James F. Hancock

Editor

Temperate Fruit Crop Breeding

Germplasm to Genomics

 Springer

James F. Hancock
Michigan State University
Dept. Horticulture
East Lansing MI 48824
USA
hancock@pilot.msu.edu

ISBN 978-1-4020-6906-2   e-ISBN 978-1-4020-6907-9

Library of Congress Control Number: 2007939760

Cover photograph by: Joseph D. Postman at the US National Clonal Germplasm Repository, Corvallis,
Oregon

Printed on acid-free paper

9 8 7 6 5 4 3 2 1

springer.com

Preface

This book is intended to be a brief compilation of the information available on the breeding of temperate fruit crops. The goal is to provide overviews on the evolution of each crop, the history of domestication, the breeding methods employed and the underlying genetics. A serious effort is made to fully integrate conventional and biotechnological breeding approaches. A discussion is also provided on licensing and patenting.

It is hoped that this book can be used as a springboard for breeders desiring an update, horticulturalists who wonder what the fruit breeders are doing and genomicists who are searching for a way to contribute to fruit breeding efforts. By far the fastest progress can be made when we all talk the same language.

This manuscript is in many regards an update of the information found in Fruit Breeding, Volumes 1 and 2, edited by J. Janick and J. N. Moore (1996). The major difference is that much more molecular information is now available on fruit crops. Molecular linkage maps have been produced for many of the commercial species and the first quantitative trait loci are being tagged and selected through marker assisted breeding. Regeneration and transformation systems are available for many of the fruit crops and potentially useful genes have been cloned and characterized. Fruit breeders will soon have all the tools in their tool box that the grain breeders have had for over a decade.

The atmosphere revolving around fruit breeders has also changed in another significant way since Janick and Moore's book. While fruit breeding has long been conducted most prominently in the public domain, private companies have become more active in the breeding of fruit crops, as large marketing organizations decide that they want to be sure of having high quality varieties in the future. In fact, most public breeders have now become private breeders, as they patent and license their cultivars to these marketing agencies.

I want to publicly thank some of the people who made this book possible. I must first acknowledge the help of Gustavo Lobos, whose willingness to tackle the references and final format was a huge help. I also need to thank Rex Brennen, John Clark, Ross Ferguson, Craig Ledbetter, Amy Iezzoni, Chris Owens and especially Chad Finn, for their agreeing to spearhead chapters, when it became clear to me that I would never finish this project without help. Susan Brown, Jim Luby, Paul Lyrene, Bill Okie, Ralph Scorza, Tom Sjulin and Nick Vorsa should also be acknowledged as coauthors who gave chapters credibility which they would not have had otherwise. Very special thanks also needs to go to my wife, Ann Hancock, who continues to be supportive of all my endeavors, sane or crazy.

East Lansing, MI, USA   J.F. Hancock

Contents

Contributors

R.M. Brennan
Scottish Crop Research Institute, Invergowrie, Dundee DD2 5DA, Scotland, UK
rex.brennan@scri.ac.uk

S.K. Brown
Department of Horticultural Sciences, 630 W. North Street, Cornell University, Geneva, New York 14456, USA
skb3@nysaes.cornell.edu

J.R. Clark
Department of Horticulture, 316 Plant Science, University of Arkansas, Fayetteville, Arkansas 72701, USA
jrclark@uark.edu

A.R. Ferguson
The Horticulture and Food Research Institute of New Zealand Ltd (HortResearch), Mt Albert Research Centre, Private Bag 92 169, Auckland, New Zealand
rferguson@hortresearch.co.nz

C.E. Finn
USDA-ARS, Horticultural Crops Research Unit, 3420 NW Orchard Avenue, Corvallis, Oregon 97330, USA
finnc@hort.oregonstate.edu

J.F. Hancock
Department of Horticulture, 342C Plant and Soil Sciences Building, Michigan State University, East Lansing, Michigan 48824, USA
hancock@msu.edu

A.E. Iezzoni
Department of Horticulture, 342B Plant and Soil Sciences Building, Michigan State University, East Lansing, Michigan 48824, USA
iezzoni@msu.edu

R.J. Jondle
Patent and Trademark Attorney, Jondle and Associates, P.C., 858 Happy Canyon Road,
Suite 230, Castle Rock, Colorado 80108, USA
rjondle@jondlelaw.com

C.A. Ledbetter
USDA-ARS, 9611 S. Riverbend Avenue, Parlier, California 93648, USA
cledbetter@fresno.ars.usda.gov

G.A. Lobos
Laboratorio de Ecofisiologia Vegetal, Escuela de Agronomia, Universidad de Talca,
Chile
globosp@utalca.cl

J.J. Luby
Department of Horticultural Sciences, 329 Alderman Hall, University of Minnesota,
St Paul, Minnesota 55413, USA
lubyx001@umn.edu

P.A. Lyrene
Horticultural Sciences, 2135 Fifield Hall, University of Florida, Gainesville, Florida
32606, USA
lyrene@ufl.edu

W.R. Okie
USDA-ARS, SE Fruit and Tree Nut Research Lab, 21 Dunbar Rd, Byron, Georgia,
31008 USDA
william.okie@ars.usda.gov

C.L. Owens
USDA-ARS, Grape Genetics Research Unit, 308 Sturtervant Hall, Cornell University,
Geneva, New York 14456, USA
chris.owens@ars.usda.gov

R. Scorza
USDA-ARS, 45 Wiltshire Rd, Kearneysville, West Virginia 25430, USA
Ralph.Scorza@ARS.USDA.GOV

A.G. Seal
The Horticulture and Food Research Institute of New Zealand Ltd (HortResearch),
Te Puke Research Centre, RD2, Te Puke, New Zealand
ASeal@hortreseach.co.nz

T.M. Sjulin
Driscoll Associates, 629 Carpenteria Road, Aromas, California 95004, USA
TomSjulin@driscolls.com

N. Vorsa
The Philip E. Marucci Center for Blueberry and Cranberry Research and Extension,
125A Lake Oswego Rd, Chatsworth, NJ 08019, USA
vorsa@aesop.rutgers.edu

Chapter 1
Apples

J.F. Hancock, J.J. Luby, S.K. Brown and G.A. Lobos

Abstract The overall objectives of modern apple breeding programs are to increase the marketability of fruit and reduce production costs. Developing well adapted cultivars with resistance to major pests is also a focus of all breeding programs. The apple is generally grown as a composite tree with a rootstock and a fruiting scion, making rootstock breeding as important as the development of scion cultivars. Genetic resistance has been found for a number of the major pests of apple. Engineering resistance to apple scab and fire blight has been the focus of many of laboratories. Most of the traits associated with adaptation and productivity have been shown to be quantitatively controlled, including chilling requirement, cold hardiness, plant vigor, season of flowering and duration of the juvenile period. Many of the traits associated with fruit quality are also quantitatively inherited including flavor, skin color, shape, size and texture. Several cDNA libraries have been developed to identify genes associated with pollination and apple fruit development. A number of apple linkage maps have been published using several different sets of parents and molecular markers have been linked to a number of monogenic traits. Mining of existing apple EST information promises to expand our knowledge of many genes important in the genetic improvement of apple.

1.1 Introduction

Apples are cultivated all across the temperate world. Their adaptive range extends from the extreme cold of places such as Siberia and North China to the much warmer environs of Columbia and Indonesia. More than 60 countries produce over 1000 or more metric tons of apples, with China, U.S.A., Turkey, Iran, France, Italy, Poland and Russia being the leading producers. World production now exceeds 57,000 million metric tons (FAOSTAT, 2004).

J.F. Hancock
Department of Horticulture, 342C Plant and Soil Sciences Building, Michigan State University, East Lansing, Michigan 48824, USA
e-mail: hancock@msu.edu

J.F. Hancock (ed.), *Temperate Fruit Crop Breeding*,
© Springer Science+Business Media B.V. 2008

Apples are an extremely versatile crop. They can be eaten directly from the tree or stored for up to a year in controlled atmospheres. They can be processed into juice, sauce and slices, and are a favorite ingredient in cakes, pies and pastries. The juice can be consumed fresh or fermented into cider, wine or vinegar. The ornamental crab apples are also known for their floral display and attractive foliage.

There are over 6,000 regionally important cultivars and land races across the world, but a few major cultivars now dominate world fruit production (O'Rourke 2003). 'Delicious' is the most important cultivar grown, followed by 'Golden Delicious', 'Granny Smith', 'Fuji' and 'Gala'. These varieties represent over 60% of the world's production. Emerging varieties include 'Cripps Pink' (often sold under the trademark Pink Lady®), 'Honeycrisp' (sold in Europe as Honeycrunch®) (Fig. 1.1), 'Scifresh' (fruit marketed under the trademark Jazz®), 'Delblush' (fruit sold as Tentation®), 'Civni' (fruit marketed as Rubens®), 'Corail' (fruit marketed as Pinova® or Pinata®) and 'Ariane'.

The genetic base of the cultivated apple has greatly eroded over time as regional cultivars have been replaced. This has been compounded by the loss of many public apple breeding projects and their associated apple cultivar collections (Brooks and Vest 1985). Forsline and his group at the USDA Germplasm Repository at Cornell University has worked hard to counter this trend by actively collecting and cataloging native apple germplasm and making it available to apple breeders (Forsline et al. 1994, Hokanson et al. 1997).

Fig. 1.1 Fruit of the Honeycrisp apple cultivar (photographed at the University of Minnesota Horticultural Research Center, Excelsior, Minnesota, U.S.A.)

1.2 Evolutionary Biology and Germplasm Resources

The genus of apples, *Malus*, belongs to the subfamily Pomoideae of the Rosaceae family. Another important fruit tree, pear (*Pyrus*), belongs to the same subfamily. There are over 30 primary species of apple and most can be readily hybridized (Korban 1986, Way et al. 1991). The cultivated apple is likely the result of initial domestication followed by inter-specific hybridization (Harris et al. 2002). Its primary wild ancestor is *M. sieversii* whose range is centered at the border between western China and the former Soviet Union. Apples are the main forest tree there and display the full range of colors, forms and tastes found in domesticated apples across the world (Forsline et al. 1994, Hokanson et al. 1997). The domesticated apple has been referred to with the epithet *Malus × domestica* (Korban and Skirvin 1984), although recently Mabberley et al. (2001) proposed that *Malus pumila* should properly refer to the domesticated apple and its presumed wild relative *M. sieversii*. Other species of *Malus* which contributed to the genetic background of the apple likely include: *M. orientalis* of Caucasia, *M. sylvestris* from Europe, *M. baccata* from Siberia, *M. mandshurica* from Manchuria, and *M. prunifolia* from China. It is likely that these species hybridized with domesticated apples as they were spread by humans (Harris et al. 2002).

The bulk of the apple species are $2n = 2x = 34$ (Table 1.1), although higher somatic numbers of 51, 68 and 85 exist; several of the cultivated types are triploid (Chyi and Weeden 1984). It is possible that the high chromosome number of apple represents an ancient genomic duplication, since there are several other Rosaceous fruit species with lower haploid chromosome numbers of $n = 8$ and 9. Based on cytology and analysis of morphological characters, the *Maloideae* likely have a polyploid origin (Phillips et al. 1991). Isozyme studies in *Malus* support an allopolyploid origin based on the presence of duplicated gene systems, allele segregation and fixed heterozygosities (Chevreau et al. 1985, Weeden and Lamb 1987, Dickson et al. 1991). An allotetraploid origin involving ancestral *Spiroideae* (mostly $x = 9$) and *Amygdaloideae* ($x = 7$) was proposed by Sax (1931) and is supported by flavonoid chemistry (Challice 1974, Challice and Kovanda 1981) and morphological traits (Phillips et al. 1991). Apples are largely self-incompatible and some are apomictic. They are propagated vegetatively, usually as composites with a separate rootstock and scion.

Apples were certainly one of the earliest fruits to be gathered by people, and their domestication was probably preceded by a long period of unintentional planting via garbage disposal. It is difficult to determine exactly when the apple was first domesticated, but the Greeks and Romans were growing apples at least 2,500 years ago. They actively selected superior seedlings and were budding and grafting 2,000 years ago (Janick et al. 1996). People in Central Asia where *M. sieversii* is native still save desirable trees when the forest is cleared for agriculture (Ponomarenko 1983) and commonly graft and plant desirable *M. sieversii* from the forest into their gardens. Planting desirable trees from root suckers may also have been a common practice prior to grafting, as *M. sieversii* trees sucker freely. Conversely, people may have cloned and moved some of their horticulturally desirable trees to areas where they

Table 1.1 Distribution of selected apple species in subsection Pumilae and their chromosome numbers

Species	Chromosome number (2n)	Distribution
M. asiatica Nakai	34	N. & N.E. China, Korea
M. baccata (L.) Borkh.	34, 68	N. & N.E. China
M.× domestica	34, 51, 68	Worldwide
M. floribunda (Siebold) ex. Van Houtte	34	Japan
M. halliana Koehne	34	Japan
M. hupehensis (Pamp.) Rehder	51	Central China
M. mandshurica (Maxim.) Kom. ex Skvortsov	34	Manschuria
M. micromalus Makino	34	S.E. China, Korea
M. orientalis Uglitzk.	?	Caucasia
M. prunifolia (Willd.) Borkh.	34	N. & N.E. China, Korea
M. pumila Mill.	34	Europe
M. sieversii (Ledeb.) M. Roem.	?	N.W. China
M. spectabilis (Aiton) Borkh.	34, 68	China
M. sikkimensis (Wenz.) Koehne ex C. K. Schneid.	51	Himalaya
M. sylvestris (L.) Mill.	34	Europe

Adapted from Way et al. 1991

seasonally grazed their animals. These trees or their open pollinated descendants may be among the horticulturally elite specimens observed in some of the forests today.

The most likely beginning of cultivation was in the region between the Caspian and Black seas (Vavilov 1949–1950); apple cultivation had reached the Near East by 3,000 B. P. (Zohary and Hopf 1993). The Romans spread the apple across Europe during their invasions and it was dispersed to the New World by European settlers during the sixteenth century.

The passage of trade routes from China to the Middle East and Europe through Central Asia probably facilitated repeated short and long distance dispersal to the east and west, either intentionally or unintentionally, of *M. sieversii* and its hybrid derivatives. The *M.× domestica* Borkh complex may have arisen through hybridization with species native to China including *M. prunifolia* (Willd.) Borkh.,

M. baccata (L.) Borkh., *M. mandshurica* (Maxim.) Kom. ex Skvortsov, and *M. sieboldii* (Regel) Rehder. To the west, hybridization with the local species *M. sylvestris* (L) Mill. and *M. orientalis* Uglitzk. is conjectured (Ponomarenko 1983, Morgan and Richards 1993, Hokanson et al. 1997, Juniper et al. 1999).

During the late nineteenth and twentieth centuries, *M.× domestica* cultivars found or bred in Europe, Russia, North America, New Zealand, Japan, and Australia were introduced throughout the world and form the basis for most current commercial apple production (Way et al. 1991, Janick et al. 1996). Several species are known to have contributed to the *M.× domestica* complex in modern breeding programs including *M. floribunda* Siebold ex Van Houtte, *M.× micromalus* Makino, *M.× atrosanguinea* (hort ex Späth) C.K. Schneid., *M. baccata* (L.) Borkh., *M. zumi* (Matsum.) Rehder, and *M. sargentii* Rehder (Ponomarenko 1983, Way et al. 1991, Janick et al. 1996).

In southern and eastern Asia, Nai, or the Chinese soft apple, *M.× asiatica* Nakai, was the primary cultivated apple for over 2000 years until *M.× domestica* was introduced in the late nineteenth and early twentieth centuries (Morgan and Richards 1993, Zhang et al. 1993, Watkins 1995, Zhou-Zhi 1999). *Malus ×asiatica* is likely a hybrid complex derived primarily from *M. sieversii* with *M. prunifolia* and perhaps other species.

Prehistoric remains and historical records, reviewed by Morgan and Richards (1993), provide evidence of the cultivation, dispersal, and human use of the apple in the Asia and Europe over the last several thousand years. Archaeological remains of apple that dated to about 6500 BC were found in Anatolia, though it is impossible to know the source of this fruit or whether it was cultivated. Historical evidence referring to apple cultivation dates to the second millennium BC from Anatolia, and northern Mesopotamia. By 500 BC, the apple likely was cultivated widely throughout the Persian Empire as fruit orchards are prominently featured in writings from the period. When Alexander the Great conquered the Persians around 300 BC, the cultivation of fruits was dispersed throughout the Greek world. By this time, the Greek philosopher, Theophrastus, distinguished the sweet cultivated apple from astringent wild forms.

The ascendance of the Roman Empire spread cultivation of the domesticated apple north and west through Europe where it supplanted and likely hybridized with, the native crab apple, *M. sylvestris*. Multiple varieties were recorded by the Roman writer, Pliny, and they attained an important place in Roman cuisine, medicine, and aesthetics by the first century AD. The Roman goddess Pomona was revered as the deity associated with apple and other fruits. With the rise and spread of Christianity and Islam over the next several centuries, apples were carefully maintained, even through wars and difficult times, in the abbey gardens throughout Europe and the orchards of Iberia. These apparently replaced the native crab apples that had a place in the diet of early Celts, Gauls, Franks, Scandinavians and other peoples of northern Europe in fermented, dried, or cooked forms. Maintenance of fruit gardens was encouraged as a basic monastic skill and many abbeys developed large orchards with many *M.× domestica* cultivars. Likewise in the Muslim world of the eastern

Mediterranean and Iberia, fruit growing was revered in keeping with Koranic teachings and skills of grafting, training and pruning became highly developed.

Today, the largest collection of apple germplasm is held at the Plant Genetics Resource Unit at Cornell University, Ithaca New York, where there are almost 4,000 accessions being maintained (http://www.ars.usda.gov/main/site_main.htm? modecode=19100500). Many of these genotypes were collected from the apples center of diversity in Central Asia (Hokanson et al. 1997, Forsline 2003).

1.3 History of Improvement

From the thirteenth century, apples became more and more widely planted throughout Europe in gardens of royalty and commoners. Raw apples were occasionally consumed, but they were more greatly prized when cooked and sometimes blended with spices and sugar or honey. Fermented juice, or cider, like beer, was preferred to the sometimes questionable local water supply. By the seventeenth century there were at least 120 cultivars described in western Europe. The rise and spread of Protestantism, which saw the apple as the special fruit of God, is credited with expanding apple cultivation across northern and eastern Europe after beginning in Germany in the early seventeenth century. By the end of the eighteenth century, many hundreds of cultivars were recognized throughout Europe. The Royal Horticultural Society of England acknowledged at least 1200 in 1826. The eighteenth and nineteenth centuries saw apple cultivars recognized and classified based on their suitability for their end uses. Aromatic dessert apples were more widely appreciated by this time, while good cooking types were still appreciated for puddings and pastries. Flavorful varieties with moderate acid and tannin levels were prized for cider production. The late nineteenth and early twentieth centuries represented the maximum of diversity in apple cultivation in Europe with hundreds of locally popular varieties being grown in thousands of small orchards. In the twentieth century, the rise of imported fruit from the Americas, New Zealand, Australia, and South Africa forced European orchards to increase in size and decrease in number and, to a large extent, to adopt the very same cultivars that were developed in, and imported from the New World.

Apples were established in the 1650s near Cape Town in South Africa to sustain settlers and to supply the ships of the Dutch East India Company. The commercial orchard district in the Western Cape apple was started by Cecil Rhodes and his associates in the late nineteenth and early twentieth century to replace a faltering wine industry.

Apples were introduced to Australia, on the island of Tasmania and at the present site of Sydney, in 1788. Orchards were established by settlers in Tasmania and New South Wales by the early 1800s. Significant production areas were eventually developed in Tasmania and the southeastern mainland. In 1814, English missionaries brought apples from Australia to New Zealand where two large apple production districts became established in the districts of Hawkes Bay and Nelson during the nineteenth and twentieth centuries.

Beginning in the sixteenth and seventeenth centuries, European colonists brought apples to the Americas. Spanish priests introduced them to their missions in Chile and California. Spanish and Portuguese settlers introduced apples to their settlements in suitable temperate climate zones of South America. European settlers brought apple seeds to establish orchards in the eastern United States and Canada. Apples grew well from northern Georgia through eastern Canada and, as in Europe, were soon highly prized for food and drink, and as a source of sugar and alcohol. The first orchards in New England were recorded in the 1620s and 1630s and became important components of the New England farmstead. Likewise, they became important on the large plantations of the mid-Atlantic colonies by the mid-1700s, including those of the early United States presidents, George Washington and Thomas Jefferson. Jefferson, an astute horticulturist, acquired and carefully trialed dozens of cultivars for his Monticello gardens in Virginia.

In Canada, French colonists established orchards in the seventeenth century along the St. Lawrence Valley. Settlers also established orchards around Lake Ontario, and in the milder valleys of Nova Scotia and New Brunswick.

As settlers moved westward in the United States, apple orchards were a requirement of homesteading throughout the territories of the Ohio River Valley. Jonathan Chapman, known as Johnny Appleseed, devoted his later life, from 1806 to 1847, to helping settlers establish thousands of apple trees on their new farms in the Ohio River drainage. The Great Lakes region of the United States, especially the states of New York, Michigan, and Ohio, continues to be a major apple production area.

In 1847, as settlers moved in to the productive valleys of western Oregon, Washington and northern California, Henderson Llewelling brought 700 trees with his family on the Oregon Trail and eventually established the first fruit nursery in the Pacific Northwest. As irrigation schemes were eventually developed in the Pacific Northwest, especially in the basin of the Columbia River and its tributaries west of the Cascade Mountains and extending to the Okanagon River valley in British Columbia, this region became one of the preeminent apple production areas of the world.

By the early twentieth century, the United States and Canada were the two largest apple producing nations. Later in the century, the Soviet Union also became important. At the beginning of the twenty-first century, China become the largest apple producer, with a large proportion of the crop being exported as concentrated juice. Major southern hemisphere production, much of it for export to northern hemisphere countries during their spring and summer, occurs in South Africa, Chile, Argentina, New Zealand, and Australia. As previously mentioned, production is currently dominated by strains of just a few cultivars: 'Delicious', 'Golden Delicious', 'McIntosh', and 'Jonagold' from North America; 'Braeburn' and 'Gala' from New Zealand; 'Granny Smith' from Australia; and 'Fuji' from Japan. Though many other cultivars remain locally important, these dominate current production and are also widely used in breeding programs around the world.

From its origins among the millions of wild *M. sieversii* trees in the mountains of central Asia (Fig. 1.2), and from the early development of thousands of local

Fig. 1.2 Photo showing the diversity of fruit collected from wild *Malus sieversii* apple trees in the Tarbagatai mountains of eastern Kazakhstan

cultivars in Europe and America, the domesticated apple, as cultivated in twenty-first century, has shrunk drastically in diversity.

1.4 Current Breeding Efforts

The overall objectives of modern breeding programs are to increase the marketability of fruit and reduce production costs. Apples are sold fresh, juiced and processed in numerous ways, but the largest overall market involves fresh fruit. Numerous apple species are also important as ornamentals (Fiala 1994).

Dessert apples are sold primarily based on appearance (size, color, shape and freedom from blemishes) and quality (taste and texture) (Janick et al. 1996; Laurens 1999; Brown and Maloney 2003). There is considerable regional variation in taste preferences, from a desire for tartness in Europe and the U.S. Midwest, to a preference for sweetness and low acidity in Asia. Low allergenic apples have become a priority in Europe. Favored colors range widely from solid green, yellow to red and bicolors of many combinations. In general, apples are expected to be blemish free, large (>70 mm in diameter), and ovate or conic shaped. Storage life is also a critical parameter, as most apples are stored for long periods of time. Resistance to apple scab and powdery mildew are common breeding goals. Niche markets are also arising for improved nutritional aspects such as higher antioxidants.

The attributes needed for processed fruit depends on their final market. Some of the most important markets are for cider, sauce and slices (Crassweller and

Green 2003). Less browning is a particularly critical parameter in the fresh cut and slices market.

Production costs are greatly reduced by maximizing yields, increasing picking efficiency and incorporating disease and pest resistance. Adaptation is a key parameter associated with yield, particularly in marginal climates with extreme winters or low chilling hours. Pest and disease resistance is critical to productivity in areas where other means of control are not available or undesirable. This applies particularly to growers interested in producing 'organic' apples.

The apple is generally grown as a composite tree with a rootstock and a fruiting scion, making rootstock breeding as important as the development of scion cultivars. There are a number of important attributes of rootstocks including; ease of propagation, clean upright stems, easy to bud or graft, well anchored root systems, no suckering, and good stock-scion compatibility (Janick et al. 1996). Rootstocks should also offer a range of tree size control from dwarfing to vigorous, induce early, heavy cropping, tolerance to cold and wet or dry soils, and be resistant the prevailing pests and diseases (Webster and Wertheim 2003).

Brown and Maloney (2003) reviewed current breeding programs and activities throughout the world. In the US, a new program was started in 1994 at Washington State University. Other programs in the US include the PRI (Purdue University, Rutgers University and the University of Illinois) cooperative that has concentrated on developing scab resistant cultivars, the University of Minnesota (of 'Honeycrisp' fame), and Cornell University, best known for 'Empire' and 'Jonagold'. The New Zealand program has been an innovator in the licensing and restricted availability of selections from their program. Increasingly, programs are partnering with private industry, examples include the collaboration between breeders, nurseries and packers in France (Laurens and Pitiot 2003).

1.5 Genetics of Economically Important Traits

1.5.1 Pest and Disease Resistance

Genetic resistance has been found for a number of the major pests and diseases of apple including: Fire blight (*Erwinia amylovora*), Alternaria blotch (*Alternaria mali*), apple blotch (*Phyllosticta solitaria*), apple canker (*Nectria galligena*), apple scab (*Venturia inaequalis*), cedar apple rust (*Gymnosporangium juniperi-virginianae*), crown rot (*Phytophthora cactorum*), powdery mildew (*Podosphaera leucotricha*), wooly apple aphid (*Eriosoma lanigerum*), rosy apple aphid (*Dysaphis plantaginea*) and rosy curling aphid (*D. devecta*) (Table 1.2).

Resistance to Alternaria blotch, crown rot, wooly apple aphid, rosy apple aphid and rosy curling aphid are regulated by a single dominant gene. RFLP markers have been found for resistance to rosy leaf curling aphid (Roche et al. 1997). Apple blotch and apple rust resistance are regulated by two dominant genes, and a

Table 1.2 Genetics of pest and disease resistance in apple

Pest or disease	Observations and source
Bacterial	
Fire blight *Erwinia amylovora*	Immunity is present in some *Malus* species (Janick et al. 1996); QTLs for resistance reported by Khan et al. (2006)
Fungi	
Alternaria blotch *Alternaria mali*	Resistance is controlled by a single dominant gene (R^{alt}) which is epistatic to a dominant gene (Alt) controlling susceptibility (Saito and Niizeki 1988)
Apple blotch *Phyllosticta solitaria*	Susceptibility is regulated by two dominant genes (Ps_1 and Ps_2) with duplicate recessive epistatic interaction between gene pairs (Mowry and Dayton 1964)
Apple canker *Nectria galligena*	Highly resistant cider apples and rootstocks have been identified (Moore 1960)
Apple scab *Venturia inaequalis*	Both quantitative and qualitative resistance exists; major source is a dominant gene, V_f; multiple resistance genes at the V_f locus have been found in several species; the resistance of V_f is enhanced by polygenes (Dayton and Williams 1968, Crosby et al. 1992, Bus et al. 2002); numerous QTL and markers identified (Tartarini and Sansavini 2003, Durel et al. 2004, Bus et al. 2005a,b, Hemmat et al. 2004)
Cedar apple rust *Gymnosporangium juniperi-virginianae*	Regulated by two, dominant genes (Gy-a and Gy-b) and perhaps other modifying genes; resistance mechanisms vary (Mowry 1964, Aldwinckle et al. 1977, Chen and Korban 1987)
Crown rot *Phytophthora cactorum*	Regulated by a single dominant gene (P_c), but polygenes are important (Alston 1970, Watkins and Werts 1971)
Powdery mildew *Podosphaera leucotricha*	Regulated by several dominant genes (Pl_1 and Pl_2) and polygenes; resistance may be enhanced by polygenes (Alston 1977, Korban and Dayton 1983, Gallott et al. 1985); quantitative resistance will be needed for durable resistance (Caffier and Parisi 2007); markers identified for the dominant alleles Pl1, Pl2, Pl-d, Pl-w, Pl-m (Markussen et al. 1995, Durel et al. 2002, Evans and James 2003)
Insects	
Wooly apple aphid *Eriosoma lanigerum*	Regulated by a single dominant gene (*Er*); 'Northern Spy' has high level of resistance, along with several other cultivars; resistance gene is closely linked to incompatibility gene (Knight et al. 1962, Knight 1962, Cummins et al. 1981); markers identified for *Er1* and *Er3* (Sandanayaka et al. 2003)
Rosy apple aphid *Dysaphis plantaginea*	Regulated by a single dominant gene (Sm_h); resistance found in open pollinated selection of *M. robusta* (Alston and Briggs 1970)
Rosy curling aphid *D. devecta*	Regulated by a single dominant gene; four different resistance genes have been identified in 'McIntosh' (SD_{pr}) 'Cox's Orange Pippin' (Sd_1), 'Northern Spy' (Sd_2) and *Malus robusta* (Sd_3); a precursor gene must be present for effective resistance (Alston and Briggs 1977); markers identified to *Sd1* (Roche et al. 1997, Cevik and King 2000)

number of polygenes. Genes for resistance to powdery mildew have been identified and markers have been developed: Pl_1 (Markussen et al. 1995), Pl_2 (Dunemann et al. 1999), Pl-d (James et al. 2004), Pl-w (Evans and James 2003). A marker has yet to be developed for $Pmis$ from Mildew immune seedling (MIS). Resistant *Malus* genotypes have also been identified for fire blight and apple canker, although the genetics have not been elucidated. 'Delicious' has fairly good resistance to fire blight, but immunity is only found outside of the cultivated species of apple (Janick et al. 1996).

At least 10 different resistance genes have been identified for apple scab (Bus et al. 2002), and one of them, *Vf* from the ornamental crabapple *M. floribunda 821*, has been used all over the world to create new scab resistant cultivars (Laurens 1999). This gene has been cloned and shown to confer scab resistance to a transgenic cultivated variety 'Gala' (Belfanti et al. 2004). Scar markers have been identified for *Vbj* from *Malus baccata jackii* (Gygax et al. 2004) and *Vm* from *M. atrosanguinea* 804 (Cheng et al. 1998). Patocchi et al. (2005) used genome scanning to identify a microsatellite tightly linked to *Vm*. Numerous other QTL have been identified for scab resistance, which will be described in the section on genetic mapping of apple. Gessler et al. (2006) review scab resistance in apple.

In breeding for multiple disease or pest resistance a balance of resistance and commercial fruit quality may be difficult to achieve. Adequate levels of resistance to the major fungal pathogens (scab and powdery mildew) coupled with resistance to fire blight requires large populations and great attention to other secondary problems such as leaf spot, moldy core and summer fruit pathogens.

Attempts have been made to associate specific enzyme activities with resistance to superficial storage scald. The total activities of guanacol-dependent peroxidase (POX), superoxidismutase (SOD) and catalase (CAT) were not found to be significantly associated with susceptibility to scald in a segregating population of 'White Angel' × 'Rome Beauty'; however, there were associations with the presence and absence of individual isozymes (Kochhar et al. 2003).

Several genes have been isolated that are related to disease resistance. A cDNA has been cloned from fruit of 'Fuji' that encodes a pathogenesis-related 5/thaumatin-like protein (PR5/TL) that was named Mdt 1 (*Malus domestica* thaumatin-like protein) (Oh et al. 2000). A salicylate-inducible PR-10 gene (designated as *APa*) was found to be expressed during infection of a compatible vs. a non-compatible race of *V. inaequalis* (Poupard et al. 2003). Eighteen genes were identified as having higher expression levels during infection of 'Golden Delicious' by *Penicillium expansum* (Sáchez-Torres and González-Candelas 2003). Two of these genes likely encoded SS-glucosidase and phosphatase 2C.

A number of apple sequences have been identified that are similar to the R (resistance) genes of other plants that contain a nucleotide binding site (NBS). NBS-containing genes are the most common class of resistance genes found in plants. Over 20 families of NBS-containing genes have been identified in apple that include the two major groups described in dicot plants, one lacking a toll-interleukin element and one containing it (Baldi et al. 2004, Calenge et al. 2005). A cluster of receptor-like genes has been identified in bacterial artificial clones derived from the

Vf scab resistance locus that are similar to the *Cladosporium fulvum* (Cf) resistance gene family of tomato (Vinatzer et al. 2001).

1.5.2 Morphological and Physiological Traits

Most of the traits associated with adaptation and productivity have been shown to be quantitatively controlled, including chilling requirement, cold hardiness, plant vigor, season of flowering and duration of the juvenile period (Table 1.3).

Several aspects of plant habit have been shown to be regulated by single genes in inheritance studies (Alston et al. 2000). A dominant gene regulates the columnar habit in the 'Wijcik' clone of 'McIntosh' (Lapins and Watkins 1973, Lapins 1974). A series of recessive alleles have been identified that regulate dwarfing (Decourtye 1967, Alston 1976). Recessive genes have been associated with the spur-habit in sports of 'Golden Delicious', 'Redspur' and 'Starkrimson' (Decourtye and Lantin 1969, Alston and Watkins 1973). RAPD markers have been identified for terminal bearing, initial bud break, root sucker formation (Weeden et al. 1994) and columnar tree habit (Hemmat et al. 1997).

Several cDNA libraries have been developed to identify genes associated with pollination and apple fruit development (Dong et al. 1997, 1998b, Sung et al. 1998; Yamada et al. 1999). *Mdh3* encoding a Phalaenopsis O39-like homeodomain protein was found to be expressed in apple ovules and may initiate the program of ovule development (Dong et al. 1999). A homologue of mammal *DAD1* (defender against cell death 1) was cloned that is expressed after flower pollination and during senescence of leaves, petals and fruit (Dong et al. 1998c). A gene encoding polygalacturonase-inhibiting protein (PGIP) was isolated that has two peaks of expression during apple maturity and is activated by wounding and fungal infection (Yao et al. 1999).

A number of MADS-box genes have been cloned and characterized from apple (Table 1.4). These genes produce transcription factors which play an important regulatory role in the development of floral meristems in all plants. To date, one of the most interesting MADS-box genes that has been cloned is a mutation of *MdP1* caused by a retrotransposon insertion, which abolishes gene expression and leads to parthenocarpic fruit development (Yao et al. 2001).

Homeobox genes have also been identified in apple that encode homeodomain proteins which are transcription factors that regulate a number of developmental processes. Watillon et al. (1997) identified three KNOTTED1 (*kn1*)-like homeobox genes, *KNAP1–3*. Transcripts from *KNAP3* accumulated in a wide rang of vegetative and reproductive organs, while mRNAs from *KNAP1* and *KNAP2* were present primarily in elongated parts of stems. Sakamoto et al. (1998) isolated two additional (*kn1*) -like homeobox genes, *APHB1* and *APHB2*. *APHB1* are expressed in shoot apical tissues, stems and flowers but not mature leaves and fruit. *APHB2* is expressed in all organs involving mature leaves and developing fruit. Another homeobox gene, *MDH1*, was isolated from developing fruit, flowers and leaves of apple that has a homeodomain similar to BEL1 which is involved in ovule development in *Arabidopsis* (Dong et al. 2000).

Table 1.3 Genetics of adaptation, productivity, plant habit and fruit quality in apple

Attribute	Observations and source
Adaptation	
Chilling requirement	Generally quantitatively inherited, although the low chill requirement of 'Anna' is thought to be controlled by one major gene and a number of minor ones (Hauagge and Cummins 1991, Labuschagne et al. 2002); bud break number is highly heritable (Labuschagne et al. 2003)
Cold hardiness	Quantitatively inherited, largely additive (Watkins and Spangelo 1970)
Growth rate	Quantitatively inherited; QTL identified (Conner et al. 1998)
Vigor	Quantitatively inherited, largely additive (Watkins and Spangelo 1970, Durel et al. 1998)
Season of flowering	Quantitatively inherited, largely additive (Janick et al. 1996); QTL identified (Liebhard et al. 2003a)
Harvest date	Quantitatively inherited; QTL identified (Liebhard et al. 2003a)
Productivity	
Flower number	Quantitatively inherited; QTL identified (Liebhard et al. 2003a)
Fruit number	Quantitatively inherited; QTL identified (Liebhard et al. 2003a)
Juvenile phase length	Quantitatively inherited; QTL identified (Liebhard et al. 2003a)
Incompatibility	Numerous S-alleles exist, are semi-compatible combinations (Broothaerts et al. 2004a,b)
Plant habit	
Compact	Single dominant gene (*Co*) identified for compact columnar habit in 'Wijcik' clone of 'McIntosh'; genetics more complex in other sources of the compact branching habit (Lapins and Watkins 1973, Lapins 1974); markers identified (Conner et al. 1997, Kim et al. 2003a,b, Tian et al. 2005)
Dwarfing	Several sources of dwarfing exist that are regulated by recessive alleles (d_1–d_4) (Decourtye 1967, Alston 1976)
Internode length and N°	Quantitatively inherited; QTL identified (Conner et al. 1998)
Spur bearing habit	Generally quantitatively inherited, but spur-habit in sports of 'Golden Delicious', 'Redspur' and 'Starkrimson' result from a single recessive gene (Decourtye and Lantin 1969, Alston and Watkins 1973)
Russett	Single dominant gene identified with many modifying polygenes (Alston and Watkins 1973, Durel et al. 1998); some russeted clones do not transmit russet to offspring
Fruit quality	
Acidity	Quantitatively inherited; QTL identified for malic acid gene, *Ma* (Maliepaard et al. 1998, Liebhard et al. 2003a)
Bioactive compounds	High variability among cultivars in ascorbic acid and phenolic compounds (Schmitz-Eiberger et al. 2003); QTL identified by Davey et al. (2006)
Firmness	Quantitatively inherited; QTL identified (Seymour et al. 2002, Liebhard et al. 2003a)

Table 1.3 (continued)

Attribute	Observations and source
Skin color	Quantitative but may be regulated by a few major genes; one possibility is that three dominant, major genes produce color (Lespinasse et al. 1988, Brown 1992); marker identified for a fruit color gene (*Rf*) regulating the red/yellow dimorphism (Cheng et al. 1996); EST research indicated that myb transcription factors are important (Espley et al. 2007, Takos et al. 2006)
Shape	Quantitatively inherited with a low genotype by environment interaction (Currie et al. 2000)
Size/weight	Quantitatively inherited (Durel et al. 1998); highly heritable (Volz et al. 2001); QTL identified (Liebhard et al. 2003a)
Storage disorders	High heritability for soft scald and superficial scald; moderate heritability for water core; low heritability for external pit, internal pit, brown heart, breakdown and chilling injury (Volz et al. 2001)
Sugar content	Quantitatively inherited; QTL identified (Liebhard et al. 2003a)
Texture	Quantitatively inherited (Durel et al. 1998); QTL identified (Seymour et al. 2002); *Md-ACS1* found to be closely associated with fruit softening (Costa et al. 2005)

1.5.3 Fruit Quality

Many of the traits associated with fruit quality are quantitatively inherited including flavor, shape, size and texture (Table 1.3). Skin color is also quantitatively inherited, but the number of major genes regulating it may be limited. Anthocyanin stripes are regulated by a single dominant gene in 'Cox's Orange Pippin' (Klein 1958). It has been proposed that three, dominant major genes regulate color (A, B and C) (Lespinasse et al. 1988). Yellow is produced by one dominant allele, red if more than two dominant alleles are present. The yellow cream flesh color of 'Cox's Orange Pippin' is dominant (Alston 1981). Russett is regulated by a single dominant gene, with numerous interacting polygenes (Alston and Watkins 1973), yet some russetted clones must only have the mutation in the L1 as they do not transmit this trait to their offspring. A RAPD marker has been identified for fruit skin color (Cheng et al. 1996). Recently the myB transcription factor has been suggested to regulate apple red fruit color (Takos et al. 2006, Espley et al. 2007).

Genes have been identified and cloned that influence fruit quality. An allele of the 1-methycyclopropene softening slower gene (*Md-ACS1*) was found that was significantly associated with softening (Oraguzie et al. 2004 and 2007).

Genes associated with anthocyanin biosynthesis were cloned from apple fruit skin, and cDNAs were identified that encode flavanone 3-hydroxylase (F3H), dihydroflavonol reductase (DFR), anthocyanidin synthase (ANS), and UDP-glucose: flavonoid 3-O-glucosyltransferase (UFGT). Each gene was found to be a member of a multigene family. The mRNAs of these genes were detected preferentially in the skin tissue and were light induced. The transcripts were more abundant in the skins

Table 1.4 MADS-box genes cloned and characterized in apple

Gene	Observations and source
AFL1	Expressed only in the floral bud during the transition from vegetative to reproductive growth. Orthologue of *Arabidopsis FL* (Wada et al. 2002)
AFL2	Expressed in vegetative shoot apex, floral buds, floral organs and root. Orthologue of *Arabidopsis FL* (Wada et al. 2002)
MdMADS1	Expressed in all floral organs and young fruits, but not in leaves. Expression highest in early stages of flower and fruit development. Shows significant sequence homology with *Arabidopsis* AGL2 (Sung and An 1997)
MdMADS2	Transcribed in all four floral organs and at all stages of flower development. Member of SQUA subfamily of snapdragon. Transgenic tobacco expressing it flowered early and had shorter bolts, but showed no homeotic changes in the floral organs (Sung et al. 1999)
MdMADS3	Expressed in the inner three whorls of the floral primordium, but not in fruit. Showed high sequence homology with *Arabidopsis* AGL2 and AGL4 (Sung et al. 2000)
MdMADS4	Expressed ubiquitously in inflorescence meristem, floral meristem, fruit and seeds. Showed high sequence homology with *Arabidopsis* AGL2 and AGL4 (Sung et al. 2000)
MdMADS5	More strongly expressed in fruit than flower buds; not expressed in leaves. Highest expression in young fruit. Most strongly expressed in cortex and skin, little expression in core. Showed high sequence homology with *Arabidopsis AP1* (Yao et al. 1999b). Expressed specifically in sepals during flower bud formation (Kotoda et al. 2000). Transgenic *Arabidopsis* with this gene flowered early (Kotoda et al. 2002)
MdMADS6	More strongly expressed in fruit than flower buds. Highest expression in young fruit. Most strongly expressed in cortex and skin, but significant expression in core. Showed high sequence homology with *PrMADS1* and *MdMADS7* (Yao et al. 1999)
MdMADS7	More strongly expressed in fruit than flower buds. Highest expression in older fruit. Most strongly expressed in cortex and skin, but significant expression in core. Showed high sequence homology with *PrMADS1* (Yao et al. 1999)
MdMADS8	More strongly expressed in fruit than flower buds. Highest expression in young fruit. Most strongly expressed in core and cortex; weak expression in skin. Showed high sequence homology with *AGL2, AGL4, MdMADS1* and *MdMADS9* (Yao et al. 1999)
MdMADS9	More strongly expressed in fruit than flower buds. Highest expression in younger fruit. Most strongly expressed in core and cortex; weak expression in skin. Showed high sequence homology with *Arabidopsis AGL2* and *AGL4* (Yao et al. 1999)
MdMADS10	More strongly expressed in fruit than flower buds. Highest expression in young fruit. Only expressed in core. Showed high sequence homology with *Arabidopsis AGL11* (Yao et al. 1999)
MdMADS11	High expression in both fruit and flower buds. Preferentially expressed in fruit after pollination. Highest expression in young fruit. Evenly expressed in all three fruit tissues. Showed high sequence homology with *Arabidopsis AGL6* (Yao et al. 1999)

Table 1.4 (continued)

Gene	Observations and source
MdMADS12	Isolated from leaf tissue but has significant homology with the *Arabidopsis* floral identity gene *AP1*. Expressed at similar levels in leaves, vegetative shoots and floral tissues. May play a role in the transition from juvenile to adult stage (van der Linden et al. 2002)
MdMADS13	Isolated from leaf tissue but has significant homology with the *Arabidopsis* floral identity gene *AP3*. Mainly expressed in petals and stamens (van der Linden et al. 2002, Kitahara et al. 2004)
MdMADS14	Isolated from leaf tissue but has significant homology with the *Arabidopsis* floral identity gene *AGAMOUS*. Preferentially expressed in carpels. Possible orthologue of *SHATTERPROOF* (van der Linden et al. 2002)
MdMADS15	Isolated from leaf tissue but has significant homology with the *Arabidopsis* floral identity gene *AGAMOUS*. Highly expressed in stamens and carpels (van der Linden et al. 2002)
MdPI	Shows high amino acid sequence identity with *Arabidopsis PI*. A retrotransposon insertion was identified that abolished expression of the gene and probably led to parthenogenic fruit development (Yao et al. 2001)

of cultivars with red skin than non-red, indicating that these genes have major roles in determination of apple skin color.

The expression of six genes (*PAL, CHS, CHI, F3H, DFR* and *ANS*) involved in anthocyanin production was also studied during flower development (Dong et al. 1998a). Maximum accumulation of all 6 RNAs was highest during early flower development and dropped drastically after petal expansion. Blocking of UV or natural light greatly reduced expression of these six genes and inhibited anthocyanin production, and after re-exposure to light, white flowers were not able to resynthesise anthocyanins.

In a study of the genetics of commonly found storage disorders, Volz et al. (2001) found high heritability for soft scald and superficial scald, moderate heritability for water core, and low heritability for external pit, internal pit, brown heart, breakdown and chilling injury. Two of the genes that are likely associated with superficial scald have been cloned from apple, *hmg1* and *hmg2* (Rupasinghe et al. 2001, Pechous and Whitaker 2002). These genes encode 3-hydoxy-3-methylglutaryl coenzyme A reductase (HMGR) which catalyses the synthesis of mevalonate from HMG-CoA. Superficial scab is thought to be caused by the oxidation of α-farnesene in apple skin and α-Farnesene is produced in the mevlonate pathway. A gene encoding (E,E)-α-farnesene synthase gene (*AFS1*) has also been cloned that uses farnesyl diphosphate as a substrate. (Pechous and Whitaker, 2004).

Boss et al. (1995) identified the gene for a full length polyphenol oxidase (*pAPO5*) from a 'Granny Smith' fruit peel cDNA library whose expression was induced by wounding and was elevated in peel with superficial scald. Polyphenoloxidase (PPO) is thought to play an important role in browning after the wounding of apples. Kim et al. (2001) cloned *pPPO5* from 'Fuji' and identified another full length PPO gene, *pMD-PPO2*, which shared about 55% identity. *MD-PPO2* was expressed in all stages of flower development, while the *APO5* transcript was detectable only in late anthesis. Both genes were expressed during early fruit ripening; however, only *APO5* was significantly induced by wounding.

Overall, flavor is a quantitative trait, but some of its individual components have been found to be regulated by single genes. The distinct aroma of 'Cox's Orange Pippin' is the result of a single dominant allele (Alston and Watkins 1973). A dominant allele also determines moderate to high acidity, with the specific levels being inherited quantitatively (Nybom 1959, Brown and Harvey 1971). Resistance to bitter pit was reported to be regulated by two dominant alleles (Korban and Skirvin 1984). Souleyre et al. (2005) have isolated an alcohol acyltransferase that produces esters involved in flavor.

A number of recent studies have shown that the antioxidant capacity of apples can have wide ranging health effects including the inhibition of colon- and liver-cancer cells (Eberhardt et al. 2000, Schirrmacher and Schempp 2003). Schmitz-Eiberger et al. (2003) found high variability among cultivars in ascorbic acid and phenolic compounds. 'Topaz', 'Berlepsch', 'AW 93', 'Golden Delicious', 'Rubinette', 'Braeburn' and 'Honeycrisp' had among the highest levels of ascorbic acid, while the highest levels of phenolics were found in 'Scesterimuher', 'Bortlinger', 'Bohnapfel' and 'Dulmener Rosenapfel'. Lee et al. (2003) found 'Rhode Island Greening' to have unusually high levels of antioxidants, while Lata (2007) documented the effect of cultivar and seasonal variation. Davey et al. (2006) identified QTL affecting vitamin C in apple using the mapping population of 'Braeburn' × 'Telamon'.

Two types of genes associated in hormone biosynthesis have been cloned and isolated. Kusaba et al. (2001) isolated a cDNA encoding gibberellin (GA) 20-oxidase that was mainly expressed in immature seeds. Wegrzyn et al. (2000) cloned an α-amylase gene from apple fruit that was transiently upregulated during low temperature exposure. Stanley et al. (2002) also isolated several α-amylase genes from apple and *Arabidopsis* that they suggested might be targeted to different compartments within the cell (cytosol, secretory pathway and plastid).

Genes for ACC-synthase, ACC-oxidase and polygalacturonadase (PG) have been cloned and characterized from apple (Dong et al. 1991 and 1992, Castiglione et al. 1998) and the promoter sequences of the genes for ACC-oxidase and PG have been characterized (Atkinson et al. 2005). Castiglione et al. (1998) examined restriction products across 12 *Malus* species and found two allelic forms of a gene for ACC-oxidase but very little variability in a gene for ACC-synthase. They suggested that the two allelic forms of ACC-oxidase might control the rate of ethylene synthesis and could be used in marker assisted selection. Harada et al. (2000) found a specific allele of ACC-synthase (*Md-ACS1-2*) that was associated with low levels of ethylene production in a screen of 35 cultivars. Atkinson et al. (1998) examined the expression of PG and ACC-oxidase mRNAs and detected them earlier in 'Royal Gala' apples relative to internal ethylene concentration than 'Braeburn' and 'Granny Smith'.

Tao et al. (1995) identified a cDNA encoding NADP-dependent sorbitol-6-phosphate dehydrogenase (S6PDH) from apple and found that protein levels were highest in mature fruit. Sorbitol is the end product of photosynthesis in apple. Tao et al. (1995) incorporated the gene into tobacco and found that sorbitol levels were positively correlated with S6PDH activity levels. Yamada et al. (1998) identified a full-length cDNA for NAD-dependent sorbitol dehydrogenase (NAD-SDH) and showed that the mRNA is expressed in mature apple fruit.

A number of plant-derived allergens have been identified and placed into specific groups, including pathogenesis-related proteins (PR), seed storage proteins and structural proteins (Hoffmann-Sommergruber 2002). Representatives of three families of PR genes have been cloned and characterized in apples including: (1) *Mal d 2*, producing a thaumatin-like protein (Krebitz et al. 2003), (2) *Ypr10*, producing a intercellular protein with unknown enzymatic action (Puhringer et al. 2000), and (3) *Mal d 3*, producing a lipid transfer protein (LTP) (Diaz-Perales et al. 2002). The promoter of *Ypr10* is both stress- and pathogen-inducible, and the product of *Mal d 2* has anti-fungal properties. Gao et al. (2005a,b) cloned and mapped *Mal d 1* and also mapped *Mal d 2* and *4*. The location of allergens was studied by Marzban et al. (2005). *Mal d 1* and *2* were distributed in peel and flesh, while *Mal d 3* is restricted to the peel.

Breeding programs are assessing low allergenicity as a breeding objective. Gao et al. (2005a) cloned and mapped the major apple allergen *Mal d 1*, and then studied *Mal d 3* (2005b). Carnes et al. (2006) demonstrated differences in antigenic and allergenic profiles for 10 different apple varieties and found significant variation in content of *Mal d 3*.

1.6 Crossing and Evaluation Techniques

1.6.1 Breeding Systems

Most apples require cross-pollination; in the orchard, pollination is carried out primarily by bees. Self-fertility is limited in apple by gametophytic self-incompatibility, where the growth of self pollen tubes is prevented by cytotoxic proteins that are produced in the stigmatic tissue. Attack is avoided if specific inhibitors of these proteins are expressed. The style-encoded toxic proteins are RNases which are produced by the *S-gene*. The pollen-expressed inhibitors have not been identified. Allele-specific PCR primers have been developed to selectively amplify and identify individual S-alleles, and 28 S-alleles have been cloned and identified in 150 diploid and triploid European, American and Japanese cultivars (Broothaerts et al. 2004a). Three S-alleles (S_2, S_3 and S_9) are very common and seven are very rare (S_4, S_6, S_8, S_{16}, S_{22}, S_{23} and S_{26}). The allelic composition of the most widely grown cultivars are: Delicious ($S_9 S_{28}$), Golden Delicious ($S_2 S_3$), Granny Smith ($S_3 S_{23}$), Fuji ($S_1 S_9$) and Gala ($S_2 S_5$). S-RNase analysis has been used to identify the parents of Japanese cultivars (Kitahara et al. 2005, Matsumoto et al. 2006). Broothaerts (2003) suggests that some S-alleles be renumbered based on new research.

1.6.2 Pollination and Seedling Culture

The blossom is typically composed of five petals, five sepals, about 20 stamens and a pistil with five styles. The flowers are borne in cymose clusters on short pedicels. Each ovary generally has five carpels with two ovules each, resulting in a total of 10 seeds (although some varieties can have up to 18 seeds).

For pollen collection, flowers are gathered at the balloon stage before petal expansion. Blossoms can be collected from rooted plants in the field or greenhouse, or flowers can be forced in the greenhouse by cutting flowering shoots and holding them in water. Anthers are first removed from flowers by passing them over a screen and then allowed to dehisce overnight in containers such as Petri plates or 'paper boats'. The dry pollen is ready to be used directly in crosses and remains viable for several days at room temperature. It can remain viable for several weeks if refrigerated under low relative humidity. For long term storage, pollen can be held for at least a year at $-15°C$ in loosely stoppered vials in a desiccators with calcium chloride.

To emasculate flowers, fingernails or scissors are generally used to remove sepals, petals and stamens at the balloon stage. Pollinations are generally made soon after emasculation, although flowers are sometimes re-pollinated one day later if conditions are thought to be too cool for normal pollen germination and tube growth. Flowers can be successfully fertilized over a period of several days. Generally, two flowers per cluster are emasculated and the rest are removed. There is no need to cover the emasculated flowers after pollination, as insects do not visit flowers without stamens and petals (Visser 1951). Some breeders do not emasculate at all and rather rely on the self incompatibility system to prevent self fertilization. In this case, flowers are bagged to prevent contamination. Keulemans et al. (1994) discussed the effect of number of flowers pollinated on fruit set in crosses.

Pollen is often placed on the stigmas after dipping a small brush into vials or Petri plates of pollen. Pencil erasers or fingertips are also sometimes used to transfer pollen. After each cross, the pollination vehicle is washed or dipped into 95% alcohol and allowed to dry to prevent cross contamination.

Several other techniques are sometimes employed to eliminate the emasculation step. Small trees can be enclosed in a bee-proof structure with a bouquet of open flowers of the desired parent and honeybees. Bouquets of potential parents with dominant marker genes can also be placed adjacent to several recipients in a field, and the desired hybrids can be identified in the seedling blocks. Seed can also be collected from orchards that contain two cultivars of interest.

Most breeders attempt to get at least 200–300 seeds per cross, although thousands of seeds are sometimes generated of crosses that are thought to have great commercial potential or when seedling screening is planned using inoculation or markers. About 50–100 pollinations are typically required to generate a few hundred seeds, but if flowers are emasculated, the general rule is that on average one flower produces one seed due to damage from emasculation. The number of crosses made varies greatly depending on program objectives, available resources and philosophy of the breeder, but can range from 5 to 50. If both parents are heterozygous for pale green lethal (Way et al. 1976), dwarfing genes (Alston 1976) or sub-lethals linked to the *Vf* gene (Gao and van de Weg 2006) then greater numbers are needed, due to the expected 25% loss due to lethals and dwarfs.

Apple seeds must be stratified in the cold for successful germination. Seeds can be left in the fruit at slightly above freezing temperatures to naturally after-ripen, but when this is done, molds often become a problem. More commonly, seeds are harvested just before the fruit reach maturity, but late enough for the seed coats to have become dark-brown. The seeds are then held in the cold at 3–5°C for 60–80 days in plastic bags containing moist filter paper or peat moss. Length of stratification may vary depending on the genetic background. Thomsen and Eriksen (2006) found that two *Malus* species (*Malus sargentii* and *M. sieboldii*) differed in their response to pretreatments and stratification temperatures. When their radicals begin to emerge, the seeds are transplanted into pots or trays at 1–2 cm depth. They are maintained under greenhouse conditions for about 60 days until they become 30–45 cm tall, and then they are planted in a nursery or moved outside in larger pots. Plants are also sometimes pre-selected before field planting for plant vigor, large mature-phase leaf type and growth habit/architecture. Often seedlings are inoculated with scab in the greenhouse, particularly if at least one parent is known to be resistant; this technique can reduce the progeny population by 50–80% (Janick et al. 1996).

1.6.3 Evaluation Techniques

The juvenile period without fruit varies from 3 to 10 years, depending on genotype and growth environment. A number of techniques have been used to shorten the duration of the juvenile period, including shoot pruning, root pruning and bark ringing (Janick et al. 1996); however, most of these are tedious and difficult to utilize with large populations. Probably the most helpful approach is to maintain active growth in the greenhouse before planting and throughout the entire evaluation period. Many programs will graft seedlings onto dwarfing rootstocks (M9 most common, also Bud 9 and EMLA 27) either in the first or second year. Dwarfing rootstock promotes more precocious flowering and saves space, but the use of rootstocks adds to the cost of the program. The rootstocks must be from virus-indexed stock to avoid infecting scions.

Field selection is generally performed in three stages: *Phase 1* – Genotypes are replicated only once whether grafted or not. If they are on their on own roots, spacing is usually 1.5–2.0 m in row and 5.0–7.0 m between rows. If they are on dwarfing rootstocks, they are usually spaced 0.6–1.0 m in the row and 4.0–5.0 m between rows. *Phase 2* – the most promising seedlings are cloned by grafting on dwarfing rootstock (M9 most common); usually 4–6 trees are produced and are planted as a single unit or split into two replications. The replications are usually evaluated at only one location but sometimes they are planted at two sites. *Phase 3* – Pre-commercial testing is conducted with multiple trees (10–50 is common) placed at many locations across the world (10–20 would not be uncommon). Virus indexing (and thermotherapy if needed) is usually performed at this stage on the most promising selections.

More breeding programs are starting to evaluate what quality traits are most important to its consumers. Fruit quality, aroma, consistent and high soluble solids, a range of acidity, high juiciness, crispness and non-browning flesh are desirable as are methods to quantify these traits. Sensory testing is becoming a part of many programs, as are studies of quality components (Harker et al. 2006).

1.7 Biotechnological Approaches to Genetic Improvement

1.7.1 Genetic Mapping and QTL Analysis

Molecular markers have been linked to a number of monogenic traits in apple (Tartarini and Sansavini 2003). The most work has been done on the *Vf* gene for scab resistance, where over 40 markers have been identified. Markers for the other scab resistance genes have also been developed by many groups and include *Vh* from Russian seedling R12740-7A of *M. sieversii* (Hemmat et al. 2002, Bus et al. 2005a,b, Boudichevskaia et al. 2006), *Vm* (Cheng et al. 1998, Patocchi et al. 2005), *Va* and *Vb* (Hemmat et al. 2004, Erdin et al. 2006), *Vd* (Tartarini et al. 2004), *Vbj* (Gygax et al. 2004) and *Vg* (Durel et al. 2000, Calenge et al. 2005). Gessler et al. (2006) reviewed the literature in this area from type of resistance through gene pyramiding.

Markers have also been linked to the pest resistance genes *Sd1* for *Dysaphis devecta*, and *Er1* and *Er2* for *E. lanigerum* (Table 1.2). A few markers have also been linked to genes regulating morphological traits including the columnar habit (*Co*), fruit color (*Rf*) and fruit acidity (*Ma*) (Table 1.3). Recently a cDNA/AFLP approach was used to identify a gene that contributes to lowering of fruit acidity (Yao et al. 2007).

A number of apple linkage maps have been published using several different sets of parents: 'White Angel' × 'Rome Beauty' (Hemmat et al. 1994), 'Wijcik McIntosh' × NY 75441-58 (Conner et al. 1997), 'Prima' × 'Fiesta' (Maliepaard et al. 1998), 'Iduna' × A679-2 (Gianfranceschi et al. 1998), 'Fiesta' × 'Discovery' (Liebhard et al. 2003b; Silfverberg-Dilworth et al. 2006) and 'Telamon' (a columnar genotype) × 'Braeburn' (Kenis and Keulemans 2005). The one with the greatest genome coverage and marker density is that of Liebhard et al. (2003b), with 475 AFLPs, 235 RAPDs, 129 SSRs and 1 SCAR marker. Two parental maps were constructed that spanned 1,140 and 1,450 cM, respectively. While their map was composed primarily with normally segregating markers, several linkage groups were found to carry groups of markers with the same distorted ratios. The highly transferable SSR frame of this map will make it a useful starting point for future *Malus* mapping projects.

In the first QTL study in apple, randomly amplified polymorphic DNAs (RAPDs) were used to locate genes associated with juvenile tree growth and development in the cross between the columnar mutant 'Wijcik McIntosh' and a standard form, disease resistant selection NY 75441-58 (Conner et al. 1998). One to eight QTL were identified for a number of traits including height increment, internode number, internode length, base diameter, branch number and leaf break. The amount of variation explained by regression on individual loci ranged from 3.9 to 24.3, with an average of 7%. Most QTL were significantly associated with a trait in only one or two years.

Several groups in Europe have been especially busy mapping the QTL associated with resistance to apple scab into various linkage groups (LGs). The cross of 'Prima' × 'Fiesta' and other related F_1 progenies have been used to identify major genes associated with resistance in the D.A.R.E. project (Durable Apple Resistance in Europe) (Durel et al. 2002, 2003). The major genes for scab resistance Vg were found on LG 12. Several different NBS-type resistance gene analogues were clustered at bottom of LG 5 and at the top of LG 10. Numerous QTL for partial scab resistance were identified that mapped to four genomic regions. Most of these QTL were race specific with a few exceptions that included a QTL on LG 2 for resistance to races 6 and 7, and a QTL on LG 17 for resistance to races 1 and 6. A major non-race-specific QTL was identified near an NBS-analog cluster on linkage group LG 10. Three major genes for powdery-mildew resistance were also identified by bulked segregant approaches, and one of them on LG 2 was located in the same region as scab resistance.

Vinatzer et al. (2004) used the inverse polymerase chain reaction and simple sequence repeats to identify BAC clones containing the apple scab resistance gene Vf and found the gene in scab-resistant accessions of *Malus micromalus* and 'Golden Gem' of *M. prunifolia*., which were previously not known to carry this gene. They also found a mistake in the published pedigree of the Vf cultivar 'Florina' by comparing SSR patterns of its presumed progenitors to characterized ones.

Five apple progenies were used in the D.A.R.E. project to identify QTL with broad spectrum of resistance towards a wide range of strains of the fungus (Durel et al. 2004). It was verified that four major genomic regions exist that carry resistance to multiple strains of the fungus, with a QTL region on LG 17 carrying the widest spectrum of resistance. Several other linkage groups carry QTL or major resistance genes to specific isolates.

Resistance to apple scab was also mapped in the cross 'Fiesta' × 'Discovery' (Liebhard et al. 2003a and c). Eight genomic regions were identified in this study, with six conferring resistance to leaf scab and two to fruit scab. The amount of variation attributed to the various genes ranged from 4% to 23%, with all but one of the QTL being present across multiple years and locations. Two of the scab resistance QTL reported by Durel et al. (2002) were located on the same linkage groups (LG 10 and 17) and two were not. While 'Discovery' showed more resistance to scab in the field, the most QTL were identified in the more susceptible parent 'Fiesta'. This may indicate that the resistance genes in 'Discovery' are largely homozygous and can not be detected because they do not segregate.

Resistance gene homologues have been mapped in two segregating populations, 'Fiesta' × 'Discovery' (Baldi et al. 2004) and 'Discovery' × TN10-8 (Calenge et al. 2005). The gene homologues are widely distributed across the genome, but often reside in clusters. A high number of the markers mapped close to major genes or QTL for resistance to scab and mildew. Research on nucleotide binding site (NBS)-encoding resistance gene homologs (RGHs) among the Rosaceae revealed synteny of a genomic region that encompasses powdery mildew resistance locus among *Malus*, *Prunus* and *Rosa* (Xu et al. 2007).

Progeny from the cross of 'Prima' × 'Fiesta' were used to detect QTL associated with physical and sensory descriptors related to fruit flesh firmness (King et al. 2000, 2001). Significant QTL were identified on nine linkage groups that were associated with firmness, stiffness, slow breakdown, crispness, granularity, hardness, juiciness, sponginess and overall liking. Considerable variability was noted across years and sites for penetrometer and acoustic resonance readings, and the presence of the QTLs associated with these traits was also highly variable. A highly significant QTL was detected on LG 16 for firmness, crispness, juiciness, sponginess and overall liking. QTL for penetrometer or acoustic resonance measures were not detected in this region, although it did map with *Ma*, the malic acid gene (Maliepaard et al. 1998). Several significant QTL associated with firmness and juiciness on linkage group LG 1 are found in proximity to the locus *Vf*, originating from the scab resistant crab apple *Malus floribunda*.

1.7.2 Regeneration and Transformation

The first apple transformation was done by James et al. (1989) using *Agrobacterium tumefaciens*. Since this seminal study, much work has been conducted to improve the efficiency of gene transfer and regeneration and a wide array of cultivars have been transformed and regenerated (Hammerschlag and Liu 2000, Brown and Maloney 2003). Particle bombardment of apple leaf explants has received much less emphasis, although Gercheva et al. (1994) developed protocols for 'Royal Gala'.

Several genes have been inserted into apple to provide resistance to fungal diseases, although their efficacy in generating pathogen resistance has not yet been published. The stilbene synthase gene from grapes and polygalacturonase-inhibiting protein from kiwi were transferred to 'Holsteiner Cox' and 'Elstar' (Szankowski et al. 2003). The antimicrobial peptide gene *A1-AMP* was incorporated into 'Jonagold' (Broothaerts et al. 2000a). Transgenic lines of 'Orin' and 'JM 7' have been selected with genes encoding chitinase, glucanase and sarcotoxin (Soejima et al. 2000).

Engineering resistance to apple scab has been the focus of several laboratories. Belfanti et al. (2004) isolated the *HcrVf2* gene for apple scab resistance from wild *Malus floribunda* and found it confers resistance in transgenic 'Gala'. Bolar et al. (2001) found that expression of endochitinase from the biocontrol fungus *Trichoderma atroviride* increased resistance to apple scab in transgenic 'Marshall McIntosh'. In other work, this group inserted genes for both endochitinase and exochitinase from *T. atroviride*, and found they had a synergistic activity against the pathogenic fungi *V. inaequalis* (Bolar et al. 2000). Chevreau et al. (2001) inserted the gene for puroindoline-b from wheat in both susceptible and resistant apple cultivars, but did not report on whether resistance was achieved.

A considerable amount of work has been undertaken to develop fire blight resistant apples through genetic engineering (Aldwinckle et al. 2003, Norelli et al. 2003). Initial efforts focused around transferring genes for anti-microbial proteins to apple, including attacin E, avian lysozyme and the cecropin analogs, SB-37 and Shiva-1. The highest levels of resistance were found in the attacin-transgenics, but the cecropin- and avian lysozyme analogues were also effective. In long term field trials, fruit from transgenic lines of 'Royal Gala' and 'Galaxy' containing the antimicrobial genes was indistinguishable from the fruit of non-transformed trees.

In the most recent attempts to engineer resistance to fire blight, antimicrobials proteins have been introduced into apple that act directly against the pathogen, *E. amylovora* (Norelli et al. 2003). Two genes are being investigated: (1) the harpin gene (*hrpN*) from *E. amylovora*, which produces an effector molecule that induces resistance when applied to apple flowers, and (2) genes for DspE-interacting kinases, which interfere with the DspE pathogenesis factor from *E. amylovora*. The gene for NPR1 protein (*MpNPR1*) has also been studied, whose homologue in *Arabidopsis* is thought to be a key regulator in the induction of disease resistance.

In other work on pest resistance, Viss et al. (2003) developed transgenic 'Jonagold' that was resistant to crown gall disease, by inserting genes designed to express double-stranded RNA from the *iaaM* and *ipt* sequences of the ocogenes of *Agrobacterium tumefaciens*. These genes are responsible for the excessive hormone production that leads to gall formation. Markwick et al. (2003) found that apple plants of 'Royal Gala' expressing biotin-binding proteins were resistant to the lightbrown apple moth. Yao et al. (1995) produced 'Royal Gala' plants resistant to the herbicide Glean™ by transforming them with pKILI110, a mutant of the *Arabidopsis* acetolactate synthase gene.

Several transgenes have been shown to have significant effects on apple growth and development. Early flowering was induced in transgenic apple when Bp-MADS4 from silver birch (*Betula pendula* Roth.) was overexpressed (Flachowsky et al. 2007). Holefors et al. (2000) found the *Arabidopsis* phytochrome B gene to reduce shoot, root and plant dry weights in transformed M26 rootstock. Bulley et al. (2005) isolated an apple GA 20-oxidase gene and inserted it into 'Greensleeves' in the sense and antisense orientations and produced dwarf lines with both constructs. Application of GA_3 restored the internode length and number of these transgenic lines, and the scion remained dwarfed after grafting to normal rootstocks. Atkinson et al. (2002) found overexpression of polygalaturonase in transgenic 'Royal Gala' led to a range of novel phenotypes including silvery colored leaves and premature leaf shedding. Mature leaves also had malformed stomata that effected water relations and lead to brittle leaves.

The role A, B, C genes from *Agrobacterium rhizogenes* are known to influence hormone metabolism and root development during infection, and as such have been tested as a means of influencing apple growth and development. The integration of the rolA gene into the genome of apple rootstock A2 reduced plant height and shortened internodes (Zhu et al. 2001a). The transformation of apple rootstock M.9/29 with the rolB gene reduced node number and stem length, but not relative growth rate (Zhu et al. 2001b). Root percentage and root number was increased in shoots of Jork 9 rootstock and the apple scion 'Florina' through the insertion of *rolB* (Sedira et al. 2001, Radchuk and Korkhovoy 2005). Introduction of *rolC* into the apple rootstock 'Marubakaidou' produced four different phenotypes in transformants: a group with reduced height and shortened intervals, a group with reduced height but normal internode lengths, a group with normal height with shortened intervals and a group that was phenotypically similar to control plants (Igarashi et al. 2002).

To reduce browning, Murata et al. (2000) produced transgenic 'Orin' apples carrying the antisense of polyphenol oxidase (PPO). The approach worked, as some of the transgenics had significantly lower levels of PPO than non-transgenic shoots and less browning. Broothaerts et al. (2000b) developed a spectrophotometric assay to rapidly screen PPO activity in apple.

Transgenic apple trees have been produced that possess extra copies of the endogenous S-gene controlling self-incompatibility in apple to induce self-fertility (Broothaerts et al. 2004a,b). In controlled self and outcrosses over a 3-year-period, the transgenic lines had normal levels of fruit and seeds after selfing, while the control plants had significant reductions. The self-fertile transgenic type was associated with an absence of pistil S-RNase proteins.

Gilissen et al. (2005) silenced the major allergen *Mal d* 1 using the RNA interference approach. Allergen levels were reduced but not eliminated.

Transgenic approaches are adding to our knowledge of flavor and ethylene responses. Dandekar et al. (2004) examined the effect of down-regulation of ethylene biosynthesis on fruit flavor complex in apple fruit. In related studies, the relationship of ethylene biosynthesis to volatile production, related enzymes, and precursor availability in apple peel and flesh tissues was studied by Defilippi et al. (2005a,b) who found that alcohol acyltransferase, a rate limiting step for ester biosynthesis important in aroma, is regulated by ethylene.

James et al. (2001) and his group (Gittins et al. 2000, 2001, 2003) have studied the ability of a number of heterologous and homologous promoters to drive expression of β-glucuronidase in tissues of apple. They found the ribulose-1,5-bisphosphate carboxylase/oxygenease small-subunit promoter (*RBCS3C*) from tomato and *SRS1P* from soybean to primarily drive activity in vegetative tissues of apple that had chloroplasts. The *SRS1P* promoter was regulated by light, while *RBCS3C* was not. They also found the *extA* promoter from rape to be very active in all apple tissues, even though its activity is root-specific in its own species. The vascular tissue promoters, *rolC* from *Arabidopsis rhizogenes* and *COYMV* from the *Commelina* yellow mottle virus were found to have localized expression in structural tissues. The group has also been active in identifying ethylene inducible promoters from apple.

Cisgenic approaches, using genes from apple in transformation, is discussed by Jacobsen and Schouten (2007).

1.7.3 Genomic Resources

Mining of existing apple EST information, such as the studies of Newcomb et al. (2006) and Park et al. (2006), and the use of microarrays (Lee et al. 2007, Pichler et al. 2007) promises to expand our knowledge of many genes important in the genetic improvement of apple. The development of public databases such as the GDR (*Genome Database for the Rosaceea*; Jung et al. 2004) and the European HIDRAS AppleBreed (Antofie et al. 2007) also offer excellent prospects for enhanced collaboration amongst breeders, bioinformatics researchers and those involved in molecular biology. In the GDR database alone, over 50,000 ESTs are available from several species, tissues and developmental stages.

References

Aldwinckle HS, Borejsza-Wysocka EE, Malnoy M, Brown SK, Norelli JL, Beer SV, Meng X, He SY, Jin Q-L (2003) Development of fire blight resistant apple cultivars by genetic engineering. Acta Hort 622:105–111

Aldwinckle H, Lamb R, Gustafson H (1977) Nature and inheritance of resistance to *Gymnosporangium juniperi-virginianae* in apple cultivars. Phytopath 67:259–266

Alston FH (1970) Resistance to collar not, *Phytophthora cactorum* (Leb. and Cohn) Schroet in apple. Rep E Malling Research sta. 1969, 143–145

Alston FH (1976) Dwarfing and lethal genes in apple progenies. Euphytica 25:505–514

Alston FH (1977) Practical aspects of breeding for mildew (*Podosphaera leucotricha*) resistance in apples. Proc Eucarpia Symp Tree Fruit Breed 4–13

Alston FH (1981) Breeding high quality high yielding apples. In: Goodenough P, Atkin R (eds) Quality in stored and processed vegetables and fruit. Academic Press, New York, pp 93–102

Alston FH, Briggs JB (1970) Inheritance of hypersensitivity to rosy apple aphid *Dysaphis plantaginea* in apple. Can J Genet Cytol 12:257–258

Alston FH, Briggs JB (1977) Resistance genes in apple and biotypes of *Dysaphis devecta*. Ann Applied Biol 87:75–81

Alston FH, Phillips K, Evans K (2000) A *Malus* gene list. Acta Hort 538:561–570

Alston FH, Watkins R (1973) Apple breeding at East Malling. Proc Eucarpia Fruit Breed Symp 14–29

Antofie A, Lateur M, Oger R, Patocchi A, Durel CE, Van de Weg WE (2007) A new versatile database created for geneticists and breeders to link molecular and phenotypic data in perennial crops: the *AppleBreed Database*. Bioinformatics 23:882–891

Atkinson CJ, Nestby R, Ford YY, Dodds PAA (2005) Enhancing beneficial antioxidants in fruits: a plant physiological perspective. Biofactors 23:229–234

Atkinson RG, Bolitho KM, Wright MA, Iturriagagoitia-Bueno T, Reid SJ, Ross GS (1998) Apple ACC-oxidase and polygalacturonase: ripening-specific gene expression and promoter analysis in transgenic tomato. Plant Mol Biol 38:449–460

Atkinson RG, Schröder R, Hallett IC, Cohen D, MacRae E (2002) Overexpression of polygalacturonase in transgenic apple trees leads to a range of novel phenotypes involving changes in cell adhesion. Plant Physiol 129:122–133

Baldi P, Patocchi A, Zini E, Toller C, Velasco R, Komjanc M (2004) Cloning and linkage mapping of resistance gene homologues in apple. Theor Appl Genet 109:231–239

Belfanti E, Silfverberg-Dilworth E, Tartarini S, Patocchi A, Barbieri M, Zhu J, Vinatzer B, Gianfranceschi L, Gessler C, Sansavini S (2004) The *HcrVf2* gene from a wild apple confers scab resistance to a transgenic cultivated variety. Proc Natl Acad Sci USA 101:886–890

Beuning L, Bowen J, Persson H, Barraclough D, Bulley S, MacRae E (2004) Characterisation of *Mal d 1*-related genes in *Malus*. Plant Mol Biol 55:369–388

Bolar JP, Norelli JL, Harman GE, Brown SK, Aldwinckle HS (2001) Synergistic activity of endochitinase and exochitinase from *Trichoderma atroviride* (*T. harzianum*) against the pathogenic fungus (*Venturia inaequalis*) in transgenic apple plants. Transgenic Res 10: 533–543

Bolar JP, Norelli JL, Wong K-W, Hayes CK, Harman GE, Aldwinckle HS (2000) Expression of endochitinase from *Trichoderma harzianum* in transgenic apple increases resistance to apple scab and reduces vigor. Phytopath 90:72–77

Boss P, Gardner R, Janssen B, Ross G (1995) An apple polyphenol oxidase cDNA is up-regulated in wounded tissues. Plant Mol Biol 27:429–433

Boudichevskaia A, Flachowsky H, Peil A, Fischer C, Dunemann F (2006) Development of a multiallelic SCAR marker for the scab resistance gene *Vr1/ Vh4/Vx* from R12740-7A and is utility for molecular breeding. Tree Genet Genomes 2:186–195

Brooks HJ, Vest G (1985) Public programs on genetics and breeding of horticultural crops in the United States. HortScience 20:826–830

Broothaerts W (2003) New findings in apple-S genotype analysis resolve previous confusion and request the re-numbering of some S-alleles. Theor Appl Genet 106:703–714

Broothaerts W, De Clubber K, Zaman S, Coppens S, Keulemans J (2000a) The feasibilty of fungal disease resistance in apple by expression of antimicrobial peptide genes. Acta Hortic 521:91–94

Broothaerts W, Keulemans J, van Nerum I (2004a) Self-fertile apple resulting from S-RNase gene silencing. Plant Cell Rep 22:497–501

Broothaerts W, McPherson J, Li BC, Randall E, Lane WD, Wiersma PA (2000b) Fast apple (*Malus × domestica*) and tobacco (*Nictotiana tobacum*) leaf polyphenol oxidase activity assay for screening transgenic plants. J Agric Food Chem 48:5924–5928

Broothaerts W, van Nerum I, Keulemans J (2004b) Update and review of incompatibility (S-) genotypes of apple cultivars. HortScience 39:943–947

Brown SK (1992) Genetics of apple. Plant Breed Rev 9:333–365

Brown AG, Harvey DM (1971) The nature and inheritance of sweetness and acidity in the cultivated apple. Euphytica 20:68–80

Brown SK, Maloney KE (2003) Genetic improvement of apple: breeding, markers, mapping and biotechnology. In: Ferree D, Warrington I (eds) Apples: botany, production and uses. CAB International, Wallingford, UK, pp 31–59

Bulley SM, Wilson FM, Hedden P, Phillips AL, Croker SJ, James DJ (2005) Modification of gibberellin biosynthesis in the grafted apple scion allows control of tree height independent of the rootstock. Plant Biotech J 3:215–223

Bus VG, Alspach PA, Hofstee ME, Brewer LR (2002) Genetic variability and preliminary heritability estimates of resistance to scab (*Venturia inaequalis*) in an apple genetics population. NZ J Crop Hort Sci 30:83–92

Bus VGM, Lauren FND, van de Weg WE, Rusholme RL, Rikkerink EHA, Gardiner SE, Basset HCM, Kodde LP, Plummer KM (2005a) The *Vh8* locus of a new gene-for-gene interaction between *Venturia inaequalis* and the wild apple *Malus sieversii* is closely linked to the *Vh2* locus in *Malus pumila* R12740-7A. New Phytologist 166:1035–1049

Bus VGM, Rikkerink EHA, van de Weg WE, Rusholme RL, Gardiner SE, Basset HCM, Kodde LP, Parisi L, Laurens FND, Meulenbroek EJ, Plummer KM (2005b) The *Vh2* and *Vh4* scab resistance genes in two differential hosts derived from Russian Seedling R12740-7A map to the same linkage group of apple. Mol Breed 15:103–116

Caffier V, Parisi L (2007) Development of powdery mildew on sources of resistance to *Podosphaera leucotricha*, exposed to an inoculum virulent against the major resistance gene *Pl-2*. Plant Breeding 126:319–322

Calenge F, Van der Linden CG, Van der Weg E, Schouten HJ, Van Arkel G, Denancé C, Durel CE (2005) Resistance gene analogues identified through the NBS-profiling method map close to major genes and QTL for disease resistance in apple. Theor Appl Genet 110:660–668

Carnes J, Ferrer A, Fernandez-Caldas E (2006) Allergenicity of 10 different apple varieties. Ann Allergy Asthma Immunol 96:564–570

Castiglione S, Pirola B, Sala F, Ventura M, Pancaldi M, Sansavini S (1998) Molecular studies of ACC synthase and ACC oxidase genes in apple. Acta Hortic 484:305–309

Cevik V, King GJ (2000) Molecular genetic analysis of the *Sd1* aphid resistance locus in *Malus*. Acta Hortic 538:553–559

Challice JS (1974) Rosaceae chemotaxonomy and the origins of the Pomoideae. Bot J Linn Soc 69:239–259

Challice JS, Kovanda M (1981) Chemotaxonomic studies in the family Rosaceae and the evolutionary origins of the subfamily Maloideae. Preslia 53:289–304

Chen H, Korban SS (1987) Genetic variability and the inheritance of resistance to cedar-apple rust in apple. Plant Path 36:168–174

Cheng FS, Weeden NF, Brown SK (1996) Identification of co-dominant RAPD markers tightly linked to fruit skin color in apple. Theor Appl Genet 93:222–227

Cheng FS, Weeden NF, Brown SK, Aldwinckle HS, Gardiner SE, Bus VG (1998) Development of a DNA marker for V_m, a gene conferring resistance to apple scab. Genome 41:208–214

Chevreau E, Lespinasse Y, Gallet M (1985) Inheritance of pollen enzymes and polyploid origin of apple (*Malus domestica* Borkh.). Theor Appl Genet 71:268–277

Chevreau E, Dupuis F, Ortolan C, Parisi L (2001) Transformation of apple for durable scab resistance: expression of a puroindoline gene in susceptible and resistant (Vf) genotypes. Acta Hortic 560:323–326

Chyi Y, Weeden N (1984) Relative isozyme band intensities permit the identification of the 2n gamete parent of triploid apple cultivars. HortScience 19:818–819

Conner PJ, Brown SK, Weeden NF (1997) Randomly amplified polymorphic DNA-based genetic linkage maps of three apple cultivars. J Am Soc Hortic Sci 122:350–359

Conner PJ, Brown SK, Weeden NF (1998) Molecular-marker analysis of quantitative traits for growth and development in juvenile apple trees. Theor Appl Genet 96:1027–1035

Costa F, Stella S, Van de Weg E, Guerra W, Cecchinel M, Dallavia J, Koller B, Sansavini S (2005) Role of the genes *Md-ACO1* and *Md-ACS1* in ethylene production and shelf life of apple (*Malus domestica* Borkh.). Euphytica 141:181–190

Crassweller R, Green G (2003) Production and handling techniques for processing apples. In: Ferree DWI (ed) Apples: botany, production and uses. CAB International, Wallingford, UK, pp 615–633

Cummins JN, Forsline PL, Mackenzie JD (1981) Woolly apple aphid colonization on *Malus* cultivars. J Am Soc Hortic Sci 106:26–30

Currie AJ, Ganeshanandam S, Noiton DA, Garrick D, Shelbourne CJ, Oraguzie N (2000) Quantitative evaluation of apple (*Malus domestica* Borkh.) fruit shape by principal component analysis of Fourier descriptors. Euphytica 111:221–227

Crosby JA, Janick J, Pecknold PC, Korban SS, O'Conner PA, Ries SM, Goffreda S, Voordeckers A (1992) Breeding apples for scab resistance:1945–1990. Acta Hortic 317:43–90

Dandekar AM, Teo G, Defilippi BG, Uratsu SL, Passey AJ, Kader AA, Stow JR, Colgan RJ, James DJ (2004) Effect of down-regulation of ethylene biosynthesis on fruit flavor complex in apple fruit. Transgenic Res 13:373–384

Davey MW, Kenis K, Keulemans J (2006) Genetic control of fruit vitamin C. Plant Physiol 142:343–351

Dayton DF, Williams EB (1968) Independent genes in *Malus* for resistance to *Venturia inaequalis*. Proc Am Soc Hortic Sci 92:89–94

Decourtye L (1967) A study of some characters under simple genetical control in apple (*Malus* sp.) and pear (*Pyrus communis*) (in French). Ann Amel Plantes 17:243–265

Decourtye L, Lantin B (1969) Contriution a la connaissance des mutants spurs de pommier heredite du caractere. Ann Amel Plantes 19:227–238

Defilippi BG, Dandekar AM, Kader AA (2005a) Relationship of ethylene biosynthesis to volatile production, related enzymes, and precursor availability in apple peel and flesh tissues. J Agric Food Chem 53:3133–3141

Defilippi BG, Kader AA, Dandekar AM (2005b) Apple aroma: alcohol acyltransferase, a rate limiting step for ester biosynthesis, is regulated by ethylene. Plant Sci 168:1199–1210

Diaz-Perales A, Garcia-Casado G, Sanchez-Monge R, Garcia-Selles FJ, Barber D, Salcedo G (2002) cDNA cloning and heterologous expression of the major allergens from peach and apple belonging to the lipid-transfer protein family. Clin Exp Allergy 32:87–92

Dickson EE, Kresovich S, Weeden NF (1991) Isozymes in North American *Malus* (*Rosaceae*): hybridization and species differentiation. Syst Bot 16:363–375

Dong, Y-H, Yao J-L, Atkinson RG, Putterill JJ, Morris BA, Gardner RC (2000) *MDH1*: an apple homeobox gene belonging to the *BEL1* family. Plant Mol Biol 42:623–633

Dong J, Kim W, Yip W, Thompson GA, Li L, Bennett AB, Yang S (1991) Cloning of a cDNA encoding 1-aminocyclopropane-1-carboxylase synthase and expression of its mRNA in ripening apple fruit. Planta 185:38–45

Dong J, Olson D, Silverstone A, Yang S (1992) Sequence of a cDNA coding for a 1-Aminocyclopropane-1-carboxylate oxidase homolog from apple fruit. Plant Physiol 98:1530–1531

Dong Y, Beuning L, Davies K, Mitra D, Morris B, Koostra A (1998a) Expression of pigmentation genes and photo-regulation of anthocyanin biosynthesis in developing Royal Gala apple flowers. Aust J Plant Physiol 25:245–252

Dong Y, Janssen BJ, Bieleski LR, Atkinson RG, Morris BA, Gardner RC (1997) Isolating and characterizing genes differentially expressed early in apple fruit development. J Am Soc Hortic Sci 122:752–757

Dong Y, Kvarnheden A, Yao J, Sutherland P, Atkinson R, Morris B, Gardner R (1998b) Identification of pollination-induced genes from the ovary of apple (*Malus domestica*). Sexual Plant Reprod 11:277–283

Dong Y-H, Yao J-L, Atkinson RG, Morris BA, Gardner RC (1999) *Mdh3* encoding a *Phalaenopsis O39*-like homeodomain protein expressed in ovules of *Malus domestica*. J Exp Bot 50: 141–142

Dong Y-H, Zhan X-C, Kvarnheden A, Atkinson R, Morris B, Gardner R (1998c) Expression of a cDNA from apple encoding a homologue of *DAD1*, an inhibitor of programmed cell death. Plant Sci 139:165–174

Dunemann F, Bräcker G, Markussen T, Roche P (1999) Identification of molecular markers for the major mildew resistance gene *Pl₂* in apple. Acta Hort 484:411–416

Durel C, Calenge F, Parisi W, van de Weg W, Kodde L, Liebhard R, Kodde LP, Dunemann F, Gennari F, Thiermann M, Lespinasse Y (2004) An overview of the position and robustness of scab resistance QTLs and major genes by aligning genetic maps of five apple progenies. Acta Hortic 663:135–140

Durel CE, Laurens F, Fouillet A, Lespinasse Y (1998) Utilization of pedigree information to estimate genetic parameters from large unbalanced data sets in apple. Theor Appl Genet 96:1077–1085

Durel CE, Lespinasse Y, Calenge F, van der Weg WE, Koller B, Dunemann F, Thiermann M, Evans K, James C, Tartarini S (2002) Four years of the European D.A.R.E. Project: numerous results on apple scab and powdery mildew resistance gene mapping. In: Plant, Animal & Microbe Genomes X Conference, San Diego, California. http://www.intl-pag.org

Durel CE, Parisi L, Laurens F, Van de Weg WE, Liebhard R, Jourjon MF (2003) Genetic dissection of partial resistance to race 6 of *Venturia inaequalis* in apple. Genome 46:224–234

Durel CE, Vande Weg WE, Venisse JS, Parioi L (2000) Localization of a major gene for scab resistance on the European genetic map of the Prima × Fiesta cross. 10BC/WPRS Bull. 23:245–248

Eberhardt MV, Lee CY, Liu RH (2000) Nutrition: antioxidant activity of fresh apples. Nature 405:903–904

Erdin N, Tartarini S, Broggini GAL, Gennari F, Sansavini S, Gessler C, Patocchi A (2006) Mapping of the apple scab-resistance gene *Vb*. Genome 49:1238–1245

Espley RV, Hellens RP, Putterill J, Stevenson DE, Kutty-Amma S, Allan AC (2007) Red colouration in apple fruit is due to the activity of the MYB transcription factor, MdMYB10. Plant J 49:414–427

Evans K, James C (2003) Identification of SCAR markers linked to *Pl-w* mildew resistance in apple. Theor Appl Genet 106:1178–1183

FAOSTAT (2004) Food and agricultural organization of the United Nations. http://faostat.fao.org/site/336.default.asp

Fiala JL (1994) Flowering crabapples: The genus *Malus*. Timer Press, Portland, Oreqon

Flachowsky H, Peil A, Sopanen T, Elo A, Hanke V (2007) Overexpression of *BpMADS4* from silver birch (*Betula pendula* Roth.) induces early-flowering in apple (*Malus × domestica* Borkh.). Plant Breed 126:137–145

Forsline PL (2003) Collection, maintenance, characterization and utilization of wild apples of central Asia. Hortic Rev 29:1–61

Forsline P, Dickson E, Djangalieu A (1994) Collection of wild *Malus, Vitis* and other fruit species genetic resources in Kasakhstan and neighboring republics. HortScience 29:433 (abstract)

Gallott JC, Lamb RC, Aldwinckle HS (1985) Resistance to powdery mildew from some small-fruited *Malus* cultivars. HortScience 20:1085–1087

Gao ZS, van de Weg WE (2006) The *Vf* gene for scab resistance in apple is linked to sub-lethal genes. Euphytica 151:123–132

Gao ZS, van de Weg WE, Schaart JG, Schouten, HJ, Tran DH, Kodde LP, van der Meer, IM, van der Geest AHM, Kodde J, Breiteneder H, Hoffmann-Sommergruber K, Bosch D, Gilissen LJWJ (2005a) Genomic cloning and linkage mapping of the *Mal d 1 (PR-10)* gene family in apple (*Malus domestica*). Theor Appl Genet 111:171–183

Gao ZS, van de Weg WE, Schaart JG, van der Meer IM, Kodde LP, Laimier M, Breiteneder H, Hoffmann-Sommergruber K, and Gilissen LJWJ (2005b) Linkage map positions and allelic diversity of two *Mal d 3* (non-specific lipid transfer protein) genes in the cultivated apple (*Malus domestica*). Theo Appl Genet 110:479–491

Gercheva P, Zimmerman RH, Owens LD, Berry C, Hammersclag FA (1994) Particle bombardment of apple leaf explants influences adventitious shoot formation. HortScience 29:1536–1538

Gessler C, Patocchi A, Sansavini S, Tartarini S, Gianfranceschi L (2006) *Venturia inaequalis* resistance in apple. Crit Rev Plant Sci 25:473–503

Gianfranceschi L, Seglias N, Tartarini R, Komjanc M, Gessler C (1998) Simple sequence repeats for the genetic analysis of apple. Theor Appl Genet 96:1069–1076

Gilissen LJ, Bolhaar ST, Matos CI, Rouwendal GJ, Boone MJ, Krens FA, Zuidmeer L, van Leeuwen A, Akkerdaas J, Hoffmann-Sommergruber K, Knulst AC, Bosch D, van de Weg E, van Ree R (2005) Silencing the major apple allergen *Mal d* 1 by using the RNA interference approach. J Allergy Clin Immunol 115:364–369

Gittins JR, Hiles ER, Pellny TK, Biricolti S, James DJ (2001) The *Brassica napus extA* promoter: a novel alternative promoter to *CaMV 35S* for directing transgene expression to young stem tissues and load bearing regions of transgenic apple trees (*Malus pumila* Mill.). Mol Breed 7:51–62

Gittins JR, Pellny TK, Biricolti S, Hiles ER, Passey AJ, James DJ (2003) Transgene expression in the vegetative tissues of apple driven by the vascular-specific *rolC* and *CoYMV*. Transgenic Res 12:391–402

Gittins JR, Pellny TK, Hiles ER, Rosa C, Biricolti S, James DJ (2000) Transgene expression driven by heterologous ribulose-1,5-bisphosphate carboxylase/oxygenase small-subunit gene promoters in the vegetative tissues of apple (*Malus pumila* Mill.). Planta 210: 232–240

Gygax M, Gianfranceschi L, Liebhard R, Kellerhals M, Gessler C, Patocchi A (2004) Molecular markers linked to the apple scab resistance gene *Vbj* derived from *Malus baccata jackii*. Theor Appl Genet 109:1702–1709

Hammerschlag FA, Liu Q (2000) Generating apple transformants free of *Agrobacterium tumefaciens* by vacuum infiltrating explants with an acidified medium and with antibiotics. Acta Hort 530:103–111

Harada T, Sunako T, Wakasa Y, Soejima J, Satoh T, Nitzeki M (2000) An allele of the 1-aminocyclopropane-1-carboxylate synthase gene (*Md-ACS1*) accounts for the low level of ethylene production in climacteric fruits of some apple cultivars. Theor Appl Genet 101: 742–746

Harker FR, Amos RL, Echeverría G, Gunson FA (2006). Influence of texture on taste: insights gained during studies of hardness, juiciness, and sweetness of apple fruit. J Food Sci 71:S77–S82

Harris SA, Robinson JP, Juniper BE (2002) Genetic clues to the origin of the apple. Trends Genet 18:426–430

Hauagge R, Cummins J (1991) Genetics of length of dormancy period in *Malus* vegetative buds. J Am Soc Hortic Sci 116:121–126

Hemmat M, Brown SK, Weeden NF (2002) Tagging and mapping scab resistance genes from R12740-7A apple. J Am Soc Hortic Sci 127:365–370

Hemmat M, Weeden NF, Conner PJ, Brown SK (1997) A DNA marker for columnar growth habit in apple contains a simple sequence repeat. J Am Soc Hortic Sci 122:347–349

Hemmat M, Weeden N, Manganaris A, Lawson D (1994) Molecular marker linkage map for apple. J Hered 85:4–11

Hoffmann-Sommergruber S (2002) Pathogenesis-related (PR)-proteins identified as allergens. Biochem Soc Trans 30:930–935

Hokanson SC, McFerson JR, Forsline PL, Lamboy WF, Luby J, Djangaliev A, Aldwinckle H (1997) Collecting and managing wild *Malus* germplasm in its center of diversity. HortScience 32:173–176

Holefors A, Xue Z-T, Zhu L-H, Welander M (2000) The *Arabidopsis* phytochrome B gene influences growth of the apple rootstock M26. Plant Cell Reports 19:1049–1056

Igarashi M, Ogasawara H, Hatsuyama Y, Saito A, Suzuki M (2002) Introduction of rolC into Marubakaidou [*Malus prunifolia* Borkh. var. Ringo Asami Mo 84-A] apple rootstock via *Agrobacterium tumefaciens*. Plant Sci 163:463–473

Jacobsen E, Schouten HJ (2007) Cisgenesis strongly improves introgression breeding and induced translocation breeding of plants. Trends Biotech 25:219–223

James CM, Clarke JB, Evans KM (2004) Identification of molecular markers linked to the mildew resistance gene *Pl-d* in apple. Theor Appl Gen 110:175–181

James DJ, Gittins JR, Massiah AJ, Pellny TK, Hiles ER, Biricolti S, Passey AJ, Vaughn SP (2001) Using heterologous and homologous promoters to study tissue specific transgene expression in fruit crops. Acta Hortic 560:55–62

James DJ, Passey AJ, Barbaro DJ, Bevan MW (1989) Genetic transformation of apple (*Malus pumila Mill.*) using a disarmed Ti-binary vector. Plant cell Rep 7:658–661

Janick J, Cummins JN, Brown SK, Hemmat M (1996) Apples. In: Janick J, Moore J (eds) Fruit Breeding, vol. I: Tree and Tropical Fruits. John Wiley & Sons, New York, pp 1–77

Juniper BE, Watkins R, Harris SA (1999) The origin of the apple. Acta Hortic 484:27–33

Jung S, Jesudurai C, Staton M, Du Z, Ficklin S, Cho I, Abbott A, Tomkins J, Main D (2004) GDR (Genome Database for Rosaceae): integrated web resources for Rosaceae genomics and genetics research. BMC Bioinform 5 http://www.biomedcentral.com/1871-2105/5/130

Kenis K, Keulemans V (2005) Genetic linkage maps of two apple cultivars (*Malus × domestica* Borkh) based on AFLP and microsatellite markers. Mol Breeding 15:205–219

Keulemans J, Eyssen R, Colda G (1994) Improvement of seed set and seed germination in apple. In: Schmidt H, Kellerhals M (eds) Progress in temperate fruit breeding. Kluwer Academic Publishers, The Netherlands, pp 225–228

Khan MA, Duffy B, Gessler C, Patocchi A (2006) QTL mapping of fire blight resistance in apple. Mol Breed 17:299–306

Kim MY, Song KJ, Hwang JH, Shin YU, Lee HJ (2003a) Development of RAPD and SCAR markers linked to the *Co* gene conferring columnar growth habit in apple (*Malus pumila* Mill.). J Hortic Sci Biotech 78:512–517

Kim JY, Seo YS, Kim JE, Sung SK, Song KJ, An GH, Kim WT (2001) Two polyphenol oxidases are differentially expressed during vegetative and reproductive development and in response to wounding in the Fuji apple. Plant Sci 161:1145–1152

Kim S-H, Lee J-R, Hong S-T, Yoo Y-K, An G, Kim S-R (2003b) Molecular cloning and analysis of anthocyanin biosynthesis genes preferentially expressed in apple skin. Plant Sci 165:403–413

King GJ, Lynn JR, Dover CJ, Evans KM, Seymore GB (2001) Resolution of quantitative trait loci for mechanical measures accounting for genetic variation in fruit texture of apple (*Malus pumila* Mill.). Theor Appl Genet 100:1227–1235

King GJ, Maliepaard C, Lynn JR, Alston FH, Durel CE, Evans KM, Griffon B, Laurens F, Manganaris AG, Schrevense E, Tartarini S (2000) Quantitative genetic analysis and comparison of physical and sensory descriptors relating to fruit flesh firmness in apple (*Malus pumila* Mill.). Theor Appl Genet 100:1074–1084

Kitahara K, Matsumoto S, Yamamoto T, Soejima J, Kimura T, Komatsu H, Abe K (2005) Parent identification of eight apple cultivars by S-RNase analysis and simple sequence repeat markers. HortScience 40:314–317

Kitahara K, Ohtsubo T, Soejima J, Matsumoto S (2004) Cloning and characterization of apple class B MADS-box genes including a novel *AP3* homologue *MdTM6*. J Jap Soc Hortic Sci 73:208–215

Klein LG (1958) The inheritance of certain fruit characters in the apple. Proc Am Soc Hortic Sci 72:1–14

Knight RL (1962) Fruit breeding. Annu Rpt E Malling Res Sta for 1961, 102–105

Knight RL, Briggs JB, Massee AM, Tydeman HM (1962) The inheritance of resistance to wooly aphid, *Eriosoma lanigerum* (Hausmn.) in the apple. J Hortic Sci 37:207–218

Kochhar S, Watkins C, Conklin P, Brown S (2003) A quantitative and qualitative analysis of antioxidant enzymes in relation to susceptibility of apples to superficial scald. J Amer Soc Hortic Sci 128:910–916

Korban S (1986) Interspecific hybridization in *Malus*. HortScience 21:41–48

Korban S, Dayton D (1983) Evaluation of *Malus* germplasm for resistance to powdery mildew. HortScience 18:219–220

Korban SS, Skirvin RM (1984) Nomenclature of the cultivated apple. HortScience 19:177–180

Korban SS, Swiader JM (1984) Genetic and nutritional status in bitter pit-resistant and susceptible apple seedlings. J Amer Soc Hortic Sci 109:428–432

Kotoda N, Wada M, Komori S, Kidou S-I, Abe K, Masuda T, Soejima J (2000) Expression pattern of homologues of floral meristem identity genes *LFY* and *AP1* during flower development in apple. J Amer Soc Hortic Sci 125:398–403

Kotoda N, Wada M, Kusaba S, Kano-Murakami Y, Masuda T, Soejima J (2002) Overexpression of *MdMADS5*, an *APETALA1*-like gene of apple, causes early flowering in transgenic *Arabidopsis*. Plant Sci 162:679–687

Krebitz M, Wagner B, Ferreira F, Peterbauer C, Campillo N, Witty M, Kolarich D, Steinkellner H, Scheiner O, Breiteneder H (2003) Plant-based heterologous expression of *Mal d 2*, a thaumatin-like protein and allergen of apple (*Malus domestica*), and its characterization as an antifungal protein. J Mol Biol 329:712–730

Kusaba S, Honda C, Kano-Murakami Y (2001) Isolation and expression analysis of gibberellin 20-oxidase homologous gene in apple. J Exp Bot 52:375–376

Labuschagne IF, Louw JH, Schmidt K, Sadie A (2003) Selection for increased budbreak in apple. J Am Soc Hortic Sci 128:363–373

Labuschagne I, Lowu J, Schmidt K, Sadie A (2002) Genetic variation in chilling requirement in apple progeny. J Am Soc Hortic Sci 127:663–672

Lapins KO (1974) Spur type growth habit in 60 apple progenies. J Am Soc Hortic Sci 99:568–572

Lapins KO, Watkins R (1973) Genetics of compact growth p.136. Ann Rpt E Malling Res Sta for 1972

Lata B (2007) Relationship between apple peel and the whole fruit antioxidant content: year and cultivar variation. J Agric Food Chem 55:663–671

Laurens F (1999) Review of the current apple breeding programs in the world: objectives for scion cultivar development. Acta Hortic 484:163–170

Laurens F, Pitiot C (2003) French apple breeding program: a new partnership between INRA and the nurserymen of NOVADI. Acta Hortic 622:575–582.

Lee KW, Kim YJ, Kim D-O, Lee HJ, Lee CY (2003) Major phenolics in apple and their contribution to the total antioxidant capacity. J Agric Food Chem 51:6516–6520

Lee YP, Yu GH, Seo YS, Han SE, Choi YO, Kim D, Mok IG, Kim WT, Sung SK (2007) Microarray analysis of apple gene expression engaged in early fruit development. Plant Cell Rep. Feb. 9 (Epub ahead of print)

Lespinasse Y, Fouillet A, Flick JD, Lespinasse JM, Delort F (1988) Contributions to genetic studies in apple. Acta Hort 224:99–108

Liebhard R, Kellerhals M, Pfammatter W, Jertmini M, Gessler C (2003a) Mapping quantitative physiological traits in apple (*Malus × domestica* Borkh.). Plant Mol Biol 52:511–526

Liebhard R, Koller B, Gianfranceschi L, Gessler C (2003b) Creating a saturated reference map for the apple (*Malus × domestica* Borkh.) genome. Theor Appl Genet 106:1497–1508

Liebhard R, Koller B, Patocchi A, Kellerhals M, Pfammatter W, Jermini M, Gessler C (2003c) Mapping quantitative field resistance against apple scab in a 'Fiesta' × 'Discovery' progeny. Phytopath 93:493–501

Mabberley DJ, Jarvis CE, Juniper BE (2001) The name of the apple. Telopea 9:421–430

Maliepaard C, Alston F, van Arkel G, Brown L, Chevreau E, Dunemann F, Evans KM, Gardiner S, Guilford P, van Heusden AW, Janse J, others (1998) Aligning male and female linkage maps of apple (*Malus pumila* Mill.) using multi-allelic markers. Theor Appl Genet 97: 60–73

Markwick NP, Docherty LC, Phung MM, Lester MT, Murry C, Yao JL, Mitra DS, Cohen D, Beuning LL, Kutty-Amma S, Christeller JT (2003) Transgenic tobacco and apple plants expressing biotin-binding proteins are resistant to two cosmopolitan insect pests, potato tuber moth and lightbrown apple moth, respectively. Transgenic Res 12:671–681

Markussen T, Krüger J, Schmidt H, Dunemann F (1995) Identification of PCR-based markers linked to the powdery-mildew-resistance gene *Pl* (1) from *Malus robusta* in cultivated apple. Plant Breed 114:530–534

Marzban G. Puehringer H, Dey R, Brynda S, Ma Y, Martinelli A, Zaccarini M, van der Weg E, Housley Z, Kolarich D, Altmann F, Laimer M (2005) Localisation and distribution of the major allergens in apple fruits. Plant Sci 169:387–394

Matsumoto S, Kitaharah K, Komatsu H, Abe K (2006) Cross-compatibility of apple cultivars possessing S-RNase alleles of similar sequence. J Hortic Sci Biotech 81:934–936

Moore M (1960) Apple rootstocks susceptible to scab, mildew and canker for use in glasshouse and field experiments. Plant Path 9:84–87

Morgan J, Richards A (1993) The book of apples. Ebury Press LTD, London

Mowry JB, Dayton DF (1964) Inheritance of susceptibility to apple blotch. J Hered 55:129–132

Mowry JB (1964) Inheritance of susceptibility to *Gymnosporangium juniperi-virginianae*. Phytopathology 54:1363–1366

Murata M, Haruta M, Murai N, Tanikawa N, Nishimura M, Homma S, Itoh Y (2000) Transgenic apple (*Malus × domestica*) shoot showing low browning potential. J Agric Food Chem 48:5243–5248

Newcomb RD, Crowhurst RN, Gleave AP, Rikkerink EHA, Allan AC, Beuning LL, Bowen JH, Gera E, Jamieson KR, Janssen BJ, Laing WA, McArtney S, Nain B, Ross GS, Snowden KC, Souleyre EJF, Walton EF, Yauk Y-K (2006) Analyses of expressed sequence tags from apple. Plant Physiol 141:147–166

Norelli JL, Jones AL, Aldwinckle HS (2003) Fire blight management in the twenty-first century. Plant Dis 87:756–765

Nybom N (1959) On the inheritance of acidity in cultivated apples. Hereditas 45:332–350

Oh DH, Song KJ, Shin YU, Chung W-I (2000) Isolation of a cDNA encoding a 31-kDa, pathogenesis-related 5/thaumatin-like (PR5/TL) protein abundantly expressed in apple fruit (*Malus domestica* cv. Fuji). Biosci Biotechnol Biochem 64:355–362

Oraguzie NC, Iwanami H, Soejima J, Harada T, Hall A (2004) Inheritance of the *Md-ACS1* gene and its relationship to fruit softening in apple (*Malus × domestica* Borkh.). Theor Appl Genet 108:1526–1533

Oraguzie NC, Volz RK, Whitworth CJ, Bassett HCM (2007) Influence of Md-ACS1 allelotype and harvest season within an apple germplasm collection on fruit softening during cold air storage. Postharvest Biol Tech 44:212–219

ORourke D (2003) World production, trade, consumption and economic outlook for apples. In: Ferree DC, Warrington I (eds) Apples: botany, production and uses. CAB International, Wallingford, UK, pp 15–29

Park SC, Sugimoto N, Larson MD, Beaudry R, van Nocker, S (2006) Identification of genes with potential roles in apple fruit development and biochemistry through large-scale statistical analysis of expressed sequence tags. Plant Physiol 141:811–824

Patocchi A, Walser M, Tartarini S, Broggini GA, Gennari F, Sansavini S, Gessler C (2005) Identification by genome scanning approach (GSA) of a microsatellite tightly associated with the apple scab resistance gene Vm. Genome 48:630–636

Pechous SW, Whitaker BD (2002) Cloning and bacterial expression of a 3-hydroxy-3-methylglutaryl-CoA reductase cDNA (*HMG1*) from peel tissue of apple fruit. J Plant Physiol 159:907–916

Pechous SW, Whitaker BD (2004) Cloning and functional expression of an (E, E)-α-farnesene synthase cDNA from peel tissue of apple fruit. Planta 219:84–94

Pichler FB, Walton EF, Davy M, Triggs C, Janssen B, Wünsche JN, Putterill J, Schaffer RJ (2007) Relative developmental, environmental, and tree-to-tree variability in buds from field-grown apple trees. Tree Genet Genomics 3:1614–2942

Phillips JB, Robertson KR, Rohrer JR, Smith PG (1991) Origins and evolution of subfam. Maloideae (Rosaceae). Syst Bot 16:303–332

Ponomarenko W (1983) History of apple *Malus domestics*. Both origin and evoluation. Bot Zh USSR 76:10–18

Poupard P, Parisi L, Campion C, Ziadi S, Simoneau P (2003) A wound- and ethephon-inducible *PR-10* gene subclass from apple is differentially expressed during infection with a compatible and an incompatible race of *Venturia inaequalis*. Physiol Mol Plant Path 62:3–12

Puhringer H, Moll D, Hoffman-Sommergruber K, Watillon B, Katinger H, Camara-Machado ML (2000) The promoter of an apple *Ypr 10* gene, encoding the major allergen *Mal d 1*, is stress- and pathogen-inducible. Plant Sci 152: 35–50

Radchuk V, Korkhovoy V (2005) The *rolB* gene promotes rooting in vitro and increases fresh root weight in vivo of transformed apple scion cultivar 'Florina'. Plant Cell Tissue Organ Cult 81:203–212

Roche P, Alston F, Maliepaard C, Evans K, Vrielink R, Dunemann F, Markussen T, Tartarini S, Brown LM, Ryder C, King GJ (1997) RFLP and RAPD markers linked to the rosy leaf curling aphid resistance gene Sd_1 in apple. Theor Appl Genet 94:528–533

Rupasinghe HP, Almquist KC, Paliyath G, Murr DP (2001) Cloning of *hmg1* and *hmg2* cDNAs encoding 3-hydroxy-3-methylglutaryl coenzyme A reductase and their expression and activity in relation to α-farnesene synthesis in apple. PlantPhysiol Biochem 39: 933–947

Sáchez-Torres P, González-Candelas L (2003) Isolation and characterization of genes differentially expressed during the interaction between apple fruit and *Penicillium expansum*. Mol Plant Path 4:447–457

Saito KI, Niizeki M (1988) Fundamental studies on breeding of apple. XI. Genetic analysis of resistance to alternaria blotch (*Alternaria mali* Roberts) in the interspecific crosses. Bulletin of the Faculty Agriculture, Hirosaki University 50:27–34

Sakamoto T, Kusaba S, Kano-Murakami Y, Fukumoto M, Iwahori S (1998) Two different types of homeobox genes exist in apple. J Jap Soc Hort Sci 67:372–374

Sandanayaka WRM, Bus VGM, Connolly P, Newcomb R (2003) Characteristics associated with woolly apple aphid *Eriosoma lanigerum*, resistance of three apple rootstocks. Entomologia Experimentalis et Applicata 109:63–72

Sax K (1931) The origins and relationships of the Pomoideae. J Arn Arbor 12:3–22

Schirrmacher G, Schempp H (2003) Antioxidative potential of flavonoid-rich extracts as new quality marker for different apple varieties. J Appl Bot 77:163–166

Schmitz-Eiberger M, Weber V, Treutter D, Baab G, Lorenz J (2003) Bioactive components in fruits from different apple varieties. J Appl Bot 77:176–171

Sedira M, Holefors A, Welander M (2001) Protocol for transformation of the apple rootstock Jork 9 with the *rolB* gene and its influence on rooting. Plant Cell Rep 20:517–524

Seymour GB, Manning K, Ericksson EM, Popovich AH, King GJ (2002) Genetic identification and genomic organization of factors affecting fruit texture. J Exper Bot 53: 2065–2071

Silfverberg-Dilworth E, Matasci CL, Van de Weg WE, Van Kaauwen MPW, Walser M, Kodde LP, Soglio V, Gianfranceschi L, Durel CE, Costa F, Yamamoto T, Koller B, Gessler C, Patocchi A (2006) Microsatellite markers spanning the apple (*Malus × domestica* Borkh.) genome. Tree Genet Genomes 2:202–224

Soejima J, Abe N, Kotoda N, Kato H (2000) Recent progress of apple breeding at the Apple Research Center in Morioka. Acta Hortic 538:211–214

Souleyre EJF, Greenwood DR, Friel EN, Karunairetnam S, Newcomb RD (2005) An alcohol acyl transferase from apple (cv. Royal Gala), MpAAT1, produces esters involved in apple fruit flavor. FEBS J 272:3132–3144

Stanley D, Fitzgerald AM, Farnden KJ, MacRae EA (2002) Characterization of putative α-amylases from apple (*Malus domestica*) and *Arabidopsis thaliana*. Biologia Bratislava 57:137–148

Sung SK, An G (1997) Molecular cloning and characterization of a MADS-Box cDNA clone of the Fuji apple. Plant Cell Physiol 38:484–489

Sung SK, Jeong DH, Nam J, Kim SH, Kim SR, An G (1998) Expressed sequence tags of fruits, peels, and carpels and analysis of mRNA expression levels of the tagged cDNAs of fruits from the Fuji apple. Mol Cells 8:565–577

Sung SK, Yu GH, An G (1999) Characterization of *MdMADS2*, a member of the *SQUAMOSA* subfamily of genes, in apple. Plant Physiol 120:969–978

Sung SK, Yu GH, Nam J, Jeong DH, An G (2000) Developmentally regulated expression of two MADS-box genes, *MdMADS3* and *MdMADS4*, in the morphogenesis of flower buds and fruits in apple. Planta 210:519–528

Szankowski I, Briviba K, Fleschhut J, Schönherr J, Jacobsen H-J, Kiesecker H (2003) Transformation of apple (*Malus domestica* Borkh.) with the stilbene synthase gene from grapevine (*Vitis vinifera* L.) and a PGIP gene from kiwi (*Actinidia deliciosa*). Plant Cell Rep 22:141–149

Tao R, Uratsu SL, Dandekar AM (1995) Sorbitol synthesis in transgenic tobacco with apple cDNA encoding NADP-dependent sorbitol-6-phosphate dehydrogenease. Plant Cell Physiol 36:525–532

Takos AM, Jaffé FW, Jacob SR, Bogs J, Robinson SP, Walker AR (2006) Light induced expression of a MYB gene regulates anthocyanin biosynthesis in red apples. Plant Physiol 142:1216–1232

Tartarini S, Sansavini S (2003) The use of molecular markers in pome fruit breeding. Acta Hort 622:129–140

Tartarini S, Gennari F, Pratesi D, Palazzetti C, Sansavini S, Parisi L, Fouillet A, Fouilet V, Durel CE (2004) Characterisation and genetic mapping of a major scab resistance gene from the old Italian apple cultivar 'Durello di Forli'. Acta Hortic 663:129–133

Thomsen KA, Eriksen EN (2006) Effect of temperatures during seed development and pretreatment on seed dormancy of *Malus sargentii* and *M. sieboldii*. Seed Sci Tech 34:215–220

Tian YK, Wang CH, Zhang JS, James C, Dai HY (2005) Mapping Co, a gene controlling the columnar phenotype of apple, with molecular markers. Euphytica 145:181–188 van der Linden CG, Vosman B, Smulders MJ (2002) Cloning and characterization of four apple MADS box genes isolated from vegetative tissue. J Exper Bot 53:1025–1036

Vavilov N (1949–1950) The origin, variation, immunity and breeding of cultivated crops. Chronica Botanica, Waltham, Massachusetts

Vinatzer B, Patocchi A, Gianfranceschi L, Tartarini S, Zhang H, Gessler C, Sansavini S (2001) Apple contains receptor-like genes homologous to the *Cladosporium fulvum* resistance gene family of tomato with a cluster of genes cosegregating with *Vf* apple scab resistance. Mol Plant Microbe Interact 14:508–515

Vinatzer B, Patocchi A, Tartarini S, Gianfranceschi L, Sansavini S, Gessler C (2004) Isolation of two microsatellite markers from BAC clones of the *Vf* scab resistance region and molecular characterization of scab-resistant accessions in *Malus* germplasm. Plant Breed 123:321–326

Viss WJ, Pitrak J, Humann J, Cook M, Driver J, Ream W (2003) Crown-gall-resistant transgenic apple trees that silence *Agrobacterium tumefaciens* oncogenes. Mol Biol 12:283–295

Visser T (1951) Floral biology and crossing technique in apples and pears. Meded Dir Tuinb 14:707–726

Volz RK, Alspach PA, White AG, Ferguson IB (2001) Genetic variability in apple fruit storage disorders. Acta Hort 553:241–244

Wada M, Cao Q-F, Kotoda N, Soejima J-I, Masuda T (2002) Apple has two orthologues of *FLOR-ICAULA/LEAFY* involved in flowering. Plant Mol Biol 49:567–577

Watillon B, Kettmann R, Boxus P, Burny A (1997) *Knotted1*-like homeobox genes are expressed during apple tree (*Malus domestica* [L.] Borkh) growth and development. Plant Mol Biol 33:757–763

Watkins R (1995) Apple and pear. In: Smartt J, Simmonds NW (eds) Evolution of Crop Plants. Longman, London

Watkins R, Spangelo LP (1970) Components of genetic variance for plant survival and vigor of apple trees. Theor Appl Genet 40:195–203

Watkins R, Werts JM (1971) Preselection for *Phytophthora cactorum* (Leb. & Con) Schroet. resistance in apple seedlings. Ann Appl Biol 67:153–156

Way R, Aldwinckle H, Lamb R, Rejman A, Sansavini S, Shen T, Watkins R, Westwood M, Yoshida Y (1991) Apples (*Malus*). In: Moore J, Ballington J (eds) Genetic resources in temperate fruit and nut crops. Acta Hortic 290:3–46

Way RD, Lamb RC, Pratt C, Cummins JN (1976) Pale green lethal gene in apple clones. J Am Soc Hortic Sci 101:679–684

Webster AD, Wertheim SJ (2003) Apple rootstocks. In: Ferree DC, Warrington I (eds) Apples: botany, production and uses. CAB International, Wallingford, UK, pp 91–124

Weeden NF, Hemmat M, Lawson DM, Lodhi M, Bell RL, Manganaris AG, Reish BI, Brown SK, Ye GN (1994) Development and application of molecular marker linkage maps in woody fruit crops. Euphytica 77:71–75

Weeden NF, Lamb RC (1987) Genetics and linkage analysis of 19 isozyme loci in apple. J Am Soc Hortic Sci 112:865–872

Wegrzyn T, Reilly K, Cipriani G, Murphy P, Newcomb R, Gardner R, MacRae E (2000) A novel α-amylase gene is transiently upregulated during low temperature exposure in apple fruit. Eur J Biochem 267:1313–1322

Xu Q, Wen X, Deng X (2007) Phylogenetic and evolutionary analysis of NBS-encoding genes in Rosaceae fruit crops. Mol Phylogenet Evol 44:315–324

Yamada K, Mori H, Yamaki S (1999) Identification and cDNA cloning of a protein abundantly expressed during apple fruit development. Plant Cell Physiol 40:198–204

Yamada K, Oura Y, Mori H, Yamaki S (1998) Cloning of NAD-dependent sorbitol dehydrogenase from apple fruit and gene expression. Plant Cell Physiol 39:1375–1379

Yao C, Conway WS, Ren R, Smith D, Ross GS, Sams CE (1999a) Gene encoding polygalacturonase inhibitor in apple fruit is developmentally regulated and activated by wounding and fungal infection. Plant Mol Biol 39:1231–1241

Yao J-L, Dong Y-H, Kvarnheden A, Morris B (1999b) Seven MADS-box genes in apple are expressed in different parts of the fruit. J Am Soc Hortic Sci 124:8–13

Yao J-L, Cohen D, Atkinson R, Richardson K, Morris B (1995) Regeneration of transgenic plants from the commercial apple cultivar Royal Gala. Plant Cell Rep 14:407–412

Yao J-L, Dong Y-H, Morris B (2001) Parthenocarpic apple fruit production conferred by transposon insertion mutations in a MADS-box transcription factor. Proc Nat Acad Sci USA 98:1306–1311

Yao Y-X, Li M, Liu Z, Hao Y-H, Zhai H (2007) A novel gene, screened by cDNA-AFLP approach, contributes to lowering the acidity of fruit in apple. Plant Physiol Biochem 45:139–145

Zhang WB, Zhang JR, Hu XL (1993) Distribution and diversity of *Malus* resources in Yunnan, China. HortScience 28:978–980

Zhu L-H, Ahlman A, Li X-Y, Welander M (2001a) Integration of the *rolA* gene into the genome of the vigorous apple rootstock A2 reduced plant height and shortened internodes. J Hortic Sci Biotech 76:758–763

Zhu L-H, Holefors A, Ahlman A, Xue Z-T, Welander M (2001b) Transformation of the apple rootstock M.9/29 with the *rolB* gene and its influence on rooting and growth. Plant Sci 160:433–439

Zohary D, Hopf M (1993) Domestication of Plants in the Old World: the Origin and Spread of Cultivated plants in West Asia, Europe and the Nile Valley, 2nd. Clarendon Press, Oxford

Zhou-Zhi Q (1999) The apple genetic resources in China: the apple species and their distributions, informative characteristics and utilization. Genet Res Crop Evol 46:599–609

Chapter 2
Apricots

C.A. Ledbetter

Abstract Several dozen publicly-sponsored breeding programs around the world are developing new fresh market and processing apricot cultivars. Apricots have a more limited environmental range than other tree fruits, and therefore, many breeders are interested in broadening adaptations for specific growing regions. Plum Pox Virus resistance is a widely pursued objective and there are ongoing efforts to identify molecular markers that are closely linked to disease resistance. Fruit sugars, acids, pigments and volatile aromatic compounds are being quantified in newly bred and historically important cultivars. Researchers have identified and characterized several stylar ribonucleases associated with self-unfruitfulness. Molecular phylogenetic studies are examining the dispersion routes of apricot germplasm from its centers of origin to those cultivars currently in production. Although several linkage maps have been developed using diverse parents and a wide variety of molecular markers from apricot and other *Prunus* crops, the scarcity of documented monogenic characters in apricot limits the effectiveness of marker assisted selection for economically important traits.

2.1 Introduction

Prunus armeniaca L. is not a true native to the plains of Armenia, but it has been continuously cultivated there since at least the first century AD. It was brought to Armenia from a more eastern center of origin much earlier as evidenced by archeological excavations at pre-Christian sites. Since those early times, Armenian foods, traditions and folklore have been influenced by the presence of apricot in the region. Perhaps due to its early ripening season, its unique and pleasant aroma, or its high nutritive content and ability to be processed into a non-perishable sustaining ration, early explorers and conquerors brought apricot with them to foreign

C.A. Ledbetter
USDA-ARS, 9611 S. Riverbend Avenue, Parlier, California, USA
e-mail: cledbetter@fresno.ars.usda.gov

lands. No attempt will be made here to convey what are currently accepted as the dissemination routes of apricot from its natural centers of origin, as an excellent review article on this subject was published recently (Faust et al. 1998).

Throughout the world apricot is considered to be among the most delectable of all fruits, with flowers, fruit and tree playing parts in various traditions of diverse human cultures. Fruit are used in both fresh and dry form, canned or otherwise preserved as jam and marmalade or pulp. Wines and distillates made from both cultivated and non-domesticated apricot are traditional beverages in parts of both Europe and Asia (Joshi et al. 1990, Genovese et al. 2004).

Since the early 1990s, both fruit tonnage and orchard area have been increasing in African and Asian countries, whereas European and South American countries have realized increased apricot production on fewer hectares of orchard. Fifty countries are listed by FAO as having annual production in excess of 1,000 Mt with Turkey being the largest current apricot producer (370,000 Mt). Three other countries (Iran, Italy and Pakistan) now have annual production in excess of 200,000 Mt. Half of the world's orchard area and nearly half of all apricot production comes from Asiatic countries. Fruit production and orchard area are both declining in North American and Oceanic growing regions. Taken as a whole, apricot fruit production and harvested orchard area are both increasing on a worldwide basis, with 2005 levels of fruit tonnage and orchard area standing at 2.8 million Mt and 434,000 ha, respectively (FAOSTAT, 2006).

2.2 Evolutionary Biology and Germplasm Resources

2.2.1 Taxonomy

Botanists in Western countries have historically placed apricots within the plant family *Rosaceae*, subfamily *Prunoideae*, tribe *Pruneae* and the genus *Prunus*. Depending on the botanical authority, opinions have been mixed on whether apricot should be placed within the sub-genera *Prunophora* or *Amygdalus*, as apricot shares some morphological and pomological characteristics of both (Zielinski 1977). Leaves emerging from dormant buds are open and in a whorl, or *convolute*, as described by Bailey (1916) for the plums, prunes and apricots of the *Prunophora*, whereas the leaves of almonds and peaches in the sub-genus *Amygdalus* have *conduplicate* leaves – folded along the midrib as they emerge from dormant buds. Genetic linkage maps based on several types of molecular markers have shown a high degree of colinearity between an F_1 progeny population from 'Polonais' × 'Stark Early Orange' apricots and an almond × peach F_2 population, indicating a very similar genomic structure between *Prunophora* and *Amygdalus* (Lambert et al. 2004). A recent investigation into the overall genetic diversity of *Prunus* based on random amplified polymorphic DNA (RAPD) analyses place apricots well within the sub-genus *Prunophora* and apart from the sub-genus *Amygdalus* (Shimada et al. 2001).

Early botanical descriptions of the different apricot species were based primarily on leaf shape and pubescence, and these characters were not always consistent between specimens. Bailey's (1916) categorical distinctions of apricot species and botanical varieties used leaf characteristics. The classification by Rehder (1940) distinguished plums (Sections Euprunus and Prunocerasus) from apricots (Section Armeniaca) on the basis of ovary pubescence, being absent or glabrous in the plums and present or pubescent in apricots. *P. brigantina* Vill. (syn. *P. brigantiaca*), a glabrous apricot, was a noted exception to the Rehder (1940) scheme. Table 2.1 provides a comparison of the classification of apricots by Bailey (1916) and Rehder (1940).

The taxonomy of apricots by Chinese investigators was also based mainly on leaf characteristics. China's immense size and varied topography, as well as its numerous geographic and climatic zones provided enormous genetic diversity in many plant families to Chinese botanists that were unknown to their counterparts in Western countries (Hou 1983). Apricot classification in China parallels that of Western taxonomists to the subfamily level (*Prunoideae*). At this point, the *Prunoideae* is divided into nine genera: *Prinsepia, Pygeum, Maddenia, Amygdalus, Armeniaca* (apricots), *Prunus, Cerasus, Padus* and *Laurocerasus* (Gu et al. 2003). These authors point out the complexity of taxonomy within *Rosaceae*, and the fact that some of the listed genera within the *Prunoideae* have been grouped together by other authorities. The genus *Armeniaca* is divided into 10 species (Lingdi and Bartholomew 2003), with mention made of an 11th species that is not present in

Table 2.1 Comparison of apricot classification schemes as suggested by Bailey (1916) and Rehder (1940)	Bailey scheme	Rehder scheme
	Prunus (genus)	*Prunus* (genus)
	Prunophora (sub-genera)	*Prunophora* (sub-genera)
	(plums, prunes & apricots)	(sections)
		Euprunus (European/Asian plums) Prunocerasus (N. American plums) Armeniaca (apricots)
	P. armeniaca L.	*P. brigantina* Vill.
	Var. *pendulata* Dipp.	*P. mandshurica* Maxim.
	Var. *variegata* Hort.	*P. sibirica* L.
	Var. *sibirica* Koch	*P. armeniaca* L.
	Var. *mandshurica* Maxim.	pomological varieties:
	Var. *Ansu* Maxim.	*P. a. variegata* Schneid.
	P. mume Sieb. & Zucc.	*P. a. péndula* Jaeg.
	Var. *Goethartiana* Koehne.	*P. a. Ansu* Maxim.
	Var. *albo-plena* Hort.	*P. mume* Sieb. & Zucc.
	other forms:	pomological varieties:
	laciniata Maxim.	*P. m. alba* Rehd.
	microcarpa Makino	*P. m. Alphandii* Rehd.
	viridicalyx Makino	*P. m. albo-plena* Bailey
	cryptopetala Makino	*P. m. Péndula* Sieb.
	P. brigantiaca Vill.	*P. m. tonsa* Rehd.
	P. dasycarpa Ehrh.	*P. dasycarpa* Ehrh.

China (*P. brigantiaca*). Five of the 10 listed species (Table 2.2) were not described by either Bailey (1916) or Rehder (1940).

The Desert apricot (*P. fremontii* S. Wats.) also deserves mention among the listed apricot species even though it is not mentioned by any of the above listed authorities. First described in 1880 during a geological survey of California, it was probably unknown or of no interest to Bailey (1916) and Rehder (1940) purposefully excluded North American trees and shrubs from subtropical and 'warmer temperate' regions in his classification key. Being a native to the Mohave and Sonoran deserts, it was naturally not mentioned by Gu et al. (2003) among the *Rosaceae* of China. *P. fremontii* has been represented in some recent molecular studies (Bortiri et al. 2001, 2002) where it has been classified within section Penarmeniaca, along with the desert dwelling species *P. andersonii* A. Gray. While *P. fremontii* differs in many ways from the other mentioned apricot species, it can hybridize freely with them and has morphological characteristics that resemble other apricot species.

From the breeding perspective, more important than the apricot species' placement in any particular classification key are their relevant characteristics that might

Table 2.2 Classification of Chinese apricot germplasm by Lingdi and Bartholomew (2003)

Rosaceae (family)
  Prunoideae (subfamily)
    Armeniaca (genus) – apricots
      A. vulgaris L.
        Var. *vulgaris* L.
        Var. *zhidanensis* Qiao & Zhu
        Var. *ansu* Maxim.
        Var. *meixianensis* Zhang
        Var. *xiongyueensis* Li

      A. limeixing Zhang & Wang
      A. sibirica L.
        Var. *sibirica* L.
        Var. *pubescens* Kostina
        Var. *multipetala* Liu & Zhang
        Var. *pleniflora* Zhang

      A. holosericea Batal.
      A. hongpingensis Li
      A. zhengheensis Zhang & Lu
      A. hypotrichodes Cardot
      A. dasycarpa Ehrh.
      A. mandshurica Maxim.
        Var. *mandshurica* Maxim.
        Var. *glabra* Nakai

      A. mume Sieb. & Zucc.
        Var. *mume* Sieb.
        Var. *pallescens* Franc.
        Var. *cernua* Franc.
        Var. *pubicaulina* Qiao & Shen

make them useful in an apricot improvement program. Key traits and general geographic origins are listed by apricot type in Table 2.3 without regard to their particular classification as distinct species or pomological/botanical varieties.

2.2.2 Eco-Geographical Groups of Apricot

Plant exploration throughout Asia by scientists of the former Soviet Union during the early part of the 20th century led to the discovery of three centers of origin for apricot. One was located in the mountainous regions and adjacent lands of central and western China known as the 'Chinese Center'. Apricot was among the more than 30 temperate fruit-producing crops listed for this region. Nine *Prunus* species including *P. armeniaca* and *P. mume* Sieb. & Zucc. were identified, as were quinces, walnuts, pecans and hazelnut species. A second region with apricots, the 'Inner-Asiatic Center' was defined by the approximate boundaries of northwestern India, Afghanistan, Tajikistan, Uzbekistan and the western Tien-Shan mountains. Smaller in land area than the Chinese center, the Inner-Asiatic center still contained many fruit and nut crops. Besides *P. armeniaca*, this region is known as the center of origin to species of *Vitis, Pistacia, Pyrus, Malus* and *Juglans*. The last center of origin for apricot described by Vavilov (1992) was known as 'Asia Minor Center'. The region is defined as the lands of Transcaucasia, Iran and Turkmenistan. More than 15 genera of fruit crop plants were identified in this region.

The Transcaucasian lands were exemplified as being particularly rich in diversity of fruit crops, in all stages of evolutionary development. Forests composed almost entirely of wild fruit trees could be found throughout the region. When clearing forested areas for agricultural development or timbering, the most horticulturally valuable wild fruit trees would be left in place, and local growers were known to graft the most valuable forms onto less desirable seedling fruit trees. Particularly promising local selections were saved and sometimes named (i.e. 'Shalah' apricot) as their popularity increased and they eventually found their course into more mainstream horticulture (Mirzaev and Kuznetsov 1984).

A result of the Russian exploration and collection expeditions was the establishment of a large apricot collection from the different centers of origin at the Nikiti Botanical Garden near Yalta, Ukraine. Over 700 apricot accessions were established there for evaluation and breeding of better adapted types. Kostina (1936) developed a classification key that could characterize any given apricot accession on seed taste (sweet or bitter), type of skin (pubescent or glabrous), stone separation (cling or freestone), flesh color (white and/or cream or yellow and/or orange) and fruit size (small, medium or large). In studying the apricot collection at the Botanical Garden, Kostina (1936) originally described three distinct eco-geographical groups of apricot based on discrete fruit characteristics. Further work by Kostina (1969) re-divided the diverse apricot germplasm into the now well known four eco-geographical groups and 13 sub-groups (Table 2.4).

Table 2.3 Key botanical traits and geographic origin of currently accepted apricot species

Species	Tree stature	Leaf specifics	Bloom specifics	Fruit specifics	Geographic origin
ansu					
Ansu apricot	Bush-like	Broad & elliptic with characteristic cuneate base, highly glabrous	Usually pink petals	Sometimes cultivated, deep & distinct suture, sweet gray-brown flesh, freestone	1000–1500 m in Central China, humid climate
armeniaca					
Common apricot	5–12 m	Ovate to round-ovate, sl. cordate at base, sharply pointed	Variable bloom date	Small to v. large, variable flesh color, glabrous & pubescent forms, bitter to sweet kernels	0–3000 m in Central & West China, Mongolia Central Asia, Japan & Korea
brigantiaca					
Alpine apricot	Small bush to 3 m	Broad-oval to ovate, abruptly short-pointed	Light pink petals, clusters of 2 to 5 blooms	Barely edible, white to yellow skin, glabrous, kernel oil for cosmetics	Foothills of the French Alps
dasycarpa					
Black apricot	Small stature tree, 4–7 m	Elliptic-ovate with closely serrate margin, smaller than common apricot	Large, showy long-stalked flowers	Pubescent purple skin, plum-like with soft/sour flesh, rarely cultivated	A natural plum-apricot hybrid, perhaps only native to Manchuria
fremontii					
Desert apricot	deciduous shrub or small tree, 1–4 m	Roundish to ovate (2 cm) with short petioles, glabrous	Complex branching habit yields intense bloom, flowers small (1 cm) with white petals	Round to elliptic and small (15 mm dia.), yellow/orange with dry flesh, freestone	below 1200 m on dry slopes & canyons of Southern California deserts

Table 2.3 (continued)

Species	Tree stature	Leaf specifics	Bloom specifics	Fruit specifics	Geographic origin
holosericea					
Tibetan apricot	Small tree, 4–5 m	Plum-like form, large & ovate with short petiole, heavy red pubescence on leaf mid-rib underside	Currently unknown	Densely pubescent with a small amount of flesh, medium fruit size, non-dehiscent, large pitted stone	Qinghai, Shaanxi, Sichuan and Yunnan provinces of China 700–3300 m
hongpingensis	Up to 10 m	Elliptic to elliptic-ovate with densely pubescent petioles	Currently unknown	Edible and cultivated locally, 3 to 4 cm fruit width	Western Hubei and Hunan Chinese providences
hypotrichodes	Shrub up to 3 m	Long lanceolate blades with short glabrous petioles, small lanceolate & caducous stipules	Single flowers, glabrous pedicle, blooms before leaf emergence	Style base & ovary are velvety, – other fruit specifics are unknown	Calcareous mountains of China's Chongqing prov. approx. 1400 m
limeixing	Small spreading tree of 3–4 m	Elliptic to obovate-elliptic, glabrous petioles, pubescence in abaxial vein axes	Solitary or groups to 3, 1.5–2.5 cm dia., white petals	Variable skin & flesh color, deep suture, sweet-sour flesh, clingstone	Natural plum-apricot hybrid cultivated in northeastern China, now unknown in the wild

Table 2.3 (continued)

Species	Tree stature	Leaf specifics	Bloom specifics	Fruit specifics	Geographic origin
mandshurica					
Manchurian apricot	5–15 m	Round with long sharp tip, double serrate margin	Typically more pink than white	Generally edible, but flesh sometimes sour, bitter or dry, with apricot aroma	200–1000 m in N.E. China, Korea & Eastern Russia
mume					
Japanese apricot	4–10 m, variable bark color	Narrow to round-ovate, long pointed tip, leaves crinkle naturally	Smaller than common apricot, variable color, double-flower varieties, very early bloom	Generally small with dry flesh, clingstone, highly pitted stones	Below 3100 m in Western Sichuan & Yunnan prov. of China and in Tiawan
siberica					
Siberian apricot	Bush-like or small tree, 5 m	Small, ovate to round, glabrous, but sometimes with pubescence on lower surface	Early and profuse, white to pink color	Small, dry & typically splitting, generally inedible	400–2500 m in diverse habitat, North & Eastern China, Mongolia, Transbaikal Steppe
zhengheensis	Erect form, 35–40 m	Elliptic or oblong with red glabrous petioles, 2–6 nectaries mid-petiole	White petals, 3 cm dia., flowers open before leaves emerge	Typically yellow with possible red blush, succulent & sweet flesh, clingstone & non-dehiscent	Mountainous regions of China's Fujian provi.

Table 2.4 Eco-geographical groups and regional sub-groups of ordinary apricot as defined by Kostina (1969)

Eco-geographical group	Regional sub-group
Central Asian	Fergana
	Horezm
	Kopet-Dag
	Samarkand
	Sharessyabz
	Verhnezeravshan
Irano-Caucasian	Dagestan
	Irano-Transcaucasian
European	Eastern European
	Western European
	Ukrainian
Dzhungar-Zailij	Dzhungar
	Zailij

On an evolutionary timescale, the Dzhungar-Zailij group is said to be the most primitive whereas the European group is believed to be the most recently developed and the product of apricot dispersion from the other eco-geographical groups. Apricots from Central Asia are certainly the most diverse group in their fruit, vegetative and phenological characters. Central Asian apricot trees are generally very long-lived, and they have a longer juvenile period prior to fruit production. Kostina (1936) initially subdivided the Central Asian apricots into two regional sub-groups (Fergana and Samarkand) and evaluations of nearly 300 apricots types from these regions in 1928–1929 exemplified the general diversity present in the apricots of these regions (Table 2.5). Two glabrous skinned Central Asian apricot accessions from the Vavilov Research Institute's Central Asian Station in Tashkent, Uzbekistan became available to US apricot breeders upon their release from quarantine in 1993 (Fig. 2.1). Apricots from the Irano-Caucasian group are also very diverse, but are generally shorter-lived than those from Central Asia. 'Shalah' (syn. 'Erevani'), a widely grown apricot from the Irano-Caucasian group, survived plant protective quarantine in the United States and became available to US breeders and interested growers in the late 1990s. Extremely late ripening apricot forms are also present in the Irano-Caucasian germplasm. 'Levent' apricot, from the Anatolia region of Turkey, is said to have a fruit development period of 190–200 days (Asma and Ozturk 2005). Importation and utilization of this germplasm in breeding programs would undoubtedly assist in the extension of the fruit maturation period.

2.3 History of Improvement

2.3.1 Historical Breeding/Selection Efforts

Selection of apricots with superior qualities and their clonal propagation began around 600 AD in China (Faust et al. 1998), and possibly as early in other regions. This is not to say that orchard establishment through the planting of apricot seed

48   C.A. Ledbetter

Table 2.5 Differences in fruit characteristics of 273 Central Asian apricot accessions from the Fergana and Samarkand regional sub-groups of Uzbekistan as evaluated during 1928–1929

Fruit character	Percentages of sub-group with this character	
	Fergana sub-group	Samarkand sub-group
Glabrous skin	5	38
Pubescent skin	95	62
Sweet kernel	97	96
Freestone pit	90	85
Clingstone pit	10	15
Large fruit size (>35 g)	3	21
Medium fruit size (20–35 g)	55	57
Small fruit size (<20 g)	42	22

disappeared at this time, as seed-propagated apricot orchards are still commonplace today in some East Asian (Geuna et al. 2003) and North African regions (Khadari et al. 2006).

Apricot breeding perhaps began accidentally after the development of grafting and budding. An astute grower might have selected a few superior trees from seed-propagated orchards, and then passed them on to friends and/or neighbors who clonally propagated them. If this were to happen either simultaneously, or even over the course of many years within a geographical region, an individual grower might find numerous and distinct selected clones within his/her orchard area. Given that these early orchards probably contained self-incompatible trees, fruit within the orchard would have arisen from cross-pollinations only. Without any knowledge of plant

Fig. 2.1 Compared with 'Patterson' apricot (*lower right*), the glabrous skin of the other Central Asian apricot accessions is quite evident. F₁ hybrids between glabrous and pubescent skinned apricots are typically pubescent

breeding, the utilization of seed from the fruits in this sort of orchard would lead to growing out of seedlings from cross-pollination of selected clones. Since it has now been clearly shown that the selection of phenotypic superior parental choices leads to significant genetic gain in apricot (Couranjou 1995, Bassi et al. 1996), the next generation of trees from seed-propagated orchards should have yielded new and variable trees worthy of selection and further propagation.

Named cultivars of apricot began appearing in the European written record during the 1600s, although apricot had been introduced to these areas many centuries before. It appears that these named cultivars were the product of selection only, from seed propagated orchards, or by chance seedlings that developed on their own. Nonetheless, some of these apricots have been important in various regions since their discovery, and have now been used extensively as parents in planned hybridizations. A listing of some of the more important historical apricot cultivars is presented in Table 2.6.

2.3.2 Current Breeding Efforts

Breeding efforts on stone and pome fruits have historically been conducted by public institutions (Table 2.7), with European breeding programs accounting for the majority of apricot improvement. Nikita Botanical Gardens in Yalta, Ukraine, is the longest ongoing apricot breeding program, beginning in 1925, while the majority of the other breeding programs began their work on apricot between the 1960s and 1980s. New cultivar introductions from these programs have numbered 150 during the last 15 years. Besides improved fruit quality traits, environmental adaptation and resistance to diseases are major objectives in many breeding programs. The extension of the fruit ripening season is also a current breeding objective for several programs. Hybridizations between locally adapted apricot accessions and Central Asian germplasm are being performed in several programs to attain that goal (Benedikova 2004, Ledbetter and Peterson 2004).

2.3.3 Repositories and Research Institute Holdings of P. armeniaca Germplasm

Considerable amounts of apricot germplasm are being held in repositories for research purposes and conservation of the species. A recent (May 2006) search of the International Plant Genetic Resources Institute (IPGRI) database revealed 62 separate research locations with holdings of *P. armeniaca* germplasm (Table 2.8). Over 6,000 accessions (with duplications) reside at these institutions in the 30 listed countries.

Table 2.6 Notable apricot cultivars in recorded history

Cultivar	Year selected or discovered	Remarks	Reference
Roman	ancient Rome	Most widely grown until 'Moor Park'	Faust et al. 1998
Shalah	unknown	Progenitor of numerous later cultivars, landrace from Armenia, still widely planted in Ararat valley	Mirzaev and Kuznetsov 1984
Nancy	1755	Disc. near Nancy, France, progenitor of numerous later cultivars, many synonyms inc. 'Peach-Apricot'	Bordeianu et al. 1967
Moor Park	1760	Superior to all previously grown apricots, sel. by Admiral Lord Anson near Watford, Herefordshire, England	Faust et al. 1998
Royal	1808	French origin, disc. by M. Hervy, seedling of 'Nancy', named by King Louis XVIII, France	Bordeianu et al. 1967
Blenheim	Bef. 1830	Syn. 'Shipley', intro. by Miss Shipley Blenheim daughter to gardener of Duke of Marlborough Blenheim, England	Hedrick 1925
Luizet	1838	Chance sdlng. found by G. Luizet, widely adapted to Europe & N. Africa	Löschnig and Passecker 1954
Hungarian Best	1868	Disc. and named by E. Lucas in Enyed, Hungary	Löschnig and Passecker 1954
Bergeron	1820	Chance sdlng. of exceptional flavor, from seed obtained at St-Cyr au Mount d'Or, Rhône, France, sel. by M. Bergeron	Lichou and Audubert 1989
Stark Earli-Orange	1920	Disc. in Grandview, Washington by W. Roberts, late-blooming apricot used extensively for resistance to sharka	Brooks and Olmo 1972
Scout	1937	Intro. by Dominion Expt. Sta. in Morden, Manitoba, Canada, sel. from seed sent by Expt. Sta. of Eastern Siberian Railway, Echo, Manchuria	Brooks and Olmo 1952
Perfection	1937	Orig. in Waterville, Washington U.S.A., unknown parents, sel. from seed planted in 1911, progenitor of many N. American cultivars	Hesse 1952

Table 2.7 Current public–sponsored apricot breeding programs throughout the world[1]

Institution & Location	Year Program Began	Named apricot cultivars since 1991	Current Breeding Objectives
South Australian Research & Development Institute Loxton, South Australia	1982	Rivergem (1995), River Ruby (2005), Riverbrite (2005), Rivergold (2005)	For fresh, dry & processing markets, fruit quality traits (flavor, size, fruit color & firmness, high TSS)
Byelorussian Research Institute for Fruit Growing Samokhvalovitchy, Minsk Region, Belarus	1935	Znahodka (1995), Govorukhin's Memory & Memory Loyko (2004), Spadchyna (2005)	First objective is extreme winter-hardiness, tolerance to *Cladosporium carpophilum* & *Monilinia laxa*
Liaoning Institute of Pomology Yingkou, P. R. China	2000	Luotuo Huang (1995), Chuanzhi Hong (1997), Fengren & Guoren (2000)	Fruit quality traits for fresh and drying markets (firm flesh, strong aroma, attractiveness, freestone, high TSS)
Research & Breeding Institute of Pomology Holovousy Ltd., Horice, Czech Rep.	1972	Darina (1999), Kompakta (1999), M-HL-1 rootstock (2002)	New cvs. for fruit quality and appearance, resistance to late frosts and brown rot, compact growth.
Mendel Agriculture and Forestry University in Brno. Horticulture Faculty of Lednice, Lednice, Czech Rep.	1977	Leala & Lebela (1995), Ledana, Legolda, Lejuna, Lemeda, Lenova & Lesorka (1999), Marlen, Minaret, Palava & Svatava (since 2000)	First priority is PPV resistance, also frost hardiness & fruit quality traits (fruit size, firmness, attractiveness, high TSS)
National Agricultural Research Foundation – Pomology Institute, Naoussa, Greece	1982	Lito & Pandora (1991), Neraida, Niobe, Nomia, Nastasia, Nina, Nausika, Nefele, Nostos & Nereis (2001), Tyrbe (2002)	New cvs. for canning & fresh market, PPV resistance is 1st selection criteria Self-compatibility, local adaptation & fruit quality (size, flavor, firmness, color)
Instituto Sperimentale per la Frutticoltura – Sezione de Caserta, Caserta, Italy	1986	No introduced cultivars yet	New cvs. for fresh & processing markets, extended fruit ripening season, high & regular productivity, high quality, Sharka & *Monilinia* resistant.

Table 2.7 (continued)

Institution & Location	Year Program Began	Named apricot cultivars since 1991	Current Breeding Objectives
Instituto de Coltivazioni Arboree – University of Milan, Milano, Italy	1980	Cora & Ninfa (1993), Boreale (1995), Bora (2002), Ardore & Pieve (2004), Priscilla (2006)	Environmental adaptation (rain crack resistant, frost tolerance), PPV & *Monilinia* resistance, self fertility, fruit quality, wide ripening season
Dipartimento de Coltivazione e Difesa Delle Specie Legnose, University of Pisa, Pisa, Italy	1980	Dulcinea & Pisana (1992), Milady (1997), Ardenza, Bona, Cabiria, Kinzica, Maharani, Piera & Salambo (2001), Angela, Caludia, Gheriana & Silvana (2005)	Improved eating quality with good postharvest characters, extension of ripening period, late flowering & environmental adaptation, Sharka & *Monilinia* resistance.
National Agriculture and Food Research Organization, National Institute of Fruit Tree Science, Tsukuba / Ibaraki, Japan	1970	Hachirou & Kagajizou (1997) (these are Japanese apricot – *P. mume*)	*P. armeniaca*: High eating quality, disease & freeze resistance, longevity, self-compatibility. *P. mume*: Self-compatibility, disease resistance, late bloom, early fruit maturity, processing ability.
The Botanical Gardens of the University of Latvia, Riga, Latvia	Late 1940s	Lasma, Daiga & Velta (1999), Jausma & Rasa (2004)	Winter hardiness (late bloom & deep dormancy), fruit quality (freestone, large size, early harvest & attractiveness), tolerance to *Monilinia* & leaf spot
Horticulture and Food Research Institute of New Zealand Ltd., Havelock North, Hawke's Bay, New Zealand	1976	Cluthastar (1991), Cluthalate & Cluthasun (1992), Cluthaearly (1993), Alex, Benmore, Dunstan, Gabriel & Vulcan (1997), Cluthafire & Mascot (1998)	New cvs. with large size, attractive & with good eating quality, good adaptation, precocity & productivity, expansion of ripening season (both early and late).

Table 2.7 (continued)

Institution & Location	Year Program Began	Named apricot cultivars since 1991	Current Breeding Objectives
Baneasa Research & Development Station for Fruit Tree Growing, Bucharest, Romania	1967	Comandor, Excelsior, Favorit & Olimp (1994), Carmela, Dacia, Rares, Sirena & Viorica (2002), Adina, Andrei, Nicusor & Valeria (2004)	Variety development for fresh & industry, disease and pest resistance, climatic adaptability & productivity, extension of fruit ripening season
Irkursk State University, Botanical Gardens Irkursk, Russia	1968	Lubímii (1996), Solnishko (1998), Four advanced selections now in registration process	Cold hardiness & local adaptation are prime objectives, high fruit quality, late bloom, dwarf tree stature
Russian Academy of Science Main Botanical Garden, Moscow, Russia	1957	Aisberg, Alyosha, Favorit, Grafinya, Lel, Monastyrsky, Tsarsky & Vodoley (2005)	New variety development (fresh & processing) for climate of Moscow, reliable long-lived rootstocks
Research Breeding Station, Vesele Piestany, Slovak Rep.	1964	Vesna, Vegama, Veharda, Velbora & rootstock MY-VS-1 (1991), Vesprima & Barbora (1996), Vestar (1997), Veselka, Vemina & Velita (1999)	Resistance to spring frosts, late blooming, high fruit quality, extended fruit season, flesh firmness, processing suitability, disease resistance (*Monilinia*, *Gnomonia*, PPV, ESFY)
Agricultural Research Council of South Africa, Stellenbosch, South Africa	1940s	Ladisun (1991), Charisma (2005)	New variety development for fresh markets (enhanced postharvest quality), canning and drying
Centro de Edafología y Biología Aplicada del Segura. Consejo Superior de Investigaciones Científicas, Murcia, Spain	1986	Rojo Pasión (2001), Selene (2002), Murciana & Dorada (2003)	Self-compatible cultivars of high fruit quality and productivity, early ripening, Sharka resistance
Instituto Valenciano de Agrarias Investigaciones, Valencia, Spain	1993	Two advanced selections are currently in the registration process	Resistance to PPV of prime consideration, early season fruit, fruit quality traits (size, blush, firmness, attractiveness, Brix : Acid ratio)

Table 2.7 (continued)

Institution & Location	Year Program Began	Named apricot cultivars since 1991	Current Breeding Objectives
Estación Experimental de Aula Del Consejo Superior de Investigaciones Científicas, Zaragoza, Spain	1998	No introduced cultivars yet	Rootstock breeding, interspecific hybridization to obtain graft-compatible stocks adapted to heavy & calcareous soils
Institut National de Recherche Agronomiques de Tunis, Tunis-Ariana, Tunisia	1955	Asli, Atef, Fakher, Meziane, Ouafer & Raki (1995)	Combining early-ripening with fruit quality traits (color, firmness, size, sugar & aroma)
Alata Horticultural Research Institute, Mersin, Turkey	1944	Alata Yıldızı, Çağrıbey, Çağataybey, Dr. Kaşka, Şahinbey	New cultivars for fresh market, combine early-ripening with high fruit quality, *Capnodis* resistance
Apricot Research & Application Center of Inonu University, Malatya, Turkey	1996	No introduced cultivars yet	Both fresh and drying types, extended fruit ripening season, Sharka resistance
Mustafa Kemal University, Antakya, Hatay, Turkey	1995	No introduced cultivars yet	Combining superior fruit quality from 'Sakit' population with early – ripening (earlier than 'Ninfa')
Nikita Botanical Gardens National Scientific Center Yalta, Crimea, UKRAINE	1925	Burevestnik (1991), Forum (1992), Krympsk Amur (1993), Aviator (1995) Autok, Alyanc, Divnee Zorkee, Krokus, Pamyati Arevoy & Shedevr (2005)	Introduction of diverse germplasm for hybridization & selection in creating highly adaptable new varieties
Department of Plant Biology & Pathology Rutgers University New Brunswick, NJ UNITED STATES	1955	SunGem (1994), Earlyblush & NJA82 (1995), NJA97 (1996), NJA150 (2006)	Improved cold hardiness, bacterial resistance, Sel. for high quality & attractiveness, extend the ripening season, novel characters (cream flesh & glabrous skin)
USDA / Agricultural Research Service San Joaquin Valley Agricultural Sciences Center parlier, CA UNITED STATES	1955	Helena (1994), Robada (1997), Lorna (1998), Apache (2002), Nicole (2003), Kettleman (2005)	Fresh and processing markets Fruit quality is 1st criteria, wide adaptation, novel fruit characters (modified sugar profile, white flesh, glabrous skin), increased ripe season

[1] Apricot breeders at 27 public-sponsored programs around the world responded to a short query to gather information on new cultivars and breeding program objectives. Other public-sponsored breeding programs might certainly exist, but information is available from only those programs where a query response was received.

Table 2.8 *Prunus armeniaca* L. germplasm resource holdings at national repositories and research institutes as listed by the International Plant Genetic Resources Institute (IPGRI)

Country	No. Accessions	Last Updated[1]
Albania	28	September 1991
Argentina	50	May 2003
Australia	693	October 1990
Brazil	4	May 1999
Canada	294	January 1994
Chile	19	October 1990
Ecuador	7	August 1990
France	406	May 2002
Greece	18	February 2003
Hungary	472	February 1995
India	28	October 1990
Israel	132	March 1995
Italy	1358	March 1995
Macedonia	56	October 1990
Mexico	200	April 1999
Morocco	68	October 1990
Netherlands	4	August1994
Norway	2	April 2002
Pakistan	32	July 1994
Poland	76	March 2003
Portugal	97	November 1994
Serbia & Montenegro	65	April1995
Slovakia	319	April 2002
South Africa	73	October 1990
Spain	212	July 2002
Switzerland	87	April 1995
Turkey	109	May 2002
Ukraine	873	August 1995
United Kingdom	2	April 1995
United States	417	April 2002

[1]Indicates date when IPGRI was last contacted by respective country. IPGRI repository database was queried during May 2006 for accession counts. Current database queries are found at: http://web.ipgri.cgiar.org/germplasm/default.asp

2.4 Problems of Genetic Significance

2.4.1 Fruit Quality

Enhanced fruit quality is the universal goal of all tree fruit breeders. Fruit quality must be sub-divided and specific characteristics evaluated by the breeder in order to measure genetic gain from planned hybridizations. Individual characters that collectively comprise fruit quality include fruit size and the degree of flesh firmness, aroma and flavor characters, color of flesh, skin and overcolor (blush) and fruit juiciness. Each of these characters can be measured objectively with appropriate instrumentation. Couranjou (1995) demonstrated that good genetic gain is possible in apricot breeding by choosing parents based on fruit phenotype. Thus, parental apricots used in hybridizations that are markedly superior in specific aspects of fruit quality (high overcolor, strong aroma & flavor, or large fruit size) generally pass along those quality characteristics to the next generation of seedlings (Fig. 2.2). Breeding programs based on apricots from the European eco-geographic group could benefit substantially in the development of higher quality fruit by utilizing germplasm from the other eco-geographical groups.

In a principal component analysis of 55 European apricot cultivars, Badenes et al. (1998) demonstrated a significant negative correlation between harvest season and fruit acidity. The lack of sweetness in early season apricots is a common consumer complaint, and a fact that can limit repeat sales of apricots later in the season. Similarly, fruit cracking was also found to be most common in the early maturing apricots. The lack of appropriate parental choices for these characteristics among European apricot clones limits genetic gain. Fruit with lower acidity and a lower potential for cracking in the early harvest season will be common only when parental germplasm with these potentials are identified and utilized in the breeding program's hybridizations.

2.4.2 Self-Compatibility

The self-(in)compatibility status of a tree is an important consideration for both breeders and producers. Opinions are divided with regard to the utility of this trait. Fully self-compatible cultivars can be grown as a monocultural system, eliminating potential problems at bloom and during the harvest period(s) that one might have growing two or more self-incompatible varieties. However, excessive fruit set can

Fig. 2.2 Utilizing apricot accession 'Habiju' (Central Asian germplasm) in hybridizations with California adapted 'Lorna' apricot, and the effect on fruit size. The 'Hibiju' fruit weigh 14 g, while the fruit of 'Lorna' weigh 117 g and the F_1 hybrids weigh 80 g

sometimes be a problem in a self-compatible orchard, and thinning costs can reduce the producer's profit margin significantly. At the same time, fruit set might be ensured in an orchard with a self-compatible cultivar during bloom periods when poor weather conditions limit bee pollination. Through trial and error, fruit set in self-incompatible apricot varieties can be manipulated by the relative number and distribution of pollinator trees in the orchard; however, weather conditions must allow adequate bee visitation.

Self-compatibility in apricot is determined by a single allele, S_c, in a multiallelic series of a monofactorial system (Burgos et al. 1997), analogous to the well-defined system in *P. dulcis* (Mill) D.A. Webb. The locus is found on linkage group 6 (Vilanova et al. 2003). The status of self-compatibility in a given tree can be determined by numerous means, and methods have grown increasingly more complex with advanced methodology. Pre-anthesis bagging of blooming branches with insect-proof bags was probably the first method employed as a means of identifying those trees capable of self-pollination. Self-pollinations and fluorescence microscopy have been utilized very effectively (Burgos et al. 1993), in particular when poor weather conditions might question the validity of bagging studies in the field.

Cross-incompatibility in apricot was first detected amongst three American apricot cultivars all having 'Perfection' apricot in their parentage (Egea and Burgos 1996). Being the first incompatibility group described in apricot, these three cultivars ('Goldrich', 'Hargrand' and 'Lambertin-1') received an identical genotype with the allelic designation $S_1 S_2$. This information is used as a starting point for further testing to find other alleles for self-incompatibility.

2.4.3 Bloom Period and Frost Tolerance

The early bloom period of apricot has limited its cultivation in some areas where it is safe to grow other stone fruits. Freezing temperatures of only a few hours in the late spring can diminish the chance for an economic yield. Breeding programs in regions where this is a problem typically have late blooming periods as major breeding priorities. Bloom date for a given cultivar is determined by both its chilling requirement and its necessary heat unit accumulation after the chilling requirement is fulfilled (Brown 1957, Cesaraccio et al. 2004). Germplasm that is consistently productive in a region where late spring frosts are problematic typically have high heat unit requirements. Selections made from native seedling populations in a region where frequent late frosts occur would undoubtedly be good starting points in hybridization programs.

Apricot germplasm collected from the Hunza region of northern Pakistan was brought to the United States in 1988 for evaluation and breeding (Thompson 1998). While not well adapted to the hot dry conditions present in California's San Joaquin Valley, the imported Hunza apricot accessions did flower significantly later than California adapted apricots. The full bloom date of some of the Hunza apricots averaged 30 days later than that of apricots typical to California (Ledbetter and Peterson 2004). Hybridizations between California adapted apricots and the Hunza types yielded F_1 trees that segregated widely in bloom date.

An extended bloom period is another means of achieving fruit set in regions plagued by less than optimal spring weather (Benedikova 2004). Sufficient variability in bloom period exists in specific germplasm within some of the eco-geographical groups such that exploitation through breeding could benefit apricot producers in regions where late frosts are ever-present. Irano-Caucasian apricot germplasm collected from Anatolia, Turkey varied greatly in average bloom date, with approximately one month difference between early and late-blooming cultivars (Asma and Ozturk 2005). The Erzincan plain of Turkey is also said to have large native seedling apricot populations from which late blooming forms can be selected (Ercisli 2004).

2.4.4 Disease Resistance

Numerous diseases plague apricot trees in the various growing regions of the world, and the development of resistance to these diseases is a major goal for many breeding programs. Some of these diseases are very widespread, while others are restricted to specific growing regions. *Monilinia laxa* Honey (Brown rot) is perhaps the most widespread and damaging fungal disease for apricot and numerous cultivars have been noted from the different eco-geographical groups that tolerate or resist the disease. A new brown rot fungus, *Monilinia mumecola* Harada, Sasaki & Sano, was recently isolated and characterized from *P. mume* trees infected in Oita Prefecture, Kyushu, Japan (Harada et al. 2004). The seriousness of this new *Monilinia* outside of its point of discovery, and its effect on *P. armeniaca* cultivars is not yet known. Powdery mildew (*Podosphaera tridactyla* DeBary) is also a widespread disease with far fewer sources of resistance or tolerance available. *Xanthomonas campestris* pv. *pruni* Young et al. is responsible for bacterial spot, a disease affecting both foliage and fruit. 'Harcot' and 'Harglow' are two apricots from Ontario, Canada that are said to be resistant to both foliar and fruit infections (Layne 1984). *P. salicina* × *P. mume* hybrids 'PM-1-1' and 'PM-1-4' have also been described as tolerant of bacterial spot (Kyotani et al. 1988). Shothole (*Stigmina carpophila* Ell.) is a prevalent fungal disease in Eastern Europe, and field observations of an apricot collection under disease pressure have shown wide variability in symptom expression (Smykov 1978). The viral disease plum pox or Sharka is becoming increasingly more important in apricot growing regions, as it is disseminated to previously Sharka-free regions by unknowing nursery persons or careless producers. Bacterial canker (*Pseudomonas syringae* van Hall) and Eutypa dieback (*Eutypa lata* Tul.) are two other serious diseases capable of killing trees with a single infection. While both diseases are limited geographically in distribution, there are few cultivars currently known that adequately resist infection.

'Blackheart' of apricot, caused by the fungus *Verticillium dahliae* Kleb., is particularly troubling in orchards where the land was previously occupied by susceptible agronomic crops. Several apricot species (*P. armeniaca, P. ansu, P. mandshurica* and *P. siberica*) have shown susceptibility to this soilborne fungus in controlled greenhouse tests (Gathercole et al. 1987). However, apricot orchards can be easily protected from *Verticillium dahliae* through the use of widely available plum (resistant) rootstocks. *P. armeniaca* is generally regarded as being uniformly immune to root knot (*Meloidogyne* sp.) nematode species and several selections of *P. mume* and *P. dasycarpa* have also shown resistance in field trials (Day 1953, Yoshida 1981). In addition, the use of *P. mume* as a rootstock protects against crown gall (*Agrobacterium tumefaciens*), whereas the *P. armeniaca* seedling rootstock 'Manicot GF 1236' has been proven to be very susceptible (Lichou and Audubert 1989).

2.5 Genetics of Important Traits

2.5.1 Male Sterility

Evidence presented by Burgos and Ledbetter (1994) indicated that male sterility is controlled by a single recessive gene, as in peach. Seedling populations segregating for this character demonstrated that the fresh market cultivar 'Helena' was heterozygous for male-sterility. The Spanish cultivars 'Gitano' and 'Pepito' (syn. 'Pepito del Rubio') have also been shown to be heterozygous for male sterility in controlled crosses (Burgos et al. 1998). Similar to the male sterile peach cultivar 'J.H. Hale', apricot cultivars 'Arrogante' and 'Colorao' have been found to be male sterile and require cross pollination with a pollen compatible male fertile apricot (García et al. 1988).

Male sterility is considered by most breeders to be a fatal flaw for an otherwise superior apricot clone. Since heterozygous individuals have fully functional pollen and are indistinguishable from homozygotes, crossing amongst heterozygotes might be quite common in some breeding programs. Seedling progenies might be left unscreened for male sterility as other duties during the bloom period could have priority over examining whole progenies for this visually apparent character. It might be only at a time near variety release that a breeder discovers that an elite clone is pollen sterile. The fact that male sterility is a discrete trait discernible only after the tree becomes reproductive makes it an excellent candidate for marker assisted selection in a breeding program. Bulk segregant analysis (BSA) was used by Badenes et al. (2000) on a segregating seedling population of apricot in an attempt to identify RAPD markers linked to the male sterility trait. Out of 228 primers used in the analysis, only primer 'M4-950' (Operon Technologies) was found to be loosely linked to male fertile trees.

2.5.2 Amygdalin Content of Seed

Amygdalin content of apricot seed was evaluated by Gómez et al. (1998) in a study designed to examine correlation between phenotypic expression (sweet or bitter seed) and actual amygdalin content. The chromatographic data revealed large numeric differences between sweet and bitter seeded apricot cultivars. Among the bitter phenotypes, significant differences did exist in actual amygdalin content, indicating that perceived bitterness could be influenced by factors other than amygdalin concentration. However, the extent of the numeric differences in amygdalin content between sweet and bitter seeded accessions indicated discrete classes or apricot seed, and not continuous variation. The inheritance of apricot seed taste had been studied previously by Kostina (1969), who determined it to be a simply inherited single gene trait with sweet kernel being dominant to bitter kernel.

2.5.3 Sugars, Acids and Nutrient Composition

Sucrose is the primary sugar present in apricot fruit. Several other sugars such as glucose, fructose, maltose, sorbitol and raffinose are also present to lesser and varying degrees (Witherspoon and Jackson 1995). Collectively, the combined concentrations of the sugars present in apricot are known as the sugar profile, and while absolute concentrations of each sugar in a given accession changes year to year, their relative ratios remain quite constant within any given variety (Bassi et al. 1996). The glucose: fructose ratio has been suggested as an indicator of juice/pulp authenticity for apricot, and glucose: fructose ratios higher than 3.3 suggest adulteration with other less expensive juice/pulp additives (Lo Voi et al. 1995). However, this study was conducted with pulp from 11 apricot varieties common to Italy. Authentic samples of 'Lorna' apricot, developed in Central California, have higher fructose levels that boost the glucose: fructose ratio to 4.6 (Ledbetter et al. 2006).

Malic and citric acids are both typically present in apricot fruit, but the predominant acid is dependent on the particular apricot accession. Gurrieri et al. (2001) studied the patterns of sugars and acids in fruit from 51 diverse apricot varieties and found that they differed greatly with respect to the levels of malic and citric acids. Malic acid predominated in 14 of the 51 sampled varieties, and no significant correlations were noted between the levels of malic and citric acids. These authors suggested that taste panels should be used in conducting correlation studies between organoleptic quality and both the levels and kinds of sugars and acids present in fruit. Apricot breeders could also exploit the observed diversity in sugar and acid contents through the employment of appropriate instrumentation as a part of the fruit evaluation procedures.

The consumption of carotenoids in the diet is associated with a degree of protection against cancers and cardiovascular diseases. Total carotenoid content of fruit was associated with the general flesh color class of the apricots (white, yellow, light orange or orange), with light orange and orange apricots having significantly more carotenoid than the white or yellow fleshed accessions sampled in the study (Ruiz et al. 2005a). The absolute determination of carotenoids in fruit requires precise extraction procedures and a diode array detector equipped HPLC. These analyses can be both time consuming and expensive for a breeding program. Ruiz et al. (2005a) however was able to find a very strong correlation between fruit flesh hue angle and carotenoid content. In a study involving 37 diverse apricot accessions, white fleshed apricots (hue angle $\sim 88°$) were found lowest in total carotenoids (2,450 mg/100 g fresh fruit). Apricots of the orange flesh class (hue angle $\sim 72°$) were highest in total carotenoids (12,750 mg/100 g fresh fruit). Given these reported findings, it seems reasonable that the apricot breeder can utilize color meter readings to identify those apricot accessions most rich in carotenoid content.

A similar study attempted to correlate apricot fruit color coordinates with the absolute phenolic content of the fruit. Unlike carotenoids, phenolic content of fruit was not related to flesh color. Neither total phenolics nor any specific class of phenolic compounds (procyanidins, hydroxycinnamic acid derivatives, flavonols or anthrocyanins) could be correlated with flesh hue angle or other color coordinates (Ruiz et al. 2005b). Therefore, if the breeder's intention is to identify apricots with particularly high or low levels of phenolic compounds, direct extractions of these compounds are the only reliable means of determining their specific quantity. While extraction procedures for phenolics are not difficult or involved, analysis of phenolic extracts requires authentic samples of the phenolics in question, and a HPLC equipped with diode array detection capability. Levels of both rutin (quercetin-3-O-rutinoside) and astragalin (kaempferol-3-O-glucoside) were found to differ significantly among apricot accessions in both mature and meristematic leaves harvested and extracted during the active growth season (Ledbetter et al. 2000). Rutin and astragalin are both important diagnostic phenolic compounds in the authentication of jams, marmalades and nectars produced from apricot fruit (Tomás-Lorente et al. 1992, Dragovic-Uzelac et al. 2005).

2.5.4 Aroma and Flavor in Apricot

Tang and Jennings (1967 and 1968) were among the first to conduct research on the aroma profiles of apricot. Several methods of extraction were used by these researchers so that chromatographic profiles could be compared and volatile artifacts detected that were products of any given extraction procedure. Headspace analysis of volatiles from intact fruit as well as simultaneous vacuum steam distillation-extraction of fresh fruit slurries have been used by other researchers to identify and quantify the compounds responsible for typical apricot aroma (Takeoka et al. 1990, Gómez 1993).

The specific methods used for extracting volatiles from apricot determine what will be separated by the gas chromatogram. Heating of the fruit slurry at any time during the extraction procedure increases detection of low-boiling point compounds whereas solvent elutions (without heat) of trapped headspace volatiles favor higher boiling point constituents. Regardless of the extraction methods used, researchers are in agreement that natural apricot aroma is complex, and the profile of volatile constituents is composed of dozens of compounds from many different classes of chemicals. A wide variety of hydrocarbons, ketones, alcohols, aldehydes, esters and lactones have been identified from both apricot headspace gasses (Gómez and Ledbetter 1993) and solvent extraction-distillation procedures (Guichard and Fournier 1990). Further, there is no clear consensus of the exact mixture of aroma constituents that is responsible for a 'typical' apricot aroma (Guichard 1990).

Varietal differences in apricot aromatic profiles have been documented by comparing apricots grown with the same cultural management/environment and extracted under similar conditions (Guichard 1995, Ledbetter et al. 1996a). When comparisons of aromatic profiles are made between fruit varieties, careful consideration must be taken for having fruit of equal maturity, as many of the key volatile constituents responsible for apricot aroma increase dramatically as fruit approach full maturity (Gómez and Ledbetter 1997). From the perspective of apricot breeding, Couranjou (1995) estimated heritability of fruit aroma for apricot at $h^2 = 0.603$, similar to that of fruit size, fruit firmness and flesh color. Thus, appropriate parental choices of apricots based on their specific phenotype (i.e. high perceived aroma) generally leads to overall improvement of that selected characteristic in the successive seedling population. As a specific example, Gómez et al. (1993) observed paternal transmission of specific volatile constituents from apricot to plum × apricot progeny. Levels of g-decalactone and g-dodecalactone were quantitatively high in plum × apricot seedlings' fruit when the apricot parent's fruit was also high in these important apricot volatiles.

2.5.5 Plum Pox Virus Resistance

Plum pox virus (PPV) or sharka disease has been devastating to the stone fruit industry in Europe during the 20th century. It was originally described in Bulgaria around 1918 and spread throughout Eastern Europe. Two major isolates or forms of PPV, Dideron (D-type) and Marcus (M-type), have been described and characterized in Europe (Candresse et al. 1994), and four other forms are now known to exist. D-type isolates of PPV have recently been identified in apricot and plum accessions at a germplasm repository in Kazakhstan (Spiegel et al. 2004), as well as from a commercial apricot/plum orchard in San Juan Providence, Argentina (Zotto et al. 2006). A mixed infection of PPV (PPV-D and PPV-Rec) has recently been detected in an orchard from Pakistan's Hunza region (Baltistan District) and characterized by ELISA and RT-PCR (Kollerová et al. 2006). The disease is naturally vectored by aphids in the orchard environment.

Because of the ease of spread and severity of sharka disease on economic losses to European growers, much emphasis has been placed on control measures as well as on developing new varieties that resist the virus. Resistance was present in some North American apricot cultivars, and Karayiannis and Mainou (1994) cite Syrgiannidis (1979) for being the first to identify 'Stark Early Orange' and

'Stella' as type-M PPV resistant apricot cultivars. Poor fruit quality and a high chill requirement of these varieties prevented their adoption into traditional European growing regions. However, they were soon used as progenitors in breeding programs to develop new PPV resistant apricot varieties.

The initial breeding efforts in the development of new PPV resistant apricot varieties were tedious and expensive. Quantities of PPV infected GF305 seedling peach rootstock would be needed for each individual seedling coming from planned hybridizations between PPV resistant and PPV susceptible apricots. Each seedling would be budded, many times in replicate, onto PPV infected GF305. The budded stocks would be placed in a darkened cold chamber (7°C) for approximately two months to simulate a dormancy period. Upon return to a greenhouse environment, the budded rootstocks would begin to grow, and symptoms *could* then develop on the emerging leaves. After several months in the greenhouse, plants would many times be pruned back and returned to the cold chamber for another round of simulated dormancy. A subsequent second cycle of growth could then stimulate symptom development in budded seedlings that had not shown symptoms in the first cycle. ELISA could be used to confirm the presence of the virus, and it would generally be employed after symptom development. Seedlings that demonstrated characteristic PPV symptoms and tested positive for ELISA would naturally be scored as PPV susceptible. PPV tolerant seedlings would be those showing no visible symptoms after a given number of growth cycles on PPV infected GF305 rootstock, but testing positive by ELISA. Resistant seedlings would neither test positive in an ELISA nor demonstrate visible symptoms after repeated growth cycles.

Current research on PPV resistance in apricot follows several paths. As evidenced in the current breeding objectives column of Table 2.7, many European programs are attempting to develop PPV resistant varieties that are adapted to their local conditions and tastes. Variety development populations are also used by some researchers to assist in developing inheritance models for PPV resistance as well as in molecular mapping studies for targeting the location of PPV resistance gene(s) in the apricot genome.

Numerous PPV resistant apricot varieties were discovered through large field screenings at the Pomology Institute's orchards in Naoussa, Greece (Karayiannis and Mainou 1994). Natural transmission by aphids spread the virus from infected peach orchards to the adjacent replicated apricot plots. After at least four years of growth, the resistant cultivars were evident amongst the mostly susceptible apricot germplasm. Resistance of both 'Stark Early Orange' and 'Stella' was re-confirmed, and apricot cultivars 'Goldrich', 'Harlayne', 'Henderson', 'NJA2', 'Sunglo' and 'Veecot' were deemed resistant to PPV through a lack of symptoms in field trials as well as subsequent ELISA. Concurrently, hybridizations had been undertaken at this Institute with PPV resistant Stark Early Orange and the traditional Greek cultivar 'Tirynthos'. Selection from the seedlings of this population led to two new PPV resistant apricots 'Lito' and 'Pandora', both introduced in 1991.

It has been of considerable interest that the original sources of PPV resistance in apricot came from North American cultivars. Since the PPV susceptible European apricot cultivars lack molecular markers common to the Asian apricot germplasm, it has been suggested that perhaps PPV resistance in the North American apricots came from *P. mandshurica* germplasm that was used in the distant pedigrees of North American cultivars (Badenes et al. 1996). A single accession of *P. mandshurica* was used in hybridizations with 'Currot' (PPV susceptible) to determine the worthiness of *P. mandshurica* in transmitting resistance to seedlings (Rubio et al. 2003). All seedlings from the cross were found to be PPV susceptible. However, given the diversity of this species in its center of origin and the unknown nature of the single examined accession, it is still quite possible the PPV resistance does reside within this botanical form.

Different isolates of PPV have complicated some analyses, and at least one case of differential resistance in apricot cultivars has been reported. 'Harcot' was determined PPV susceptible in Greek field tests with the predominant 'M' isolate (Karayiannis and Mainou 1994) whereas this same cultivar was found PPV resistant when the 'D' type Spanish isolate was employed under greenhouse conditions (Martínez-Gómez and Dicenta 2000). A survey of popular *Prunus* rootstocks was also conducted recently to identify those resistant to the Spanish D-isolate. After artificial inoculation and four complete cycles of growth/artificial dormancy under controlled conditions, *Prunus* rootstocks 'GF677' and 'Myrobalan 29C' were found free of PPV symptoms as well as being negative in ELISA and PCR assays. Rootstocks 'Marianna 2624' and 'Nemaguard' were both PPV susceptible based on symptom expression and laboratory assays (Rubio et al. 2005). In a test with six different PPV isolates, apricot cultivars 'Harlayne' and 'Betinka' were shown to be highly resistant or immune to all isolates during the three year examination period (Polák et al. 2005).

There are several published accounts of the inheritance mode for resistance to PPV in apricot. All associate resistance with the presence of one or more dominant genes. Nearly 300 seedlings segregating for PPV resistance and susceptibility from 20 different cross combinations led Dicenta et al. (2000) to believe PPV resistance was controlled by a single dominant gene. Symptoms were recorded after one or two cycles of growth/artificial dormancy and corroborated with laboratory ELISA. Vilanova et al. (2003) used a 76 seedling population from the self-pollination of 'Lito' apricot for molecular mapping of SSRs and AFLPs as well as for establishing the segregation of resistant: susceptible seedlings. These researchers observed a 46:30 ratio (resistant : susceptible) which deviated significantly from a 1:1, but fit a 9:7 ratio that could be expected if resistance were controlled by two dominant genes.

A three dominant gene model was proposed recently by Salava et al. (2005) using 'Stark Early Orange' as a PPV resistance donor. This study differed from the two previously mentioned ones in that the more aggressive M-type PPV isolate was used. Salava et al. (2005) allowed at least 3 complete cycles of active growth/artificial dormancy prior to final scoring of segregation ratios. Resistant plants were only considered as those with visual symptoms and either positive ELISA or PCR during the last three growth cycles. This study documented changes over time in the ratio of resistant to susceptible seedlings. Higher numbers of resistant (symptomless) plants were observed after the first growth cycle as compared to after the third cycle.

As was discussed by these researchers, there might be variability in the amount of time necessary for any particular genotype to express PPV symptoms. Furthermore, a 'false susceptible' might be a result of an insufficient time period after inoculation, prior to the plant's recovery and elimination of the virus. The combined results of three specific crosses with over 200 segregating seedlings yielded a 1:7 segregation ratio (resistant : susceptible), indicating a tri-genic mode of inheritance.

Linkage of a molecular marker to PPV resistance would be a huge benefit for apricot breeding programs in sharka infested areas or even in areas where the disease is not yet present. PPV resistance has been mapped with a diversity of molecular markers in several studies (see section on genetic mapping and QTL analysis). Soriano et al. (2005) has also characterized apricot resistance gene analogs (RGAs) for the development of specific AFLPs that are tightly linked to PPV resistance. An RGA marker, SEOBT101, has recently been identified as an amplification product only in PPV resistant apricots. The marker was present in the six tested PPV resistant accessions ('Stark Early Orange', 'Lito', 'Pandora', 'Stella' and two breeding selections from the Department of Tree Culture, University of Bologna, Italy) and failed to amplify in the 10 examined PPV susceptible apricot cultivars (Dondini et al. 2004).

2.6 Crossing and Evaluation Techniques

2.6.1 Pollen and Seed Management

Pollen is typically collected from flowers in the 'balloon' stage, prior to the unfurling of petals and anther dehiscence. This is best done when the flowers are dry, after any morning dew has evaporated. The harvested flowers can be brought back to the laboratory in small paper bags, and dozens of samples can be collected from the orchard and stored in a small cold ice chest prior to laboratory handling. A coarse metal-wire kitchen sieve is used to render the anthers from the bulk of the floral tissues. The anthers are collected on clean paper as the flowers are carefully rubbed through the sieve. The anthers are then dried overnight at room temperature. A 60–100 watt incandescent lamp placed approximately 30 cm above the sample aids in the drying process.

Dried anthers are then placed in a smaller nylon fine-mesh sieve to remove any dry floral tissues and to break open the anthers. The pollen and anthers are again collected on clean paper. Rubbing the dry sample is seldom necessary as the sample can be easily separated with a few light taps to the side of the sieve. Pollen/anthers are stored in appropriately labeled vials in the freezer.

Viability of pollen can be easily examined with a germination test. Petri dishes are prepared for this purpose with a 12% sucrose solution and 0.5% gelling agent. When dishes are cool, pollen can be distributed over the medium by gently tapping a pollen coated brush on the dish's edge to release and distribute the pollen on the surface of the medium. These dishes are stored under refrigeration (2–7°C) and

Fig. 2.3 Emasculated apricot flowers receive pollen in a planned hybridization. Given favorable weather conditions, emasculated flowers are receptive to pollen for approximately one week

the samples are scored in 12–36 hours with the aid of a dissecting microscope. Pollen samples are kept in the freezer when not in use and in a cold ice chest when being transported to and from the orchard. In this manner, samples can normally be used with good results for two seasons. Pollen can be applied successfully with the fingertip (Fig. 2.3), a small paint brush or with other small applicators.

Fruits containing hybridized seed are handled in a manner similar to the other stone fruits. There are many variations in the specifics of handling seed of *Prunus* (Grisel 1974), and procedures are typically modified to suit the individual program and its resources. In our situation, we prefer to harvest fruit with hybridized seed while still a bit under-ripe. This is actually for sanitation, to reduce the amount of free sugar available for contaminant bacteria and fungi. Seed are cut from the pits, taking care not to cut the seed open or damage the seed coat. They are then surface disinfested with a cleaning agent and rinsed thoroughly with sterile water to remove any residue. Seed are stored in zip-closure bags containing a pre-moistened/autoclaved filter paper to provide moisture during the stratification period. The bags are then held in a common household refrigerator (1–2° C) and checked periodically during the stratification period for contamination and whether they require more moisture. A high percentage of the seed germinate during the stratification period.

The early fruit ripening period of apricot can benefit to the breeder as it is possible to perform hybridizations, collect and stratify the seed and plant the seedlings in the same calendar year. However, one or more things might limit this possibility, with a large part depending on a program's resources and length of the growing season. For example, in Central California where we harvest our hybridized seed during May and June, it is possible to have seed ready for greenhouse planting by September 1. However, certain constraints and responsibilities limit our field transplanting possibilities until after the bloom period. Therefore, we choose to hold stratified seed under refrigeration until November. Seed planted in the greenhouse at this time produce healthy and vigorous seedlings, and are ready to field transplant in the March time frame. The greenhouse planting of seed prior to November leads to larger and more root-bound seedlings that can complicate the field transplanting procedure.

Seed are generally planted in the greenhouse when most of their roots extend 2–5 cm. The seed are planted at the soil level in flats with a soil depth of approximately 10 cm. Typically, each seedlot is allowed to soak in a shallow pan of water before it is planted. Seed coats are removed prior to planting, and the soaking period greatly facilitates this procedure. Apricot seed coats have been shown to have an inhibitory effect on seed germination (Chao and Walker 1966) however, they are primarily removed to allow easier emergence of the elongating shoots. Seed are commonly planted on a 2.5 cm grid within the flat and initially drenched with a fungicide to reduce pressure from damping off organisms. The flats are watered deeply and infrequently, and allowed to dry down prior to re-wetting.

Apricots exhibit hypogeal germination, and the cotyledons remain in place at the soil surface as the seedling begins to develop. When the seedlings attain approximately 15 cm in height, they are pruned back to their 4th or 5th true leaf. This is done to strengthen the small stem and to allow a higher rate of survival upon field transplanting. The seedlings will typically produce multiple shoots with this treatment, and they are pruned individually to their single strongest shoot. The general pruning treatment is performed again after the majority of the seedlings have again attained 15–20 cm in height. The pruning cut is made to allow just a node or two to grow above the level of the first pruning cut. This cycle can occur from four to six times prior to field transplanting, with the seedling stem diameter increasing with each pruning cycle.

Recent advances in breeding for earlier ripening apricots has led to a situation where there is a noted reduction in viability for seed from very early ripening cultivars. Seed from cultivars such as 'Apache' and 'Poppy' appear normal compared to seed from later-ripening cultivars, and will germinate after sufficient stratification. However, seedling emergence is reduced, and many of the emerged seedlings of very early-ripening cultivars die in the seedling flat. The in vitro culture of apricot embryos is a solution to this problem (Burgos and Ledbetter 1993).

2.6.2 Fruit Evaluation

The unlinking of visual appeal and fruit taste can be accomplished through the use of a formal tasting panel. For apricots and the other stone fruits, taste can be ranked by participants on the panel in specific descriptive categories: sweet, sour, astringency, flavor, texture and juiciness. Visual appeal and all the descriptive categories of taste in apricot can be directed and improved through selective breeding.

The evaluation of fresh fruit quality in new apricot accessions requires a knowledge of the quality characteristics of competing cultivars, or those that would be available during the same maturity season. The competing cultivars should be grown with similar orchard conditions and cultural practices, and harvest maturity of the different apricot accessions must be very similar in order to have valid comparisons. Flesh firmness is often used as a measure of harvest maturity, as it can be measured objectively and quickly with widely available instruments. Thus, when apricot samples from two different accessions do not differ significantly in flesh firmness, other quality characteristics such as Brix, acidity and flesh color can be compared validly with appropriate measures. Prior to the release of 'Lorna' apricot (selection K505-50), its fruit were compared objectively with fruit from 'Katy', a cultivar that ripens during the same period. Fruit from K505-50 had significantly higher Brix, significantly lower juice acidity and flesh with a significantly lower hue angle (deeper orange flesh) than fruit from 'Katy' in samples that did not differ significantly in flesh firmness (Ledbetter et al. 1996a).

There are other quality characteristics that are important for processing apricots, depending on the particular industrial use. Because large quantities of apricots are typically processed for any given industrial use, the percentage of usable flesh in a given apricot shipment is of primary importance to the processor. Brix, acidity and juice pH are also important measured parameters for apricots at any industrial starting point. It is extremely important to examine the particular industrial product and evaluate quality during what would be considered a normal storage period.

Fruit softening during storage is a major problem in canned apricots, and citrates present in the apricot juice have been implicated in chelation reactions that lead to an unacceptably soft canned product (French et al. 1989). Developing new apricots specifically for canned product might therefore involve a closer examination of the acids present in newly selected accessions. Citric and malic acids are most predominant in apricot fruit, and within a collection of apricot accessions, large differences exist in the levels of each acid per each accession. However, the ratio of malic to citric acids remains relatively constant year to year in a given accession, allowing for selection of germplasm with the desired contents of specific acids (Bassi et al. 1996). Hence, fruit evaluation of new apricot material for canned product might involve measurements of malic and citric acids in order to identify those types less prone to fruit softening after canning.

In dried apricot, color retention during storage is a major concern. Storage at a higher temperature exasperates the problem, but even during cold storage, darkening of the product affects marketability (Ledbetter et al. 2002). The degree of darkening during storage is influenced by the particular apricot accession, and one discovers whether or not an accession is suitable for drying by actually putting it through the drying process. Immediately after drying, baseline color coordinates such as

Luminosity and Chroma can be established with one of the available tristimulus colorimeters, and the dry product can then be sampled periodically during storage to establish rates of change. Stone freeness is certainly of equal importance in selecting new apricots for drying quality. Processing speed and efficiency is severely affected on both mechanical and hand-cut lines if the apricot stone clings to any of the cavity flesh.

2.6.3 Trialing and Variety Introduction

Trialing of advanced selections with a commercial grower is extremely important to fully realize a clone's performance. The trial of experimental apricot selection(s) and competing cultivars should be sized realistically with both the grower and breeder in mind. In some cases the trial grower might choose to top work the experimental selection into trees of an existing orchard, perhaps within the orchard of the competing cultivar. Doing so provides the opportunity to observe both the new selection and the competing cultivar throughout the year, and eases comparisons of growth habit, as well as seasons of bloom, fruit maturity and fall senescence. New orchards for trialed varieties can also be established, and in this case evaluations can be conducted on trees grown on the same rootstock, an important consideration for self-incompatible varieties where bloom matching with another variety is critical. Whatever the makeup of the trial orchard, it must be sufficiently large such that commercial quantities of fruit can be treated in an identical manner with other varieties. When it is probable that a new variety's destiny is predominantly for large scale producers or export markets, then it is important to have sufficient fruit to be convinced that the new selection performs adequately during the harvest and packing operation. Pre-conditioning and/or cold storage, and perhaps quarantine treatments might be applied to determine the effects on the new apricot selection(s). If the new variety's destiny is intended for smaller growers who would market fruit locally, then a smaller-scale orchard trial might be more appropriate. Fruit of greater maturity is typical of the local 'farmers markets', so a breeder could get an indication the bruising potential of new selections as well as customer opinions on fruit quality, by conducting smaller-scale trials where fruit of higher maturity is handled.

If there is insufficient fruit from the trialed trees for marketing, the grower will not be motivated to harvest the fruit, or work the trees in a manner similar to the existing varieties that are being commercially productive. Hence, the breeder must allow trials to be large enough such that the producer has the opportunity for a successful commercial harvest. A single flaw in an experimental selection can make the apricots unworthy for market, so trials should not be so large as to produce an unnecessary financial burden on the producer. A large failure with a grower who is new at trialing experimental selections might sour the grower's opinion of providing such assistance to the breeder again in the future.

2.7 Biotechnological Approaches to Genetic Improvement

2.7.1 Cultivar Fingerprinting and Phylogenetic Studies

The analysis of plant isozymes was an early technique used for hybrid verification (Byrne and Littleton 1989a) and to characterize apricot germplasm (Byrne and Littleton 1989b), but low numbers of useful (segregating) loci limited the effectiveness of the technique in establishing precise relationships among closely related accessions of a given species (Badenes et al. 1996). DNA fragment based analyses have proven more useful in discerning similarities or differences between apricot cultivars or between accessions in different eco-geographical groups. There are many examples of apricot cultivars having different names that have been successfully 'fingerprinted' to demonstrate their genetic origin. Other examples of the technique's usefulness are the ability to identify mistakes in a cultivar's pedigree, or to demonstrate genetic identity of new cultivars, and thereby insist or provide evidence regarding the protection of plant breeder's rights. As a key example, Ahmad et al. (2004) utilized a set of 28 single sequence repeat (SSR) primers on a specific set of apricot, Japanese plum, plumcot and pluot^{TM} cultivars. Developed by a private plant breeding company in California, a pluot^{TM} is said to be the product of backcrossing a plumcot (*P. salicina* × *P. armeniaca*) with a plum, thereby creating a novel new fruit type with 25% apricot germplasm in its pedigree. Since their arrival in California nurseries in 1989, the pluot^{TM} cultivars have steadily increased in acreage and fruit volume to present levels of approximately 5 million 12 kg boxes (California Tree Fruit Agreement, 2002). Consumers have paid dearly at the marketplace for this novel new fruit type, but the SSR analysis made by Ahmad et al. (2004) found no alleles specific to apricot amongst any of the six tested pluot^{TM} cultivars. This research has the potential to affect the marketing of these 'novel' fruits as well as pesticide labeling information.

When genetically and geographically diverse germplasm has been compared in molecular phylogenetic studies, good separation is usually discovered between species and/or subgenera. Such was the case with a RAPD analysis conducted by Shimada et al. (2001), on 40 *Prunus* accessions representative of four subgenera (*Prunophora*, *Amygdalus*, *Lithocerasus* and *Cerasus*). In another RADP analysis of a genetically diverse group of 35 apricot cultivars that included a large group of Japanese cultivars, as well as cultivars from China, Europe, Nepal, Turkey and North America, all the Japanese germplasm grouped together and separate from apricots of other origin (Takeda et al. 1998). Apricots from China, Europe, Nepal, Turkey and North America were represented together in another single group. These researchers also found that the Turkish cultivars 'Hajihaliloulu' and 'Hasanbay' were probably identical, being inseparable in both plant morphology and RAPD analysis. In an early restriction fragment length polymorphism (RFLP) study using chloroplast DNA, Uematsu et al. (1991) successfully discriminated non-domesticated *Prunus* species (*P. mira* Wilson., *P. davidiana* Maxim.) from several diverse cultivated varieties (*P. persica* Sieb and Zucc., *P. domestica* L., *P. armeniaca* and *P. mume*).

Hagen et al. (2002) examined genetic diversity within a group of 53 apricot accessions (47 diverse cultivars and 6 related apricot species accessions) with AFLP markers. *Prunus ansu*, *P. mume*, *P. dasycarpa* and *P. brigantiaca* were well separated from *P. armeniaca* in their analysis. All cultivars had unique AFLP profiles and segregated into four clusters. In another AFLP study of 118 apricot accessions representing Europe, China and North America, Geuna et al. (2003) provided evidence that the North American cultivars were created from a complex blend of germplasm from both Europe and China, although only 17% of the total observed variation was described by the first three principal components. In agreement with the AFLP-based study by Hagen et al. (2002), they found the cultivars 'Erevani' ('Shalah'), 'Stella' and 'Veecot' to be located in close proximity, but they differed in the placement of 'Goldrich' and 'Harcot'. These two cultivars were clustered together by Hagen et al. (2002), whereas they found themselves in completely different clusters in the study by Geuna et al. (2003). In a much smaller AFLP-based study designed specifically to examine genetic variability within the progenitors of a breeding program with the objective of developing sharka resistance, 'Goldrich' and 'Harcot' clustered together amongst other sharka resistant accessions (Hurtado et al. 2002). In another unrelated study 'Goldrich' and 'Harcot' were placed in different clusters by Khadari et al. (2006) in an AFLP study examining the uniqueness of Tunisian apricot germplasm relative to cultivars from other geographical regions.

Peach and cherry SSRs were used to study a group of 48, 40 and 74 apricot accessions by Hormaza (2002), Romero et al. (2003) and Zhebentyayeva et al. (2003), respectively. Hormaza (2002) studied European and North American apricot germplasm, along with a single Chinese cultivar, 'Piu Sha Sin'. Romero et al. (2003) focused their analysis on a somewhat geographically wider collection including apricots from Europe, North America, and Central Asian origin, as well as Central Asian × European apricot hybrids. Zhebentyayeva (2003) compared 74 apricot accessions representing both domesticated and wild forms from all four eco-geographic groups. The French traditional cultivar 'Bergeron' was the only apricot accession common to all three studies. The total number of SSR primers analyzed in these studies were 20, 16 and 14 for the studies by Hormaza (2002), Romero et al. (2003) and Zhebentyayeva et al. (2003), with only a single SSR (98–406) common to all three studies. While these studies separated most cultivars successfully, there were some dissimilarities. In the Hormaza (2002) study, 'Bergeron' and 'Gönci Magyar' appeared quite similar whereas 'Canino' and 'Pandora' were very different. This contrasts with results from Romero et al. (2003) where 'Bergeron' and 'Gönci Magyar' were similar and 'Canino' and 'Pandora' resided as neighbors on the dendrogram.

An even larger SSR phylogenetic study was conducted in by Maghuly et al. (2005) with newly developed SSRs from the apricot genome. Over 130 apricot cultivars representing Europe, Central Asia, Irano-Caucasian region and North America were screened with the broad goal of grouping the accessions by their eco-geographic origin. Ten of the 120 then known apricot-derived SSRs (Lopes et al. 2002, Messina et al. 2004) were used to elucidate the similarities and/or differences between the accessions. The UPGMA dendrogram presented is complex, and the authors broadly state that the position of Central Asian cultivars in the dendrogram supports the notion that most of the tested cultivars are of Asiatic origin. Based on their analysis, numerous cases of synonymous cultivars could be identified ('Alberna' = 'Andormaktájai Magyar kajszi' = 'Crvena ungarska' = 'Gönci Magyar kajszi' = 'Naggyümölcsü vagyar kajszi'; 'Kalasek' = 'Krasnoshchokijiz Nikolajeva 1486'; 'Cacansko zlato' = 'Magyar kajszi 235'; 'Chershonskij 1469' = 'Paksi Magyar kajszi'; 'Ceglédi óriás' = 'Ligeti óriás' = 'Szegedi mamut' and 'Kecskemet early' = 'Rosensteiner'); however, other reportedly synonymous cultivars did not group as genetically identical ('OrangeRed' and 'Bahrt', 'Erevan' and 'Shalah').

Other noteworthy placements of cultivars in the UPGMA dendrogram of Maghuly et al. (2005) are of practical breeding interest. 'Morden 604' and 'Kletnice' were placed as virtual outliers to all others in the chart. While 'Kletnice' is of Czech origin and of unknown parentage, 'Morden 604' was developed in Canada, from the cross 'Scout' × 'McClure'. 'Scout' was selected from seed of Siberian origin, perhaps being *P. manchurica* or *P. sibirica* (Brooks and Olmo 1952). The broadly adapted cultivar 'Goldrich' clusters with six other cultivars including the French cultivar 'Bergeron', two seedling selections and 'San Castrese' from Italy, and the Eastern European apricots 'Marille Bauer' and 'Spätblühende Koch'.

Overall, the researchers working on SSR-based phylogenetic studies of apricot germplasm have concluded that European cultivars have a narrower genetic base compared to the other eco-geographic groups. Most have also pointed out the need for diversification in apricot breeding programs. Cultivars and/or landraces from the other eco-geographic groups (Central Asian, Dzhungar-Zailij, Irano-Caucasian) need to be used in variety development (North American or European) in order to obtain wider climactic adaptation, disease resistance, lengthened fruit developmental periods and other diverse and interesting characters. These same suggestions were put forward by Kostina (1936) in her comprehensive whole plant studies of the 1920s and 1930s. Central Asian apricot germplasm has been utilized to breed higher quality California-adapted apricots since the early 1990s (Ledbetter and Peterson 2004, Ledbetter et al. 2006). High Brix, long fruit development period, strong fruit attachment, white flesh and glabrous skin are all available to the breeder with access to this germplasm. Fruit sizes of many Central Asian apricots are small, but good progress can be made in a single round of hybridization with a European parent.

2.7.2 Stylar Ribonuclease Characterization

Beyond bagging trials at bloom time to exclude bee visitation and fluorescent microscopy to examine pollen tube growth in planned self-pollinations, the next advancement in analyzing self-(in)compatibility in apricot involved characterizing stylar ribonucleases. Emasculated flowers were gathered from numerous apricot cultivars and evaluated on polyacrylamide gels for ribonucleases activity. Among the initial examined cultivars were 'Goldrich', 'Hargrand' and 'Lambertin-1'.

Non-equilibrium pH gradient electrofocusing was used to separate the stylar proteins, and after staining identical banding patterns were observed for 'Goldrich', 'Hargrand' and 'Lambertin-1' (Burgos et al. 1998). Stained bands migrated differently for other cultivars, and it was possible to assign other allelic designations on the basis of these tests. In other studies, the male-sterile cultivar 'Colorao' was determined to carry alleles for self-compatibility, although it was incapable of effecting self-pollination. When it was used as a female parent in a cross with self-compatible 'Pepito', several seedlings were obtained that proved to have the $S_c S_c$ genotype (Alburquerque et al. 2002). These authors pointed out the general interest of apricot breeders in possessing germplasm homozygous for self-compatibility.

The self-(in)compatibility locus has been examined and characterized in a number of *Prunus* species including almond, both sweet and sour cherry, and Japanese apricot. Similarities exist in each of these species: stylar ribonucleases genes all contain two variable sized introns, five conserved regions (C1, C2, C3, RC4 and C5) and a hypervariable region (RHV) that is putatively recognized as the recognition site for the S-determinant in a pollen tube. An area known as the S-locus F-box (SFB) gene, is tightly linked with the stylar ribonucleases gene such that the two genes are inherited as a single unit. Recombination between the two is thought to be suppressed by a high content of repetitive sequences present between them (Ushijima et al. 2003). SFB genes are expressed specifically in the developing pollen.

The identification of self-(in)compatibility alleles from leaf tissue (as opposed from floral organ tissues) would aid breeders by allowing selection for self- compatibility (or any specific known allelic structure) at a very early stage of development. To achieve this, stylar ribonucleases DNA sequences specific to certain self- incompatibility groups needed to be characterized so that PCR-based primers could be developed for further identification of new self-incompatible alleles. Romero et al. (2004) characterized three stylar ribonucleases alleles (S_1, S_2 and S_4) from 'Goldrich' and 'Harcot' apricots and confirmed that they were linked closely to SFB genes as had been found in other *Prunus* species. Amino acid identity amongst these three alleles averaged 75.3%, indicating a high level of sequence diversity. Using other cultivars, Vilanova et al. (2005) characterized the other four known self-incompatibility alleles (as well as S_c) with consensus primers developed from stylar ribonucleases genomic sequences of apricot and sweet cherry. These developed protocols are now a tool available to the apricot breeder, and self-compatibility can be determined in the seedling flat with meristematic tissues.

2.7.3 Genetic Mapping and QTL Analysis

Several genetic linkage maps of have been generated for apricot. Hurtado et al. (2002) used AFLP, RAPD, RFLP and SSR markers to map 81 F_1 individuals from the cross of 'Goldrich' × 'Valenciano'. A total of 132 markers were placed on eight linkage groups of 'Goldrich', with a coverage of 511 cM. A total of 80 markers were placed into seven linkage groups of 'Valenciano' that defined 467.2 cM. Two codominant markers were located on linkage group 2 that flanked sharka resistance. Vilanova et al. (2003) generated a map using 76 individuals from a self-pollination of 'Lito' ('Stark Early Orange' × 'Tyrinthos'). A total of 212 markers (180 AFLPs, 29 SSRs and two agronomic traits) were assigned to 11 linkage groups spanning 602 cM. Plum pox resistance was mapped to linkage group 1 and the self-incompatibility trait to linkage group 6. Twenty two loci were held in common with other *Prunus* maps and most of them showed the same linkage relationships. Lambert et al. (2004) examined 142 F_1 apricot hybrids from the cross 'Polonais' × 'Stark Early Orange' using 83 AFLPs, 88 RFLP and 20 SSRs from the *Prunus* reference map of almond 'Texas' × peach 'Earlygold' (T × E). A total of 110 markers were placed on a map of 'Polanais', covering 538 cM, and 141 markers were located on a map of 'Stark Early Orange', defining a length of 699 cM. Almost all markers could be aligned with those from the T × E map. Salava et al. (2007) developed an integrated genetic linkage map with 316 molecular markers (290 AFLPs, 26 SSRs) using a backcross progeny of 'LE-3246' × 'Vestar'. They assigned markers to 8 linkage groups covering 574 cM and found several markers linked to the *PPVres1* locus conferring resistance to PPV. Vilanova et al. (2006) hybridized sixteen SSRs to a BAC library and were able to identify clones belonging to the G1 linkage group.

2.7.4 Micropropagation/Plantlet Production

Literature reports of apricot micropropagation first appeared in the late 1970s (Skirvin et al. 1979). Pérez-Tornero et al. (1999a) developed successful techniques for meristem tip culture as a means of eliminating the persistent endophytic bacteria typically found in field-grown trees. Snir (1984) utilized growth chamber grown 'Canino' apricot shoots rather than field grown materials to avoid the heavy infestation problems of the later. Snir's work was focused on simply developing an effective in vitro rooting technique for apricot, as both softwood and hardwood cuttings of *P. armeniaca* L. are difficult to root, and other research at that time had demonstrated that fruit trees grown on their own root had better nutrient uptake and higher productivity (Couvillon 1982, Thibault and Herman 1982). Snir (1984) observed that MS medium was completely unsuccessful in supporting vegetative growth of 'Canino' apricot, but Woody Plant Medium (WPM) (Lloyd and McCown 1980) allowed a high percentage (70%) of the buds to elongate into shoots. Rooting of the elongated shoots was effectively accomplished on $1/2$ strength MS medium supplemented with 0.5 mg L-1 1-naphthaleneacetic acid (NAA).

Marino et al. (1993) studied proliferation and rooting ability of apricot cultivars San Castrese and Portici in modified MS medium. Specifically, these researchers examined differences in the in vitro growth rate as related to different carbon sources. Sorbitol as a carbon source enhanced the proliferation rate of both cultivars, as did increasing the 6-benzyladenine (BA) concentration. However, high levels of BA (8.8 mM) in the proliferation medium caused hyperhydricity in the explants, in particular when sucrose was the medium's carbon source. While sorbitol enhanced the proliferation rates of both apricot cultivars, lower percentages of rooting were reported when sorbitol was used as the carbon source in the rooting medium. A 70% success rate of rooted plantlets was reported by Marino et al. (1993) using a modified MS rooting medium with the inclusion of indolebutyric acid (IBA). Hyperhydricity in proliferating in vitro cultures has been a reported problem in other *Prunus* species (Rugini and Verma 1982, Ledbetter et al. 1996b). Basal cooling for several weeks during proliferation has been an effective treatment in reducing hyperhydricity among cultured apricot genotypes. The level of hyperhydric explants in specific apricot cultivars is also influenced by the medium's gelling agent (Pérez-Tornero et al. 2001).

Further work on the traditional Spanish apricot cultivar 'Canino' was carried out by Pérez-Tornero et al. (1999a and 2000b) to identify a more suitable nutrient medium to support healthy growth and enhance proliferation. Working on four apricot cultivars, they found meristem survival in culture was significantly affected by cultivar, as well as interactions of the cultivar with BA concentration, and the interaction between BA and gibberellic acid (GA). Establishment medium prepared without BA prevented all cultivars from developing rosettes of leaves and elongating into usable explant shoots (Pérez-Tornero et al. 1999a). A subsequent study by Pérez-Tornero and Burgos (2000) involved the development of a medium designated as 'M3' that provided a significantly superior number of proliferated shoots and total shoot length as compared to MS, QL (Quoirin and Lepoivre 1977) and WPM. Optimum medium BA concentrations were found to be highly cultivar dependent. In vitro productivity of 'Búlida', 'Helena' and 'Lorna' apricot explants was highest with a concentration of 4.44 mM BA in the medium whereas 'Canino' proliferation was optimal when BA concentration was 1.78 mM (Pérez-Tornero and Burgos 2000). The inclusion of a low BA concentration in the rooting medium was beneficial in alleviating apical necrosis of the proliferated explants; however, lower percentages of rooted explants were also associated with the BA inclusion. As a remedy, a sterile 22–44 mM BA solution was used as a dip treatment for explant shoot apices prior to insertion into the BA-free rooting medium.

2.7.5 Regeneration

Early research studies on vegetative regeneration in apricot utilized embryonic tissues. Pieterse (1989) observed that Stage 2 embryos (50 percent fill) produced the most regeneration buds, and that a MS medium modified with 2, 4-dichlorophenoxyacetic acid (2, 4-D) and BA could provide the stimulus for cultured embryos to produce shoots, some of which could spontaneously root in the culture medium. Similar results were obtained by Goffreda et al. (1995), again working with MS medium supplemented with either BA or thidiazuron (TDZ), where Stage 2 embryos (30–60 percent fill) from the cultivars 'Zard' and 'NJA82' produced shoot primordia. Transgenic regeneration of *P. armeniaca* L. was first reported by Laimer da Câmara Machado et al. (1992), with the successful insertion into 'Kecskemeter' of a marker gene, ß-glucuronidase (GUS), and the coat protein gene for Plum Pox Virus (PPV). Cotyledonary tissues were the actual explants used in this work, and regeneration rates were highest for embryos harvested and utilized between 68 and 89 days after full bloom.

At the turn of this century, the Department of Breeding at the Centro de Edafología y Biología, Aplicada del Sugura (CEBAS) in Murcia, Spain began investigations on improving the efficiency of apricot regeneration. Studies were conducted on the establishment of explants in vitro and storage conditions that would allow high rates of regeneration after prolonged semi-dormant storage (Pérez-Tornero et al. 1999b). Numerous factors significantly affected the rate of shoot regeneration in apricot: leaf age and leaf position on the explant source, light and darkness regime during culture, specific gelling agent of the medium and plant growth regulator regime. As was found with the proliferation phase of micropropagation, apricot genotypes responded differently to culture conditions and medium composition (Pérez-Tornero et al. 2000a). TDZ at 9.0 mM was the most effective growth regulator at improving regeneration rates, across genotypes. Silver thiosulphate (STS) at 30–60 mM increased regeneration rates significantly in 'Helena' and 'Canino', whereas incorporating 8.6–17.1 mM kanamycin in the culture medium increased significantly the regeneration rate in 'Helena', but not in 'Canino' apricot. Utilizing STS and a low level of kanamycin in the regeneration medium resulted in regeneration rates 200% over what had previously been reported (Burgos and Alburquerque 2003).

The first successful genetic transformation in apricot from clonal vegetative material was reported by Petri et al. (2004) with 'Helena'. These CEBAS researchers succeeded in inserting a marker gene (green fluorescent protein – *gfp*) into 'Helena' through *A. tumefaciens* mediated transformation. It was no accident that cultivar 'Helena' was used as the test subject as this genotype responded favorably to in vitro systems. Subsequent work from this research group refined further the culture conditions necessary to increase transformation events during regeneration and select transgenic explants from regenerating cultures. Four day pulses of 2, 4-D and spermidine/STS in the regeneration medium increased stable *gfp*-producing calli in 'Helena' apricot (Petri et al. 2005a). Paromomycin has been recently suggested as an improved antibiotic alternative to kanamycin for selecting transformed explants in regenerating cultures (Petri et al. 2005b).

References

Ahmad R, Potter D, Southwick SM (2004) Identification and characterization of plum and pluot cultivars by microsatellite markers. J Hortic Sci Biotechnol 79:164–169

Alburquerque N, Egea J, Pérez-Tornero O, Burgos L (2002) Genotyping apricot cultivars for self-(in)compatibility by means of RNases associated with S alleles. Plant Breeding 121:343–347

Asma BM, Ozturk K (2005) Analysis of morphological, pomological and yield characteristics of some apricot germplasm in Turkey. Genet Resour Crop Evol 52:305–313

Badenes ML, Asins MJ, Carbonell EA, Llácer G (1996) Genetic diversity in apricot (*Prunus armeniaca* L.) aimed at improving resistance to plum pox virus. Plant Breeding 115:133–139

Badenes ML, Martínez-Calvo J, Llácer G (1998) Analysis of apricot germplasm from the European ecogeographical group. Euphytica 102:93–99

Badenes ML, Hurtado MA, Sanz F, Archelos DM, Burgos L, Egea J, Llácer G (2000) Searching for molecular markers linked to male sterility and self-compatibility in apricot. Plant Breeding 119:157–160

Bailey LH (1916) *Prunus*. In: The standard cyclopedia of horticulture, vol. V. P–R. Mount Pleasant Press, J. Horace McFarland Co., Harrisburg, PA, pp 2822–2845

Bassi D, Bartolozzi F, Muzzi E (1996) Patterns and heritability of carboxylic acids and soluble sugars in fruits of apricot (*Prunus armeniaca* L.). Plant Breeding 115:67–70

Benedikova D (2004) The importance of genetic resources for apricot breeding in Slovakia. J Fruit Orn Plant Res 12:107–113

Bordeianu T, Constantinescu N, Stefan N (1967) Pomologia Republicii Socialiste Romania. V. Caisul – Piersicul. (in Romanian) Bucuresti, Romania

Bortiri E, Oh SH, Jiang J, Baggett S, Granger A, Weeks C, Buckingham M, Potter D, Parfitt DE (2001) Phylogeny and systematics of *Prunus* (Rosaceae) as determined by sequence analysis of ITS and the chloroplast *trnL-trnF* spacer DNA. Syst Bot 26:797–807

Bortiri E, Oh SH, Gao FY, Potter D (2002) The phylogenetic utility of nucleotide sequences of sorbitol 6-phosphate dehydrogenase in *Prunus* (Rosaceae). Am J Bot 89:1697–1708

Brooks RM, Olmo HP (1952) Apricots. In: Registry of New Fruit and Nut Varieties 1920–1950. University of California Press, Berkeley, CA, pp 24–28

Brooks RM, Olmo HP (1972) Apricots. In: Registry of New Fruit and Nut Varieties, 2nd edn. University of California Press, Berkeley, CA, pp 120–134

Brown DS (1957) The rest period of apricot flower buds as described by a regression of time of bloom on temperature. Plant Physiol 32:75–85

Burgos L, Berenguer T, Egea J (1993) Self- and cross-compatibility among apricot cultivars. HortScience 28:148–150

Burgos L, Ledbetter CA (1993) Improved efficiency in apricot breeding: effects of embryo development and nutrient media on in vitro germination and seedling establishment. Plant Cell Tiss Organ Cult 35:217–222

Burgos L, Ledbetter CA (1994) Observations on inheritance of male sterility in apricot. HortScience 29:127

Burgos L, Ledbetter CA, Pérez-Tornero O, Ortín-Párraga F, Egea J (1997) Inheritance of sexual incompatibility in apricot. Plant Breeding 116:383–386

Burgos L, Pérez-Tornero O, Ballester J, Olmos E (1998) Detection and inheritance of stylar ribonucleases associated with incompatibility alleles in apricot. Sex Plant Reprod 11:153–158

Burgos L, Alburquerque N (2003) Ethylene inhibitors and low kanamycin concentrations improve adventitious regeneration from apricot leaves. Plant Cell Reports 21:1167–1174

Byrne DH, Littleton TG (1989a) Interspecific hybrid verification of plum × apricot hybrids via isozymes analyses. HortScience 24:132–134

Byrne DH, Littleton TG (1989b) Characterization of isozymes variability in apricots. J Am Soc HortSci 114:674–678

California Tree Fruit Agreement (2002) Press Release 5/9/2002. Tree fruit industry examines interspecific issues. http://eatcaliforniafruit.com/ppn/media/prDetail.asp?prID=15

Candresse T, Mac Quaire G, Lanneau M, Busalem T, Wetzel T, Quiot-Douine L, Quiot JB, Dunez J (1994) Detection of Plum Pox potyvirus and analysis of its molecular variability using immunocapture-PCR. Eur Plant Prot Organ Bull 24:585–595

Cesaraccio C, Spano D, Snyder RL, Duce P (2004) Chilling and forcing model to predict bud-burst of crop and forest species. Agric Forest Meteorology 126:1–13

Chao L, Walker DR (1966) Effects of temperature, chemicals and seed coat on apricot and peach seed germination and growth. Proc Amer Soc Hort Sci 88:232–238

Couranjou J (1995) Genetic studies of 11 quantitative characters in apricot. Scientia Hort 61: 61–75

Couvillon GA (1982) Leaf elemental content comparisons of own-rooted peach cultivars to the same cultivars on several peach seedling rootstocks. J Amer Soc Hort Sci 107:555–558

Day LH (1953) Rootstocks for stone fruits. University of California, California Agricultural Experiment Station Extension Service Bull 736. 74p

Dicenta F, Martínez-Gómez P, Burgos L, Egea J (2000) Inheritance of resistance to plum pox potyvirus (PPV) in apricot, *Prunus armeniaca*. Plant Breeding 119:161–164

Dondini L, Costa F, Tataranni G, Tartarini S, Sansavini S (2004) Cloning of apricot RGAs (Resistant Gene Analogs) and development of molecular markers associated with Sharka (PPV) resistance. J Hortic Sci Biotechnol 79: 729–734

Dragovic-Uzelac V, Delonga K, Levaj B, Djakovic S, Pospisil J (2005) Phenolic profiles of raw apricots, pumpkins and their purees in the evaluation of apricot nectar and jam authenticity. J Agric Food Chem 53:4836–4842

Egea J, Burgos L (1996) Detecting cross-incompatibility of three North American apricot cultivars and establishing the first incompatibility group in apricot. J Amer Soc Hort Sci 121: 1002–1005

Ercisli S (2004) A short review of the fruit germplasm resources of Turkey. Genet Resour Crop Evol 51:419–435

FAOSTAT data (2006) Last accessed February 2006. http://faostat.fao.org/faostat/

Faust M, Surányi D, Nyujtó F (1998) Origin and dissemination of apricot. In: Janick J (ed) Horticultural Reviews, vol 22. John Wiley & Sons, Inc., pp 225–266

French DA, Kader AA, Labavitch JM (1989) Softening of canned apricots: a chelation hypothesis. J Food Sci 54:86–89

García JE, Egea J, Egea L, Berenguer T (1988) The floral biology of certain apricot cultivars in Murcia. Adv Hort Sci 2:84–87

Gathercole FJ, Wachtel MF, Magarey PA, Stevens KM (1987) Resistance of potted apricot and plum rootstocks to *Verticillium dahliae* (Kleb.). Aust Plant Path 16:88–91

Genovese A, Ugliano M, Pessina R, Gambuti A, Piombino P, Moio L (2004) Comparison of the aroma compounds in apricot (*Prunus armeniaca* L. cv Pellecchiella) and apple (*Malus pumila* L. cv Annurca) raw distillates. Ital J Food Sci 16:185–196

Geuna F, Toschi M, Bassi D (2003) The use of AFLP markers for cultivar identification in apricot. Plant Breeding 122:526–531

Goffreda JC, Scope AL, Fiola JA (1995) Indole butyric acid induces regeneration of phenotypically normal apricot (*Prunus armeniaca* L.) plants from immature embryos. Plant Growth Regul 17:41–46

Gómez E, Ledbetter CA (1993) Transmission of biochemical flavor constituents from apricot and plum to their interspecific hybrid. Plant Breeding 111:236–241

Gómez E, Ledbetter CA, Hartsell PL (1993) Volatile compounds in apricot, plum and their interspecific hybrids. J Agri Food Chem 41:1669–1676

Gómez E, Ledbetter CA (1997) Development of volatile compounds during fruit maturation: characterization of apricot and plum × apricot hybrids. J Sci Food Agric 74:541–546

Gómez E, Burgos L, Soriano C, Marín J (1998) Amygdalin content in the seeds of several apricot cultivars. J Sci Food Agric 77:184–186

Grisel TJ (1974) *Prunus* L. Cherry, peach and plum. pp 658–673 In: Seeds of Woody Plants in the United States (Schopmeyer CS, Technical Coordinator) Agriculture Handbook No. 450. Forest Service, USDA, Washington DC

Gu C, Li C, Lu L, Jiang S, Alexander C, Bartholomew B, Brach AR, Boufford DE, Ikeda H, Ohba H, Robertson KR, Spongberg SA (2003) Rosaceae. In: Wu CY, Raven PH (eds) Flora of China, vol 9. (Pittosporaceae through Connaraceae). Science Press, Beijing, and Missouri Botanical Garden Press, St. Louis, pp 46–434

Guichard E, Fournier N (1990) Dosage des composés volatils présents dans différentes variétés d'abricots et corrélation avec la typicité d'arôme. 9° Colloque sur les recherches fruitières. Avignon, France, 4–6 December 1990. p 229–237

Guichard E, Schlich P, Issanchou S (1990) Composition of apricot aroma: correlations between sensory and instrumental data. J Food Sci 55:735–738

Guichard E (1995) Chiral g-lactones, key compounds to apricot flavor. Sensory evaluation, quantification and chirospecific analysis in different varieties. In: Rouseff RL, Leahy MM (eds) Fruit

Flavors: Biogenesis, Characterization and Authentication, American Chemical Society, Oxford University Press, New York, NY USA pp 258–267

Gurrieri F, Audergon JM, Albagnac G, Reich M (2001) Soluble sugars and carboxylic acids in ripe apricot fruit as parameters for distinguishing different cultivars. Euphytica 117:183–189

Hagen LS, Khadari B, Lambert P, Audergon JM (2002) Genetic diversity in apricot revealed by AFLP markers: species and cultivar comparisons. Theor Appl Genet105:298–305

Harada Y, Nakao S, Sasaki M, Sasaki Y, Ichihashi Y, Sano T (2004) *Monilia mumecola*, a new brown rot fungus on *Prunus mume* in Japan. J Gen Plant Path 70:297–307

Hedrick UP (1925) Varieties of Apricots. In: Bailey LH (ed.) Systematic Pomology. The Macmillan Company, New York, pp 313–319

Hesse CO (1952) Apricot Culture in California. California Agricultural Experiment Station Extension Service Circular 412

Hormaza JI (2002) Molecular characterization and similarity relationships among apricot (*Prunus armeniaca* L.) genotypes using simple sequence repeats. Theor Appl Genet 104: 321–328

Hou HY (1983) Vegetation of China with reference to its geographical distribution. Ann Missouri Bot Gard 70:509–548

Hurtado MA, Romero C, Vilanova S, Abbott AG, Llácer G, Badenes ML (2002) Genetic linkage maps of two apricot cultivars (*Prunus armeniaca* L.), and mapping of PPV (sharka) resistance. Theor Appl Genet 105:182–191

Hurtado MA, Westman A, Beck E, Abbott GA, Llácer G, Badenes ML (2002) Genetic diversity in apricot cultivars based on AFLP markers. Euphytica 127:297–301

Joshi VK, Bhutani VP, Sharma RC (1990) The effect of dilution and addition of nitrogen source on chemical, mineral and sensory qualities of wild apricot wine. American J Enol Vit 41:229–231

Karayiannis I, Mainou A (1994) Resistance to plum pox potyvirus in apricots. Bulletin OEPP 24:761–765

Khadari B, Krichen L, Lambert P, Marrakchi M, Audergon JM (2006) Genetic structure in Tunisian apricot, *Prunus armeniaca* L., populations propagated by grafting: a signature of bottleneck effects and ancient propagation by seedlings. Genet Resour Crop Evol 53:811–819

Kollerová E, Nováková S, Subr Z, Glasa M (2006) Plum Pox Virus Mixed Infection Detected on Apricot in Pakistan. Plant Dis 90:1108

Kostina KF (1936) The Apricot. (in Russian) Supplement No. 83 to the bulletin of applied botany, genetics and plant breeding. Lenin Academy of Agricultural Sciences, Institute of Plant Industry, Leningrad, Russia

Kostina KF (1969) The use of varietal resources of apricots for breeding. (in Russian) Trud Nikit Bot Sad 40:45–63

Kyotani H, Yoshida M, Yamaguchi M, Ishizawa Y, Kozono T, Nishida T, Kanato K (1988) Breeding of plum-mume parental lines 'PM-1-1' and 'PM-1-4', interspecific hybrids of Japanese Plum (*Prunus salicina* Lindl.) and Mume (*P. mume* Sieb. et Zucc.). (in Japanese). Bull Fruit Tree Res Sta (Ministry of Agriculture, Forestry and Fisheries). Series A, 15:1–10

Laimer da Câmara Machado M, da Câmara Machado A., Hanzer V, Weiss H, Regner F, Steinkellner H, Mattanovich D, Plail R, Knapp E, Kalthoff B, Katinger H (1992) Regeneration of transgenic plants of *Prunus armeniaca* containing the coat protein gene of Plum Pox Virus. Plant Cell Rep 11:25–29

Lambert P, Hagen LS, Arus P, Augerdon JM (2004) Genetic linkage maps of two apricot cultivars (*Prunus armeniaca* L.) compared with the almond Texas × peach Earlygold reference map for *Prunus*. Theor Appl Genet 108:1120–1130

Layne REC (1984) 'Harglow' apricot. HortScience 19:136–137

Ledbetter CA, Gómez E, Burgos L, Peterson S (1996a) Evaluation of fruit quality of apricot cultivars and selections. J Tree Fruit Prod 1:73–86

Ledbetter CA, Peterson S, Palmquist D (1996b) In vitro tolerance of six clonally propagated *Prunus* accessions. J Genet Breed 50:1–6

Ledbetter CA, Obenland D, Palmquist D (2000) Rutin and astragalin in dried apricot leaves as affected by leaf type, apricot accession and leaf harvest date. J Genet Breed 54:41–47

Ledbetter CA, Aung LH, Palmquist DE (2002) The effect of fruit maturity on quality and colour shift of dried 'Patterson' apricot during eight months of cold storage. J Hortic Sci Biotechnol 77:526–533

Ledbetter CA, Peterson SJ (2004) Utilization of Pakistani apricot (*Prunus armeniaca* L.) germplasm for improving Brix levels in California adapted apricots. Plant Genet Resour Newsl 140:14–22

Ledbetter CA, Peterson S, Jenner J (2006) Modification of sugar profiles in California adapted apricots (*Prunus armeniaca* L.) through breeding with Central Asian germplasm. Euphytica 148:251–259

Lichou J, Audubert A (1989) L'abricotier. Centre Technique Interprofessionnel des Fruits et Légumes. (CTIFL). ISBN: 2-901002-69-2

Lingdi L, Bartholomew B (2003) Armeniaca. In: Wu CY, Raven PH (eds.), Flora of China, vol 9 (*Pittosporaceae* through *Connaraceae*). Science Press, Beijing, and Missouri Botanical Garden Press, St. Louis. pp 396–401

Lloyd G, McCown B (1980) Commercially-feasible micropropagation of mountain laurel, *Kalmia latifolia*, by use of shoot-tip culture. Proc Internat Plant Propagators Soc 30:421–427

Lopes MS, Sefc KM, Laimer M, Da Camara Machado A (2002) Identification of microsatellite loci in apricot. Mol Ecology Notes 2:24–26

Löschnig HJ, Passecker DF (1954) Die Marille (Aprikose) und ihre Kultur. (in German) Österreichischer Agrarverlag Druck – Austrian Agrarian Publishing Company, Vienna, Austria

Lo Voi A, Impembo M, Fasanaro G, Castaldo D (1995) Chemical characterization of apricot puree. J Food Composit Anal 8:78–85

Maghuly F, Fernandez EB, Ruthner S, Pedryc A, Laimer M (2005) Microsatellite variability in apricots (*Prunus armeniaca* L.) reflects their geographic origin and breeding history. Tree Genet Genom 1:151–165

Marino G, Bertazza G, Magnanini E, Altan AD (1993) Comparative effects of sorbitol and sucrose as main carbon energy sources in micropropagation of apricot. Plant Cell Tissue Organ Cult 34:235–244

Martínez-Gómez P, Dicenta F (2000) Evaluation of resistance of apricot cultivars to a Spanish isolate of plum pox potyvirus (PPV). Plant Breed 119:179–181

Messina R, Lain O, Marrazzo MT, Cipriani G, Testolin R (2004) New set of microsatellite loci isolated in apricot. Mol Ecology Notes 4:432–434

Mirzaev MM, Kuznetsov VV (1984) Apricot in Uzbekistan: Biology, Varieties, Selection and Agricultural Techniques. (in Russian) Central Asian Branch of the Science-Investigation Agricultural Laboratory, Scientific Investigation Institute of Horticulture, Grape Growing & Winemaking. FAN Publishing House, Tashkent, Uzbekistan, pp 22–101

Pérez-Tornero O, Burgos L, Egea J (1999a) Introduction and establishment of apricot in vitro through regeneration of shoots from meristem tips. In Vitro Cell Develop Biol 35: 249–253

Pérez-Tornero O, Ortín-Párraga F, Egea J, Burgos L (1999b) Medium-term storage of apricot shoot tips in vitro by minimal growth method. HortScience 34:1277–1278

Pérez-Tornero O, Burgos L (2000) Different media requirements for micropropagation of apricot cultivars. Plant Cell Tiss Organ Cul 63:133–141

Pérez-Tornero O, Egea J, Vanoostende A, Burgos L (2000a) Assessment of factors affecting adventitious shoot regeneration from in vitro cultured leaves of apricot. Plant Sci 158:61–70

Pérez-Tornero O, López JM, Egea J, Burgos L (2000b) Effect of basal media and growth regulators on the in vitro propagation of apricot (*Prunus armeniaca* L.) cv. Canino. J Hortic Sci Biotechnol 75:283–286

Pérez-Tornero O, Egea J, Olmos E, Burgos L (2001) Control of hyperhydricity in micropropagated apricot cultivars. In vitro Cell Develo Biol – Plant 37:250–254

Petri C, Alburquerque N, García-Castillo S, Egea J, Burgos L (2004) Factors affecting gene transfer efficiency to apricot leaves during early *Agrobacterium*-mediated transformation steps. J Hortic Sci Biotechnol 79:704–712

Petri C, Alburquerque N, Pérez-Tornero O, Burgos L (2005a) Auxin pulses and a synergistic interaction between polyamines and ethylene inhibitors improve adventitious regeneration from apricot leaves and *Agrobacterium*-mediated transformation of leaf tissues. Plant Cell Tiss Organ Cult 82:105–111

Petri C, Alburquerque N, Burgos L (2005b) The effect of aminoglycoside antibiotics on the adventitious regeneration from apricot leaves and selection of *npt*II-transformed leaf tissues. Plant Cell Tiss Organ Cult 80:271–276

Pieterse RE (1989) Regeneration of plants from callus and embryos of 'Royal' apricot. Plant Cell Tiss Organ Cult 19:175–179

Polák J, Krska B, Pívalová J, Svoboda J (2005) Apricot cultivars 'Harlayne' and 'Betinka' were proved to be highly resistant to the six different strains and isolates of plum pox virus (PPV). Phytopath Poland 36:53–59

Quoirin M, Lepoivre P (1977) Etude de milieux adaptes aux cultures in vitro de *Prunus*. Acta Hortic 78:437–442

Rehder A (1940) Manual of cultivated trees and shrubs hardy in North America, exclusive of the subtropical and warmer temperate regions, 2nd revised and enlarged edition. Macmillan, New York, NY, USA

Romero C, Pedryc A, Muñoz V, Llácer G, Badenes ML (2003) Genetic diversity of different apricot geographical groups determined by SSR markers. Genome 46:244–252

Romero C, Vilanova S, Burgos L, Martínez-Calvo J, Vicente M, Llácer G, Badenes ML (2004) Analysis of the S-locus structure in *Prunus armeniaca* L. Identification of S-haplotype specific S-RNase and F-box genes. Plant Mol Biol 56:145–157

Rubio M, Dicenta F, Martínez-Gómez P (2003) Susceptibility to sharka (Plum pox virus) in *P. mandshurica* × *P. armeniaca* seedlings. Plant Breeding 122:465–466

Rubio M, Martínez-Gómez P, Pinochet J, Dicenta F (2005) Evaluation of resistance to sharka (Plum pox virus) of several *Prunus* rootstocks. Plant Breeding 124:67–70

Rugini E, Verma DC (1982) Micropropagation of difficult-to-propagate almond (*Prunus amygdalus*, Batsh) cultivar. Plant Sci Letters 28:273–281

Ruiz D, Egea J, Tomás-Barberán F, Gil M (2005a) Carotenoids from new apricot (*Prunus armeniaca* L.) varieties and their relationship with flesh and skin color. J Agric Food Chem 53:6368–6374

Ruiz D, Egea J, Gil M, Tomás-Barberán F (2005b) Characterization and quantitation of phenolic compounds in new apricot (*Prunus armeniaca* L.) varieties. J Agric Food Chem 53:9544–9552

Salava J, Polák J, Krska B (2005) Oligogenic inheritance of resistance to plum pox virus in apricots. Czech J Genet Plant Breeding 41:167–170

Salava J, Polák J, Krska B, Lalli DA, Abbott AG (2007) Construction of a genetic map for apricot with molecular markers and identification of markers associated with plum pox virus resistance. Acta Hort 738:657–661

Shimada T, Hayama H, Nishimura K, Yamaguchi M, Yoshida M (2001) The genetic diversities of 4 species of subg. *Lithocerasus* (*Prunus*, *Rosaceae*) revealed by RAPD analysis. Euphytica 117:85–90

Skirvin RM, Chu MC, Rukan H (1979) Tissue culture of peach, sweet and sour cherry and apricot shoot tips. Trans Illi State Horti Soc 113:30–38

Smykov VK (1978) Biology of apple and apricot, and principals of formation of industrial varieties. (in Russian). Moldovan Ministry of Agriculture. Scientific Research Institute of Moldova for Horticulture, Grape Growing and Winemaking. Sheentsa Publishing House. Kishinev, Moldova. p 128–131

Snir I (1984) In vitro propagation of 'Canino' apricot. HortScience 19:229–230

Soriano JM, Vilanova S, Romero C, Llácer G, Badenes ML (2005) Characterization and mapping of NBS-LRR resistance gene analogs in apricot (*Prunus armeniaca* L.). Theor Appl Genet 110:980–989

Spiegel S, Kovalenko EM, Varga A, James D (2004) Detection and partial molecular characterization of two plum pox virus isolates from plum and wild apricot in southeast Kazakhstan. Plant Disease 88:973–979

Syrgianndis G (1979) Research on the sensitivity of apricot varieties to sharka (plum pox) virus disease (in Greek). Georgike Ereuna 3:42–48

Takeda T, Shimada T, Nomura K, Ozaki T, Haji T, Yamaguchi M, Yoshida M (1998) Classification of apricot varieties by RAPD analysis. J Japan Soc Hort Sci 67:21–27

Takeoka GR, Flath RA, Mon TR, Teranishi R, Guentert M (1990) Volatile constituents of apricot (*Prunus armeniaca*). J Agric Food Chem 38:471–477

Tang CS, Jennings WG (1967) Volatile components of apricot. J Agric Food Chem 15:24–28

Tang CS, Jennings WG (1968) Lactonic compounds of apricot. J Agric Food Chem 16:252–254

Thibault B, Herman L (1982) Culture of Bartlett on its own roots: comparisons with quince and French seedling rootstocks. Acta Hort 124:21–26

Thompson MM (1998) Plant quarantine: a personal experience. Fruit Var J 52:215–219

Tomás-Lorente F, García-Viguera C, Ferreres F, Tomás-Barberán FA (1992) Phenolic compounds analysis in the determination of fruit jam genuineness. J Agric Food Chem 40:1800–1804

Uematsu C, Sasakuma T, Ogihara Y (1991) Phylogenetic relationships in the stone fruit group of *Prunus* as revealed by restriction fragment analysis of chloroplast DNA. Japan J Genet 66:59–69

Ushijima K, Sassa H, Dandekar AM, Gradziel TM, Tao R, Hirano H (2003) Structural and transcriptional analysis of the self-incompatibility locus of almond: identification of a pollen-expressed F-box gene with haplotype-specific polymorphism. Plant Cell 15:771–781

Vavilov NI (1992) The phyto-geographical basis for plant breeding. In: Dorofeyev VF (ed) Origin and Geography of Cultivated Plants. Cambridge University Press, Cambridge, UK, pp pp 316–366

Vilanova S, Romero C, Abbott AG, Llácer G, Badenes ML (2003) An apricot (*Prunus armeniaca* L.) F_2 progeny linkage map based on SSR and AFLP markers, mapping plum pox virus resistance and self-incompatibility traits. Theor Appl Genet 107:239–247

Vilanova S, Romero C, Llácer G, Burgos L, Badenes ML (2005) Identification of self-(in)compatibility alleles in apricot by PCR and sequence analysis. J Am Soc Hort Sci 130:893–898

Vilanova S, Soriano JM, Lalli DA, Romero C, Abbott AG, Llácer G (2006) Development of SSR markers located in the G1 linkage group of apricot (*Prunus armeniaca* L.) using a bacterial artifical chromosome library. Mol Eco Notes 6:789–791

Witherspoon JM, Jackson JF (1995) Analysis of fresh and dried apricot. In: Linskens HF and Jackson JF (eds) Modern methods of plant analysis, vol 18. Springer-Verlag, Berlin, Germany, pp 111–131

Yoshida M (1981) Breeding of peach rootstocks resistant to root knot nematode. I. Root knot nematode resistance in peaches and plums. (in Japanese). Bulletin of the Fruit Tree Research Station (Ministry of Agriculture, Forestry and Fisheries). Series A, 8:13–30

Zhebentyayeva TN, Reighard GL, Gorina VM, Abbott AG (2003) Simple sequence repeat (SSR) analysis for assessment of genetic variability in apricot germplasm. Theor Appl Genet 106:435–444

Zielinski QB (1977) Apricots. In: Modern Systematic Pomology. Pomona Books, Ontario, Canada, pp 127–131

Zotto AD, Ortego JM, Raigón JM, Caloggero S, Rossini M, Ducasse DA (2006) First report in Argentina of plum pox virus causing Sharka disease in *Prunus*. Plant Disease 90:523

Chapter 3
Blackberries

C.E. Finn

Abstract Few crops are as genetically diverse as blackberry, with several *Rubus* species in its background. This diversity is being used by plant breeders to fill a myriad of marketing opportunities. Several strategies are being employed to broaden the climatic zone of blackberries including the development of primocane fruiting plants, mixing eastern and western germplasm and reducing chilling requirements. Efforts are also being made to improve harvest efficiency through appropriate architecture and thornlessness to keep hand and machine picking costs as low as possible. New knowledge about basic fruit chemistry and the volatiles associated with flavor is giving breeders opportunities they have never had in the past to improve fruit quality. While blackberries are more disease tolerant than raspberries, breeders are still working to develop resistance to major fungal diseases, and most recently, several virus diseases. Molecular tools have been developed to understand taxonomy and to explore the potential of marker assisted selection, although traditional plant breeding approaches are still the primary methods being used to develop new cultivars.

3.1 Introduction

Blackberries belong to the genus *Rubus,* which bear aggregate fruit consisting of a number of fleshy drupelets, each containing a single seed (pyrene) around the central torus or receptacle. The central torus remains attached when the fruit is picked. Species in this genus, which also includes red and black raspberries, have a tremendous diversity in plant form from prostrate plants to bushes over 5 m tall (Clark et al. 2007). All *Rubus* species have perennial roots and crowns. Some species have annual or biennial fruiting canes or woody perennial growth.

C.E. Finn

USDA-ARS, Horticultural Crops Research Unit, 3420 NW Orchard Avenue, Corvallis, Oregon 97330, USA

e-mail: finnc@hort.oregonstate.edu

J.F. Hancock (ed.), *Temperate Fruit Crop Breeding,*
© Springer Science+Business Media B.V. 2008

Blackberry fruits are eaten fresh or processed. The most common primary products from processing are individually quick-frozen (IQF), canned, pureed, juiced, and freeze dried fruit. These products are in turn used to produce a myriad of products found throughout all sections of a grocery store and in institutional food service product lines.

Three main types of blackberries have been developed into commercial crops: the trailing, erect, and semi-erect blackberries. Several hybrids between raspberry and trailing blackberry have been produced including 'Logan', 'Tayberry', and 'Boysen'. Since these hybrids grow and are harvested like a trailing blackberry they are usually included with that group.

Trailing blackberries (Fig. 3.1) have a complex genetic background that includes red raspberry and eastern blackberries. However the predominant species in the background of these types is *R. ursinus* Cham. & Schltdl, (western dewberry), which is native along the Pacific Coast from British Columbia (Canada) to California (U.S.A.) and inland to Idaho (U.S.A.) in the mountains. Trailing blackberry plants are characterized by their production of vigorous primocanes from a single crown that lie on the ground. The floricanes thus need to be trained onto a support trellis for good fruit production. Fruit from trailing blackberries tend to have an excellent, aromatic flavor, with less noticeable seeds than many eastern North American and

Fig. 3.1 a) Fruiting lateral of the trailing black raspberry cultivar Siskiyou. **b)** Blackberry geek examining blackberry fruit closely (photos compliments of the USDA-ARS)

European species. The primary cultivars of this type include 'Marion', 'Olallie', 'Pacific', 'Waldo', 'Black Diamond' and 'Obsidian', and were primarily developed by the USDA-ARS in Oregon.

Another species of trailing blackberry, *R. laciniatus* Willd. (the cutleaf or evergreen blackberry), a European native, was imported and led to what is still a strong industry in Oregon. 'Evergreen' and another European native, 'Himalaya' (*R. armeniacus* Focke [= *R. procerus* auct.]), have naturalized throughout much of the western, Pacific states and are considered noxious weeds.

Blackberry/red raspberry hybrids have largely been developed unintentionally. Most have been found in plots or in the wild where red raspberry has been grown with the dioecious *R. ursinus*. Despite their purple to red fruit color, they are technically blackberries as the receptacle/torus picks with the fruit and they have a trailing blackberry growth habit. While 'Boysen' and 'Logan' have historically been important in the commercial industry, the commercial acreage of these, particularly 'Logan', has declined over the last 10 years.

Erect blackberries were developed from blackberries native to the eastern U.S. They are characterized by plants that produce stiff upright canes that are 1–4 m tall. While cultivars such as 'Eldorado' trace back to the 1880s, these blackberries really developed as a crop when the University of Arkansas began breeding them in the 1960s. This led to a series of high quality cultivars with Native American tribal names, some examples include 'Cheyenne' and 'Cherokee' released in the 1970s, 'Shawnee' and 'Navaho' in the 1980s, 'Kiowa', 'Apache' and 'Chickasaw' in the 1990s, and 'Ouachita' in the 2000s. In general, the new cultivars of this type are thornless.

Primocane fruiting/fall bearing erect blackberries are a recent development from the University of Arkansas. Plants flower and fruit very late in the season on the new growth. The plants are cut to the ground in winter simplifying management.

Semi-erect blackberries were developed from a very similar background as the erect types. Plants are characterized as being thornless, with very vigorous, large erect canes that will grow 4–6 m long from a crown and arch to the ground. Their fruit is similar in quality to the erect blackberries and they are often incredibly productive. The cultivars released included 'Smoothstem', 'Thornfree', 'Black Satin' and 'Dirksen Thornless' in the 1960s, 'Chester Thornless' and 'Hull Thornless' in the 1970s, and 'Triple Crown' and 'Loch Ness' in the 1980s, and 'Loch Tay' in the 2003.

Blackberries are grown in many areas of the world, but they are most productive in regions with mild winters and long, moderate summers. Blackberry production is rapidly increasing (Strik 1992, Clark 2005a, 2007, Strik et al. 2007). Strik et al. (2007) estimated that there were 20,035 ha of blackberries planted and commercially harvested in 2005 and that this area produced about 140,292 Mg.

Europe leads the world in area planted in blackberries (7,692 ha) and North America has the greatest production (59,123 Mg). European production is concentrated in Serbia (69%) although a number of countries have significant production. In North America, the U.S., particularly Oregon, is the major producer.

However, Mexican production has been rapidly increasing, particularly in Michoacán and Jalisco in the past 5 years. California and Arkansas are the only other states in the U.S. with over 1,000 Mg in production. Central American production (1620 ha) is predominantly located in Costa Rica and Guatemala where in addition to cultivated stands, a great deal is harvested from feral blackberry stands. South American production (1,597 ha) is predominantly from Ecuador and Chile.

Asian production is really beginning to develop with over 1550 ha of new plantings predominantly in China. Chinese production is concentrated in Jiangsu, although Liaoning, Shandong and Hubei are increasing their production. The area planted in Oceania is very low (259 ha), and is concentrated in New Zealand. African production is reportedly only found in South Africa. In the Pacific Northwest U.S., Serbia, and China the bulk of the fruit is grown for processing applications whereas elsewhere fresh market sales are the focus of the industry.

3.2 Evolutionary Biology and Germplasm Resources

The domesticated blackberries are found in the subgenera *Rubus* of the genus *Rubus,* which is divided into 12 sections. The cultivated types are predominantly derived from representatives of the sections *Allegheniensis*, *Arguti*, *Rubus* and *Ursini*, all of which have temperate distributions, although the equatorial, but not tropical, Andean blackberry (*R. glaucus* Benth) and many genotypes in the pedigrees of cultivated blackberry are in the *Idaeobatus*.

Blackberry cultivars generally have complex backgrounds and are often composed of species at several ploidy levels. Four diverse groups of blackberries have been domesticated (Clark et al. 2007): (1) the European blackberries that were derived from a group of diploid and polyploidy species ($2n = 28$, 42, and 56); the background of the European cultivars is so mixed that the designation *R. fruticosus* L. agg. is sometimes used (Daubeny 1996) (2) erect blackberries and trailing dewberries domesticated from mostly diploid and tetraploid species ranging across eastern America, and (3) trailing blackberries generated from only polyploid species from western America, predominantly *R. ursinus* at $2n = 56$, 84, with an infusions of 4x blackberry and 2x red raspberry through various intersectional hybrids such as 'Logan' and 'Tayberry' ($2n = 42$), 'Boysen' and 'Young' ($2n = 49$); the trailing western blackberries cultivars can be found at $2n = 42$, 49, 56, 63, 72, and 80 along with various aneuploids such as 'Aurora' ($2n = 58$) and 'Santiam' ($2n = 61$) (Thompson 1997, Meng and Finn 2002).

A wide range of blackberry species have been identified as carrying valuable genes (Table 3.1). High inter-fertility generally exists between homoploids of the same subgenera, and many hybrids can be formed between different ploidies within subgenera and even between subgenera. Crosses among the higher ploidy (6x–12x) blackberries and with the 4x blackberries, usually have some fertility and can have complete fertility. Very successful cultivars have been released with odd,

Table 3.1 Important sources of germplasm in blackberry breeding ($2x = 2n = 14$)

Subgenus, Section	Species	Ploidy	Location	Useful characteristics
Rubus, Alleghaniensis	R. allegheniensis Porter	2x	E. N. America	Erect, large, long shaped fruit
Rubus, Arguti	R. argutus Link	2x, 4x	E. N. America	Winter hardiness, adaptation to heavy soils, erect habit, large fruit, spinelessness
Rubus, Arguti	R. frondosus Bigel	4x	E. N. America	Low chilling requirement, adaptation to high temperatures, erect habit
Rubus, Caesii	R. caesius L.	4x	Europe, Tajikistan	Winter hardiness, primocane fruiting, glaucous fruit
Rubus, Canadenses	R. canadensis L.	2x, 3x	E. N. America	Winter hardiness, semi-thornlessness, erect habit, fruit quality
Rubus, Cuneifolii	R. cuneifolius Pursh.	2x, 3x, 4x	E. N. America	Low chilling requirement, erect habit, small seed size, late ripening
Rubus, Flagellares	R. baileyanus Britt.	4x	E. N. America	Good fruit quality
Rubus, Flagellares	R. flagellaris (Willd.)	4x, 8x, 9x	E. N. America	Flavor, improved hardiness in trailing type, early ripening, drought tolerant
Rubus, Rubus	R. armeniacus Focke (= R. procerus auct.)	4x	Europe	Double blossom rosette, leaf spot Verticillium wilt resistance
Rubus, Rubus	R. bartonii Newton syn. 'Ashton Cross'	4x	Europe	Earliness

Table 3.1 (continued)

Subgenus, Section	Species	Ploidy	Location	Useful characteristics
Rubus, Rubus	*R. caucasicus* Focke	4x	Russia, Caucus Mountains	Productive, vigorous, excellent set, disease resistance, excellent flavor
Rubus, Rubus	*R. echinatus* Lindl.	4x	Europe	Cane and leaf rust resistance
Rubus, Rubus	*R. laciniatus* Willd.	4x	Europe	Drought resistance, anthracnose resistance, high yield
Rubus, Rubus	*R. nitidioides* Wats.	4x	Europe	Early ripening, large fruit with good quality
Rubus, Rubus	*R. praecox* Bertol. *armeniacus* (= *R. procerus* Muell)	4x	Europe	Vigor, resistance to double blossom and verticillium wilt
Rubus, Rubus	*R. ulmifolius* Schott *inermis* (Willd.) Focke	2x	Europe	Spinelessness, late ripening
Rubus, Rubus	*R. ulmifolius* Schott.	2x	Europe	Vigor, wide adaptation
Rubus, Ursini	*R. ursinis* Cham. et Schlecht	7–13x	W. N. America	Good fruit quality, early ripening, resistance to verticillium wilt, blackberry rust, and *Phytophthora* root rot
Rubus, Verotriviales	*R. trivialis* Michx (= *R. rubrisetus* Rydb.)	2x	E. N. America	Low chilling requirement, adaptation to high temperatures, humidity, and drought, early ripening, double blossom rosette resistance
Idaeobatus	*R. idaeus* L.	2x	Europe	Primocane fruiting, disease and pest resistance

Table 3.1 (continued)

Subgenus, Section	Species	Ploidy	Location	Useful characteristics
Idaeobatus	R. strigosus Michx.	2x	N. America	Winter hardiness, primocane fruiting, disease and pest resistance
Idaeobatus	R. glaucus Benth.	4x	S. America	Low chilling requirement, excellent fruit quality, large fruit size, small seeds and drupelets, extended production season
Lampobatus	R. bogotensis Kunth	2x	N. S. America	Low chilling, high vigor
Lampobatus	R. nubigenus Kunth (= R. macrocarpus Benth.)	6x	S. America	Good fruit quality, large size
Lampobatus	R. roseus Poir.	6x	S. America	Low chilling, large fruit size, small seed size
Hybrid species	R. ×thyrsiger Banning & Focke	4x	Europe	Fruit quality

Sources: Jennings et al. 1991, Thompson 1995a, 1997, Daubeny 1996, Jennings 1988, Finn et al. 2002a,b

high ploidy levels such as 'Kotata' (7x), 'Siskiyou' (7x) and 'Black Pearl' (9x) (Thompson 1995b, Finn et al. 1999, 2005c, Meng and Finn 2002).

3.3 History of Improvement

Theophrastus in the 3rd century BC mentions that hedgerows of blackberries were used to keep out invading forces as long as two millennia ago (Jennings 1988). In the late 17th century, English gardening books began to mention blackberries and one of the most important European blackberry species, *R. laciniatus* was probably domesticated around this time. Selections of this species were imported to the Northwestern coast of North America before 1860 where it, and another later introduced European blackberry 'Himalaya' (*R. armeniacus*), became noxious weeds (Jennings 1988).

The earliest cultivars of eastern North American blackberries were selected in the 1800s, as forests were actively cleared and numerous wild species spread and hybridized (Darrow 1937). According to Hedrick (1925), 'Dorchester' was the first named blackberry cultivar in 1841. 'New Rochelle' was the first cultivar to be widely planted after it was named in 1854. The American Pomological Society chose to rename 'New Rochelle' as 'Lawton' in a contentious debate (Hedrick 1925). 'Dorchester', 'Lawton' and 'Texas Early' ('Crandall') were natural hybrids of *R. allegheniensis* Porter × *R. frondosus* Bigelow and played a key role in the domestication of the crop.

Judge James H. Logan in the 1880s was the first to release a blackberry cultivar from a breeding program (Logan 1955). Logan's crowning achievement was the discovery and development of the intersectional hybrid most likely between a pistillate *R. ursinus* selection 'Aughinbaugh' and 'Red Antwerp' red raspberry. Later studies showed that the Loganberry was an allohexaploid derived from a reduced gamete of an octoploid *R. ursinus* and an unreduced gamete of diploid *R. idaeus* (Crane 1940, Thomas 1940, Waldo and Darrow 1948, Jennings 1981). Logan's effort encouraged Luther Burbank in California to develop 'Phenomenal' in 1905 and Byrnes Young, who could grow neither 'Logan' nor 'Phenomenal' on his southern Louisiana farm, then crossed 'Phenomenal' with the adapted 'Austin Mayes' to produce 'Youngberry', which was released in 1926 (Darrow 1918, 1925, 1937, Clark et al. 2007). Other inter-specific polyploidy hybrids were selected in the late 1800s and early 1900s including the 'Laxtonberry' and 'Boysenberry'.

'Boysenberry' is still an important cultivar in the marketplace. The precise origin of 'Boysenberry' is unknown and often debated (Darrow 1937, Thompson 1961, Jennings 1988, Hall et al. 2002b); however, Wood et al. (1999) presented an excellent case for its historical origins. They traced 'Boysenberry' from its discovery on John Lubben's farm in Napa county (California) to southern California, to its commercialization by Walter Knott who went on to develop the Knott's Berry Farm empire based on his success with 'Boysenberry'.

The first public breeding program was begun in 1908 by the Texas Agricultural Experiment Station (College Station) where the emphasis was on developing blackberries and hybrid berries with a low chilling requirement that were adapted to warm climates (Darrow 1937). The John Innes Horticultural Institute in England, the New York State Agricultural Experiment Station, and the USDA-ARS in Georgia (later moved to Beltsville, Maryland) and Oregon programs began in succession soon thereafter. The USDA-ARS program in Corvallis Oregon, begun in 1928, is the oldest continuously active breeding program in the world.

While there have been many programs worldwide over time, few are still active (Finn and Knight 2002). The major concerted breeding efforts are with the University of Arkansas, the USDA-ARS in Oregon and the private program run by Driscoll's Strawberry Associates (Watsonville). Smaller sized and very productive programs are active elsewhere as demonstrated by the recent development of 'Tupy' from Brazil (Clark and Finn 2002), 'Loch Ness' and 'Loch Tay' from the United Kingdom (Jennings 1989, Clark and Finn 2006), and 'Čačaka Bestrna' ('Čačak Thornless') from Serbia (Clark and Finn 1999, Stanisavljevic 1999).

In North America, the eastern (Beltsville, Md.) and western (Corvallis, Ore.) USDA-ARS programs had the greatest early impact. The Beltsville program took advantage of the 'Merton Thornless' genotype developed at the John Innes Institute in the U.K. with its recessive spineless gene(s) (Crane 1943). While 'Merton Thornless' itself was not a very good cultivar, Scott and Ink (1966) were able to use it to develop 'Smoothstem' and 'Thornfree' which were impressive for the time. John W. 'Jack' Hull began his thornless blackberry program at USDA-ARS Carbondale, Illinois (Hull 1968) in the early 1960s and collaborated with Beltsville (Finn 2006). Over the next 20 years a whole series of excellent thornless semi-erect blackberries were developed including, 'Black Satin' (Brooks and Olmo 1972), 'Chester Thornless' (Galletta et al. 1998a), 'Dirksen Thornless' (Brooks and Olmo 1972), 'Hull Thornless' (Galletta et al. 1981), and 'Triple Crown' (Galletta et al. 1998b).

George M. Darrow, who is primarily known for the impact he had as head of the USDA-ARS small fruit lab in Beltsville (Md.), actually spent a few years establishing the small fruit breeding program in Oregon in the late 1920s and early 1930s. In 1932, George F. Waldo who was as the USDA-ARS Small Fruit Lab in Beltsville swapped positions with Darrow (Finn 2006). 'Loganberry', 'Youngberry', 'Mammoth', 'Himalaya', and wild selections of R. ursinus (e.g. 'Zielinski') predominated in the germplasm Waldo assembled, along with perfect-flowered cultivars (e.g. 'Santiam'/'Ideal'), where the maternal parent was R. ursinus and the paternal parent was unknown but suspected to be 'Loganberry' (Waldo 1968). Very quickly Waldo released cultivars that had a commercial impact including 'Pacific' and 'Cascade' (Waldo and Wiegand 1942), 'Chehalem' (Waldo 1948), and 'Olallie' (Waldo 1950). His next generation of crosses produced 'Marion' (Waldo 1957), which is still the major cultivar in the Northwest (Finn et al. 1997). From the late 1960s to the 1980s, while important cultivars continued to be released such as 'Kotata' (Lawrence 1984), it was the development of a germplasm pool that used the 'Austin Thornless' source of thornlessness (S_f) that took precedence.

This work lead to the release of 'Waldo', the first thornless, trailing blackberry (Lawrence 1989), and subsequently to the release of the thornless 'Nightfall', 'Black Pearl', and 'Black Diamond', which are having a significant commercial impact in the processing industry (Finn et al. 2005a,c,d, Yorgey and Finn 2005). These are in addition to 'Metolius' and Obsidian' that extend the fresh season earlier (Finn et al. 2005b,d) 'Black Diamond' provides an excellent example of germplasm exchange coming full circle. The USDA-ARS program provided significant amounts of germplasm to the New Zealand HortResearch Inc. program when it was begun in 1980. Over the years the programs had exchanged selections and seed lots and a 1991 cross between NZ 8610L163 and 'Kotata' made in New Zealand was grown out in Oregon and from it 'Black Diamond' was selected.

James N. Moore, who ran the USDA-ARS Beltsville program for a couple of years in the early 1960s moved to the University of Arkansas and began a blackberry program (Clark 1999, Finn 2006). He combined his knowledge of the Beltsville germplasm with that of the relatively nearby program Hull was running in southern Illinois and began to develop the erect blackberries. 'Cherokee', 'Comanche', and 'Cheyenne' were the first erect blackberries with high yield and very good fruit quality (Moore et al. 1974c,d, 1977). These were followed by cultivars such as 'Shawnee' (Moore et al. 1985) that had much better fruit quality and 'Choctaw' (Moore and Clark 1989a) that filled a different market season. While several of these cultivars were excellent, the release of 'Navaho' in 1989 significantly raised the standards for all fresh market blackberries. 'Navaho' was the first thornless, erect blackberry and had excellent fruit quality in the wholesale fresh market where it could be shipped internationally with good quality (Moore and Clark 1989b, 2000). The program is still active and has released many valuable cultivars in the 1990s and early 2000s, however their most recent releases 'Prime-Jan'[TM] and 'Prime-Jim'[TM] have marked another milestone (Clark et al. 2005). These two cultivars are the first primocane fruiting blackberry cultivars. As with the primocane fruiting raspberries, it is expected that these types of blackberries will revolutionize parts of the fresh market blackberry industry.

While continued funding for the New Zealand HortResearch Inc. program is currently uncertain, this program has had a significant impact over the past 25 years particularly in the areas of germplasm development. The New Zealand program, begun in 1980, has had several objectives, but the most important has been the development of new 'Boysenberry-like' cultivars (Hall et al. 2002b). The program was initiated by Harvey K. Hall and has been primarily based in Riwaka (HortResearch, Nelson Region). Initially the breeding program relied on available commercial cultivars and germplasm from the USDA-ARS (Oregon) and the Scottish Crop Research Institute (Dundee). In addition to the germplasm that was developed from mixing and selection within this group, the New Zealand program developed a new source of spinelessness, the 'Lincoln Logan' source (S_{fL}) (Hall et al. 1986a,b,c) and incorporated genotypes from the tetraploid blackberry germplasm pools (Hall and Stephens 1999). The most important recent releases from this program have been 'Ranui', 'Waimate', 'Karaka Black', and 'Marahau' (Hall and Stephens 1999, Clark and Finn 2002, Hall et al. 2003).

Private breeding programs are becoming increasingly important in the commercial industry. While the program of Driscoll's Strawberry Associates (Watsonville, Calif.) is the oldest and most productive, many private companies in North and South America and Europe are developing breeding programs. Other than information published in patents, there is little information on any of the cultivars developed in these programs as they are usually kept within the company. Companies like Driscoll's are primarily growing cultivars developed within their company.

3.4 Current Breeding Efforts

In 2001, an assessment found that there were roughly 30 (now 38) active *Rubus* breeding programs, and of these, 15 programs had some blackberry efforts (Finn and Knight 2002). However, an overall rise in importance of blackberries in the marketplace has led to a recent increase in breeding in the public and private sectors (Clark and Finn 2007). The largest programs are now at the USDA-ARS in Oregon, with emphasis on trailing types, and the University of Arkansas, with emphasis on erect types (Finn and Knight 2002). All *Rubus* breeding programs emphasized developing cultivars with high quality fruit, good yields, suitability for shipping if fresh market, machine harvestability and suitability for processing, adaptation to the local environment and improved pest and disease resistance. Numerous new blackberry cultivars have been developed over the last 30 years that incorporated many of these traits (Table 3.2).

3.5 Genetics of Important Traits

3.5.1 Disease and Pest Resistance

Blackberries are generally much more resistant to fungal diseases than raspberries (Jennings et al. 1991). In maritime and Mediterranean climates, cane Botrytis (*B. cinerea*), cane spot (*Septoria rubi* Westend), purple blotch (*Septocyta ruborum* [Lib.] Petr.) and spur blight are the most common cane diseases. In these climates, fruit ripening often occurs in the driest part of the year and therefore fruit rots are often less of a problem. In years or climates, with warm wet springs, downy mildew (*Peronospora sparsa* Berk.) can be a serious problem and lead to severe crop loss particularly on raspberry-blackberry hybrids such as 'Boysen' (Clark et al. 2007).

In more continental climates such as the eastern and Midwestern U.S.A., anthracnose (*E. veneta*), Botrytis fruit rot (*B. cinerea*), Botryosphaeria cane canker (*Botryosphaeria dothidea* (Moug.: Fr.) Ces. & De Not, and, occasionally, *Colletotrichum* spp. (Clark et al. 2007) are common problems. Two diseases that can be devastating in this climate include orange rust [*Gymnoconia peckiana* (Howe) Trott.] and rosette (double blossom) [*Cercosporella rubi* (Wint.) Plakidas]. With the exception of the very successful 'Navaho', many eastern erect and semi-erect blackberries are resistant to orange rust. While many cultivars have some resistant

Table 3.2 Blackberry/hybridberry cultivars released since the 1970s

Location of releasing program	Cultivar
Australia	Murrindindi, Silvan
Brazil	Brazos, Caiguangue, Guarani, Tupy, Xavante
Hungary	Fertodi botermo
New Zealand	Kaiteri, Karaka Black, Lincoln Logan, Mahana, Mapua, McNicol's Choice, Ranui, Riwaka's Choice, Riwaka Tahi, Taranaki, Tasman, Waimate
Poland	Orkan, Gazda
Serbia	Cacanska Bestrna
Sweden	Douglas (different from Douglass patented by B. Douglass)
U.K.	
England	Adrienne, Helen, Malling Sunberry
Scotland	Loch Ness, Loch Tay, Tayberry, Tummelberry
Canada-Quebec	Per Can, Perron's Black
U.S.A.	
Arkansas	Apache, Arapaho, Chickasaw, Choctaw, Kiowa, Navaho, Ouachita, Prime Jan (APF-8), Prime Jim (APF-12), Shawnee
California	Driscoll Carmel, D. Cowles, D. Eureka, D. Sonoma, Pecos, Sleeping Beauty, Sonoma, Zorro
Illinois	Everthornless, Illini Hardy
Indiana	Doyle's Thornless
Maryland	Chester Thornless, Hull Thornless, Triple Crown, Chesapeake
Oregon	Black Butte, Black Diamond, Black Pearl, Douglass, Kotata, Metolius, Nightfall, Obsidian, Siskiyou, Waldo
Texas	Clark Gold
West Virginia	Cox's Miracle Berry

to double blossom, in the Deep South where the environment is especially conducive to this disease, most blackberries cannot be grown. While double blossom is usually found in continental climates, it has been associated with 'Boysenberry decline' in New Zealand (Wood et al. 1999). Blackberry rust (*Phragmidium violaceum* [C.F. Schulz] G. Wint.), which was used as a biocontrol to limit the spread of European blackberries such as *R. armeniacus,* has been recently introduced to the western U.S. (Bruzzese 1986, Evans et al. 1998, Osterbauer et al. 2005). While it appears that in the long run it will have a minimal impact on feral *R. armeniacus* populations, in years conducive to the disease it has been devastating to the 'Thornless Evergreen' (*R. laciniatus*) growers in the Pacific Northwest. Should this become a problem in commercial fields for genotypes other than those derived from *R. laciniatus*, sources of resistance have been identified primarily in the trailing blackberry cultivars (Bruzzese and Hasan 1987, Evans et al. 2005).

Historically, virus diseases were not considered a major problem in blackberry. In recent years, the pollen borne *Raspberry bushy dwarf virus* (RBDV) has been found in 'Navaho', 'Boysen', and 'Marion' and found to reduce yield, however, it has yet to become a substantial commercial problem (Wood 1995, Wood and Hall 2001, Strik and Martin 2003). While symptomless, *Tobacco streak virus* (TSV) is another common pollen borne virus in blackberry (Converse 1991, Finn and Martin 1996). In addition, new viruses are being identified in blackberry that appear to be serious pests in the commercial industry (Martin et al. 2004). Reports of virus-like symptoms have become more common in blackberry plantings in the southeastern U.S. as this industry has grown. A recent survey of numerous cultivars across the regions found single and mixed infections of *Impatiens necrotic spot virus* (INSV), RBDV, *Tomato ringspot virus* (ToRSV) and *Tobacco ringspot virus* (TRSV) (Guzmán-Baeny 2003, Martin et al. 2004). Virus-free planting stock is an important component of control; however, genetic resistance is the only long term viable control option (Converse 1991). No one has examined resistance to virus diseases in blackberry, however, if these viruses were to become a significant problem, sources of resistance might be identified within blackberry germplasm or the broader *Rubus* germplasm pool, as has been the case in red raspberry.

Insect and mite problems are usually specific to regions or environments. In monocultures, insecticides/acaricides are often applied as needed for specific problems such as raspberry crown borer (*Pennisetia marginata* [Harris]), red-necked caneborer (*Agrilus ruficollis* [Fabricius]), redberry mite (*Acalitus essigi* Hassan), strawberry weevil (*Anthonomus signatus* Say), brown and green stink bugs (*Euschistus* spp. and *Acrosternum hilare* Say, respectively), Japanese beetle (*Popillia japonica* Newman), thrips (eastern and western flower thrips, *Frankliniella tritici* Fitch and *F. occidentalis* Pergande, respectively), grass grub (*Costelytra zealandia* White), raspberry fruitworms (*Byturus tomentosus* Degeer in Europe and *B. unicolor* Say in North America), root weevils (*Otiorhynchus singularis* L., *O. sulcatus* Fab, *O. ovatus* L. and *Sciopithes obscurus* Horn), and foliar nematode (*Aphelenchoides ritzemabosi* [Schwartz] Steiner). Insecticides are also applied as a 'knockdown' to remove insects such as orange tortrix (*Argyrotaenia citrana* Fernald) that can contaminate machine harvested fruit (Clark et al. 2007). New Zealand HortResearch identified resistance in *R. occidentalis* to bud moth (*Eutorna phaulacosma* Meyerick) and several leafroller species (e.g. *Cnephasia jactatana* Walker, *Ctenopseustis obliquana* Walker, *C. herana* Felder and Rogenhofer, *Epiphyas postivittana* Walter, *Planotortrix exessana* Walker, and *P. octo* Dugdale) and has tried to move that resistance into blackberry (Clark et al. 2007). While virus vector resistance has been identified in red raspberry and could presumably be moved into the trailing blackberries, it has not been attempted (Jennings 1988). While there is anecdotal evidence of differences in susceptibility among blackberry genotypes, there is no breeding effort to pursue resistance as these pests are not consistently a problem and they usually can be controlled with cultural techniques. Red berry mite is an example where extension publications will tell you that 'Chester Thornless' and 'Thornless Evergreen' are prone to this disease but no breeding program has developing resistance to this pest as a primary objective.

3.5.2 Environmental Adaptation

Lack of winter cold tolerance limits the range of successful blackberry cultivation in the continental climates of central and eastern Europe, and eastern and central North America (Moore 1984, Warmund et al. 1986, 1988, 1989, Warmund and George 1990, Daubeny 1996). While blackberries are less winter hardy than raspberries, there is still a wide range in the hardiness of cultivars and species (Hall 1990). Within the erect/semi-erect germplasm pool, the European cultivar 'Merton Thornless' is one of the least tolerant genotypes, and its use as a source of spinelessness initially led to reductions in hardiness in this pool (Jennings 1989, 1991). While blackberry breeding programs active in Poland, Nova Scotia (Canada) and Russia are selecting in climates with severe winters, by-and-large breeding for cold hardiness has been discontinued in the United States. Where cold hardiness is an essential trait, the trait is treated as any other quantitative trait, with the most hardy parents being intercrossed and the seedlings evaluated in cold environments (Galletta et al. 1980, Clark et al. 2007).

The recent development of the primocane fruiting cultivars 'Prime-Jim' and 'Prime Jan' by the University of Arkansas may make winter hardiness less of an issue in blackberry, as the canes can be removed after harvest and not be subject to winter cold (Clark et al. 2005). This trait could allow for a significant expansion of blackberry into more northern climates; however there are concerns remaining about reduced crown hardiness, which may negate this advantage (Clark, pers. comm.).

Blackberry production has expanded rapidly in Mexico, and to a lesser extent Guatemala and Spain. In all of these locations, chilling is generally 300 hours or less below 7° C and many blackberries would require greater chilling in order to develop flower and fruit normally. The industry was initially built around 'Brazos' and rapidly changed over to the higher quality 'Tupy' along with 'Choctaw' and 'Sleeping Beauty', in the early 2000s (Clark et al. 2007, Finn and Clark 2004). These cultivars have a lower chilling requirement or are amenable to cultural manipulation including defoliation, pruning, and growth regulator applications to force the plants to flower and produce good crops (J. Lopez-Medina, pers. commun.).

The EMBRAPA program in Pelotas, Brazil, and to a lesser extent the University of Arkansas, is actively breeding for genotypes with low chilling requirements and have released several low-chill cultivars including 'Ebano' (Bassols and Moore 1981), 'Tupy' (Clark and Finn 2002), 'Guarani' (Clark and Finn 2002), 'Caigangue' (Raseira 2004) and 'Xavante' (Moore et al. 2004). The cultivars developed by the University of Arkansas have a range of chilling requirements with 'Prime-Jim' and 'Kiowa' at approximately 200–300 hours, 'Arapaho', and 'Shawnee' at 400–500 hours, and 'Navaho' at 800–900 hours (Drake and Clark 2001, Yazzetti et al. 2002, Carter and Clark 2003, Warmund and Krumme 2005). 'Marion', the primary trailing cultivar, has a requirement of about 300 hours and the standard semi-erect 'Chester Thornless' about 800 hours (Takeda et al. 2002). Earlier, 'Chester Thornless' and 'Hull Thornless' were determined to break buds more sporadically due to lack of chilling in California compared to southern U.S. and trailing cultivars including 'Silvan', 'Kotata' 'Marion', 'Boysen', 'Navaho', and 'Arapaho' (Fear and Meyer 1993). With the increased emphasis on a low chilling requirement in the industry, combined with the apparently readily available genetic variability for this trait, more cultivars should be forthcoming.

In general, blackberries are much more tolerant of high heat and intense sunlight (ultraviolet [UV] light) than are raspberries, and a wide array of resistant germplasm is available (Daubeny 1996, Stafne et al. 2000, 2001). Blackberries that are susceptible to UV damage will get white drupelets and/or sunburned fruit. Surprisingly, many of the erect cultivars that tolerate very high temperatures in the Midwest are susceptible to UV damage in the Pacific Northwest where the humidity is much lower and the nights much cooler. Blackberry species with high temperature adaptations and a low chilling requirement include *R. trivialis* Michx., *R. cuneifolius* Pursh, and *R. frondosus*. *Rubus trivialis* has been used to breed diploid trailing blackberries such as 'Oklawaha' and 'Flordagrand', wich are adapted to the hot, long summers of Florida and Texas (Daubeny 1996).

Extended harvest seasons are important all across the world, regardless of climate. Early and later ripening floricane and primocane fruiting blackberries are being developed. Season extension requires that the developmental periods of cultivars are 'in tune' with the specific climatic requirements of a region. Breeding for this trait often requires selecting in a very specific location as while there appears to be plenty of variability for this trait in blackberry there is also significant genotype × environment interaction especially among types of blackberries. A couple of examples include: (1) in Oregon, the earliest trailing blackberries are about two weeks ahead of the earliest erects but in Arkansas, given more heat units in spring, they ripen in a similar time frame and (2) the primocane fruiting cultivars Prime-Jim and Prime-Jan have much better fruit quality on their primocane crop when it ripens in the relatively cool late summer of Oregon than in the hot mid-summer in Arkansas (Clark 2005b). In some locations, early floricane-fruiting cultivars have been developed by selecting for those genotypes that rapidly flower under cool temperatures, while in more variable environments with frequent frosts, a safer strategy has been to identify genotypes with a delayed bloom but rapid ripening. In primocane fruiting types, earliness is often at a premium so that the fruit can be harvested before conditions become too cool for ripening.

3.5.3 Plant Characteristics

3.5.3.1 Plant Habit/Architecture

Blackberries have a range in habits from stiffly upright to completely procumbent. As previously mentioned, cultivated blackberries are usually classified with three cane types, trailing, semi-erect, and erect (Strik 1992). Trailing cultivars (e.g. 'Marion', 'Black Diamond', 'Thornless Evergreen', and 'Obsidian') are crown-forming and grow at or near ground level, and the canes must be bundled and tied to a trellis.

The trailing blackberry cultivars have a wide range of chromosome numbers from $6x$ to $12x$. Trailing cultivars have primarily been developed by the USDA-ARS in Oregon. Semi-erect plants also are crown-forming and require a trellis, with the mature canes growing upward approximately 1 m before arching over horizontally. The semi-erect cultivars (e.g., 'Chester Thornless', 'Loch Ness', and 'Triple Crown') are grown worldwide for the fresh market and were developed in a few breeding programs including the USDA-ARS Maryland and the Scottish Crop Research Institute. Erect blackberries (e.g., 'Navaho', 'Arapaho', and 'Ouachita'), also primarily grown for the fresh market, sucker beneath the soil line and grow more stiffly upright than the semi-erects. Erect cultivars have been primarily developed at the University of Arkansas. Support wires on either side of the row are often used to help keep the erect blackberry canes upright in the Midwestern U.S. In maritime climates, erect cultivars behave more similarly to semi-erects and grow much more vigorously upright to 1–2 m making trellising essential. While erect and semi-erect cultivars respond positively to tipping of the canes, trailing cultivars do not and this is reflected in how primocanes are managed. The erect and semi-erect cultivars have historically all been tetraploid.

All three blackberry growth habits offer production advantages and the fact that the germplasm pool for the eastern blackberries tends to be erect or semi-erect in background and that for the western trailing blackberries tends to be trailing has resulted in the continued development of different types. The emphasis on erect-caned cultivars at the University of Arkansas was guided by the objective of developing self-supporting plants that could be machine harvested. The eastern germplasm tended towards this habit and selection by the breeders accentuated this in many of the first high quality cultivars such as 'Comanche' and 'Cherokee'. Later when the thornlessness was introduced from 'Merton Thornless' new challenges were presented as 'Merton Thornless' also carried several undesirable traits including late-ripening, poor cold hardiness, tart flavor, variable drupelet fertility, and small fruit size. Over time excellent thornless cultivars such as 'Navaho' were developed (Moore and Clark 1989b).

Primocane blackberry germplasm is just beginning to be developed and while the trait is recessive, other genes are involved such that thorny non-primocane fruiting seedlings are common from crosses that theoretically should be only primocane fruiting (Ballington and Moore 1995, Lopez-Medina and Moore 1999, Lopez-Medina 2000).

The trailing habit in the western trailing blackberries offers advantages for machine harvesting and winter hardiness. New primocane growth trails along the ground beneath the potentially damaging catcher plates of the machine harvester. With the primocanes out of the way, the berries fall through the floricane vegetation more easily as the machine passes. After harvest, the primocanes can be left on the ground over the winter thereby decreasing potential winter injury (Bell et al. 1995a,b). While this reduces the risk of winter injury it also reduces yield.

The traditional delineation between the growth habit of these eastern and western types is beginning to be broken down. Recently advanced selections that are 25%

trailing to 75% erect/semi-erect or vice versa have been developed that are mixing the growth habits and fruit quality characteristics traditionally restricted to these two germplasm pools. The problems of high labor cost and poor availability are strongly pushing the development of machine harvested, processing genotypes with an erect habit.

3.5.3.2 Thornlessness

Blackberry species naturally have varying levels of 'thorns' on their canes. The term, 'thorn', which has historically been the preferred term in North America, should more correctly be referred to as spines as they are often referred to in the United Kingdom. We will refer to spines and spinelessness in blackberries as thorns and thornlessness. Clark et al. (2007) presents an extensive and complete review of breeding for thornlessness in blackberry, the highlights of which will be given here. While cultural methods have been designed to reduce thorn contamination in machine harvested blackberries, the ultimate solution to this problem is genetic (Strik and Buller 2002).

Breeding for thornlessness in blackberries is complicated by several factors including: (1) there are recessive and dominant sources (2) sources are available at the 4x and the 6x or greater ploidy levels (3) thornlessness varies as to when it becomes apparent in the seedlings (4) when the sources were identified at different times and (5) some sources seem to have linkages with undesirable traits.

While numerous thornless sports have been selected from a variety of species over the years (Hedrick 1925, Darrow 1928, 1929), four sources are of the greatest interest as they have been valuable in the development of thornless cultivars or hold the promise to do so in the near future, these include: (1) the 'Merton Thornless' recessive gene s (2) the 'Austin Thornless' dominant gene S_f (3) the 'Thornless Evergreen' S_{fTE} gene, and (4) the 'Lincoln Logan' dominant gene S_{fL} source (Clark et al. 2007).

Breeders at the John Innes Institute in the U.K. selfed and crossed thornless plants of the 4x species *R. ulmifolius* Schott and found that the trait was recessive. They went on to develop 'Merton Thornless' from this work and while this cultivar was never very useful in the U.K., at the USDA-ARS in Beltsville, Md. it was extensively used in breeding and led to the release of 'Thornfree' and 'Smoothstem' (Scott and Ink 1966, Clark et al. 2007). The 'Merton Thornless' source of thornless has been valuable in the development of commercial tetraploid erect and semi-erect blackberry cultivars. The gene is very stable and all seedlings carrying the homozygous recessives are thornless. The main limitations of this source are the slowness with incorporating new germplasm, as it requires going through a step where all of the progeny are heterozygous and thorny, and, to this point, its use has been limited to the tetraploid germplasm breeding pool. After the original Beltsville releases, many programs took advantage of the germplasm to develop a whole series of tetraploid thornless cultivars including 'Black Satin', 'Dirksen Thornless', 'Hull Thornless', 'Chester Thornless', 'Arapaho', 'Navaho', 'Apache', 'Triple Crown', 'Ouachita',

'Pecos', 'Driscoll Sonoma', 'Driscoll Cowles', 'Loch Ness', 'Loch Tay', 'Orkan', and 'Čačanska Bestrna' ('Čačak Thornless') (Moore 1997, Clark 1999, 2006, Daubeny 1995).

Early in the 20th Century, a thornless sport of 'Austin Mayes' was found and released as 'Austin Thornless' (Hedrick 1925, Butterfield 1928). The 'Austin Thornless' source has been very useful in breeding at the $6x$ and higher ploidy levels and therefore has been primarily of value for breeding the trailing blackberries. While dominant in different genetic backgrounds, this trait shows a range of expression from completely thornless to thorniness on the basal 0.3 m of the cane. Canes that are thorny on the base of their canes are thornless from a commercial standpoint as there are no thorns in the area where the fruit are picked and the catcher plates on a machine harvester are above this level. Since the plants must be over 0.3 m tall before their thornlessness status can be determined, they must be grown in pots until they are large enough to screen or all seedlings must be field planted and evaluated later. While sterility was often associated initially with the S_f gene, this was overcome with time and the effort of a number of breeding programs. Other negative traits associated with this gene have been a somewhat dwarfed plant growth habit including short fruiting laterals, as well as brittle canes and tight fruiting clusters that are more prone to Botrytis. In 1989, 'Waldo' was the first cultivar to be released with this character; since then, cultivars that are $6x$, $8x$, and $9x$ have been released with this gene including 'Adrienne', 'Black Diamond', 'Black Pearl', 'Douglass', 'Helen', 'Murrindindi', and 'Nightfall' (Douglass 1993, 1995, Finn et al. 2005c,d).

'Thornless Evergreen' is a 4x trailing blackberry that is a periclinal chimeral of the *R. laciniatus* cultivar 'Evergreen' and accounts for 15–20% of the blackberry acreage in the Pacific Northwest. The S_{fTE} gene has been investigated but does not appear to be useful in breeding or further cultivar development (Hall et al. 1986a, Clark et al. 2007). McPheeters and Skirvin (1983, 1989, 2000) harvested the epidermal, thornless layer of 'Thornless Evergreen' and through the use of cell culture developed 'Everthornless' which is a genetically thornless cultivar with the fruit characteristics very similar to 'Thornless Evergreen'. Unfortunately, 'Everthornless' was commercialized just as 'Thornless Evergreen' was falling out of favor in the industry.

Using similar techniques to those described by McPheeters and Skirvin (1983, 1989) to produce 'Everthornless', Hall et al. (1986b,c), developed a non-chimeral, dominant thornless 'Loganberry' type from a thornless clone in New Zealand (Rosati 1986, Clark et al. 2007). The S_{fL} gene is dominant and genotypes can be screened at the very young seedling stage to determine their thornlessness, which can save a breeding program a tremendous amount of money. Initially this source of thornlessness tended to produce blackberry plants with a somewhat raspberry type plant where the canes were semi-erect but brittle and fruit that was soft, purple and small. While an ongoing process, many of these negative associations have been broken. To this point, no commercial cultivars have been released with this thornless gene, however advanced selections from New Zealand HortResearch Inc. and USDA-ARS Oregon programs are in commercial trial.

3.5.4 Fruit Quality

The critical traits associated with high fruit quality in blackberry include physical characteristics such as size, shape, color, firmness, skin strength, texture, seed (botanically pyrene) size, and ease of harvest and chemical characteristics such as flavor, soluble solids, pH, titratable acidity, and nutritional/nutraceutical content. Obviously, whether the fruit is being grown for the fresh or processing market determines which traits rise or drop in importance. For instance, while machine harvested caneberries need to maintain their shape and integrity throughout the harvesting process, they are frozen or otherwise processed within hours of harvest and therefore do not need the same level of firmness that is essential for fruit for fresh shipping. Fruit that is processed needs high soluble solids, titratable acidity levels, and relatively low pH in order that the products that they are made into have greater shelf stability. Similarly, since many times fruit for processing is only a small portion of a product, it is essential that they have intense flavor and color.

Breeding for fruit size has seen an interesting evolution over the past century (Moore et al. 1974a). Blackberry breeders have increased fruit size tremendously compared to the wild germplasm. While large fruit size has never been that important in processed berries, especially once machine harvesting became the standard practice, it has been a common objective to select for large fruit size. However, beginning in the late 20th Century, it became apparent that for fruit to fit in the plastic 'clamshell' containers, which are the standards in the wholesale fresh market, it cannot be too large. Cultivars that are large fruited make it harder for a clamshell to make the stated weight and very large fruit are subject to being injured when the clamshell is closed. While blackberry breeders were beginning to develop cultivars that averaged 13–15 g fruit (Hall 1990, Finn et al. 1998), the industry really was only adapted to handle 8–10 g fruit.

While fruit firmness is a result of multiple factors, in blackberries it is primarily related to skin toughness and drupelet size, all things being equal, the larger the drupelet the more likely it is to be damaged. In general, blackberry fruit with smaller drupelets and higher drupelet number have greater firmness but they can appear seedier to the consumer if there are too many tiny drupelets. While Strik et al. (1996) examined variability among blackberry genotypes for drupelet number and size, very little additional work has been done on blackberry. Although many blackberry cultivars have been selected for hand-picking, selecting for machine harvest ability has been essential since the advent of viable commercial harvesters in the 1950s (Hall et al. 2002a).

Generally, blackberries harvest fairly efficiently (Clark et al. 2007) as they form a good, clear abscission zone at the base of the receptacle (Hall et al. 2002a). There are exceptions, as some genotypes that appeared to be easy to harvest when hand tested did not machine harvest well because the pedicel separated from the plant instead of separating at the base of the torus.

Blackberry seeds can be offensive when chewed. Perception of 'seediness' is due to a combination of seed shape, endocarp thickness and the capsule of flesh which surrounds the seed when it is removed from the drupelet (Takeda 1993).

Trailing blackberry seeds are generally ellipsoidal and smaller than those of eastern semi-erect blackberries that are 'clam shaped' (Takeda 1993). This shape difference as well as the fact that trailing blackberries tend to have flat pyrenes with a soft, thin endocarp leads to a perception of low seediness in trailing blackberries. Moore et al. (1975) found that seed size was quantitatively inherited with partial dominance for small size; therefore progress in crossing and selecting for small seeds in erect blackberries should be successful. Progenies derived from crosses between eastern erect and western trailing blackberries show a range of seediness. Large fruit size can be attained in cultivars with small seed size; 'Siskiyou' is an excellent example of this (Strik et al. 1996, Finn et al. 1999).

Appropriate fruit color is essential for the success of a new cultivar. Blackberry cultivars, with the exception of the hybrid berries such as 'Boysen' and 'Loganberry', must retain their black color when refrigerated or frozen. Blackberries naturally have a very intense color and high anthocyanin levels. While relatively little research had been done historically on fruit color in blackberry, the fact that anthocyanins and polyphenolics are powerful antioxidants has led to a number of investigations on the nutraceutical/antioxidant levels in all caneberries (Wang and Lin 2000, Siriwoharn et al. 2005, Cho et al. 2005, Connor et al. 2005a,b, Ding et al. 2006).

Deighton et al. (2000) measured the antioxidant properties of domesticated and wild species, along with total phenol, anthocyanin and ascorbic acid contents. Antioxidant capacities ranged from 0 to 25.3 μmol Trolox equivalents g^{-1}, with *R. caucasicus* Focke having the highest values. Ascorbic acid and anthocyanins had only a minor influence on antioxidant capacity in their study. Moyer et al. (2002) found that the western blackberry species *R. ursinus* and its derivatives were higher in antioxidant capacity than the eastern blackberry cultivars and most European *Rubus* species.

Commercial blackberry genotypes have a much greater range in flavors than do raspberry genotypes. Ideal fruit flavor represents a balance between acidity and sweetness as well as appropriate intensity with desirable aroma. In blackberry, Wrolstad et al. (1980) were among the first to examine differences in sugars and acids in a diverse group of eastern and western cultivars. The most important trailing cultivars ('Marion' and 'Thornless Evergreen') have had their flavors profiled and the variability from season to season characterized (Klesk and Qian 2003a,b, Qian and Wang 2005).

Differences in flavor among blackberry genotypes have been examined as well as what characterizes good flavor (Perkins-Veazie et al. 2000, Turemis et al. 2003, Fan-Chiang and Wrolstad 2005, Kurnianta 2005). Environmental effects on flavor are documented (Qian and Wang 2005, Wang et al. 2005). The blackberry aroma component of flavor has recently been analyzed in depth at Oregon State University in 'Marion' as well as a number of selections and cultivars grown in the Pacific Northwest (Klesk and Qian 2003a,b, Qian and Wang 2005). The same genotypes whose aromas were profiled at Oregon State University were also evaluated by a trained consumer panel in a sensory analysis study (Kurnianta 2005). In the principal component analysis in this study, PC1 separated the eastern blackberry, 'Chester

Thornless' and the *R. laciniatus* cv. Thornless Evergreen, which were described as having 'cardboard', 'cooked fruits', 'moldy', 'prunes', 'vegetal', 'vinyl', 'wheat bread' and 'woody' notes, from the trailing blackberries, which were described as having 'floral', 'raspberry', 'citrus', and 'strawberry' characters. Differences were observed among many of the genotypes but the new cultivars 'Black Pearl' and 'Black Diamond' were comparable to 'Marion' for flavor; this was encouraging as this had been the goal of the original crosses and seedling selection criteria in the breeding program.

3.6 Crossing and Evaluation Techniques

3.6.1 Breeding Systems

Blackberries have a variety of reproductive systems. Most European polyploid blackberries are either facultative or obligate apomicts (pseudogamy), and several species produce sexual and asexual progeny in varying proportions (Jennings 1988). Most of the eastern American blackberry species are sexual and self-incompatible, although some pseudogamous reproduction occurs in *R. canadensis* L. and a few other species. The western American blackberry *R. ursinus* are mostly sexual and dioecious, with perfect flowered genotypes being rare but selectable (Finn 2001).

3.6.2 Pollination and Seedling Culture

Over time, every breeding program puts their own 'personality' into their procedures as their techniques, based on described procedures and research results, are combined with their experiences at their specific location, in their climate, and with their germplasm (Ourecky 1975, Jennings 1988, Hall 1990, Daubeny 1996, Clark et al. 2007). The guidelines listed here are predominantly the experience of the USDA-ARS (Ore) program as developed and defined by Mary Peterson and Kirsten Wennstrom. In the early 1990s, we expected seed germination rates of 0–40% with an average about 25%. After tweaking the system, our germination average is now approaching 80% although there are always a few crosses that just do not work.

Blackberry flowers are typical for the *Rosaceae* and therefore the emasculation and pollination techniques are similar to those for others in the family. One initial difference in the approaches for the caneberries versus other members of the family is that the pollen parents need to tested free of viruses, since the pollen-borne viruses RBDV and TSV are an issue.

While crosses can be done either in the field or in the greenhouse, in the field there is a plethora of flowers at the correct stage and pollen production seems to be greater. However, the New Zealand HortResearch Inc. program reports much better germination on seeds from crosses made in the greenhouse (Hall, pers. comm.).

As the plants begin to bloom, buds are collected when they are expanded to show some petal, but have not yet opened to allow contamination by pollinators. Buds from each genotype are placed in a sterile Petri dish, cut in half, and dried overnight under a 25-watt bulb placed about 20–24 cm above the bench and stored in small salve tins in a refrigerated desiccator. Some programs extract pollen or anthers, but this practice is much more time consuming and in our experience is not important. The pollen will remain viable in these conditions for 1–2 weeks, but if a longer storage period is needed, pollen is dried and then frozen in a desiccator.

Emasculation of the female genotypes is most efficiently done when inflorescences have reached the stage when the primary bud on a lateral has bloomed but the secondary buds are just beginning to swell and show a bit of petal. Neither the pollen nor the stigmas are mature at this stage, so the emasculation can be done without fear of self-pollination. Typically, four inflorescences (about 16 buds) are emasculated for each cross, which generally produces enough seed to yield 100 seedlings, after the multitude of potential losses occur.

Buds are emasculated by slicing from the underside through the sepal, petal, and stamen whorls simultaneously with a single edged razor, leaving only the receptacle and gynoecium. Some programs prefer to use thumbnails, forceps or scalpels to do the emasculation and a great deal depends on the dexterity and personal preference of the emasculators. A waxed paper or glassine bag is generally secured over the inflorescence to keep the flowers dry (Finn 1996, Clark et al. 2007).

Approximately three days after emasculation the stigmas mature and become receptive to pollen. This is indicated by expansion of the cluster of styles and a change in stigma color from bright green to a pale yellowish. Pollen from the storage tin is brushed onto the emasculated buds using small paint brushes that are kept inside the tins (Fig. 3.2). As time permits, and depending on the length of time the stigmas are receptive, bags are removed for repollination in one to three days and when possible a third pollination is also done.

As the fruit ripen, they are placed into magenta boxes or baby food jars, which are filled about halfway with tap water containing 2–4 drops of pectinase. The fruit are than mashed with a fork and left overnight. The pectinase separates the flesh from the seeds and is greatly preferred to blenders as they cause too much seed damage. After the pectinase treatment, the solution is poured through a small strainer and rinsed. The cleaned seed is allowed to dry over night, placed in envelopes and these can be held at room temperature for several weeks. For long-term storage, envelopes can be kept in a refrigerated desiccator for ten or more years.

Rubus seeds generally require scarification and stratification. The size of the seed lot and the rarity of the cross determine the pre-germination and germination protocol to be used. Typical crosses, and those with >50 seeds, are treated with a 'standard' protocol. Challenging crosses (interspecific, expected sterility, etc.) and those with <50 seeds are germinated with an 'in vitro' protocol.

The standard germination-to-field protocol consists of an acid scarification (concentrated sulfuric acid), water and sodium bicarbonate rinse, a 5–6 day calcium

Fig. 3.2 Emasculated blackberry flower being pollinated (photo compliments of the USDA-ARS)

hypochlorite soak, another rinse, overnight warm stratification, 6–10 week cold stratification, 1–4 week germination and transplanting, six weeks as greenhouse plugs, one week acclimation to outdoor conditions and, finally, field planting. The target date for planting determines when the pre-germination treatments should begin; 18–22 weeks is the typical timeframe.

In preparation for scarification, seeds are placed in 100 ml test tubes. The number of seeds per tube is kept under ~300 so that the acid will be evenly distributed and the seeds will not clump together. Trailing blackberries require 1–4 hours of stratification, depending on size and semi-erect/erect blackberries might require 3–4 hours (Moore et al. 1974b). Approximately 10 ml acid is poured into each tube, and then stirred using a vortex mixer to coat the seeds. The tubes are then immediately placed in a test tube rack in an ice bath. When the designated scarification time has elapsed, ice water is poured quickly into the tubes to within 2.5 cm of the rim to dilute the acid and slow the reaction. The solution and seeds are then poured through a strainer under cold running tap water. They are then drenched with a sodium bicarbonate solution (30–50 g of baking soda in a 500 mls of water) and rinsed again with water. The seeds are finally placed into a jar of calcium hypochlorite solution (3 g L^{-1}) and held at $4°$ C for 5–6 days.

A wide variety of germination flats are used which are filled with a medium of textured vermiculite, watered, and topped with 0.6 cm sphagnum peat. Flats are placed into the mist bench overnight, the seeds are spread on the peat surface and then the flats are wrapped with clear plastic bags and stored at $4°$ C with 16 hours of light for 6–10 weeks. Stratification time varies from cross to cross depending on the genetic background, so flats should be checked every couple of weeks to make sure the seed bed is still moist and to discover germinating seedlings.

After stratification, the plastic wrap is removed and flats are moved to the mist bench and given bottom heat (24° C) and intermittent mist. Seedlings generally begin to emerge in less than a week, and germination is mostly complete within four weeks. When seedlings have developed two true leaves they are pricked out and transplanted into 50–72 cell plug trays filled with a bedding plant mix. Plugs are watered in and grown in the greenhouse at 22–24° C under 16 hours daylength. They are initially fertilized with a balanced fertilizer at 1–2x per week with 100 ppm N for 2–3 weeks, then 200 ppm N for 3–4 weeks. When roots fill the plugs and outdoor temperatures allow, the flats are moved outdoors under shade cloth for 1 week, then moved to full sun to await field planting after the last frost date.

While most seed lots are germinated using the basic procedure, in vitro procedures are used for small seed lots or for seeds from wide or challenging crosses. In addition to what we describe below, there are other procedures available (Galletta and Puryear 1983, Galletta et al. 1986, Hall 1990) The in vitro germination protocol involves surface sterilization with ethanol and bleach, a 6–10 week cold stratification, repeat surface sterilization, dissection, 1–2 week germination on media, transplanting, six weeks as greenhouse plugs and one week of acclimation prior to field planting.

The seed is surface sterilized prior to stratification using 1 minute in 70% ethanol while swirling by hand, then into a 20–25 ml solution of 10% bleach + 1–2 drops surfactant with agitation on a shaker table at 300 + RPM for 60 minutes. Seed and bleach are poured through a strainer, and seeds are then placed into the sterilized Petri dishes, sealed with Parafilm and stratified at 4° C for 6–10 weeks.

When the stratification period has been satisfied, seed is ready for seed coat removal and germination. Seed is removed from its Petri dish and surface sterilized, using the 70% ethanol for 1 min., then bleach + surfactant for 1 hour and placed into a tube of sterile water to await dissection. For embryo extraction, the radicle end of each seed is grasped with forceps and the top half of the seed containing the cotyledon tips is removed with a cut by a scalpel cleaned in 70% ethanol. The portion containing the radicle, embryo, and a piece of the cotyledon is retained and placed into test tubes containing sterile water. While embryos will show signs of germination as quickly as 2–4 hours after initial cutting, if they are left for more than 4 hours, or even overnight, they expand enough to be expelled from the seed coat, thereby separating the embryo from a major source of contamination. At this point the embryos can be poured through a strainer and transferred to the germination medium. While a variety of vessels work, we have found that 48-well (0.4 ml) sterile culture plates work well. Wells in the vessel are filled about 3/3 full with autoclaved medium of 1/2 strength MS media with 100 mg L^{-1} myoinositol, 10 mg L^{-1} sucrose, and 7 mg L^{-1} agar.

Germination occurs at room temperature on the bench top, but better results have been obtained with 16 hour daylength and maintaining temperature around 25° C. Within 5–10 days the initially white embryos will have begun to take on a green color and root growth will be observed. Once germinated, the seedlings are treated by one of two paths: (1) transfer to a test tube to allow further development before establishing in soil, or (2) remaining in the culture plate for a few extra days until the first true leaves begin to show, then transferring to small plug trays in germination mix for further development before establishing in final-size plug trays. The latter path has been more successful for us.

When transplanting from the in vitro environment it is critical that seedlings be protected from drying out. Soil should be moist prior to transplanting and new transplants must either be misted with a spray bottle periodically while the rest of the flat is filled, or given mist and plastic film laid over them to prevent exposure to drying air. Once complete, the flat is placed in the mist bench with supplemental light delivering a 16 hour daylength for a day or two, and then brought out for successively longer exposures to open air. After about a week the plants are hardened enough so that they can be moved to the greenhouse and grown under 16 hour days and temperatures ranging from 21 to 27° C.

Blackberry seedlings from either protocol can be screened in the seedling flat for thornlessness if they have the 'Lincoln Logan' source of thornlessness as well for any off types (e.g. chlorotic mutants).

3.6.3 Evaluation Techniques

The number of seedlings that are planted is obviously restricted by the objectives and the land, labor and supply resources that are available. Typically, 100 seedlings per cross are planted with each plant at 0.8–0.9 m apart within the row. In an ideal world, floricane fruiting caneberries could be evaluated about 14 months after planting. However, it is usually very difficult to get all of the seedlings large enough to fairly evaluate them in 14 months, so seedlings are usually grown for one year and cut to the ground during the dormant season to simplify management. The primocanes produced in the second year are intensively managed so that all of the seedlings can be evaluated in the third year, two years after planting. While it is not uncommon to have programs that evaluate seedlings again in the 4th year, the expense of this drives most programs to try to complete selection in the 3rd year. Primocane fruiting seedlings can be evaluated in the planting year in many climates but are often left for a second year to ensure that the assessment of the flowering and fruiting habit is accurate and not affected by any juvenility.

Typical for most fruit breeding programs, 0.5–1% of the seedlings are selected primarily based on the perceived vigor, yield and fruit quality with few notes or detailed evaluations made. These are then propagated for more critical trials. In some cases where there is a critical need, these selections are placed into screening trials such as those for root rot resistance or for adaptation for machine harvest. More typically they are placed in either single, multiple plant observation plots or in replicated trials. In both cases more detailed observations are made and the most promising are harvested for yield and often for postharvest fresh market storage or processing evaluation. For processing market applications, the fruit is frozen, pureed, and/or juiced for evaluation after the season. Ideally about 10% of the advanced selections, with excellent yield, horticultural traits and fruit quality traits are identified that are propagated further and planted with collaborators at public institutions or in the industry. Decisions about release as a cultivar are generally made 8–12 years after the initial crosses.

3.7 Biotechnological Approaches to Genetic Improvement

A wide array of molecular markers have been developed in *Rubus*, including isozymes, restriction fragment length polymorphisms (RFLPs), random amplified polymorphic DNA (RAPDs), sequence-characterized amplified regions (SCARs), simple sequence repeats (SSRs), microsatellites, and DNA sequences (Hokanson 2001, Stafne, 2005). These have been used to distinguish cultivars and hybrids (Stafne et al. 2003), estimate levels of apomixis (Kraft et al. 1996), conduct taxonomic studies (Waugh et al. 1990, Alice et al. 1997, 2001, Stafne 2005) and measure interspecific and intraspecific genetic variation (Kraft and Nybom 1995, Stafne et al. 2003, 2004, Stafne and Clark 2004). Genomic in situ hybridization (GISH) and fluorescence in situ hybridization (FISH) have been utilized to distinguish between raspberry and blackberry chromosomes, and identify translocations (Lim et al. 1998).

Graham et al. (2002) developed the first SSR markers for *Rubus* from red raspberry and successfully tested them on blackberries and blackberry × raspberry hybrids. While no mapping has been done on blackberry, primers have been identified for SSR markers that might be useful for mapping (Lewers et al. 2005, Stafne et al. 2005).

While regeneration systems have been developed for blackberries (Swartz and Stover 1996, Meng et al. 2004); no transgenic blackberries have been produced to date. The highest regeneration efficiency (70% of explants) was accomplished when leaves were incubated in TDZ pretreatment medium for three weeks before culturing them on regeneration medium (Woody Plant Medium with 5 uM BA and 0.5 uM IBA) in darkness for a week, and then transferring them to a 16 hour light photoperiod at 23 C for 4 weeks.

References

Alice LA, Eriksson T, Eriksen B, Campbell CS (1997) Intersubgeneric hybridization between a diploid raspberry, *Rubus idaeus* and a tetraploid blackberry, *R. caesius* (Rosaceae). Am J Bot 84:171 (Abstr.)

Alice LA, Eriksson T, Eriksen B, Campbell CS (2001) Hybridization and gene flow between distantly related species of *Rubus* (Rosaceae): Evidence from nuclear ribosomal DNA internal transcribed spacer region sequences. Syst Bot 26:769–778

Ballington JR, Moore JN (1995) NC 194 primocane-fruiting thorny erect tetraploid blackberry germplasm. Fruit Var J 49:101–102

Bassols MCM, Moore JN (1981) 'Ebano' thornless blackberry. HortScience 16:686–687

Bell NC, Strik BC, Martin LW (1995a) Effect of primocane suppression date on 'Marion' trailing blackberry. I. Yield components. J Am Soc Hort Sci 120:21–24

Bell NC, Strik BC, Martin LW (1995b) Effect of primocane suppression date on 'Marion' trailing blackberry. II. Cold hardiness. J Am Soc Hort Sci 120:25–27

Brooks RM, Olmo HP (1972) Blackberry. In: Brooks RM, Olmo HP (eds) Register of fruit & nut varieties. Second edition. ASHS Press, Alexandria, Va, pp 158–167

Bruzzese E (1986) Proposed introduction to Australia of the rust fungus *Phragmidium violaceum* (Schultz) Winter for the biological control of European blackberry (*Rubus fruticosus* L. Agg.). Keith Turnbull Res. Inst., Frankston, Australia

Bruzzese E, Hasan S (1987) Infection of blackberry cultivars by the European blackberry rust fungus, *Phragmidium violaceum*. J Hort Sci 62:475–479

Butterfield HM (1928) The origin of certain thornless blackberries and dewberries. J Hered 19:135–138

Carter, ME, Clark JR (2003) Chilling response of Arkansas blackberry cultivars. In: Evans MR, Karcher DE (eds) Hort Studies 2002. Ark Agr Exp Sta Res Ser 506:65–67

Cho MJ, Howard LR, Prior RL, Clark JR (2005) Flavonol glycosides and antioxidant capacity of various blackberry and blueberry genotypes determined by high-performance liquid chromatography/mass spectrometry. J Sci Food Agric 85:2149–2158

Clark JR (1999) The blackberry breeding Program at the University of Arkansas: thirty-plus years of progress and developments for the future. Acta Hortic 505:73–77

Clark JR (2005a) Changing times for eastern United States blackberries. Horttechnology 15:491–494

Clark JR (2005b) Intractable traits in eastern U.S. blackberries. HortScience 40:1954–1955

Clark JR, Finn CE (1999) In: Okie WR (ed) Register of new fruit and nut varieties Brooks and Olmo list 39: Blackberries and hybrid berries. HortScience 34:183–184

Clark JR, Finn CE (2002) Blackberry. In: Okie WR (ed) Register of new fruit and nut varieties, list 41. HortScience 37:251

Clark JR, Finn CE (2006) Blackberry and hybrid berry. In: Clark JR, Finn CE (eds) Register of new fruit and nut cultivars, list 43. HortScience 41, pp 1104–1106

Clark JR, Finn CE (2007) Trends in blackberry breeding. Acta Hortic: In press

Clark JR, Moore JN, Lopez-Medina J, Perkins-Veazie P, Finn CE (2005) Prime-Jan (APF-8) and Prime-Jim (APF-12) primocane-fruiting blackberries. HortScience 40:852–855

Clark JR, Stafne ET, Hall HK, Finn CE (2007) Blackberry breeding and genetics. Plant Breed Rev 29:19–144

Connor AM, Finn CE, Alspach PA (2005a) Genotypic and environmental variation in antioxidant activity and total phenolic content among blackberry and hybridberry cultivars. J Am Soc Hort Sci 130:527–533

Connor AM, Finn CE, McGhie TK, Alspach PA (2005b) Genetic and environmental variation in anthocyanins and their relationship to antioxidant activity in blackberry and hybridberry cultivars. J Am Soc Hort Sci 130:680–687

Converse RH (1991) Diseases caused by viruses and viruslike agents. In: Ellis MA, Converse RH, Williams RN, Williamson B (eds) Compendium of raspberry and blackberry diseases and insects. APS Press, St. Paul, Minn, pp 42–58

Crane MB (1940) Reproductive versatility in *Rubus*. I. Morphology and inheritance. J Genet 40:109–118

Crane MB (1943) United States Plant Patent: Blackberry plant, PP571. Washington DC

Darrow GM (1918) Culture of the Logan blackberry and related varieties. U.S. Deptartment of Agricultural Farmers Bulletin 998, pp 34

Darrow GM (1925) The Young dewberry, a new hybrid variety. Am Fruit Grow 45:33

Darrow GM (1928) Notes on thornless blackberries: Their chromosome number and their breeding. J Hered 19:139–142

Darrow GM (1929) Thornless sports of the Young dewberry. J Hered 20:567–569

Darrow GM (1937) Blackberry and raspberry improvement. U.S. Deptartment of Agriculture. Yearbook of Agriculture Yrbk, pp 496–533

Daubeny HA (1995). In: Cummins JN (ed) Register of new fruit and nut varieties Brooks and Olmo list 37: Blackberries and hybrid berries. HortScience 30:1136–1137

Daubeny HA (1996) Brambles. In: Janick J, Moore JN (eds) Fruit Breeding, vol. II. Vine and small fruits. John Wiley & Sons, Inc., New York

Deighton N, Brennan R, Finn CE, Davies HV (2000) Antioxidant properties of domesticated and wild *Rubus* species. J Sci Food Agr 80:1307–1313

Ding M, Feng R, Wang SY, Bowman L, Lu Y, Qian Y, Castranova V, Jiang BH, Shi X (2006) Cyanidin-3-glucoside, a natural product derived from blackberry, exhibits chemopreventive and chemotherapeutic activity. J Biol Chem 281:17359–17368

Douglass BS (1993) United States Plant Patent: Blackberry plant named 'Douglass', PP8, 423. Washington DC

Drake CA, Clark JR (2001) Chilling requirement of Arkansas thornless blackberry cultivars. In: Lindstrom JT, Clark JR (eds) Hort. Studies 2000. Ark Agr Sta Res Ser, vol. 483, pp 30–32

Evans KJ, Jones MK, Roush RT (2005) Susceptibility of invasive taxa of European blackberry to rust disease caused by the uredinial stage of Phragmidium violaceum under field conditions in Australia. Plant Pathol 54:275–286

Evans KJ, Symon DE, Roush RT (1998) Taxonomy and genotypes of the *Rubus fruticosus* L. Aggregate in Australia. Plant Protection Quart Plant Prot Q 13:152–156

Fan-Chiang HJ, Wrolstad RE (2005) Anthocyanin pigment composition of blackberries. J Food Sci 70:198–202

Fear CD, Meyer ML (1993) Breeding and variation in *Rubus* germplasm for low winter chill requirement. Acta Hortic 352:295–299

Finn CE (1996) Emasculated trailing blackberry flowers set some drupelets when not protected from cross pollination. HortScience 31:1035

Finn CE (2001) Trailing blackberries: From clear-cuts to your table. HortScience 36:236–238

Finn CE (2006) Caneberry breeders in North America. HortScience 41:22–24

Finn CE, Clark JR. (2004) Blackberry and hybrid berries. In: Okie WR (ed) Register of new fruit and nut varieties List 42, HortScience, 39, p 1509

Finn CE, Knight VH (2002) What's going on in the world of *Rubus* breeding? Acta Hortic 585:31–38

Finn CE, Lawrence FJ, Strik BC (1998) 'Black Butte' trailing blackberry. HortScience 33:355–357

Finn CE, Lawrence FJ, Strik BC, Yorgey B, DeFrancesco J (1999) 'Siskiyou' trailing blackberry. HortScience 34:1288–1290

Finn CE, Martin RR (1996) Distribution of tobacco streak, tomato ringspot, and raspberry bushy dwarf viruses in *Rubus ursinus* and *R. leucodermis* collected from the Pacific Northwest. Plant Dis 80:769–772

Finn CE, Strik BC, Lawrence FJ (1997) 'Marion' trailing blackberry. Fruit Varieties J 51:130–133

Finn CE, Swartz HJ, Moore PP, Ballington JR, Kempler C (2002a) Use of 58 *Rubus* species in five North American breeding programs-breeders notes. Acta Hortic 585:113–119

Finn CE, Swartz HJ, Moore PP, Ballington JR, Kempler C (2002b) Breeders experience with *Rubus* species. http://www.ars-grin.gov/cor/rubus/rubus.uses.html

Finn CE, Yorgey B, Strik BC, Hall HK, Martin RR, Qian MC (2005a) 'Black Diamond' trailing thornless blackberry. HortScience 40:2175–2178

Finn CE, Yorgey B, Strik BC, Martin RR, Kempler C (2005b) 'Obsidian' trailing blackberry. HortScience 40:2185–2188

Finn CE, Yorgey B, Strik BC, Martin RR, Qian MC (2005c) 'Black Pearl' trailing thornless blackberry. HortScience 40:2179–2181

Finn CE, Yorgey B, Strik BC, Martin RR, Qian MC (2005d) 'Nightfall' trailing thornless blackberry. HortScience 40:2182–2184

Galletta GJ, Draper AD, Hill RG Jr (1980) Recent progress in bramble breeding at Beltsville, Md Acta Hortic 112:95–102

Galletta GJ, Draper AD, Hill RG, Blake RC, Skirvin RM (1981) 'Hull Thornless' blackberry. HortScience 16:796–797

Galletta G, Draper A, Maas J (1998a) 'Chester Thornless' blackberry. Fruit Varieties J 53:188–122

Galletta GJ, Draper AD, Puryear RL (1986) Characterization of *Rubus* progenies from embryo culture and from seed germination. Acta Hortic 183:83–89

Galletta GJ, Maas JL, Clark JR, Finn CE (1998b) 'Triple Crown' thornless blackberry. Fruit Varieties J 52:124–127

Galletta GJ, Puryear RL (1983) A method for *Rubus* embryo culture. HortScience 18:588

Graham J, Smith K, Woodhead M, Russell J (2002) Development and use of simple sequence repeat SSR markers in *Rubus* species. Mol Ecol Notes 2:250–252

Guzmán-Baeny TL (2003) Incidence, distribution, and symptom description of viruses in cultivated blackberry (*Rubus* subgenus *Eubatus*) in the Southeastern United States. MS Thesis, North Carolina State Univ., Raleigh

Hall HK (1990) Blackberry breeding. Plant Breed Rev 8:249–312

Hall HK, Brewer LR, Langford G, Stanley CJ, Stephens MJ (2003) 'Karaka Black': Another 'Mammoth' blackberry from crossing eastern and western USA blackberries. Acta Hortic 626:105–110

Hall HK, Cohen D, Skirvin RM (1986a) The inheritance of thornlessness from tissue culture-derived Thornless Evergreen blackberry. Euphytica 35:891–898

Hall HK, Quazi MH, Skirvin RM (1986b) Isolation of a pure thornless Loganberry by meristem tip culture. Euphytica 35:1039–1044

Hall HK, Skirvin RM, Braam WF (1986c) Germplasm release of 'Lincoln Logan', a tissue culture-derived genetic thornless 'Loganberry'. Fruit Varieties J 40:134–135

Hall HK, Stephens J (1999) Hybridberries and blackberries in New Zealand – breeding for spinelessness. Acta Hortic 505:65–71

Hall HK, Stephens MJ, Alspach P, Stanley CJ (2002a) Traits of importance for machine harvest of raspberries. Acta Hortic 585:607–610

Hall HK, Stephens MJ, Stanley CJ, Finn CE, Yorgey B (2002b) Breeding new 'Boysen' and 'Marion' cultivars. Acta Hortic 585:91–95

Hedrick UP (1925) The small fruits of New York. J.B. Lyon, Albany, NY

Hokanson SC (2001) SNiPs, chips, BACs, and YACs: are small fruits part of the party mix. HortScience 36:859–871

Hull JW (1968) Sources of thornlessness for breeding in bramble fruits. Proc Am Soc Hortic Sci 93:280–288

Jennings DL (1981) A hundred years of Loganberries. Fruit Varieties J 35:34–37

Jennings DL (1988) Raspberries and blackberries: their breeding, diseases and growth. Academic Press, London

Jennings DL (1989) United States Plant Patent: Blackberry plant-Loch Ness cultivar, PP6,782, Washington DC

Jennings DL, Daubeny HA, Moore JM (1991) Blackberries and raspberries (*Rubus*). In: Moore JN, Ballington JR (eds) Genetic resources of fruit and nut crops, vol 1. International Society for Horticultural Science, Wageningen, pp 329–320

Klesk K, Qian MC (2003a) Aroma extract dilution analysis of cv. Marion (*Rubus spp.* hyb.) and cv. Evergreen (*R. laciniatus* L.) blackberries. J Agric Food Chem 51:3436–3441

Klesk K, Qian MC (2003b) Preliminary aroma comparison of Marion (*Rubus spp.* hyb.) and Evergreen (*R. laciniatus* L.) blackberries by dynamic headspace/OSME technique. J Food Sci 68:697–700

Kraft T, Nybom H (1995) DNA fingerprinting and biometry can solve some taxonomic problems in apomictic blackberries (*Rubus* subgen. *Rubus*). Watsonia 20:329–343

Kraft T, Nybom H, Werlemark G (1996) DNA fingerprint variation in some blackberry species (*Rubus* subgen. *Rubus*, Rosaceae). Plant Syst Evol 199:93–108

Kurnianta AJ (2005) Descriptive sensory analysis of thornless blackberry selections to determine sensory similarity to 'Marion' blackberry flavor. M.S. Thesis, Oregon State University, Corvallis, Ore

Lawrence FJ (1984) In: T.B. Kinney and J.R. Davis (eds.). Naming and release of blackberry cultivar Kotata. USDA-ARS Release Notice

Lawrence FJ (1989) Naming and release of blackberry cultivar 'Waldo'. U.S. Department of Agriculture, Oregon Agric Expt Sta Release Notice

Lewers KS, Styan SMN, Hokanson SC, Bassil NV (2005) Strawberry GenBank-derived and genomic simple sequence repeat [SSR] markers and their utility with strawberry, blackberry, and red and black raspberry. J Am Soc Hortic Sci 130:102–115

Lim KY, Leitch IJ, Leitch AR (1998) Genomic characterization and the detection of raspberry chromatin in polyploid *Rubus*. Theor Appl Genet 97:1027–1033

Logan ME (1955) The Loganberry. Mary E. Logan (Mrs. J.H. Logan) Publisher, Oakland, Ca

Lopez-Medina J, Moore JN (1999) Chilling enhances cane elongation and flowering in primocane-fruiting blackberries. HortScience 34:638–640

Lopez-Medina J, Moore JN, McNew RW (2000) A proposed model for inheritance of primocane fruiting in tetraploid erect blackberry. J Am Soc Hort Sci 125:217–221

Martin RR, Tzanetakis TE, Gergerich R, Fernandez GE, Pesic Z (2004) Blackberry yellow vein associated virus: A new crinivirus found in blackberry. Acta Hortic 656:137:142

McPheeters KD, Skirvin RM (1983) Histogenic layer manipulation in chimeral 'Thornless Evergreen' trailing blackberry. Euphytica 32:351–360

McPheeters KD, Skirvin RM (1989) Somaclonal variation among ex vitro 'Thornless Evergreen' trailing blackberries. Euphytica 42:155–162

McPheeters KD, Skirvin RM (2000) 'Everthornless' blackberry. HortScience 35:778–779

Meng R, Chen THH, Finn CE, Li H (2004) Improving in vitro plant regeneration for 'Marion' blackberry. HortScience 39:316–320

Meng R, Finn CE (2002) Determining ploidy level and nuclear DNA content in *Rubus* by flow cytometry. J Am Soc Hortic Sci 127:767–775

Moore JN (1984) Blackberry breeding. HortScience 19:183–185

Moore JN (1997) Blackberries. p. 161–173, The Brooks and Olmo register of fruit and nut varieties. Third ed. ASHS Press, Alexandria, Va.

Moore JN, Brown GR, Brown ED (1974a) Relationships between fruit size and seed number and size in blackberries. Fruit Varieties J 28:40–45

Moore JN, Brown GR, Lundergan CA (1974b) Effect of duration of acid scarification on endocarp thickness and seedling emergence of blackberries. HortScience 9:204

Moore JN, Brown ED, Sistrunk WA (1974c) 'Cherokee' blackberry. HortScience 9:246

Moore JN, Brown ED, Sistrunk WA (1974d) 'Comanche' blackberry. HortScience 9:245–246

Moore JN, Brown ED, Sistrunk WA (1977) 'Cheyenne' blackberry. HortScience 12:77–78

Moore JN, Clark JR (1989a) 'Choctaw' blackberry. HortScience 24:862–863

Moore JN, Clark JR (1989b) 'Navaho' erect thornless blackberry. HortScience 24:863–865

Moore JN, Clark JR (2000) 'Navaho' blackberry. Fruit Varieties J 54:162–163

Moore JN, Lundergan CA, Brown ED (1975) Inheritance of seed size in blackberry. J Am Soc Hortic Sci 100:377–379

Moore JN, Sistrunk WA, Buckley JB (1985) 'Shawnee' blackberry. HortScience 20:311–312

Moore JN, Santos AM, Clark JR, Raseira MCB, Antunes LEC (2004) Cultivar de amora-preta 'Xavante'. Simpósio Nacional do Morango e 1° Encontro de Pequenas Frutas e Frutas Nativas do Mercosul, Embrapa, Série Documentos 123:213–216

Moyer R, Hummer K, Finn C, Frei B, Wrolstad R (2002) Anthocyanins, phenolics and antioxidant capacity in diverse small fruits: *Vaccinium, Rubus* and *Ribes*. J Agric Food Chem 50:519–525

Osterbauer N, Trippe A, French K, Butler T, Aime MC, McKemy J, Bruckart WL, Peerbolt T, Kaufman D (2005) First report of *Phragmidium violaceum* infecting Himalaya and evergreen blackberries in North America. Online. Plant Health Progress doi:10.1094/PHP-2005-0923-01-BR

Ourecky DK (1975) Brambles. In: Janick J, Moore JN (eds) Advances in Fruit Breeding. Purdue University Press, West Lafayette, Indiana, pp 98–129

Perkins-Veazie P, Collins JK, Clark JR (2000) Shelflife and quality of 'Navaho' and 'Shawnee' blackberry fruit stored under retail storage conditions. J Food Qual 22:535–544

Qian MC, Wang Y (2005) Seasonal variation of volatile composition and odor activity value of 'Marion' [*Rubus* spp. *Hyb*] and 'Thornless Evergreen' [R. *laciniatus* L.] blackberries. J Food Sci 70:C13–20

Raseira MCB (2004) A pesquisa com amora-preta no Brasil. Second Simpósio Nacional do Morango e 1° Encontro de Pequenas Frutas e Frutas Nativas do Mercosul, Embrapa, Série Documentos 124:219–223

Rosati P (1986) Genetic stability of micropropagated 'Loganberry' plants. HortScience 61:34–41

Scott DH, Ink DP (1966) Origination of 'Smoothstem' and 'Thornfree' blackberry varieties. Fruit Varieties Hort Dig 20:31–33

Siriwoharn T, Wrolstad RE, Finn CE, Pereira CB (2005) Influence of cultivar, maturity and sampling on blackberry (*Rubus* L. hybrids) anthocyanins, polyphenolics, and antioxidant properties. J Agric Food Chem 52:8021–8030

Stafne ET (2005) Characterization, differentiation, and molecular marker analysis of blackberry germplasm. Ph.D. Dissertation, University of Arkansas

Stafne ET, Clark JR (2004) Genetic relatedness among eastern North American blackberry cultivars based on pedigree analysis. Euphytica 139:95–104

Stafne ET, Clark JR, Rom CR (2000) Leaf gas exchange characteristics of red raspberry germplasm in a hot environment. HortScience 35:278–280

Stafne ET, Clark JR, Rom CR (2001) Leaf gas exchange response of 'Arapaho' blackberry and six raspberry cultivars to moderate and high temperatures. HortScience 36:880–883

Stafne ET, Clark JR, Szalanski AL (2004) Genetic sequence variation in the ITS region of nine *Rubus* genotypes. HortScience 39:754 (abstr.)

Stafne ET, Clark JR, Weber CA, Graham J, Lewers KS (2005) Simple sequence repeat (SSR) markers for genetic mapping of raspberry and blackberry. J Am Soc Hortic Sci 103:722–728

Stafne ET, Szalanski AL, Clark JR (2003) Nuclear ribosomal ITS region sequences for differentiation of *Rubus* genotypes. J Arkansas Acad Sci 57:176–180

Stanisavljevic M (1999) New small fruit cultivars from Cacak: 1. The new blackberry [*Rubus* sp.] cultivar 'Cacanska Bestrna'. Acta Hortic 505:291–295

Strik BC (1992) Blackberry cultivars and production trends in the Pacific Northwest. Fruit Varieties J 46:207–212

Strik BC, Buller G (2002) Reducing thorn contamination in machine harvested 'Marion' blackberry. Acta Hortic 585:677–681

Strik B, Clark JR, Finn C, Buller G (2007) Management of primocane-fruiting blackberry to maximize yield and extend the fruiting season. Acta Hortic In press

Strik BC, Mann J, Finn CE (1996) Percent drupelet set varies among blackberry genotypes. J Am Soc Hortic Sci 121:371–373

Strik BC, Martin RR (2003) Impact of Raspberry bushy dwarf virus on 'Marion' blackberry. Plant Dis 87:294–296

Swartz HJ, Stover EW (1996) Genetic transformation in raspberries and blackberries (*Rubus* species). In: Bajaj YPS (ed) Biotechnology in Agriculture and Forestry, vol 38. Springer-Verlag, Berlin

Takeda F (1993) Characterization of blackberry pyrenes. HortScience 28:488 (abstr.)

Takeda F, Strik BC, Peacock D, Clark JR (2002) Cultivar differences and the effect of winter temperature on flower bud development in blackberry. J Am Soc Hortic Sci 127:495-501

Thomas PT (1940) Reproductive versatility in *Rubus*, II the chromosomes and development. J Gen 40:119–128

Thompson MM (1961) Cytogenetics of *Rubus* II. Cytological studies of the varieties 'Young', 'Boysen', and related forms. Am J Bot 48:667–673

Thompson MM (1995a) Chromosome numbers of *Rubus* species at the National Clonal Germplasm Repository. HortScience 30:1447–1452

Thompson MM (1995b) Chromosome numbers of *Rubus* cultivars at the National Clonal Germplasm Repository. HortScience 30:1453–1456

Thompson MM (1997) Survey of chromosome numbers in *Rubus* Rosaceae: Rosoideae. Ann Rpt Mo Bot Gard 84:128–163

Turemis N, Kafkas S, Kafkas E, Onur C (2003) Fruit characteristics of nine thornless blackberry cultivars. J Am Pomol Soc 57:161–165

Waldo GF (1948) The Chehalem blackberry. Oregon Agric Expt Sta Circ 421

Waldo GF (1950) Notice of naming and release of a new blackberry adapted to the Pacific Coast region. U.S.D.A. Release Notice

Waldo GF (1957) The Marion blackberry. Oregon Agric Expt Sta Circ 571

Waldo GF (1968) Blackberry breeding involving native Pacific Coast parentage. Fruit Varieties J 22:3–7

Waldo GF, Darrow GM (1948) Origin of the Logan and the Mammoth blackberries. J Hered 39:99–107

Waldo GF, Wiegand EH (1942) Two new varieties of blackberry the Pacific and the Cascade. Oregon Agric Expt Sta Circ 269

Wang SY, Lin HS (2000) Antioxidant activity in fruits and leaves of blackberry, raspberry and strawberry varies with cultivar and developmental stage. J Agric Food Chem 48:140–146

Wang Y, Finn CE, Qian MC (2005) Impact of growing environments on 'Chickasaw' blackberry (*Rubus* L.) aroma evaluated by gas chromatography olfactory dilution analysis. J Agric Food Chem 53:3563–3571

Warmund MR, George MF, Clark JR (1986) Bud mortality and phloem injury of six blackberry cultivars subjected to low temperatures Fruit Var J 40:144–146

Warmund MR, George MF (1990) Freezing survival and supercooling in primary and secondary buds of *Rubus* spp. Can J Plant Sci 70:893–904

Warmund MR, George MF, Cumbie BG (1988) Supercooling in 'Darrow' blackberry buds. J Amer Soc Hort Sci 113:418–422

Wrolstad RE, Culbertson JD, Nagaki DA, Madero CF (1980) Sugars and nonvolatile acids of blackberries. J Agric Food Chem 28:553–558

Warmund MR, George MF, Ellersieck MR, Slater JV (1989) Susceptibility of blackberry tissues to freezing injury after exposure to 16° C. J Am Soc Hortic Sci 114:795–800

Warmund MR, Krumme J (2005) A chilling model to estimate rest completion in erect blackberries. HortScience 1259–1262

Waugh R, Van de Ven WTG, Phillips MS, Powell W (1990) Chloroplast DNA diversity in the genus *Rubus* (Rosaceae) revealed by Southern hybridizations. Plant Syst Evol 172:65–75

Wood GA (1995) Further investigations of raspberry bushy dwarf virus in New Zealand. NZ J Crop Hortic Sci 23:273–281

Wood GA, Andersen MT, Forster RLS, Braithwaite M, Hall HK (1999) History of Boysenberry and Youngberry in New Zealand in relation to their problems with Boysenberry decline, the association of a fungal pathogen, and possibly a phytoplasma, with this disease. NZ J Crop Hortic Sci 27:281–295

Wood GA, Hall HK (2001) Source of raspberry busy dwarf virus in *Rubus* in New Zealand and the infectibility of some newer cultivars to the virus. NZ J Crop Hortic Sci 29:177–186

Yazzetti D, Clark JR, Stafne ET (2002) Evaluating the usage of stem cuttings to determine chilling requirements in six Arkansas blackberry cultivars. In: Clark JR, Evans MR (eds) Hort Studies 2001. Ark Agric Exp Sta Res Ser 494:40–41

Yorgey B, Finn CE (2005) Thornless blackberry (*Rubus* sp. L) genotypes evaluated as individually quick frozen and puree products. HortScience 40:513–515

Chapter 4
Blueberries and Cranberries

J.F. Hancock, P. Lyrene, C.E. Finn, N. Vorsa and G.A. Lobos

Abstract Most blueberry breeding activity is focused on northern highbush, southern highbush and rabbiteye types. The major objectives of blueberry breeders center on high plant vigor, improved disease resistance, flavor, longer storing fruit and expanded harvest dates. Cranberry breeders have concentrated on early maturing fruit, uniform large size, intense color, keeping quality, high productivity, disease resistance and plant vigor. Considerable variability exists in blueberry and cranberry for most of the horticulturally important traits, and while only a limited number of genetic studies have been performed, most inheritance patterns fit quantitative models. Several genes have been identified through molecular, genetic and genomic approaches that are associated with cold hardiness. Wide hybridization is commonly employed in blueberry breeding and southern highbush types were derived primarily by incorporating genes from the diploid species *Vaccinium darrowii* into the highbush background via unreduced gametes. A wide array of molecular markers has been used in blueberry for fingerprinting and linkage mapping, and a major QTL regulating the chilling requirement in diploids has been identified. Transgenic blueberries have been produced with herbicide resistance and the *Bt* gene (*Bacillus thuringiensis*) has been incorporated into cranberry. A large EST library of highbush blueberry has been produced.

4.1 Introduction

Several species of *Vaccinium* are important commercially. Most production comes from species in section Cyanococcus including cultivars of *Vaccinium corymbosum* L. (highbush blueberry) and *Vaccinium ashei* Reade (rabbiteye blueberry; syn.

J.F. Hancock
Department of Horticulture, 342C Plant and Soil Sciences Building, Michigan State University, East Lansing, Michigan 48824, USA
e-mail: hancock@msu.edu

Vaccinium virgatum Ait.), and native stands of *Vaccinium angustifolium* Ait. (low-bush blueberry). Highbush cultivars are further separated into northern or southern types depending on their chilling requirements and winter hardiness. *Vaccinium macrocarpon* Ait. (large cranberry), a member of section Oxycoccus, is also an important domesticated species. *Vaccinium myrtillus* L. (bilberry, whortleberry), *Vaccinium membranaceum* Douglas. ex Torr. (tall bilberry, big huckleberry), *Vaccinium deliciosum* Piper (Cascade bilberry or huckleberry), and *Vaccinium ovalifolium* Sm. (oval-leaved huckleberry) in section Myrtillus and *Vaccinium vitis-idaea* (lingonberry) in section Vitis-Idaea are collected primarily from the wild.

Blueberries are eaten as fresh fruit and in processed forms. About 50% of the highbush crop is marketed fresh and the remainder is processed. Individually quick frozen (IQF) fruit, pureed, juiced, dried/freeze-dried are the primary processed products and from these a myriad of products appear in grocery stores. The fruits of cranberries are very tart and for this reason are mostly processed into juices or baked goods (Eck, 1990).

Many of the wild, edible *Vaccinium* species have been harvested for thousands of years by indigenous peoples (Moermàn 1998). Native Americans in western and eastern North America intentionally burned native stands of blueberries and huckleberries to renew their vigor. The cultivation of *Vaccinium* by immigrant Europeans first began in the early nineteenth century when cranberry farmers in the Cape Cod area of Massachusetts started building dykes and ditches to control the water levels in native stands. Highbush and rabbiteye blueberries were domesticated at the end of the nineteenth century. Plants were initially dug from the wild and transplanted into New England and Florida fields.

Most of the commercial production of blueberry now comes from highbush and lowbush types, although rabbiteyes are important in the North American southeast and hybrids of highbush × lowbush (half-highs) have made a minor impact in the Upper Midwest of the U.S.A. Rabbiteye cultivars are beginning to be grown in the Pacific Northwest and Chile for their very late ripening fruit. Highbush blueberries are grown in 37 states in the U.S.A., in six Canadian provinces, and in Australia, Chile, Argentina, New Zealand and a number of countries in Europe (Strik 2005, Strik and Yarborough 2005). The largest acreages of northern highbush are in Michigan, New Jersey, North Carolina, Oregon and Washington in the U.S.A., and British Columbia in Canada. The greatest amount of southern highbush acreage is in Georgia, Florida and California. Commercial production of lowbush blueberries is mainly in Maine, Quebec, New Brunswick, and Nova Scotia (Strik 2005). While the half-high blueberries are not a major contributor to the fruit market, they are very widely used as an ornamental plant for landscaping. Cranberries are grown primarily in Wisconsin, Massachusetts, New Jersey, Washington and Oregon, with limited plantings in British Columbia, Michigan, Nova Scotia, Quebec and Germany. Lingonberries are primarily harvested from the wild in Scandinavia with a significant harvest in northeast China and a much smaller wild harvest in Newfoundland.

Over 110,000t of highbush fruit are produced annually in the United States on over 20,000ha (USDA Agricultural Statistics). The estimated area of rabbiteye production is currently about 3000ha, with half the surface planted in Georgia. The total annual production is over 5500 t. Half-high production is restricted to a few hundred hectares in Minnesota and Michigan. Annual production of lowbush blueberries ranges from 40,000 to 55,000 t on about 40,000 ha in primarily Maine and the Maritime provinces of eastern Canada. Cranberry production in North America is about 380,000 t annually on 21,700 ha.

4.2 Evolutionary Biology and Germplasm Resources

The genus *Vaccinium* is widespread, with high densities of species being found in the Himalayas, New Guinea and the Andean region of South America. The origin of the group is thought to be South American. Estimates of species numbers vary from 150 to 450 in 30 sections (Luby et al. 1991). The commercially important species are found in the sections Cyanococcus, Oxycoccus, Vitis-Idaea and Myrtillus (Table 4.1).

Species delineation has been difficult to resolve in *Cyanococcus* due to polyploidy, overlapping morphologies, continuous introgression through hybridization and a general lack of chromosome differentiation. In the first detailed taxonomy of the group, Camp (1945) described 9 diploid, 12 tetraploid and 3 hexaploid species, but Vander Kloet (1980, 1988) reduced this list to 6 diploid, 5 tetraploid and 1 hexaploid taxa. He included all the crown-forming species into *V. corymbosum* with three chromosome levels. Most horticulturists and blueberry breeders feel that the variation patterns in *V. corymbosum* are distinct enough to retain Camp's diploid *Vaccinium elliottii* Chapm. and *Vaccinium fuscatum* Ait., tetraploid *Vaccinium simulatum* Small and hexaploid *V. ashei* and *Vaccinium constablaei* A. Gray (Ballington 1990, 2001, Galletta and Ballington 1996, Lyrene 2007).

All the polyploid *Cyanococcus* are likely of multiple origin and active introgression between species is ongoing. The tetraploid highbush blueberry *V. corymbosum* has been shown to be genetically an autopolyploid (Draper and Scott 1971, Krebs and Hancock 1989), as well as an interspecific tetraploid hybrid of *V. darrowii* Camp and *V. corymbosum* (Qu and Hancock 1995, Qu et al. 1998). Wenslaff and Lyrene (2003) found considerable chromosome homology in tetraploid southern highbush × *V. elliottii* hybrids. The lowbush blueberry, *V. angustifolium* appears to be a direct descendant of *V. pallidum* Ait. × *V. boreale* Hall & Aalders, but introgression with *V. corymbosum* may have also influenced its subsequent development (Vander Kloet 1977).

The primary mode of speciation in *Vaccinium* has been through unreduced gametes, as there is a strong but not complete triploid block (Lyrene and Sherman 1983, Vorsa and Ballington 1991). The unreduced gametes are produced primarily through first division restitution (Qu and Hancock 1995, Qu and Vorsa 1999), although some second division restitution occurs (Vorsa and Rowland 1997). Embryo culture was not successful in recovering triploids of *V. elliottii* × tetraploid highbush (Munoz and Lyrene 1985).

Table 4.1 Important species of blueberries and cranberries

Section	Species	Ploidy	Location
Batodendron	*V. arboreum* Marsh	2x	S.E. North America
Cyanococcus	*V. angustifolium* Ait.	4x	N.E. North America
	V. ashei Reade.	6x	S.E. North America
	V. boreale Hall & Aald.	2x	N.E. North America
	V. constablaei Gray	6x	Mountains of SE North America
	V. corymbosum L.	2x	S.E. North America
	V. corymbosum L.	4x	E. North America
	V. darrowii Camp	2x	S.E. North America
	V. fuscatum Ait	2x	Florida
	V. myrtilloides Michx.	2x	Central North America
	V. pallidum Ait.	2x, 4x	Mid-Atlantic North America
	V. tenellum Ait.	2x	S.E. North America
	V. elliottii Chapm.	2x	S.E. North America
	V. hirsutum Buckley	4x	S.E. North America
	V. myrsinites L.	4x	S.E. North America
	V. simulatum Small	4x	S.E. North America
Oxycoccus	*V. macrocarpon* Ait.	2x	North America
	V. oxycoccos L.	2x, 4x, 6x	Circumboreal
Vitis-Idaea	*V. vitis-idaea* L.	2x	Circumboreal
Myrtillus	*V. cespitosum* Michx.	2x	North America
	V. chamissonis Bong.	2x	Circumboreal
	V. deliciosum Piper	4x	N.W. North America
	V. membranaceum Dougl. Ex Hook	4x	W. North America
	V. myrtillus L.	2x	Circumboreal
	V. ovalifolium Sm.	4x	N.W. North America
	V. parvifolium Sm.	2x	N.W. North America
	V. scoparium Leiberg ex Coville	2x	N.W. North America
Polycodium	*V. stamineum* L.	2x	Central and E. North America
Pyxothamnus	*V. consanguineum* Klotzch	2x	S. Mexico and Central America
	V. ovatum Pursh	2x	N. W. North America
	V. bracteatum Thunb.	2x	East Asia, China and Japan
Vaccinium	*V. uliginosum* L.	2x, 4x, 6x	Circumboreal

Vaccinium macrocarpon is an endemic of eastern North America and is thought to be the most primitive species in section Oxycoccus (Camp 1945). Its closest relatives are diploid, tetraploid and hexaploid races of *V. oxycoccos* L. that have a circumboreal distribution. Tetraploid *V. oxycoccos* has been shown to be genetically an autopolyploid, although it likely carries genes from *V. macrocarpon* (Mahy et al. 2000). Gene exchange is now severely limited between the species due to a disjunct distribution and a flowering period that differs by three weeks (Vander Kloet 1988). *Vaccinium myrtillus* is very similar to *V. scoparium* Leiberg and may have been derived in the Rocky Mountains of North America (Camp 1945). There has been little speculation about the origin of *V. vitis-idaea*, but it must be closely related to *V. myrtillus*, since hybrids have been discovered between these two species at numerous locations across northern Europe (Luby et al. 1991).

Interspecific hybridization within *Vaccinium* section *Cyanococcus* has played a major role in the development of highbush blueberries (Ballington 1990, Ballington 2001). Most homoploids freely hybridize and interploid crosses are frequently successful (Lyrene et al. 2003). Genotypes have been found in many blueberry species that produce unreduced gametes (Ballington et al. 1976, Cockerman and Galletta 1976, Ortiz et al. 1992), and colchicine can be used to produce fertile genotypes with doubled chromosome numbers (Perry and Lyrene 1984). Even pentaploid hybrids of diploid × hexaploid crosses have been shown to cross relatively easy to tetraploids (Jelenkovic 1973, Chandler et al. 1985a, Vorsa et al. 1987).

Numerous interspecies crosses have been made by breeders within section *Cyanococcus* including: (1) tetraploid *V. corymbosum* × tetraploid *V. angustifolium* (Luby et al. 1991), (2) tetraploid *V. myrsinites* L. × tetraploid *V. angustifolium* and *V. corymbosum* (Darrow 1960, Draper 1977), (3) colchicine-doubled diploid hybrids of *V. myrtilloides* Michx. × tetraploid *V. corymbosum* (Draper 1977), (4) diploid *V. darrowii* × hexaploid *V. ashei* (Darrow et al. 1954, Sharp and Darrow 1959) and (5) diploid *V. elliottii* × tetraploid highbush cultivars (Lyrene and Sherman 1983). Probably the most widely employed interspecific hybrid has been US 75, a tetraploid derived from the cross of diploid *V. darrowii* selection Fla 4B × the tetraploid highbush cultivar Bluecrop (Fig. 4.1). In spite of its being a hybrid of an evergreen, diploid species crossed with a deciduous, tetraploid highbush, US 75 is completely fertile and is the source of the low chilling requirement of many southern highbush cultivars (Draper and Hancock 2003).

Many of the highbush types now being released are complex hybrids. Some of the most dramatic examples are 'O'Neal' which contains genes from four species (*V. corymbosum, V. darrowii, V. ashei* and *V. angustifolium*) and 'Sierra' which possesses the genes of five species (*V. corymbosum, V. darrowii, V. ashei, V. constablaei* and *V. angustifolium*). 'Biloxi' contains the genes from five taxa [*V. corymbosum* (diploid and tetraploid), *V. darrowii, V. ashei* and *V. angustifolium*], and has fewer *V. corymbosum* than non -*V. corymbosum* genes in its genome.

Intersectional crosses have generally proved difficult, although partially fertile hybrids have been derived from *V. tenellum* Ait. and *V. darrowii* (section Cyanococcus) × *V. stamineum* L. (section Polycodium) (Lyrene and Ballington 1986), *V. darrowii* and *V. tenellum* × *V. vitis-idaea* (section Vitis-Idaea) (Vorsa 1997), *V. darrowii* × *V. ovatum* Pursh (section Pyxothamnus), *V. arboreum* Marshall (section Batodendron) and *V. stamineum* (section Polycodium) (Ballington 2001), and tetraploid *V. uliginosum* (section Vaccinium) × highbush cultivars (Rousi 1963, Hiirsalmi 1977, Czesnik 1985). Genes of *V. arboreum* have also been moved into tetraploid southern highbush using *V. darrowii* as a bridge (Lyrene 1991, Brooks and Lyrene 1998a,b). Genes from *V. ovatum* have been incorporated into ornamental highbush selections in the USDA-ARS Oregon program via NC 3048.

There are several important collections of native blueberry germplasm and hybrids (Ballington 2001). The most extensive is held at the U.S. Dept of Agriculture,

Fig. 4.1 Morphological differences in hybrids of *V. darrowii* and *V. corymbosum*; (**a**) *V. corymbosum*; (**b**) Backcross hybrid; (**c**) F₁ hybrid; (**d**) *V. darrowii*

Agricultural Research Service's National Clonal Germplasm Repository at Corvallis Oregon (http://www.ars.usda.gov/main/site_main.htm?modecode=53581500), where representatives of most species can be found along with almost all named non-patented cultivars. Jim Ballington at North Carolina State University has a particularly large collection of southern species material. Paul Lyrene at the University of Florida and Jim Hancock at Michigan State University also have large collections of southern and northern adapted material, respectively. The largest collections of cranberry germplasm are held by Nicholi Vorsa at Rutgers University and Eric Zeldin and Brent McCown at the University of Wisconsin.

4.3 History of Improvement

Blueberry breeding is a very recent development (Lyrene 1998, Hancock 2006a). Highbush breeding began in the early 1900s in New Jersey, with the first hybrid being released in 1908 by Frederick Coville of the United States Department of Agriculture (USDA). He conducted the fundamental life history studies of the blueberry that served as the basis of cultivation such as soil pH requirements, cold and day-length control of development, pruning strategies and modes of propagation. Working with Elizabeth White and others, he collected several outstanding wild clones of *V. corymbosum* and *V. angustifolium*, which he subsequently used in breeding improved types. Over 75% of the current blueberry acreage is still composed of his hybrids, most notably 'Bluecrop', 'Jersey', 'Weymouth', 'Croatan', 'Blueray', 'Rubel' and 'Berkeley' (Mainland 1998).

George Darrow assumed the USDA program after Coville died in 1937 and made important contributions on the interfertility and phylogeny of the native *Vaccinium* species in cooperation with the taxonomist W.H. Camp (Hancock 2006a). He formed a large collaborative testing network that encompassed private growers and Agricultural Experiment Station (AES) scientists in Connecticut, Florida, Georgia, Maine, Massachusetts, Michigan, New Jersey and North Carolina. From 1945 to 1961, he sent out almost 200,000 seedlings to his cooperators for evaluation.

Arlen Draper followed Darrow and focused on incorporating the genes of most wild *Vaccinium* species into the cultivated highbush background (Draper 1995, Hancock 2006b). He maintained and strengthened Darrow's collaborative network and released a prodigious number of southern and northern highbush cultivars, with improved fruit color and firmness, smaller pedicle scars and higher productivity (Hancock and Galletta 1995). His 'Duke' and 'Elliott' have been major successes, along with the newer releases 'Nelson' and 'Legacy'. Mark Ehlenfeldt assumed the USDA-ARS program in 1998.

Ralph Sharp began working in the 1950s in Florida on the development of southern highbush types in collaboration with Darrow (Sharp and Darrow 1959, Lyrene 1998). He was the first collector of *V. darrowii* for breeding, and until very recently, all southern highbush cultivars contained genes from his wild clones. Sharp, and his colleague Wayne Sherman, developed several successful cultivars, including 'Sharpblue', which was grown commercially until very recently. Paul Lyrene took over the breeding work in Florida in 1977.

Stanley Johnson at Michigan State University spent a considerable amount of time in the 1950s and 1960s improving the cold tolerance of highbush by crossing it with *V. angustifolium*. Out of this work came the 'half-high' cultivar Northland and the mostly pure highbush type 'Bluejay', which was released by his successor Jim Moulton. The program was abandoned in 1978, but was renewed in 1990 by Jim Hancock.

In the Pacific Northwest, Joseph Eberhart, in Olympia, Wash. released three cultivars, Pacific, Olympia, and Washington in the 1920s and 1930s. 'Olympia' is still widely grown today.

Outside of the U.S.A., blueberry breeding work was conducted in Australia, Germany, and New Zealand. Johnston sent open pollinated seed to D. Jones and Ridley Bell in Australia in the 1960s that generated the important cultivar 'Brigitta Blue' along with several others. Narandra Patel at HortResearch in New Zealand released the cultivars Nui, Puru and Reka from breeding material initially provided by the University of Arkansas and the USDA at Beltsville in the 1960s and 1970s. Walter Heermann in Germany, working with seed provided by Frederick Coville, released several varieties in the 1940s and 1950s including 'Blauweiss-Goldtraube', 'Blauweiss-Zukertraube', 'Heerma', 'Rekord', 'Ama' and 'Gretha'.

Rabbiteye breeding was initiated in the 1939 by George Darrow in collaboration with Otis J. Woodard at the Georgia Coastal Plain Experiment Station (Tifton, Ga.) and Emmett B. Morrow at the North Carolina Experiment Station, although a collection of wild selections from Florida and Georgia had been planted at Tifton in the 1920s (Austin 1994). This work was continued by Max Austin and then Scott NeSmith in Georgia, Gene Galletta followed by Jim Ballington in North Carolina, and Ralph Sharp, Wayne Sherman and then Paul Lyrene in Florida (Lyrene 1987). These breeding programs have resulted in significant improvements in fruit color, size, texture and appearance over the original wild selections. The most important cultivars have been 'Tifblue' (1955) and 'Brightwell' (1971) from Georgia, 'Bluegem' (1970) and 'Bonita' (1985) from Florida, and 'Powderblue' and 'Premier' (1978) from North Carolina. Rabbiteye cultivars were also bred in the New Zealand HortResearch Inc. program of Narandra Patel. Several releases came from this program in the 1990s including 'Maru' and 'Rahi'.

Lowbush blueberry breeding has generally received little attention. The primary effort has been centered with Agriculture and Agri-Foods Canada (Kentville, NS), currently overseen by Andrew Jamieson. At this station, wild selections from Maine and the Maritime Provinces were tested and crossed, resulting in a number of releases including 'Augusta', 'Blomidon', 'Brunswick', 'Chignecto', and 'Fundy'. Recently, a seed-propagated lowbush cultivar, 'Novablue', was released by Andrew Jamieson from the cross of 'Fundy' × 'Brunswick'. The hybrids have unusually large berries and spread more rapidly by rhizomes than the parent clones.

Lowbush blueberries have been hybridized with V. corymbosum to produce 'half-high' cultivars (Finn et al. 1990). The major releases of this type were 'Northland' developed by Stanley Johnston in Michigan and 'Northblue', 'Northsky', 'Northcountry', 'St. Cloud', 'Polaris' and 'Chippewa' released by Jim Luby in Minnesota. The 'half-highs' have much higher yields and larger fruit than lowbush, but have low enough stature to be protected by snow in areas with extreme winter cold.

Breeding of cranberries has been sporadic since the mid-1900s. However, during the last decade of the 20th century, much of the acreage previously planted to native selections 'Early Black', 'Howes', 'McFarlin' and 'Searles' has been renovated with first generation hybrids; the cultivar Stevens being the most widely planted. In 1929 the USDA began a major cooperative cranberry breeding project with the New Jersey, Massachusetts and Wisconsin Agriculture Experiment Stations to develop varieties resistant to false blossom disease, a phyoplasma (Chandler et al. 1947). Resistance to false bottom was based on developing varieties which would be less attractive to the blunt-nosed leaf hopper, the vector of the false blossom agent. The majority of the seedlings were planted in New Jersey because of the severity and prevalence of false bottom in the state. Out of this program came 'Pilgrim', 'Wilcox' and 'Stevens'. 'Pilgrim' was released for improved productivity, size, color (purplish red), keeping quality, productivity and resistance to the blunt-nosed leafhopper. 'Stevens' was selected and released for its improved productivity, color (deep red), firmness and resistance to softening (Dana 1983). 'Crowley' was introduced from the Washington Agriculture Station in the 1960s as a better pigmented replacement for 'McFarlin', but has lost favor due to variable and generally low productivity.

4.4 Current Breeding Efforts

The current goals of southern highbush breeders are to obtain early ripening cultivars with high plant vigor, improved disease resistance and later flowering dates (particularly in the southeastern U.S.A., where late freezes are a problem). Higher yields, better flavor and characteristics favorable for mechanical harvest are also being sought. Cultivars and advanced breeding lines are being used to breed southern highbush, along with hybrids derived from native, low-chill highbush selections from Florida and Georgia (*V. ashei*, *V. elliottii* and *V. darrowii*). Because of their low chill requirement and the influence of genes from *V. darrowii*, many southern highbush cultivars can be grown as evergreens that avoid dormancy in areas with mild winters, with a harvest season that extends for several months through the winter and early spring (Darnell and Williamson 1997, Lyrene 2007). Rabbiteye breeders hope to expand harvest dates, improve berry size and fruit quality, reduce susceptibility to rain cracking and extend storage life.

Southern highbush cultivars are being developed at several locations, including Arkansas, Australia, California, Florida, Georgia, Mississippi, Chile and Spain. Paul Lyrene at the University of Florida has the most active program dealing with very low chill genotypes and has released many high impact cultivars including 'Emerald', 'Jewel', 'Misty' and 'Star'. Jim Ballington in North Carolina has the most significant program operating at the interface between northern and southern highbush types, and has generated a number of important cultivars including 'Lenore', 'New Hanover', 'O'Neal', 'Reveille' and 'Sampson'. Jim Moore and now John Clark at the University of Arkansas have focused on mixing southern wild species with northern types and recently released 'Ozarkblue', a very high quality late type. Scott NeSmith at the University of Georgia has generated several new early varieties including 'Rebel', 'Camellia' and 'Palmetto'. He also has an active rabbiteye breeding program and his late season cultivar Ochlockonee has generated considerable interest. Steve Stringer, Arlen Draper and Jim Spiers at the USDA-ARS in Mississippi have developed a number of southern highbush types including 'Biloxi', 'Gupton' and 'Magnolia'. Several private breeding programs have also emerged that are developing southern highbush types including Atlantic Blue in Spain (Ridley Bell), Berry Blue in Michigan and Chile (Ed Wheeler), Driscoll Associates in California (Brian Caster), Mountain Blue Orchard in Australia (Ridley Bell) and Vital Berry in Chile (Jim Ballington). Berry Blue is also devoting some effort to rabbiteye types.

Northern highbush breeders are concentrating on flavor, longer storing fruit, expanded harvest dates, disease and pest resistance and machine harvestability. Established breeding lines are being used in these efforts, along with complex hybrids made up of *V. darrowii*, *V. angustifolium*, *V. constablaei* and most of the other wild species. Even though it has limited winter hardiness, *V. darrowii* has proven to be an interesting parent in colder climates, because it passes on a powder blue color, firmness, high flavor, heat tolerance and upland adaptation (Hancock 1998).

Northern highbush blueberries are currently being bred in New Jersey, Michigan, Oregon and Chile. Jim Hancock at Michigan State University is focusing on late maturing, long storing genotypes and has released three new northern highbush cultivars that show high promise, 'Aurora', 'Draper' and 'Liberty'. Mark Ehlenfeldt of the USDA program in New Jersey is focusing on identifying genotypes with high disease resistance and tolerance to winter cold, and has released several cultivars including 'Chanticleer' and 'Hannah's Choice'. Nicholi Vorsa at the Cranberry and Blueberry Research Station of Rutgers University has begun a program in New Jersey to develop locally adapted highbush cultivars with machine harvestability and high fruit quality. Chad Finn of the USDA in Oregon is active in identifying genotypes that are well suited to the Pacific Northwest. Other worldwide northern highbush breeding projects include 'Berry Blue' in Michigan and Chile, Fall Creek Farm and Nursery in Oregon, Driscoll Associates in California and Washington, the University of Talca and Vital Berry in Chile.

Danny Barney at the University of Idaho and to a lesser extent the USDA-ARS (Ore.), is selecting superior genotypes of *V. membranaceum*, *V. ovalifolium* and *V. deliciosum* that may have potential as commercial 'huckleberry' cultivars and some of these are in commercial trial. They have also attempted to cross these species with highbush blueberry with very limited success.

Cranberry breeding efforts are being focused on early maturing fruit, uniform large size, intense color (total anthocyanin content – TACy), keeping quality, high productivity, disease resistance and plant vigor. The greatest emphasis is being placed on productivity and resistance to fruit rot organisms. Cranberries are currently being bred by Nicholi Vorsa at Rutgers University in New Jersey, and Eric Zeldin and Brent McCown at the University of Wisconsin. The Wisconsin team recently released the first new cranberry cultivar in over 30 years – 'HyRed', which is distinguished by its earliness and deep red color. The Rutgers program released three cultivars in 2006, 'Crimson Queen', 'Mullica Queen' and 'Demoranville'. 'Crimson Queen' and 'Demoranville' have tested for high TACy, large fruit size and productivity. 'Mullica Queen' is being released for high production potential and improved TACy relative to 'Stevens'.

4.5 Genetics of Important Traits

The genetics of blueberry is complicated by the fact that domesticated species are polyploid, with highbush generally behaving as an autotetraploid. Rabbiteye, a hexaploid, is considered a segmental autoallopolyploid by ancestry. American cranberry, being a diploid, is under disomic inheritance.

4.5.1 Disease and Pest Resistance

Blueberries are subject to a wide array of diseases (Caruso and Ramsdell 1995, Cline and Schilder 2006). Probably the most widespread problems in highbush blueberry are mummy berry [*Monilinia vaccinii-corymbosi* (Reade)], blueberry stunt phytoplasma, *Blueberry shoestring virus, Blueberry shock virus* (BlShV), *Tomato ringspot virus* (TmRSV), *Blueberry scorch virus* (BlScV), stem blight [*Botryosphaeria dothidea* (Moug.: Fr.) Ces and de Not.], stem or cane canker (*Botryosphaeria corticis* Demaree and Wilcox), Phytophthora root rot (*Phytophthora cinnamomi* Rands), Phomopsis canker (*Phomopsis vaccinii* Shear), Botrytis (*Botrytis cinerea* Pers.: Fr.) and anthracnose fruit rots [*Colletotrichum gloeosporioides* (Penz.) Penz. and Sacc.]. Most of these diseases are widespread, although mummy berry and the virus diseases are most prevalent in areas that grow northern highbush, and stem blight, cane canker and Phytophthora root rot are most common in rainy, hot climates where southern highbush are grown. Fungal induced defoliation is also a problem in the southeastern U.S.A. Rabbiteye blueberries have somewhat different disease susceptibilities than highbush, but can be affected by Botrytis blossom and twig blight, stem blight and mummy berry, and several defoliating fungus diseases. Lowbush is most negatively impacted by Botrytis stem and twig blight, and red leaf disease caused by *Exobasidium vaccinii* (Fckl.) Wor.

Resistant or tolerant genotypes have been described for most of the above diseases in highbush blueberry, but the genetics of resistance has only been determined for *Phytophthora* root rot, *Phomopsis* canker, cane canker, and stem blight (Luby et al. 1991, Galletta and Ballington 1996). Inheritance of resistance to all these diseases is quantitative, with resistance to Phytophthora root rot being partially recessive (Clark et al. 1986). Resistant genotypes have been identified for Phytophthora root rot (Draper et al. 1972, Clark et al. 1986, Erb et al. 1987), cane canker (Ballington et al. 1993, Polashock 2006), stem blight (Creswell and Milholland 1987, Gupton and Smith 1989, Ballington et al. 1993, Polashock 2006), Fusicoccum canker (Hiirsalmi 1988, Baker et al. 1995), mummy berry blight (Stretch et al. 1995, Ehlenfeldt et al. 1996, 1997), mummy berry fruit rot (Stretch et al. 2001, Ehlenfeldt and Stretch 2002), anthracnose fruit rot (Ehlenfeldt and Stretch 2002, Polashock et al. 2005) and shoestring virus (Schulte et al. 1985, Hancock et al. 1986, Acquaah et al. 1995). Resistance to stem blight and cane canker in Florida and North Carolina is so critical that high proportions of otherwise-acceptable test clones are eliminated, because they have insufficient resistance (Lyrene 2007). BlScV has Northwest and East Coast (formerly Sheep Pen Hill Disease) strains. Cultivars show a range of responses to the Northwest strain whereas 'Jersey' is the only cultivar that appears unaffected by the East Coast strain (Bristow et al. 2000, Martin et al. 2006). BlShV is pollen borne and no resistance has yet to be identified, however, the virus spreads through blocks of cultivars at very different rates, suggesting a range of susceptibility (Martin et al. 2006).

The most serious disease problems of cranberry are the phytoplasma false blossom, and fruit rots caused by a number of organisms including blotch rot

(*Physalospora vaccinii* (Shear) Arx & E. Muller), bitter rot (*Glomerella cingulata* (Stoneman) Spauld. & H. Schrenk), end rot (*Godronia cassandrae* Peck), ripe rot (*Coleophoma empetri*), early rot or scald (*Phyllosticta vaccinii* Earle) and Botryosphaeria (*Botryosphaeria vaccinii* (Shear) Barr) (Vorsa 2004). Variability in fruit rot resistance has been observed in cranberry germplasm, although resistant types are in the minority (Johnson-Cicalese et al. 2005).

Several insects and arthropods do significant damage to highbush blueberries including blueberry maggot (*Rhagoletis pomonella* Walsh), blueberry gall midge (*Dasineura oxycoccana* Johnson), blueberry bud mite (*Acalitus vaccinii* Keifer), flower thrips (*Franklinellia* ssp.), Japanese beetle (*Popillia japonica* Newman), sharp-nosed leafhopper (stunt vector) *Staphytopius magdalensis* Prov., blueberry aphid (shoestring and blueberry scorch virus vector) (*Illinoia pepperi* Mac. G.), cranberry fruit worm (*Acrobasis vaccinii* Riley), cherry fruit worm (*Grapholita packardi* Zell), and the plum curculio (*Conotrachelus nenuphar* Herbst). Flower thrips, blueberry bud mite and the gall midge are particular problems in the southeastern U.S.A. Lowbush and rabbiteye blueberries generally suffer from fewer major pests than highbush types; however, significant damage is caused by cranberry fruitworm and stunt in rabbiteye blueberry, and maggot in lowbush. The most serious pests of cranberries are the blunt-nosed leafhopper *Scleroracus vaccinii* Van Duzee, which vectors false blossom disease, black root weevil *Otiorhynchus sulcatus* Fabr., cranberry tip worm *Dasyneura vaccinii* Smith, black-headed fireworm *Rhopobota naevana* Hübner and the cranberry fruit worm.

Little variation in resistance has been reported to most of these pests in *Vaccinium*, except for sharp-nosed leafhopper, blueberry aphid, bud mite and gall midge. Some variation exists for feeding preference of blunt-nosed leafhopper in cranberry. Most southern highbush cultivars have medium to high resistance to the blueberry gall midge and numerous cultivars exist that are resistant to the blueberry bud mite (Lyrene 2007). A wide range of densities of blueberry aphids were found on northern highbush cultivars, but no immunity was identified (Hancock et al. 1982). Ranger et al. (2006, 2007) found several *Vaccinium* species that were more resistant to aphid colonization and population growth than *V. corymbosum*. Resistance to the sharp-nosed leafhopper based on non-feeding preference can be found in *V. ashei* and *V. elliottii*, but not in wild or cultivated *V. corymbosum*. Resistance to the sharp-nosed leafhopper is quantitatively inherited in *V. ashei*, but monogenic, recessive resistance has been found in *V. elliottii* (Meyer and Ballington 1990, Ballington et al. 1993).

4.5.2 Environmental Adaptation

Most blueberry breeding programs are concerned with expanding the harvest season. Earliness is at a particular premium in the southern parts of the U.S.A, Spain, Argentina and north-central Chile, while lateness is extremely important in Michigan and the Pacific Northwest. Early maturation is also important in cranberry.

Increases in earliness have been achieved by selecting for earlier bloom dates and shorter ripening periods, although too early a bloom date makes cultivars subject to spring frost damage in both warm and cold climates. There appears to be little variation in frost tolerance of open flowers and developing fruit among rabbiteye and highbush cultivars, but frost damage can be avoided by selecting genotypes with late blooming dates (Hancock et al. 1987). Up until anthesis, southern highbush flower buds and developing flowers are noticeably more cold-tolerant than rabbiteye flower buds at similar stages. Genes for blossom frost tolerance may exist in *V. angustifolium*, *V. boreale* and *V. myrtilloides* (Luby et al. 1991).

Bloom date, ripening interval and harvest dates are highly heritable in blueberry populations (Lyrene 1985, Hancock et al. 1991), with strong genotype by environmental interactions (Finn et al. 2003). Bloom date is strongly correlated with ripening date, but early ripening cultivars have been developed that have later than average flowering dates such as 'Duke' and 'Spartan' (Hancock et al. 1987). Finn and Luby (1986) found additive genetic variation was more important than non-additive effects for date of 50% bloom, 50% ripe fruit and for length of fruit development interval in populations from hybrids between *V. angustifolium* and *V. corymbosum*. They also observed a positive relationship between ripening interval and crop load, although sufficient variability was available to obtain genotypes with high yield potential and uniform ripening (Luby and Finn 1987). A long fruit development period was not necessary for large fruit.

The native species of *Vaccinium* have considerable variability in flowering dates and ripening seasons (Ballington et al. 1984b). The earliest ripening species are southern diploid *V. corymbosum*, *V. angustifolium* and *V. pallidum*, while the latest are *V. ashei*, *V. ovatum*, *V. stamineum* and *V. tenellum* (Table 4.2). In hexaploid *Vaccinium* hybrid progenies, Ballington et al. (1986) also found significant variation in flowering date, with a high percentage of the families flowering at the same time or later than highbush blueberry.

Among the abiotic factors limiting blueberries, high pH and tolerance to mineral soils are the most important (Chandler et al. 1985b). The *Vaccinium* are 'acid-loving' and as such generally require soils below pH 5.8 for high vigor. Korcak (1986) found an inverse relationship between seedling growth on mineral soils and their proportion of *V. corymbosum* genes. Perhaps for this reason, southern highbush cultivars tend to have broader adaptations than northern ones. Erb et al. (1990, 1993, 1994) discovered several interspecific hybrids that transmitted mineral soil adaptation including hexaploid JU-11 (*V. ashei* × *V. corymbosum*), tetraploid JU-64 (tetraploid *V. myrsinites* × *V. angustifolium*) and tetraploid US 75 (*V. darrowii* × *V. corymbosum*). Scheerens et al. (1999a,b) also found that hybrids with JU 11, JU-64 and US 75 had mineral soil adaptation, along with 'Jersey', 'Sunrise' and complex hybrids of *V. elliottii*. Overall, US 75 was the best parent, as its progeny had the highest fruit quality and it has produced a number of cultivars with remarkable adaptation including 'Legacy', 'Gulfcoast', 'Georgiagem', 'Dixieblue', 'Cooper' and 'Cape Fear'. Finn et al. (1993a,b) found progenies from *V. corymbosum*, *V. angustifolium* and *V. corymbosum/V. angustifolium* hybrids to significantly vary in their pH tolerance, even though *V. angustifolium* was not generally a good source of

Table 4.2 Native species of blueberry that carry potentially useful traits (Ballington 1990, Luby et al. 1991, Galletta and Ballington 1996, Lyrene 2007)

Species	Useful characteristics
V. angustifolium	Winter hardiness, early ripening, blossom frost tolerance, adaptation to high pH, stem blight and Phytophthora root rot resistance, light blue fruit color, small scar, high soluble solids and low acidity
V. arboreum	Drought tolerance, adaptation to basic mineral soils, open flower clusters, upright bush habit, stem blight resistance, resistance to sharp-nosed leafhopper
V. ashei	Drought tolerance, low chilling requirement, upright plant habit, late ripening, long flowering to ripening period, fruit firmness, small scar, loose fruit cluster, cane canker, stem blight and Phytophthora root rot resistance, resistance to sharp-nosed leafhopper
V. boreale	Winter hardiness, blossom frost tolerance
V. bracteatum	Tolerance to high pH
V. caespitosum	Blossom frost tolerance
V. chamissonis	Winter hardiness
V. constablaei	Winter hardiness, high chilling requirement, light blue fruit color
V. consanguineum	Blossom frost tolerance
V. corymbosum – 2x	Low chilling requirement, upright plant habit, early ripening, light blue fruit color, small fruit scar
V. corymbosum – 4x	Low chilling requirement, upright plant habit, light blue and firm fruit color, small fruit scar, excellent flavor, stem canker resistance
V. darrowii	Low chilling requirement, heat tolerance, resistance to mummy berry, adaptation to high pH, tolerance to mineral soils, late flowering, late ripening, long flowering to ripening period, fruit firmness, excellent complex flavor, small scar, light blue fruit color, fruit hold well in heat, high soluble solids and low acidity, loose fruit cluster
V. deliciosum	Winter hardiness, blossom frost tolerance, light blue fruit color, excellent flavor
V. elliottii	Drought tolerance, adaptation to high pH, tolerance to mineral soils, low chilling requirement, upright plant habit, late flowering, early ripening, upright habit, small fruit scar, excellent flavor, cane canker, stem blight and Phytophthora root rot resistance, resistance to sharp-nosed leafhopper
V. fuscatum	Very low chilling requirement, upright plant habit, vigorous
V. membranaceum	Winter hardiness, internal fruit pigmentation, large fruit size, excellent flavor
V. myrtillus	Winter hardiness, blossom frost tolerance, internal fruit pigmentation, excellent flavor
V. myrtilloides	Winter hardiness, early ripening, blossom frost tolerance, resistance to mummy berry, small scar, high soluble solids and low acidity
V. myrsinites	Low chilling requirement, small scar, low acidity, firm fruit
V. ovalifolium	Firm fruit, light blue fruit color
V. ovatum	Adaptation to mineral soils, late ripening, ornamental value
V. pallidum	Adaptation to mineral soils, early ripening, small scar, high soluble solids and low acidity
V. simulatum	Large fruit, winter hardiness, adaptation to mineral soils, deep root system
V. stamineum	Drought tolerance, adaptation to mineral soils, late ripening, very high soluble solids and low acidity, large and firm fruit size, small stem scar, excellent flavor, resistance to sharp-nosed leafhopper
V. tenellum	Adaptation to mineral soils, late ripening, firm fruit
V. uliginosum	Winter hardiness, blossom frost tolerance, resistance to Fusicoccum canker and mummy berry

tolerance. They also developed an in vitro screening method for identifying geno-types that were more tolerant of higher pH root environments (Finn et al. 1991).

Most blueberry species are negatively impacted by high temperature and drought. However, rabbiteye types tolerate these conditions better than highbush, and south-ern highbush are generally superior to northern highbush (Galletta and Ballington 1996). Successful adaptation to heat and drought in blueberry may depend on how rates of CO_2 assimilation, transpiration and water use efficiency are influenced by changes in leaf temperature. Moon et al. (1987a,b) and Hancock et al. (1992) found the Fla 4B clone of *V. darrowii* to have much higher temperature tolerance than 'Bluecrop' and this tolerance was heritable in hybrid populations. Rabbiteye blue-berry has been shown to have very high water use efficiencies under non-irrigated conditions (Teramura et al. 1979, Davies and Johnson 1982) that may relate to high epicuticular wax deposition around stomatal pores (Anderson et al. 1979, Freeman et al. 1979). In his screens of wild species material, Erb et al. (1988a,b) found *V. elliottii*, *V. darrowii* and *V. ashei* to be the most drought tolerant species and this characteristic was transmitted to hybrid progeny. Finn et al. (2003) found a strong genotype × environment interaction for survival in highbush families grown in Michigan and Oregon.

Other sources of drought tolerance include the native species *V. stamineum* and *V. arboreum*. *Vaccinium stamineum* is the most drought tolerant species in the south-eastern U.S.A., but hybrids derived with species in section Cyanococcus have not been vigorous (Ballington 1980, Lyrene 2007). The use of *V. arboreum* appears to be more promising, as this species can be crossed with *V. darrowii* to produce vigorous hybrids, and these hybrids can be used as a bridge to tetraploid southern highbush types (Lyrene 1991, Brooks and Lyrene 1998a,b). *Vaccinium arboreum* is drought tolerant because it has deep tap roots in contrast to the spreading, shallow root systems of highbush blueberry. Because of their greater adaptation to varying soil environments, *V. ashei* and *V. arboreum* have been investigated as rootstocks for highbush blueberry (Galletta and Fish 1972).

Expanding the range of adaptation of the northern highbush blueberry by reduc-ing its chilling requirement has been an important breeding goal for over 50 years. This was largely accomplished by incorporating genes from the southern diploid species *V. darrowii* into *V. corymbosum* via unreduced gametes, although hybridiza-tions with native southern *V. corymbosum* and *V. ashei* have also played a role. Cultivars with an almost a continuous range of chilling requirements are now available from 0 to 1,000 hours. The genetics of the chilling requirement has not been formally determined, although segregation patterns suggest that it is largely quantitatively inherited with the low chilling requirement showing some domi-nance. There are likely variations in temperature thresholds that have not been explored.

Winter cold often causes severe damage to blueberry flower buds and young shoots in the colder production regions. Cold hardiness is a complex interaction between rate of acclimation and deacclimation, as well as mid-winter tolerance. In general, northern highbush types survive much colder mid-winter temperatures than rabbiteye and southern highbush cultivars, although considerable variability exists within groups (Hancock et al. 1997, Ehlenfeldt et al. 2003, 2006, Hanson et al. 2007). In full dormancy, northern highbush genotypes have been found to range in tolerance from –20 to –30°C, while rabbiteye genotypes range from –14 to –22°C. Few southern highbush have been evaluated, although 'Legacy' has been found to tolerate temperatures to –17°C and 'Ozarkblue' – 26°C. 'Sierra', which is composed of 50% southern germplasm has tolerated temperatures in excess of –32°C (Hancock, personal observation). US 245, a hybrid of US 75 ('Bluecrop' × *V. darrowii* 'Fla 4B') × 'Bluecrop', is tolerant to at least –24°C. The wood of half-high cultivars, such as 'Northblue', can survive to – 40°C and the flower buds can tolerate –36°C (Finn, personal observation).

The flower buds of rabbiteye and southern highbush cultivars are generally considered to acclimate more slowly in the fall than those of northern highbush cultivars, but there is considerable variability for this trait. Southern adapted material typically sets flower inflorescence buds later in the season than northern highbush types. Rowland et al. (2005) found the northern highbush 'Duke' to be the most rapid deacclimator of a mixed group of 12 cultivars, while the southern highbush 'Magnolia', the northern highbush × rabbiteye pentaploid hybrid 'Pearl River', the rabbiteye × *V. constablaei* cultivar 'Little Giant' and the half-highs 'Northcountry' and 'Northsky' were the slowest. Northern highbush 'Bluecrop' and 'Weymouth', southern highbush 'Legacy' and 'Ozarkblue', and rabbiteye 'Tifblue' were intermediate. Hanson et al. (2007) found that leaf retention in the fall was not a good predictor of rate of deacclimation, as 'Ozarkblue' and US 245 retain their leaves until the very late fall, but they are just as hardy as the mid-season standard 'Bluecrop'. Bittenbender and Howell (1975) also found no correlation between flower bud hardiness and fall leaf retention.

Little formal genetic analysis of cold tolerance of tetraploid blueberry has been performed, although Arora et al. (2000) found in diploid populations that the cold hardiness data fit a simple additive-dominance model of gene action, with the additive effects being greater than the dominance ones. Several wild species likely carry useful genes for cold hardiness including *V. angustifolium*, *V. boreale* and *V. myrtilloides* (Galletta and Ballington 1996). Ehlenfeldt and Rowland (2006) have shown that *V. constablaei* possesses extreme cold tolerance and it could be used to develop winter hardy rabbiteye types.

The cold-responsive proteins known as dehydrins appear to have a role in the cold hardiness of blueberries. Muthalif and Rowland (1994a,b) originally examined changes in protein levels in the floral buds of the cold-tolerant, northern highbush 'Bluecrop' and the cold-sensitive, rabbiteye 'Tifblue'. They found that three proteins of 65, 60 and 14 kDa increased in both cultivars in response to cold and became the predominant proteins. The highest levels of the dehydrins were found in 'Bluecrop'. This correlation has held up across a number of other cultivars with varying levels of cold tolerance (Arora et al. 1997, Panta et al. 2001). A 2.0 kb blueberry cDNA was identified that encodes the 60 kDa dehydrin (*bbdhn1*) (Levi et al. 1999). This clone was used to probe cold-hardened floral buds of 'Bluecrop' and another five dehydrins were identified (*bbdhn2–bbdhn5*) (Rowland et al. 2004).

4.5.3 Fruit Quality

In blueberries, the fruit characteristics most sought after are flavor, large size, light blue color (a heavy coating of wax), a small scar where the pedicel detaches, easy fruit detachment for hand or machine harvest, firmness and a long storage life. Other important characteristics are uniform shape (Fig. 4.2), size and color, high aroma, and ability to retain texture in storage. Considerable genetic variability has been identified for most of these traits in cultivated and wild species (Luby et al. 1991, Galletta and Ballington 1996, Ehlenfeldt 2002). Ballington et al. (1984a) found the native species with the best fruit quality to be: highest soluble solids (SS) – *V. angustifolium*, *V. pallidum* and *V. stamineum*, lowest titratable acidity (TA) – *V. ashei*, *V. darrowii*, *V. myrsinites* and *V. pallidum*, lowest SS/TA – 4× *V. corymbosum*, largest fruit size – *V. ashei* and *V. corymbosum*, smallest scar – *V. elliottii*, easiest detached fruit – 2× *V. corymbosum*, *V. darrowii*, *V. elliottii*, *V. myrtilloides* and *V. tenellum*, and most firm – *V. darrowii*, *V. tenellum* and *V. ashei* (Galletta and Ballington 1996). *Vaccinium darrowii* has been a particularly important source of powder blue color, intense flavor and fruit that remain in good condition in hot weather (Ehlenfeldt et al. 1995, Ballington 2001, Draper and Hancock 2003).

A number of studies have been conducted on the genetics of fruit characteristics. Albino fruit has been shown to be regulated by a single recessive gene in several *Vaccinium* species (Hall and Aalders 1963, Draper and Scott 1971, Lyrene 1988). Edwards et al. (1974) using mid-parent regressions found fruit size, color, firmness and scar to be highly heritable among southern highbush seedlings. Finn and

Fig. 4.2 Two clusters of 'Draper' fruit showing concentrated ripening and regularity of fruit size

Luby (1992) found that general combining ability was more important than specific combining ability for these same traits in a partial diallel mating scheme using 17 *V. corymbosum*, *V. angustifolium* and half-high parents. Interestingly, crosses of light-blue fruited genotypes often produced black fruited progeny. Finn et al. (2003) found a significant genotype × environment for picking scar but not for fruit color and firmness in highbush families grown in Michigan and Oregon. Ehlenfeldt and Martin (2002) discovered a wide range in fruit firmness measurements for blueberry genotypes and found that genotypes derived from *V. angustifolium* were softer and *V. darrowii* firmer than average. In 6x hexaploid families of *V. ashei* and *V. ashei* × *V. constablaei*, Ballington et al. (1986) found the fruit of most progeny to be acceptable for color, fruit scar, firmness and flavor. *Vaccinium constablaei* hybrids had small fruit but good potential for machine harvest.

High antioxidant capacity has become an important fruit quality parameter in blueberries. Considerable amounts of variability have been observed in this characteristic that is quantitatively inherited (Connor et al. 2002a,b). In general, blueberries are one of the richest sources of antioxidant phytonutrients among the fresh fruits, with total antioxidant capacity ranging from 13.9 to 45.9 μmol Trolox equivalents/g fresh berry (Ehlenfeldt and Prior 2001, Connor et al. 2002a,b, Moyer et al. 2002). Total anthocyanins in blueberry fruit range from 85 to 270 mg per 100 g, and species in the subgenus *Cyanococcus* carry the same predominant anthocyanins, aglycones and aglycone-sugars, although the relative proportions vary (Ballington et al. 1988). The predominant anthocyanins were delphinidin-monogalactoside, cyanidin-monogalactoside, petunidin-monogalactoside, malvidin-monogalactoside and malvidin-monoarabinoside.

In North American species in the Myrtillus, total monomeric anthocyanin (ACY) contents ranged from 101 to 400 mg·100 g^{-1} and the total phenolics (TP) from 367 to 1286 mg·100 g^{-1} (Lee et al. 2004). Populations could be identified that had high ACY or TP contents. Two of the major anthocyanins in *V. membranaceum* and *V. ovalifolium* were cyanidin glycosides and delphinidin glycosides *V. ovalifolium* also had high levels of malvidin glycosides while *V. membranaceum* levels were low. In contrast, while 'Rubel' was high in malvidin glycosides and delphinidin glycosides, it was low in cyanidin glycosides. In 'Rubel' highbush blueberry, the anthocyanins exist almost exclusively in the fruit skins and the polyphenolics are mostly in the skin with lesser amounts in the flesh and seeds. As a result, the skins are where the highest antioxidant levels are found (Lee and Wrolstad 2004).

In cranberries, highly colored fruit with a high anthocyanin content, good keeping quality and processing ability are important characteristics. Variability for all these traits has been identified among cultivars (Wang and Stretch 2001). Cranberries have total anthocyanins varying from 25 to 100 mg per 100 g fruit, with the most important anthocyanins being cyanidin-3-monogalactoside, peonidin-3-monogalactoside, cyanidin-3-monoarabinoxide, and peonidin-3-monoarabinoside. Significant amounts of qualitative and quantitative variation exist for these compounds among genotypes (Vorsa et al. 2002, 2005).

The anthocyanin pigments of the American cranberry are primarily conjugated with arabinose and galactose, rather than the more 'healthy' glucose (Vorsa and

Polashock 2005). These ratios could be improved through breeding, as *V. oxycoccos* has a higher percentage of these pigments conjugated to glucose and hybrids with the American cranberry have intermediate levels. The genetic control of glycosylation appears to be at a single locus or two closely linked ones with incomplete dominance.

Cranberry has been recognized for beneficial effects on urinary tract health. Foo et al. (2000) identified A-type proanthocyanidins to be the component in cranberry which inhibited the adherence of uropathogenic P-type *E. coli* to urinary tract cell surfaces. Quantitative genetic variation exists for overall fruit proanthocyanidin levels (Vorsa et al. 2003). Heritability estimates for fruit proanthocyanidin content exhibited significant year-to-year variation and ranged low to moderate, suggesting breeding and selection cycles could increase proanthocyanidin content, but a negative correlation with yield might impede genetic gain (Vorsa and Johnson-Cicalese 2007).

A cDNA encoding dihydroflavonol-4-reductase (DFR) was cloned from leaves of cranberry (Polashock et al. 2002). This enzyme is thought to play an important role in anthocyanin and proanthocyanin production. When expressed in tobacco using the CaMV 35S promoter, the corolla of flowers was much darker pink and the filaments became highly colored.

4.5.4 Plant Architecture

The most desirable highbush habit is one that is upright, open and vase shaped, with a bush height of 1.5–2.0 m and a modest number of renewal canes. This architecture is favored for ease of pruning, a reduced tendency to sprawl when it is carrying a heavy crop load and facilitates harvesting by over-the-row machines. The genetic background of the highbush types consists of two contrasting morphologies. The first type is found in *V. corymbosum*, which is generally tall (1.5–2.0 m) and crown forming, with varying degrees of bushiness. The other type is represented by *V. darrowii* and *V. angustifolium*, which are short (less than 0.3 m tall), rhizomatous and form large colonies. In general, plant height appears to be quantitatively inherited, although the short stature of *V. angustifolium* and *V. darrowii* is dominant to highbush in many interspecific crosses (Johnston 1946, Luby and Finn 1986, Lyrene 2007). There are likely major genes in both of these lowbush species that produce dwarf plants in hybrid populations (Draper et al. 1984, Baquerizo 2005). In fact, high percentages of dwarf plants are found in many southern highbush breeding populations. Rabbiteye breeding populations are all upright and tall growing, with most being much taller than the highbush types.

Another important architectural feature in blueberries is an open flower cluster that is easily picked. Tight clusters can lead to misshapen fruit, provide a more conducive environment for fruit diseases, and provide refuge for arthropod pests that may become machine harvest contaminants, and lead to entire clusters being harvested by machine or prevent ripe fruit from falling during machine harvest.

Long pedicels and peduncles are critical to enhancing this feature. While no formal genetic studies have been conducted on this trait, there appears to be considerable genetic variability in the primary gene pool being utilized by blueberry breeders. The native species *V. arboreum* could also be a good source of open clusters (Brooks and Lyrene 1998a,b).

In cranberry, a high density of short, upright stems with flowers is highly desirable. Fruit borne at a uniform height is also beneficial for harvesting, along with strong runnering for bed establishment. Considerable variability exists for these characteristics, although genetic studies have not been conducted (Luby et al. 1991, Galletta and Ballington 1996). In general, these morphological traits appear heritable, with significant additive genetic variation. In contrast to blueberry which has a determinate flower raceme, cranberry has a indeterminate rachis in which vegetative growth resumes after flowering. This terminal growth supports the developing fruit. Significant genetic variation exists for terminal shoot length, and is thought to significantly impact fruit set and development.

4.6 Crossing and Evaluation Techniques

4.6.1 Breeding Systems

Blueberries are all primarily outcrossing with varying levels of self-fertility, depending on species and genotype. In general, northern highbush blueberries have the highest levels of self-fertility, followed by southern highbush, and then rabbiteye. Cultivars that are not highly self-fertile display reduced fruit set and berry size when self pollinated (Morrow 1943, El-Agamy et al. 1981, Rabaey and Luby 1988, Gupton and Spiers 1994, Ehlenfeldt 2001). Highbush are generally planted in solid blocks, although having a pollinizer would be beneficial for most cultivars. All rabbiteye cultivars need pollinizers and alternate row plantings are recommended. Lowbush fruit is harvested from highly variable native stands, with abundant opportunity for cross pollination. Self infertility in blueberries has been shown to be the result of late-acting inbreeding depression (Krebs and Hancock 1988, 1990, Hokanson and Hancock 2000). Harrison et al. (1993) found that parental self fertility was not predictive of the self fertility of progeny in segregating families of half-high and highbush genotypes. Cranberries are generally self-fertile, but cross pollination can enhance seed production (Sarracino and Vorsa 1991, Galletta and Ballington 1996).

Blueberries and cranberries are asexually propagated through cuttings and tissue culture, so elite genotypes can be directly utilized without the need to develop pure lines. Self-pollinations are rarely used in *Vaccinium* breeding due to reduced seed set, germination and because seedlings from selfing tend to be weak. Most breeding programs have relied primarily on pedigree breeding where elite parents are selected each generation for inter-crossing. However, the Florida southern highbush and rabbiteye breeding programs have utilized recurrent selection (Lyrene 1981,

2005). About 150 different genotypes are used in the Florida program each year, in random pair wise combinations.

4.6.2 Pollination and Seedling Culture

Blueberry. Pollen in blueberry is shed as a tetrad with four united grains that can all germinate. It is common for breeders to bring in branches bearing flower buds that are about to open to facilitate pollen collection. The base of the stem is freshly cut to prevent vascular clogging and the branches are placed in jars of water. Pollen is dehisced through pores at the end of anthers and as a result, pollen is most easily obtained for crosses by rolling flowers between the thumb and forefingers. The pollen is collected for immediate use on a flat surface such as a thumb or Petri plate, or into a container for storage such as a Petri dish, small plastic bag, and deepwell microscope slide or glass vial. Some breeders place the pollen on a microscope slide that is then stored in a Petri dish. Pollen which is going to be used within a few days can be stored at room temperatures, but for longer periods of time it should be held in a closed container at slightly above freezing temperatures. Blueberry pollen maintained in this way can remain viable for years if relative humidity is kept below 50% with a desiccant (Galletta and Ballington 1996).

Blueberry flowers do not naturally shed pollen until the flower is open. Emasculation is accomplished by cutting or tearing away the corolla and stamens with small, sharp forceps. The filaments and stamens lie in a ring around the receptacle just below the petals. Pollination is done soon after emasculation (1–5 days) by touching the stigma with a pollen coated fingertip, thumbnail, slide, Petri dish, spatula or glass rod. The pollen transfer tools should be washed with 70% alcohol between pollinations. Field-pollinated plants need to be caged to prevent contamination or the pollinated flower clusters must be wrapped in gauze or nylon mesh or waxed paper bags.

Pollen germination generally occurs within 1–2 h after placement on the stigma and the pollen tubes grow to the styler base within 72 hrs (El-Agamy et al. 1981, Krebs and Hancock 1990). *Vaccinium* fruit contain many small seeds, ranging from 10 to 100 per fruit. Most breeders pollinate at least 25 flowers to assure adequate numbers of seeds for germination and field planting.

Seed can be extracted from accumulated ripe fruit of a single cross by placing them in a blender with a little water and subjecting them to a short burst of energy. The pulp and skin float and the seed sink, allowing the seed to be decanted onto an absorbent surface. Seed are also commonly removed by scooping them out of fruit using a spatula and smearing them on an absorbent surface. As with blackberries (Chapter 3), pectinase is a very efficient way to extract seed from fruit without the risk of damaging them during blender extraction. A few drops of pectinase is mashed with the fruit and a small quantity of water, left overnight and then the slurry is poured off leaving very clean seeds. Fruit are stored in a refrigerator until seed are extracted.

Blueberry seed can be germinated directly after extraction if they are not allowed to dry. Alternately, they can be stored dry or moist at slightly above freezing for 2–3 months and then germinated. Seed are generally placed on the surface of pots or flats containing acid peat or soil combinations of at least 1/3 acid peat. The seeds and medium must be kept moist throughout germination and growth. Light is necessary for germination, with continuous red being more effective than continuous white (Stushnoff and Hough 1968), although most breeders find that just keeping the germination flats in light is sufficient. Fluctuating diurnal greenhouse temperatures ranging from 10 to 30°C give higher germination percentages than constant temperatures, and germination is inhibited above 24°C constant temperature (Stushnoff and Hough 1968). Treatment of blueberry seeds with gibberellins can enhance germination (Ballington et al. 1976, Smagula et al. 1980). Seed generally germinate over a 6–8 week period, with different crosses emerging at different times.

Blueberry seedlings are generally transplanted when they have produced their first true leaves (2–3 cm tall). Numerous soil medium are adequate for growth as long as they have excellent drainage, a high organic content and a pH below 5.5. Under good growing conditions, the seedlings can be ready for field planting within 3–4 months after transplanting. The minimum size for high field survival is about 20–25 cm tall. Most breeding programs set the plants at spacing of about 60 cm apart in the row, although some use a higher density. The Florida breeding program does their primary selections in a 'fruiting nursery' at spacing of 10,000–15,000 seedlings in a 0.2 ha field nursery (Sherman et al. 1973, Lyrene 2005).

Cranberry. Stem cuttings with flower buds can be harvested from September to November from the field, rooted in a mist chamber, subjected to winter chilling, and grown in pots for two months and then used for crossing in late winter. Pollen is collected by tapping flowers onto finger tips or any smooth surface. The bloom dates of most cultivars overlap, so fresh pollen is generally used for pollinations, although the cranberry pollen can generally be handled as was described for blueberry pollen. Pollen may be stored for a month at 4–5°C in deep well microscope slides with a cover slip secured by tape. Emasculation is done with forceps at the full pink bud stage of development and pollinations are made 3–7 days after emasculation. In crosses made in the field, the pollinated flowers are covered with a glycine bag, gauze or cheesecloth bags to prevent contamination. Flowers produce 10–50 seeds. Seed is removed by hand or macerated in a food blender. Seed can be sown directly after extraction for germination, held in the fruit at 0–5°C for 3–5 months or cold stored dry for 6–12 weeks at 7.2°C (Galletta and Ballington 1996).

Light is necessary for cranberry seed germination (Devlin and Karczmarczyk 1975, 1977). Germination generally takes 10–14 days to begin, and seedlings can be transplanted as soon as 2 weeks later, although for some families (especially where seed were dried) the germination process can take several months. A sand and peat mixture is generally used for potting. The seedlings are held in the greenhouse at least until they begin to produce runners. The seedlings are sometimes subjected to a hardening period of 1–2 months of 4–7°C before they are placed in the field, to make sure that they are not damaged by cool temperatures. As single plant plots they are generally set in the field at 1.5 × 1.5 m spacing or 1.8 × 1.8 M. Alternatively, once chilling requirement has been fulfilled, runners of each seedling can be cut and easily rooted in late winter to generate 16–24 rooted plugs for establishing a plot.

4.6.3 Evaluation Techniques

In most northern highbush and southern rabbiteye breeding programs, evaluation begins two years after planting and selections are made over the next two years. Traditionally, the selected seedling plants were dug and moved to further spacing distances and evaluated for another year or two, before the most elite types were propagated and tested in rows of 25–50 plants for several years (Galletta and Ballington 1996). The most promising selections from this row trial were then again propagated and tested in small numbers (5–10 plants) in replicate designs across multiple sites. The whole process took from 15 to 20 years for release of a new cultivar from the original cross.

An accelerated program is now being conducted at Michigan State University where the selected plants in the original planting are propagated and tested directly in replicated plantings at multiple sites. It is expected that about 1% of the progeny plants will go into this trial. After 3–5 years, the elite types will be released as cultivars. It is hoped that this approach will speed the release time to 8–10 years, even though it will result in the final testing of a larger number of ultimately rejected genotypes.

In the close-spaced southern highbush program, the first selections are made within 12 months of planting (Stage I). Ninety percent of the seedlings are removed and the remaining plants are left in place for three more years (Stage II). Each year they are evaluated for possible advancement to Stage III, with about 300 selections being advanced each year into 15-plant plots. These plantings are observed for 10 years, with about 15 clones being selected each year and propagated for planting at multiple locations in larger plots (Stage IV). Cultivars are ultimately selected from these blocks at a rate of about one genotype from each Stage IV test (Lyrene 2007). The fastest moving varieties can go through this system in 10–12 years, although many are evaluated much longer.

One problematic issue in the selection process of highbush cultivars for large commercial plantings of one genotype is the self-fruitfulness of a selection. Inherently, breeder's trials are heterogeneous in the composition of genotypes, selections and standards being tested, which usually facilitates the opportunity for cross-pollination resulting in a quite different environment than that of commercial plantings. For this reason, the self fertility of a selection should be tested before commercial release by comparing the response of self and outcrossed pollinations.

Cranberry. Fruit on some plants can be evaluated a year after planting, but the whole selection process and second testing can take 10–15 years (Galletta and

Ballington 1996). A major characteristic that needs to be determined in cranberry is stable year-to-year high yields. Many varieties exhibit a biennial bearing habit. To assess yield and propensity of biennial bearing, a plot needs to be evaluated for a minimum of four years to assess this characteristic. Field fruit rot resistance can be assessed in growing areas having high fruit rot pressure, such as New Jersey and Massachusetts, and reducing or eliminating fungicide applications.

Sapers et al. (1983, 1986) described a systematic process of evaluation where genotypes with the highest anthocyanin content are first identified, and then tested for juice yield, pH, titratable acidity and soluble solids. Schmid (1977) calculated mean values for a number of horticulturally important parameters of 12 cranberry cultivars that included fruit size, pH, titratable acidity, glucose, fructose and totals sugars, water-, oxylate- and NaOH-soluble and total pectin's, vitamin C, benzoic acids, dry matter, anthocyanins and carotenoids. These values could be used to select elite genotypes in breeding populations (Galletta and Ballington 1996).

4.7 Biotechnological Approaches to Genetic Improvement

4.7.1 Genetic Mapping and QTL Analysis

A wide array of markers have been utilized in blueberry for fingerprinting and linkage mapping including proteins (Vorsa et al. 1988, Bruederle et al. 1991, Hokanson and Hancock 1998), RFLPs (Haghighi and Hancock 1992), RAPDs (Aruna et al. 1993, Levi et al. 1993, Qu and Hancock 1997), SSR and EST-PCR (Rowland et al. 2003a,b, Boches et al. 2005, 2006).

More limited numbers of marker studies have been conducted in cranberry, although isozymes were used to measure diversity patterns in native *V. macrocarpon* (Bruederle et al. 1996) and RAPDs were utilized to determine cultivar identity and heterogeneity in commercial beds (Novy et al. 1994). Most recently, Polashock and Vorsa (2002a,b) have used the SCARs technique to fingerprint over 500 accessions, and to estimate the degree of genetic similarity.

Rowland and Levi (1994) developed the first blueberry map using a diploid population segregating for chilling requirement. Their population was a cross between a F_1 inter-specific hybrid (*V. darrowii* × *V. elliottii*) and another clone of *V. darrowii*. They have continued to periodically add markers and at last report, the map had 72 RAPD markers mapped to 12 linkage groups, which is in agreement with the basic chromosome number of blueberry (Rowland and Hammerschlag 2005). Later, Rowland et al. (1999, 2003b) constructed RAPD-based maps of diploid *V. corymbosum* (*V. caesariense* Mack.) × *V. darrowii* hybrids crossed with other *V. darrowii* and *V. corymbosum* selections. The goal was to develop populations that were segregating for chilling requirement and cold tolerance. First RAPD and more recently EST-PCR markers were added to this map and a QTL was identified that explained about 20% of the genotypic variance associated with cold hardiness (Rowland et al. 2003, Rowland and Hammerschlag 2005).

Qu and Hancock (1997) constructed a RAPD-based genetic map of a tetraploid population resulting from the cross of US 75 × tetraploid *V. corymbosum*, 'Bluecrop'. One hundred and forty markers were mapped to 29 linkage groups. The map was essentially that of *V. darrowii*, as US 75 was produced from an unreduced gamete of *V. darrowii* and only unique markers for Fla 4B were used. Fla 4B was one of the *V. darrowii* clones used by Rowland and Levi (1994) and Rowland et al. (1999). As was previously noted, Fla 4B hybrids (in particular US 75) have been used extensively in breeding to produce low-chilling types.

Most recently, Brevis and Hancock at Michigan State University are using the SSR and EST-PCR markers of Boches et al. (2005, 2006) to develop a linkage map of the tetraploid cross 'Jewel' (southern highbush) × 'Draper' (northern highbush). The ultimate goal is to identify QTL for the chilling requirement. Polashock and Vorsa (2006) are using bulk segregant analysis to tag genes for mummy berry resistance in segregating blueberry populations with *V. darrowii* as the source of resistance.

4.7.2 Regeneration and Transformation

A number of studies have been conducted to perfect blueberry regeneration systems (Cao et al. 1998, 2002, Cao and Hammerschlag 2000, Song and Sink 2004). Overall, In-vitro derived leaves have been found to be the most useful explant source (Billings et al. 1988, Callow et al. 1989, Rowland and Ogden 1992), although Nickerson (1978) first reported on lowbush blueberry regeneration using hypocotyls and cotyledon sections.

Two groups have reported on the transformation of blueberry. Graham et al. (1996) described the transformation and regeneration of the half high blueberry 'Northcountry', using *A. tumefaciens* strain LBA4404 with a binary vector carrying an intron-containing GUS marker gene. They found 'Northcountry' to be hypersensitive to kanamycin and ticaricillin, and as a result, they did not use antibiotic selection. The regenerates were shown to be GUS-positive, but a Southern analysis was not performed to confirm transformation.

Song and Sink (2004, 2006) generated Southern-blot confirmed transgenic plants of highbush blueberry of 4 cultivars, 'Aurora', 'Bluecrop', 'Brigitta Blue', and 'Legacy' (Fig. 4.3). Two selectable marker genes, *npt*II and *bar*, were used to produce the transgenic plants. High-level tolerance to the glufosinate-herbicide Rely, 750–3000 mg l^{-1} glufosimate ammonium, was oberved in the *bar*-expressing, greenhouse plants (Song et al. 2006). Field trials showed little or no damage to plants from 4 independent transgenic events after application of a workable concentration of glufosinate ammonim, 750 mg l^{-1}, for weed control (Song, personal communication). The protocol was then used to produce herbicide-resistant transgenic plants of 'Legacy' (Song et al. 2006). A detailed step-by-step description of the transformation procedure can be found in Song and Sink (2006).

Fig. 4.3 Transformation, selection and regeneration of transgenic blueberry plants of 'Aurora' (from Song and Sink 2004). (**A**) Non-transformed leaf explants. (a) On RM, (b) On RM containing 10 mg l^{-1} Km and 250 mg l^{-1} Cx, *Bar* = 1 mm; (**B**) Production of Km-resistant buds and shoots from surface (a) or wounded postions (b) of leaf explants, *Bar* = 1 mm; (**C**) Formation of Km-resistant shoot clusters from different positions of leaf explants, *Bar* = 1 mm; (**D**) Elongation of GUS-positive shoots on RM (a) or stock culture medium (b) containing 10 mg l^{-1} Km and 250 mg l^{-1} Cx, *Bar* = 1 mm (D-a), 1 cm (D-b)

The first transgenic cranberry 'Stevens' was obtained using the particle bombardment method (Serres et al. 1992), to which a patent (United State Patent 5240839) was issued in 1993 (http://www.freepatentsonline.com/5240839.html). A vector containing the *gus*A, *npt*II and *Bt* (*Bacillus thuringiensis* Subsp. *Kurstaki* crystal protein) was used to optimize transformation protocols for 'Stevens' (Serres et al. 1992, Serres et al. 1997). Four *Bt* transclones showed a significant increase in the mortality of blackheaded fireworm (BHFW, *Rhopobota naevana*) although none of the 64 *Bt*-expressing transclones yielded significant BHFW larval mortality in in vitro feeding assays (Polashock and Vorsa 2002b). In later work, Zeldin et al. (2002) tranformed two genes, *npt*II and *bar*, into 'Pilgrim' to obtain herbicide resistant plants. One transclone with *bar* showed moderate tolerance to 500 mg/L glufosinate ammonium.

While *Agrobacterium*-mediated transformation has not been employed on cranberry to date, an efficient regeneration system has been developed (Qu et al. 2000). Preliminary studies have shown that cranberry leaf explants are susceptible to *A. tumefaciens* strain EHA105 and that leaf explants of several cultivars of cranberry are amenable to regeneration (Polashock and Vorsa 2002a,b).

4.7.3 Genomic Resources

A large EST library of highbush blueberry has been generated in the laboratory of L.J. Rowland at the USDA-ARS Fruit Laboratory at Beltsville, Maryland. Through traditional molecular genetic and genomic approaches, she and her collaborators have identified and isolated several genes associated with cold hardiness including several members of the dehydrin gene family (Muthalif and Rowland 1994a,b, Dhanaraj et al. 2003, 2005a,b). They have also conducted gene expression studies under field and cold room conditions using cDNA microarrays (Dhanaraj et al. 2006a, Alkharouf et al. 2007).

References

Acquaah T, Ramsdell DC, Hancock JF (1995) Resistance to blueberry shoestring virus in southern highbush and rabbiteye cultivars. HortScience 30:1459–1460

Alkharouf NW, Dhanaraj AL, Naik D, Overall C, Rowland LJ (2007) BBGD: an online database for blueberry genomic data, BMC Plant Biol. 7:1–6

Anderson PD, Buchanan DW, Albrigo LG (1979) Water relations and yields of three rabbiteye blueberry cultivars with and without drip irrigation. J Am Soc Hortic Sci 104:731–736

Arora R, Rowland LJ, Lehman JS, Lim CC, Panta GR, Vorsa N (2000) Genetic analysis of freezing tolerance in blueberry (*Vaccinium* section Cyanococcus). Theor Appl Genet 100:690–696

Arora R, Rowland LJ, Panta GR (1997) Cold hardiness and dormancy transitions in blueberry and their association with accumulation of dehydrin-like proteins. Physiol Plant 101:8–16

Aruna M, Ozias-Akins P, Austin ME, Kochert G (1993) Genetic relatedness among rabbiteye blueberry (*Vaccinium ashei*) cultivars determined by DNA amplification using single primers of arbitrary sequence. Genome 36:971–977

Austin ME (1994) Rabbiteye blueberries: development, production and marketing. Agscience, Inc., Auburndale, FL, p 160

Baker JB, Hancock JF, Ramsdell DC (1995) Screening highbush blueberry cultivars for resistance to *Phomopsis* canker. HortScience 30:586–588

Ballington JR (1980) Crossability between subgenus *Cyanococcus* (Gray) Klotzsch and subgenus *Polycodium* (Raf.) Sleumer in *Vaccinium*. HortScience 15:419 (Abstr.)

Ballington JR (1990) Germplasm resources available to meet future needs for blueberry cultivar improvement. Fruit Varieties J 44:54–62

Ballington JR (2001) Collection, utilization and preservation of genetic resources in *Vaccinium*. HortScience 36:213–220

Ballington JR, Ballinger WE, Mainland CM, Swallow WH, Maness EP, Galletta GJ, Kushman LJ (1984b) Ripening season of *Vaccinium* species in southeastern North Carolina. J Am Soc Hortic Sci 109:392–396

Ballington JR, Ballinger WE, Swallow WH, Galletta GJ, Kushman LJ (1984a) Fruit quality characterization of 11 *Vaccinium* species. J Am Soc Hortic Sci 109:392–396

Ballington JR, Galletta GJ (1976) Potential fertility levels in four diploid *Vaccinium* species. J Am Soc Hortic Sci 101:507–509

Ballington JR, Galletta GJ, Pharr DM (1976) Gibberellin effects on rabbiteye blueberry seed germination. HortScience 11:410–411

Ballington JR, Isenberg YM, Draper AD (1986) Flowering and fruiting characteristics of *Vaccinium ashei* and *Vaccinium ashei–Vaccinium constablaei* derivative blueberry progenies. J Am Soc Hortic Sci 111:950–955

Ballington JR, Kirkman WB, Ballinger WE, Maness EP (1988) Anthocyanin, aglycone and aglycone-sugar content in the fruits of temperate North American species of four sections in *Vaccinium*. J Am Soc Hortic Sci 113:746–749

Ballington JR, Rooks SD, Milholland RD, Cline WO, Meyer JR (1993) Breeding blueberries for pest resistance in North Carolina. Acta Hortic 346:87–94

Baquerizo D (2005) The inheritance of dwarf seedlings in southern highbush blueberries. PhD. dissertation, University of Florida, Gainesville

Billings SG, Chin CK, Jelenkovic G (1988) Regeneration of blueberry plantlets from leaf segments. HortScience 23:63–766

Bittenbender HC, Howell GS (1975) Interactions of temperature and moisture content on spring de-acclimation of flower buds of highbush blueberry. Can J Plant Sci 55:447–452

Boches P, Rowland LJ, Bassil NV (2005) Microsatellite markers for *Vaccinium* from EST and genomic libraries. Mol Ecol Notes 5:657–660

Boches P, Rowland LJ, Hummer KE, Bassil NV (2006) Microsatellite markers evaluate genetic diversity in blueberry and generate unique fingerprints. Plant and Animal Genome Conference, p 133

Bristow PR, Martin RR, Windom GE (2000) Transmission, field spread, cultivar response, and impact on yield in highbush blueberry infected with blueberry scorch virus. Phytopathology 90:474–479

Brooks SJ, Lyrene PM (1998a) Derivatives of *Vaccinium arboreum* × *Vaccinium* section Cyanococcus: I. Morphological characteristics. J Am Soc Hortic Sci 123:273–277

Brooks SJ, Lyrene PM (1998b) Derivatives of *Vaccinium arboreum* × *Vaccinium* section Cyanococcus: II. Fertility and fertility parameters. J Am Soc Hortic Sci 123:997–1003

Bruederle LP, Hugan MS, Dignan JM, Vorsa N (1996) Genetic variation in natural populations of the large cranberry, *Vaccinium macrocarpon* Ait. (Ericaceae). B Torrey Bot Club 123:41–47

Bruederle LP, Vorsa N, Ballington JR (1991) Population genetic structure in diploid blueberry *Vaccinium* section Cyanococcus (Ericaceae). Am J Bot 78:230–237

Callow P, Haghighi K, Giroux M, Hancock JF (1989) In vitro shoot regeneration on leaf tissue from micropropagated highbush blueberry. HortScience 24:373–375

Camp WH (1945) The North American blueberries with notes on other groups of *Vaccinium*. Brittonia 5:203–275

Cao X, Hammerschlag FA (2000) Improved shoot organogenesis from leaf explants of highbush blueberry. HortScience 35:945–947

Cao X, Hammerschlag FA, Douglass L (2002) A two step pretreatment significantly enhances shoot organogenesis from leaf explants of highbush blueberry cv. Bluecrop. HortScience 37:819–821

Cao X, Liu Q, Rowland LJ, Hammerschlag FA (1998) GUS expression in blueberry (*Vaccinium* spp.): factors influencing *Agrobacterium*-mediated gene transfer efficiency. Plant Cell Rep 18:266–270

Caruso FL, Ramsdell DC (1995) Compendium of blueberry and cranberry diseases. APS Press, St. Paul, MN, p 87

Chandler CK, Draper AD, Galletta GJ (1985a) Crossability of a diverse group of polyploidy inter-specific hybrids. J Am Soc Hortic Sci 110:878–881

Chandler CK, Draper AD, Galletta GJ (1985b) Combining ability of blueberry interspecific hybrids for growth on upland soil. HortScience 20:257–258

Chandler FB, Wilcox RB, Bergman HF, Dermen H (1947) Cranberry Breeding Investigation of the U.S.D.A. Cranberries-The National Cranberry Mag 12:6–9

Clark JR, Moore JN, Draper AD (1986) Inheritance of resistance to *Phytophthora* root rot in high-bush blueberry. J Am Soc Hortic Sci 111:106–109

Cline WO, Schilder A (2006) Identification and control of blueberry diseases. In: Childers NF, Lyrene PM (eds) Blueberries for growers, gardeners, promoters. Childers Publications, Gainesville, FL, pp 115–138

Cockerman LE, Galletta GJ (1976) A surevy of pollen characteristic in certain *Vaccinium* species. J Am Soc Hort Sci 101:671–676

Connor AM, Luby JJ, Tong CBS (2002a) Variability in antioxidant activity in blueberry and correlations among different antioxidant assays. J Am Soc Hortic Sci 127:238–244

Connor AM, Luby JJ, Tong CBS, Finn CE, Hancock JF (2002b) Genotypic and environmental variation in antioxidant activity, total phenolics and anthocyanin content among blueberry cultivars. J Am Soc Hortic Sci 127:89–97

Creswell TC, Milholland RD (1987) Responses of blueberry genotypes to infection to *Botryosphaeria dothidea*. Plant Dis 71:710–713

Czesnik E (1985) Investigation of F$_1$ generation of interspecific hybrids *Vaccinium corymbosum* L. × *V. uliginosum* L. Acta Hortic 165:85–91

Dana MN (1983) Cranberry cultivar list (*Vaccinium macrocarpon*). Fruit Varieties J 37:88–95

Darnell RL, Williamson JG (1997) Feasibility of blueberry production in warm climates. Acta Hortic 446:251–256

Darrow GM (1960) Blueberry breeding, past, present, future. Am Horticult Mag 39:14–33

Darrow GM, Scott DH, Derman H (1954) Tetraploid blueberries from hexaploid × diploid species crosses. Proc Am Soc Hortic Sci 63:266–270

Davies FS, Johnson CR (1982) Water stress, growth and critical water potentials of rabbiteye blueberry (*Vaccinium ashei* Reade). J Am Soc Hortic Sci 107:6–8

Devlin RM, Karczmarczyk SJ (1975) Effect of light and gibberellic acid on the germination of 'Early Black' cranberry seeds. Hortic Res 15:19–22

Devlin RM, Karczmarczyk SJ (1977) Influence of light and growth regulators on cranberry seed dormancy. J Hortic Sci 52:283–288

Dhanaraj AL, Alkharouf AL, Beard HS, Chouikha IB, Mathews BF, Wei H, Arora R, Rowland LJ (2006a) Gene expression profiles during cold acclimation in blueberries using cDNA microarrays. Proceedings from Functional Genomics of Model Organisms to Crop Plants for Global Health, Washington DC

Dhanaraj AL, Alkharouf AL, Beard HS, Chouikha IB, Mathews BF, Wei H, Rowland LJ (2005a) Monitoring gene expression profiles during cold acclimation in blueberry under field and cold room conditions using cDNA microarrays. In Vitro Biol 41:38A (abstract)

Dhanaraj AL, Slovin JP, Rowland LJ (2003) Analysis of gene expression patterns associated with cold acclimation in blueberry floral buds using expressed sequence tags. Plant Sci 166:863–872

Dhanaraj AL, Slovin JP, Rowland LJ (2005b) Isolation of a cDNA clone and characterization of expression of highly abundant, cold acclimation-associated dehydrin of blueberry. Plant Sci 168:949–957

Draper AD (1977) Tetraploid hybrids from crosses of diploid, tetraploid and hexaploid *Vaccinium* species. Acta Hortic 61:33–36

Draper AD (1995) In search of the perfect blueberry variety. J Small Fruit Vitic 3:17–20

Draper AD, Hancock JF (2003) Florida 4B: native blueberry with exceptional breeding value. J Am Pomol Soc 57:138–141

Draper AD, Chandler CK, Galletta GJ (1984) Dwarfed plants in some blueberry (*Vaccinium*) seedling populations. Acta Hortic 146:63–68

Draper AD, Scott DH (1971) Inheritance of albino seedlings in tetraploid highbush blueberry. J Am Soc Hortic Sci 96:791–792

Draper AD, Stretch AM, Scott DH (1972) Two tetraploid sources of resistance for breeding blueberries resistant to *Phytophthora cinnamomi* Rands. HortScience 7:266–268

Eck P (1990) The American cranberry. Rutgers University Press, New Brunswick, NJ, p 420

Edwards TW, Sherman WB, Sharpe RH (1974) Evaluation and inheritance of fruit color, size, scar, firmness and plant vigor in blueberry. HortScience 9:20–22

Ehlenfeldt MK (2001) Self and cross-fertility in recently released highbush cultivars. HortScience 36:133–135

Ehlenfeldt MK (2002) Postharvest research and technology in *Vaccinium*. Acta Hortic 574:31–38

Ehlenfeldt MK, Draper AD, Clark JR (1995) Performance of southern highbush blueberry cultivars released by the U.S. Department of Agriculture and cooperating state agriculture stations. HortTechnology 5:127–130

Ehlenfeldt MK, Martin RB Jr (2002) A survey of fruit firmness in highbush blueberry and species-introgressed blueberry cultivars. HortScience 37:386–389

Ehlenfeldt MK, Ogden EL, Rowland LJ, Vinyard B (2006) Evaluation of midwinter cold hardiness among 25 rabbiteye cultivars. HortScience 41:579–581

Ehlenfeldt MK, Prior RL (2001) Oxygen radical absorbance capacity (ORAC) and phenolic and anthocyanin concentrations in fruit and leaf tissues of highbush blueberry. J Agric Food Chem 49:2222–2227

Ehlenfeldt MK, Rowland LJ (2006) Cold-hardiness of *Vaccinium ashei* and *V. constablaei* germplasm and the potential for northern–adapted rabbiteye cultivars. Acta Hortic 715:77–80

Ehlenfeldt MK, Rowland LJ, Arora R (2003) Bud hardiness and deacclimation in blueberry cultivars with varying species ancestry: flowering time may not be a good indicator of deacclimation. Acta Hortic 626:39–44

Ehlenfeldt MK, Stretch AW (2002) Identifying sources of resistance to mummy berry and anthracnose in highbush, rabbiteye and species germplasm. Acta Hortic 574:63–69

Ehlenfeldt MK, Stretch AW, Brewster V (1996) Genetic and morphological factors influence mummy berry blight resistance in highblush blueberry cultivars. HortScience 31:1271–1273

Ehlenfeldt MK, Stretch AW, Lehman JS (1997) Shoot length affects susceptibility to mummy berry blight within highbush blueberry cultivars. HortScience 32:806–835

El-Agamy SZA, Sherman, WB, Lyrene PM (1981) Fruit set and seed number from self- and cross-pollinated highbush (4x) and rabbiteye (6x) blueberries. J Am Soc Hortic Sci 106:443–445

Erb W, Draper AD, Galletta GJ, Swartz HJ (1990) Combining ability for plant and fruit traits of interspecific blueberry progenies on mineral soil. J Am Soc Hortic Sci 115:1025–1028

Erb W, Draper AD, Swartz HJ (1988a) Methods of screening blueberry populations for drought resistance. HortScience 25:312–314

Erb W, Draper AD, Swartz HJ (1988b) Screening interspecific blueberry seedling populations for drought resistance. J Am Soc Hortic Sci 113:599–604

Erb W, Draper AD, Swartz HJ (1993) Relation between moisture stress and mineral soil adaptation in blueberries. J Am Soc Hortic Sci 118:130–134

Erb W, Draper AD, Swartz HJ (1994) Combining ability for seedling root system size and shoot vigor in interspecific blueberry progenies. J Am Soc Hortic Sci 119:793–797

Erb WA, Moore JN, Sterne RE (1987) Response of blueberry cultivars to inoculation with *Phytophthora cinnamomi* Rands Zoospores. HortScience 22:298–300

Finn CE, Hancock JF, Mackey T, Serce S (2003) Genotype × environment interactions in highbush blueberry (*Vaccinium* sp. L.) families grown in Michigan and Oregon. J Am Soc Hortic Sci 128:196–200

Finn CE, Luby JJ (1986) Inheritance of fruit development interval and fruit size in blueberry progenies. J Am Soc Hortic Sci 11:784–788

Finn CE, Luby JJ (1992) Inheritance of fruit quality traits in blueberry. J Am Soc Hortic Sci 117:617–621

Finn CE, Luby JJ, Rosen CJ, Ascher PD (1991) Evaluation in vitro of blueberry germplasm for higher pH tolerance. J Am Soc Hortic Sci 116:312–316

Finn CE, Luby JJ, Rosen CJ, Ascher PD (1993a) Blueberry germplasm screening at several soil pH regimes I. Plant survival and growth. J Am Soc Hortic Sci 118:377–382

Finn CE, Rosen CJ, Luby JJ, Ascher PD (1993b) Blueberry germplasm screening at several soil pH regimes. II. Plant nutrient composition. J Am Soc Hortic Sci 118:383–387

Finn CE, Luby JJ, Wildung DK (1990) Half-high blueberry cultivars. Fruit Varieties J 44:63–68

Foo LY, Lu Y, Howell AB, Vorsa N (2000) The structure of cranberry proanthocyanidins which inhibit adherence of uropathogenic P-fimbriated *Escherichia coli* in vitro. Phytochemistry 54:173–181

Freeman B, Albrigo LG, Biggs RH (1979) Cuticular waxes of developing leaves and fruit of blueberry *Vaccinium ashei* Reade cv. Bluegem. J Am Soc Hortic Sci 104:398–403

Galletta GJ, Ballington JR (1996) Blueberries, cranberries and lingonberries. In: Janick J, Moore JN (eds) Fruit breeding. Vine and small fruit crops, vol 2. John Wiley and Sons, Inc., New York

Galletta GJ, Fish AS (1972) Interspecific blueberry grafting, a way to extend *Vaccinium* culture to different soils. J Am Soc Hortic Sci 96:294–298

Graham J, Greig K, McNicol RJ (1996) Transformation of blueberry without antibiotic selection. Ann Appl Biol 128:557–564

Gupton CL, Smith BJ (1989) Inheritance of tolerance to stem blight in *Vaccinium* species. HortScience 24:748

Gupton CL, Spiers JM (1994) Interspecific and intraspecific pollination effects in rabbiteye and southern highbush blueberry. HortScience 29:324–326

Haghighi K, Hancock JF (1992) DNA restriction fragment length variability in genomes of highbush blueberry. HortScience 27:44–47

Hall IV, Aalders LE (1963) Two-factor inheritance of white fruit in the common lowbush blueberry, *Vaccinium angustifolium* Ait. Can J Genet Cytol 5:371–373

Hancock JF (1998) Using southern blueberry species in northern highbush breeding. In: Cline WO, Ballington JR (eds) Proceedings of the 8th North American Blueberry Research and Extension Workers Conference, North Carolina State University, Raleigh, pp 91–94

Hancock JF (2006a) Northern highbush breeding. Acta Hortic 715:37–40

Hancock JF (2006b) Highbush blueberry breeders. HortScience 41:20–21

Hancock JF, Erb WA, Goulart BL, Scheerens JC (1997) Blueberry hybrids with complex backgrounds evaluated on mineral soils: cold hardiness as influenced by parental species and location. Acta Hortic 446:389–396

Hancock JF, Galletta GJ (1995) Dedication: Arlen D. Draper: Blueberry Wizard. Plant Breed Rev 13:1–10

Hancock JF, Haghighi K, Krebs SL, Flore JA, Draper AD (1992) Photosynthetic heat stability in highbush blueberries and the possibility of genetic improvement. HortScience 27:1111–1112

Hancock JF, Sakin M, Callow PW (1991) Heritability of flowering and harvest dates in *Vaccinium corymbosum*. Fruit Varieties J 45:173–176

Hancock JF, Schulte NL, Siefker JH, Pritts MP, Roueche JM (1982) Screening highbush blueberry cultivars for resistance to the aphid *Illinoia pepperi*. HortScience 17:362–363

Hancock JF, Morimoto KM, Schulte NL, Martin JM, Ramsdell DC (1986) Search for resistance to blueberry shoestring virus in highbush blueberry cultivars. Fruit Varieties J 40:56–58

Hancock JF, Nelson JW, Bittenbender HC, Callow PW, Cameron JS, Krebs SL, Pritts MP, Schumann CM (1987) Variation among highbush blueberry cultivars in susceptibility to spring frost. J Am Soc Hortic Sci 112:702–706

Hanson EJ, Berkheimer SF, Hancock JF (2007) Seasonal changes in the cold hardiness of the flower buds of highbush blueberry with varying species ancestry. J Am Pomol Soc 61: 14–18

Harrison RE, Luby JJ, Ascher PD (1993) Genetic characteristics of self-fertility in highbush and half-high blueberries. Euphytica 67:79–88

Hiirsalmi H (1977) Inheritance of characters in hybrids of *Vaccinium uliginosum* and highbush blueberries. Ann Agric Fenn 16:7–18

Hiirsalmi H (1988) Small fruit breeding Finland. J Agric Sci Finland 60:223–234

Hokanson K, Hancock J (1998) Levels of allozymic diversity in diploid and tetraploid *Vaccinium* section Cyanococcus (blueberries). Can J Plant Sci 78:327–332

Hokanson K, Hancock J (2000) Early-acting inbreeding depression in three species of *Vaccinium* (Ericaceae). Sex Plant Reprod 13:145–150

Jelenkovic G (1973) Breeding value of pentaploid interspecific hybrids of *Vaccinium*. Jugoslovensko Vocarstvo 7:237–244

Johnson-Cicalese J, Vorsa N, Polashock J (2005) Evaluating cranberry germplasm for fruit rot resistance. N Amer Cranberry Res Ext Workers Ann Meeting. pp 23–25

Johnston S (1946) Observations on hybridizing lowbush and highbush blueberries. Proc Am Soc Hortic Sci 47:199–200

Korcak RF (1986) Adaptability of blueberry species to various soil types I: growth and initial fruiting. J Am Soc Hortic Sci 111:816–821

Krebs SL, Hancock JF (1988) The consequences of inbreeding on fertility in *Vaccinium corymbosum* L. J Am Soc Hortic Sci 113:914–918

Krebs SL, Hancock JF (1989) Tetrasomic inheritance of isoenzyme markers in the highbush blueberry, *Vaccinium corymbosum*. Heredity 63:11–18

Krebs SL, Hancock JF (1990) Early-acting inbreeding depression and reproductive success in highbush blueberry, *Vaccinium corymbosum* L. Theor Appl Genet 79:825–832

Lee J, Finn CE, Wrolstad RE (2004) Anthocyanin pigment and total phenolic content of three *Vaccinium* species native to the Pacific Northwest of North America. HortScience 39:959–964

Lee J, Wrolstad RE (2004) Extraction of anthocyanins and polyphenolics from blueberry processing waste. J Food Sci 69:564–573

Levi A, Panta GR, Parmentier CM, Muthalif MM, Arora R, Shanker S, Rowland LJ (1999) Complementary DNA cloning, sequencing and expression of an unusual dehydrin from blueberry floral buds. Physiol Plant 107:98–109

Levi A, Rowland LJ, Hartung JS (1993) Production of reliable randomly amplified polymorphic DNA (RAPD) markers from DNA of woody plants. HortScience 28:1188–1190

Luby JJ, Ballington JR, Draper AD, Pliszka K, Austin ME (1991) Blueberries and cranberries (*Vaccinium*). In: Moore JN, Ballington JR (eds) Genetic resources of temperate fruit and nut crops. International Society for Horticultural Science, Wageningen, The Netherlands, pp 391–456

Luby JJ, Finn CE (1986) Quantitative inheritance of plant growth habit in blueberry progenies. J Am Soc Hortic Sci 111:609–611

Luby JJ, Finn CE (1987) Inheritance of ripening uniformity and relationship to crop load in blueberry progenies. J Am Soc Hortic Sci 112:167–170

Lyrene PM (1981) Recurrent selection in breeding rabbiteye blueberries (*Vaccinium ashei* Reade). Euphytica 30:505–511

Lyrene PM (1985) Effects of year and genotype on flowering and ripening dates in rabbiteye blueberry. HortScience 20:407–409

Lyrene PM (1987) Breeding rabbiteye blueberries. Plant Breeding Rev 5:307–357

Lyrene PM (1988) An allele for anthocyanin-deficient foliage, buds and fruit in *Vaccinium elliottii*. J Hered 79:80–82

Lyrene PM (1991) Fertile derivatives from sparkleberry × blueberry crosses. J Am Soc Hortic Sci 116:899–902

Lyrene PM (1998) Ralph Sharpe and the Florida blueberry breeding program. In: Cline, WO, Ballington JR (eds) Proceedings of the 8th North American Blueberry Research and Extension Workers Conference, North Carolina State University, Raleigh, pp 1–7

Lyrene PM (2005) Breeding low-chill blueberries and peaches for subtropical areas. HortScience 40:1947–1949

Lyrene PM (2007) Breeding southern highbush blueberries. Plant Breeding Rev (in press)

Lyrene PM, Ballington JR (1986) Wide hybridization in *Vaccinium*. HortScience 21:52–57

Lyrene PM, Sherman WB (1983) Mitotic instability and 2n gamete production in *Vaccinium corymbosum* × *V. elliottii* hybrids. J Am Soc Hortic Sci 108:339–342

Lyrene PM, Vorsa N, Ballington JR (2003) Polyploidy and sexual polyploidization in the genus *Vaccinium*. Euphytica 133:27–36

Mahy G, Bruederle LP, Connors B, Van Hofwegen M, Vorsa N (2000) Allozyme evidence for genetic autopolyploidy and high genetic diversity in tetraploid cranberry, *Vaccinium oxycoccos* (Ericaceae). Am J Bot 87:1882–1889

Mainland CM (1998) Frederick Coville's pioneering contributions to blueberry culture and breeding. Proceedings of the North American Blueberry Workers Conference, Wilmington, DC

Martin RR, Bristow PR, Wegener LA (2006) Scorch and shock: emerging virus diseases of highbush blueberry and other *Vaccinium* species. Acta Hortic 715:463–467

Meyer JR, Ballington JR (1990) Resistance of *Vaccinium* species to the leafhopper *Scaphytopius magdalensis* (Homoptera: Cicadellidae). Ann Entomol Soc Am 83:515–520

Moerman DE (1998) Native American ethnobotany. Timber Press, Portland, Oregan, p 927

Moon JW, Flore JA, Hancock JF (1987a) A comparison of carbon and water vapor gas exchange characteristics between a diploid and a highbush blueberry. J Am Soc Hortic Sci 112:134–138

Moon JW, Flore JA, Hancock JF (1987b) Genotypic differences in the effect of temperature on CO_2 assimilation and water use efficiency in blueberry. J Am Soc Hortic Sci 112:170–173

Morrow EB (1943) Some effects of cross pollination versus self pollination in the cultivated blue-berry. Proc Am Soc Hortic Sci 42:469–472

Moyer R, Hummer K, Finn C, Frei B, Wrolstad R (2002) Anthocyanins, phenolics and antioxidant capacity in diverse small fruits: *Vaccinium*, *Rubus* and *Ribes*. J Agric Food Chem 50:519–525

Munoz CE, Lyrene PM (1985) In vitro attempts to overcome the cross-incompatibility between *V. corymbosum* L. and *V. elliottii* Chapm. Theor Appl Genet 69:591–596

Muthalif MM, Rowland LJ (1994a) Identification of dehydrin-like proteins responsive to chilling in floral buds of blueberry (*Vaccinium*, section Cyanococcus). Plant Physiol 104:1439–1447

Muthalif MM, Rowland LJ (1994b) Identification of chilling responsive proteins from floral buds of blueberry. Plant Sci 101:41–49

Nickerson NL (1978) In vitro shoot formation in lowbush blueberry seedling explants. HortScience 13:698

Novy RG, Kokak C, Goffreda J, Vorsa N (1994) RAPDs identify varietal misclassification and regional divergence in cranberry [*Vaccinium macrocarpon* (Ait) Pursh]. Theor Appl Genet 88:1004–1010

Ortiz R, Vorsa N, Bruederle LP, Laverty T (1992) Occurrence of unreduced pollen in diploid blue-berry species, *Vaccinium* sect. Cyanococcus. Theor Appl Genet 85:55–60

Panta GR, Rieger MW, Rowland LJ (2001) Effect of cold and drought stress on blueberry dehydrin accumulation. J Hortic Sci Biotech 76:549–556

Perry JL, Lyrene PM (1984) In vitro induction of tetraploidy in *Vaccinium darrowii*, *V. elliottii*, and *V. darrowii* × *V. elliottii* with colchicine treatment. J Am Soc Hortic Sci 109:4–6

Polashock JJ (2006) Screening for resistance to *Botryosphaeria* stem blight and *Phomopsis* twig blight. Acta Hortic 715:493–495

Polashock J, Ehlenfeldt M, Stretch A, Kramer M (2005) Anthracnose fruit rot resistance in blue-berry cultivars. Plant Dis 89:33–38

Polashock JJ, Griesbach RJ, Sullivan RF, Vorsa N (2002) Cloning of a cDNA encoding the cranberry dihydroflavonol-4-reductase (DFR) and expression in transgenic tobacco. Plant Sci 163:241–251

Polashock J, Vorsa N (2002a) Breeding and biotechnology: a combined approach to cranberry improvement. Acta Hortic 574:171–174

Polashock J, Vorsa N (2002b) Cranberry transformation and regeneration. In: Khachatourians GG, McHughen A, Scorza R, Nip W, Hui YH (eds) Transgenic plants and crops. Marcel Dekker Inc., New York, pp 383–396

Polashock J, Vorsa N (2006) Segregating blueberry populations for mummy berry fruit rot resis-tance. New Jersey Ann Veg Meeting Proc, Atlantic City, NJ

Qu L, Hancock JF (1995) Nature of 2n gamete formation and mode of inheritance in interspe-cific hybrids of diploid *Vaccinium darrowi* and tetraploid *V. corymbosum*. Theor Appl Genet 91:1309–1315

Qu L, Hancock JF (1997) RAPD-based genetic linkage map of blueberry derived from an interspe-cific cross between diploid *Vaccinium darrowi* and tetraploid *V. corymbosum*. J Am Soc Hortic Sci 122:69–73

Qu L, Hancock JF, Whallon JH (1998) Evolution in an autopolyploid group displaying pre-dominantly bivalent pairing at meiosis: genomic similarity of diploid *Vaccinium darrowi* and autotetraploid *V. corymbosum* (Ericaceae). Am J Bot 85:698–703

Qu L, Polashock J, Vorsa N (2000) A highly efficient in vitro cranberry regeneration system using leaf explants. HortScience 35:827–832

Qu L, Vorsa N (1999) Desynapsis and spindle abnormalities leading to 2n pollen formation in *Vaccinium darrowii*. Genome 42:35–40

Rabaey A, Luby J (1988) Fruit set in half-high blueberry genotypes following self and cross polli-nation. Fruit Varieties J 42:126–129

Ranger CM, Johnson-Cicalese J, Polavarapu S, Vorsa N (2006) Evaluation of *Vaccinium* spp. for resistance to *Illinoia pepperi* (Hemiptera: Aphididae) performance and phenolic content. J Econ Entomol 99:14741482

Ranger CM, Singh AP, Johnson-Cicalese J, Polavarapu S, Vorsa N (2007) Intraspecific variation in aphid resistance and constitutive phenolics exhibited by the wild blueberry *Vaccinium darrowi*. J Chem Ecol 33:711–729

Rousi A (1963) Hybridization between *Vaccinium uliginosum* and cultivated blueberry. Ann Agric Fenn 2:12–18

Rowland LJ, Hammerschlag FA (2005) *Vaccinium* spp. blueberry. In: Litz RE (ed) Biotechnology of fruit and nut crops. CABI Publishing, Wallingford, UK, pp 222–246

Rowland LJ, Levi A (1994) RAPD-based genetic linkage map of blueberry derived from a cross between diploid species (*Vaccinium darrowi* and *V. elliottii*). Theor Appl Genet 87:863–868

Rowland LJ, Mehra S, Dhanaraj A, Ogden EL, Arora R (2003b) Identification of molecular markers associated with cold tolerance in blueberry. Acta Hortic 625:59–69

Rowland LJ, Ogden EL (1992) Use of a cytokinin conjugate for efficient shoot regeneration from leaf sections of highbush blueberry. HortScience 27:1127–1129

Rowland LJ, Ogden EL, Arora R, Lim CC, Lehman JS, Levi A, Panta GR (1999) Use of blueberry to study genetic control of chilling requirement and cold hardiness in woody perennials. HortScience 34:1185–1191

Rowland L, Ogden E, Ehlenfeldt M, Vinyard B (2005) Cold hardiness, deacclimation kinetics and bud devlopment in diverse blueberry (*Vaccinium*) genotypes under field conditions. J Amer Soc Hort Sci 130:508–514

Rowland LJ, Panta GR, Mehra S, Parmentier-Line C (2004) Molecular genetic and physiological analysis of the cold-responsive dehydrins of blueberry. J Crop Improv 10:53–76

Rowland L, Smriti M, Dhanaraj A, Ehlenfeldt M, Ogden E, Slovin J (2003a) Development of EST-PCR markers for DNA fingerprinting and mapping in blueberry (*Vaccinium*, section Cyanococcus). J Am Soc Hortic Sci 128:682–690

Sapers GM, Jones SB, Kelley MJ, Phillips JG (1986) Breeding strategies for increasing the anthocyanin content of cranberries. J Am Soc Hortic Sci 111:618–622

Sapers GM, Phillips JG, Rudolf HM, DiVito AM (1983) Cranberry quality: selection procedures for breeding programs. J Am Soc Hortic Sci 108:241–246

Sarracino J, Vorsa N (1991) Self- and cross-fertility in cranberry. Euphytica 58:129–136

Scheerens JC, Erb WA, Goulart BL, Hancock JF (1999a) Blueberry hybrids with complex genetic backgrounds evaluated on mineral soils: stature, growth rate, yield potential and adaptability to mineral soil conditions as influenced by parental species. Fruit Varieties J 53:73–90

Scheerens JC, Erb WA, Goulart BL, Hancock JF (1999b) Blueberry hybrids with complex genetic backgrounds evaluated on mineral soils: flowering, fruit development, yield and yield components as influenced by parental species. Fruit Varieties J 53:91–104

Schmid P (1977) Long-term investigation with regard to the constituents of various cranberry varieties. Acta Hortic 61:241–254

Schulte NL, Hancock JF, Ramsdell DC (1985) Development of a screen for resistance to blueberry shoestring virus. J Am Soc Hortic Sci 110:343–346

Serres R, McCown B, Zeldin E (1997) Detectable β-glucuronidase activity in transgenic cranberry is affected by endogenous inhibitors and plant development. Plant Cell Rep 16:641–647

Serres R, Stang E, McCabe D, Russell D, Mahr D, McCown B (1992) Gene transfer using electric discharge particle bombardment and recovery of transformed cranberry plants. J Am Soc Hortic Sci 117:174–180

Sharp RH, Darrow GM (1959) Breeding blueberries for the Florida climate. Proc Fl St Hortic Soc 72:308–311

Sherman WB, Sharpe RH, Janick J (1973) The fruiting nursery: ultrahigh density for evaluation of blueberry and peach seedlings. HortScience 8:170–172

Smagula JM, Michand M, Hepler PR (1980) Light and gibberellic acid enhancement of low bush blueberry seed germination. J Am Soc Hortic Sci 105:816–818

Song G-Q, Roggers RA, Sink KC, Particka M, Zandstra B (2006) Production of herbicide-resistant highbush blueberry 'Legacy' by *Agrobacterium*-mediated transformation of the *Bar* gene. Acta Hortic 738:397–407

Song G-Q, Sink KC (2004) *Agrobacterium tumefaciens*-mediated transformation of blueberry (*Vaccinium corymbosum* L.). Plant Cell Rep 23:475–484

Song G-Q, Sink KC (2006) *Agrobacterium*-mediated transformation of highbush blueberry (*Vaccinium corymbosum* L.) cultivars. In: Wang K (ed) Agrobacterium Protocols (2nd Edition)-Methods in Molecular Biology 344, Humana, Totowa, NJ, pp 37–44

Stretch AW, Ehlenfeldt MK, Brewster V (1995) Mummy berry disease blight resistance in high-bush blueberry cultivars. HortScience 30:426–444

Stretch AW, Ehlenfeldt MK, Brewster V (2001) Resistance of diploid *Vaccinium* spp. to the fruit rot stage of mummy berry disease. Plant Dis 85:27–30

Strik B (2005) Blueberry – an expanding world berry crop. Chronica Hortic 45:7–12

Strik BC, Yarborough D (2005) Blueberry production trends in North America, 1992 to 2003, and predictions for growth. HortTechnology 15:391–398

Stushnoff C, Hough LF (1968) Response of blueberry seed germination to temperature, light, potassium nitrate, and coumarin. Proc Am Soc Hortic Sci 93:260–266

Teramura AH, Davies FS, Buchanan DW (1979) Comparative photosynthesis and transpiration in excised shoots of rabbiteye blueberry. HortScience 14:723–724

Vander Kloet SP (1977) The taxonomic status of *Vaccinium boreale*. Can J Bot 55:281–288

Vander Kloet SP (1980) The taxonomy of highbush blueberry, *Vaccinium corymbosum*. Can J Bot 58:1187–1201

Vander Kloet SP (1988) The genus *Vaccinium* in North America. Res Branch Agric Can Publ 1828, p 201

Vorsa N (1997) On a wing: the genetics and taxonomy of *Vaccinium* from a pollination perspective. Acta Hortic 446:59–66

Vorsa N (2004) www.hort.wisc.edu/cran/pubs_archive/proceedings/1994/brevor.pdf

Vorsa N, Ballington JR (1991) Fertility of triploid highbush blueberry. J Am Soc Hortic Sci 116:336–341

Vorsa N, Cunningham J, Roderick R, Howell AB (2002) Evaluation of fruit chemistry in cranberry germplasm: potential for breeding varieties with enhanced health constituents. Acta Hortic 574:215–219

Vorsa N, Howell AB, Foo LY, Lu Y (2003) Structure and genetic variation of cranberry proantho-cyanidins that inhibit adherence of uropathogenic P-fimbriated *E. coli*. In: Shahidi F, Ho C-T, Watanabe S, Osawa T (eds) Food factors for health promotion and disease prevention. ACS Symposium Series 851, American Chemical Society, Washington DC, pp 298–311

Vorsa N, Jelenkovic G, Draper AD, Welker WV (1987) Fertility of 4x × 5x and 5x × 4x pro-genies derived from *Vaccinium ashei/corymbosum* pentaploid hybrids. J Am Soc Hortic Sci 112:993–997

Vorsa N, Johnson-Cicalese J (2007) Variation in proathocyanidin content. In: Desjardins Y (ed) Acta Hortic, Proceeding of the first IS on Human Health Effects of F&V, 744, pp 243–250

Vorsa N, Manos PS, van Heemstra MI (1988) Isozyme variation and inheritance in blueberry. Genome 30:776–781

Vorsa N, Polashock J (2005) Genetic manipulation of cranberry fruit and leaf anthocyanin glyco-sylation. Theor Appl Genet 130:711–715

Vorsa N, Polashock J, Cunningham D, Roderick R (2005) Genetic inferences and breeding im-plications from analysis of cranberry germplasm anthocyanin profiles. J Amer Soc Hort Sci 128:691–697

Vorsa N, Rowland LJ (1997) Estimation of 2n megagametophyte heterozygosity in a diploid blue-berry (*Vaccinium darrowi* Camp) clone using RAPDs. J Hered 88:423–426

Wang SY, Stretch AW (2001) Antioxidant capacity in cranberry as influenced by cultivar and storage temperature. J Agric Food Chem 49:969–974

Wenslaff TF, Lyrene PM (2003) Chromosome homology in tetraploid southern highbush × *Vaccinium elliottii* hybrids. HortScience 38:263–265

Zeldin EL, Jury TP, Serres RA, McCown BH (2002) Tolerance to the herbicide glufosinate in transgenic cranberry (*Vaccinium macrocarpon* Ait.) and enhancement of tolerance in progeny. J Am Soc Hortic Sci 127:502–507

Chapter 5
Cherries

A.F. Iezzoni

Abstract Two major cherry species are grown for their fruit, the diploid sweet cherry and the tetraploid sour cherry. For both these species, new cultivars are needed that possess improved fruit quality and disease resistance. Genetic variation exists for most of the desired traits; however, little is known about their inheritance. The one exception is self-compatibility, where the molecular genetic basis has been elucidated in both sweet and sour cherry and molecular markers are available to identify self-compatible individuals in segregating progeny populations. Complete genetic linkage maps in cherry are just now being generated and as a result QTL analyses in cherry lag behind those in other *Prunus* species, most notably peach. Successful regeneration and transformation of cherry has been reported; however, utilizing this technology for gene function analysis and cultivar generation is still in its infancy.

5.1 Introduction

The major cherry species grown commercially for their fruit are sweet and sour cherry (*Prunus avium* L. and *Prunus cerasus* L., respectively). The ground cherry, *Prunus fruticosa* Pall., is also of considerable importance in Russia. Sour cherries (syn. tart cherries) are divided into two groups, the Amarelles and the Morellos. Amarelle cultivars have red fruits with uncolored juice. Morello types have colored skin and juice ranging from red to deep purple.

Annual world production of cherries is about 2 million tons, which is split relatively evenly between sweet and sour cherry (FAO 2005). Cherry production is limited to temperate regions that experience moderately cold winter temperatures. In colder regions, cherry production is limited by cold mid-winter temperatures. As

A.F. Iezzoni
Department of Horticulture, 342B Plant and Soil Sciences Building, Michigan State University, East Lansing, Michigan 48824, USA
e-mail: iezzoni@msu.edu

J.F. Hancock (ed.), *Temperate Fruit Crop Breeding*,
© Springer Science+Business Media B.V. 2008

a result, better cherry sites are frequently near large bodies of water that buffer the temperatures and on higher locations with adequate cold air drainage.

World production of cherries is greatest in Europe, with sweet cherries more important in Western Europe and sour cherries more important in Eastern Europe. Major cherry producing countries in Europe include Germany, Poland, and Turkey. In North America, most of the fresh market sweet cherry production is in the Pacific Northwest, while most of the processed cherries (sweet and sour) are produced in Michigan. Pockets of cherry production are also growing rapidly in Chile and Argentina, with local supplies present in a vast number of countries.

Cherries are clonally propagated either on seedling or vegetatively propagated rootstocks. Trees are propagated by inserting a bud from the scion into the rootstocks in mid-August. The next year the rootstock above the bud is removed permitting the scion bud to grow. These one year old grafted trees are dug in the fall after the first growing year and stored in the nursery cooler until planting the next spring.

Sweet cherries are best eaten fresh. Fruit have either dark red skin and flesh or yellow skin with a pink blush and yellow flesh. Yellow fleshed cherries are also used for marashino cherries. Sour cherries are excellent for cooking and are used mainly in pies, jams, juices and pastry products, although certain sour cherries are also excellent fresh such as the 'Pándy' (syn. 'Cri:ana') cherry that is grown for local markets in Eastern Europe. Dried cherries, with the pits removed, are also eaten fresh or used in salads or baked goods.

5.2 Evolutionary Biology and Germplasm Resources

Cherries have been valued since ancient times as one of the first tree fruits to ripen in temperate growing regions. The two main groups of cherries, sweet and sour cherries, are reported to have originated in an area that includes Asia Minor, Iran, Iraq, and Syria (Vavilov 1951). The first diploid *Prunus* species arose in central Asia and the sweet, sour and ground cherry (a low growing bush cherry native to Russia) were early derivatives of this ancestral *Prunus*. Sour cherry is believed to have arisen through natural hybridizations between ground cherry and sweet cherry (Olden and Nybom 1968). As sweet and sour cherries spread throughout Europe, ecotypes evolved within each species that differed in cold hardiness, tree habit and fruit and leaf characteristics (Kolesnikova 1975). These ecotypes presumably arose due to continued gene flow between sweet, sour and ground cherry.

There are more than 30 species of cherries, with most indigenous to Europe and Asia. The main species grown for their fruit are in the Section Eucerasus which consists of the sweet cherry (*P. avium* L.), sour cherry (*P. cerasus* L.) and ground cherry (*P. fruticosa* Pall.). The basic chromosome number is $x = 8$. Sweet cherry is diploid ($2n = 16$) and sour cherry and ground cherry are tetraploid ($2n = 32$). Ground cherry and sour cherry hybridize readily and some of the cultivars in commercial production in Russia and eastern Europe are actually hybrids between these two species. The Duke cherry, is presumed to have arisen from the pollination of sour cherry by unreduced $(2n)$ pollen of sweet cherry. Tree and fruit types of Duke cherry are intermediate between these two species. The most important rootstock species are *P. mahaleb* L $(2n = 16)$, wild *P. avium* commonly called mazzard, and more recently triploid hybrids between sour cherry and the diploid *P. canescens* Bois $(2n = 24)$. A likely race of sour cherry, the maraska cherry (named *P. marasca* by some authorities), is found in the wild in parts of the former Yugoslavia.

The continual gene flow between sour cherry and its progenitor species has had a major negative impact on the reproductive behavior of sour cherry. Sour cherry typically has very reduced fertility due to irregular meiosis resulting from the unbalanced genomic contributions. Cytological analyses reveal the prevalence of multivalent and quadrivalent formations at meiosis (Schuster 2000). Genetic data supports this meiotic behavior as sour cherry exhibits both disomic and tetrasomic inheritance (Beaver and Iezzoni 1993), therefore sour cherry is classified as a segmental allotetraploid. Morphologically, this promiscuity is manifest as sour cherry exhibiting an extreme amount of genetic diversity ranging from types resembling sweet cherry to those resembling the *P. fruticosa* parent (Hillig and Iezzoni 1988).

Sweet cherry exhibits a classic gametophytic self-incompatibility (GSI) system that is a common genetic mechanism promoting outcrossing in flowering plants (de Nettancourt 2001). In GSI, self-incompatibility (SI) is determined by a single multi-allelic locus, called the S-locus that contains a minimum of two genes, one controlling stylar specificity and the other controlling pollen specificity of the SI reaction. The stylar-S in cherry is a ribonuclease (S-RNase) that is expressed in the pistil and degrades the RNA of incompatible pollen (McClure et al. 1990). A similar system is found in three other families, the Solanaceae, Scrophulariaceae, and Rosaceae (Anderson et al. 1990, Boskovic and Tobutt 1996, McClure et al. 1989, Sassa et al. 1992, Xue et al. 1996). The pollen-S gene in cherry as in other *Prunus*, is believed to be an S-haplotype specific F-box gene (SFB) and has been identified in *Prunus dulcis*, *P. avium*, and *P. cerasus* (Entani et al. 2003, Ikeda et al. 2004a, Ushijima et al. 2003, Yamane et al. 2003b). All natural sweet cherry selections are reported to be SI, except 'Cristobalina', an old Spanish landrace cultivar (Wünsch and Hormaza 2004).

Like sweet cherry, sour cherry exhibits an S-RNase based GSI system (Boskovic et al. 2006, Hauck et al. 2006a, Tobutt et al. 2004, Yamane et al. 2001); however, natural sour cherry selections include both SI and self-compatible (SC) types (Lansari and Iezzoni 1990, Redalen 1984). This genotype-dependent loss of self-incompatibility in sour cherry indicates that genetic changes, and not polyploidy per se, cause the breakdown of SI. Instead, the genetic control of SI and SC in sour cherry has been shown to be regulated by the accumulation of non-functional S-haplotypes according to the 'one-allele-match model' (Hauck et al. 2006). In this model, the match between a functional pollen-S gene produced by the $2x$ pollen and its cognate functional S-RNase in the style, results in an incompatible reaction. A similar reaction occurs regardless of whether the pollen contained a single functional pollen-S gene or two different pollen-S genes. The absence of a functional match results in a compatible reaction. Thus for successful fertilization, the $2x$ pollen must contain two non-functional S-haplotypes.

Many of the naturally occurring S-haplotypes present in sweet cherry are also present in sour cherry. However, three of these S-haploptyes (S_1, S_6, and S_{13}) also have non-functional variants in sour cherry that have lost pollen or stylar function (Hauck et al. 2006a, Tsukamoto et al. 2006). Loss of function was due to structural alterations of the S-RNase, SFB or S-RNase upstream sequences.

5.3 History of Improvement

Cherry was undoubtedly an early food source for early inhabitants of Europe, as pits recovered from cave dwelling date to as early as 4000–5000 BC (Marshall 1954). Cherry was cultivated for its fruit by the time of the Greek Empire, about 300 BC, and probably for its wood several centuries earlier. The Romans are credited with introducing cherry cultivation to England in the first century.

Little is known about the development of cherries before the sixteenth century. However, early records describe cherry cultivation around this period and many old sweet cherry varieties trace back to German origins. From ancient times to the 1600s, numerous local varieties (landraces) arose that were identified with specific regions or towns where the selection had been popular. Much of the cherry germplasm available today represents collections of these old landraces. In many cases, these landraces were exchanged between communities and countries and as a result it is quite common for a landrace variety to have multiple names. In Europe, the major tonnage of sweet and sour cherry production is still from the old landrace selections: 'Schneiders' sweet cherry, syn. '0900 Ziraat' and 'Schattenmorelle' sour cherry.

In European countries with extensive landrace diversity for cherry, the breeding programs began by selecting among landraces and using these selections as parents. All of the sour cherry cultivars grown today are either landrace selections themselves or one generation removed from these landrace selections.

Early settlers brought cherry seeds and budwood to North America and pioneers moved the cherries westward with them. Most notably, in the mid 1800s the Lewellin brothers journeyed from Iowa to Oregon with numerous grafted cherry trees. Seth Lewelling became interested in growing cherry seedlings for pollenizers and for selection of promising cultivars. From these seedlings, he selected 'Bing' which is still the leading sweet cherry cultivar in North America. Other important cultivars that came from his selection process were 'Lambert' and 'Republican', which is still a leading pollenizer variety.

A major advance in sweet cherry breeding occurred with the introduction of self-fertility in sweet cherry. Lewis and Crowe (1954) used irradiated 'Napoleon' pollen to fertilize 'Emperor Francis' styles. One of the induced mutations, named S_4', renders the pollen self-compatible on any sweet cherry style. Therefore, sweet cherry varieties carrying the S_4' do not need to be planted with a pollinator variety. The sweet cherry breeding program in Summerland, British Columbia, was the first program to capitalize on the use of the S_4' with the release of the variety 'Stella'

(Lapins 1970). 'Stella' did not have suitable quality to become commercially important, however, it was used extensively as a parent in breeding programs as a source of self-fertility. Later Summerland self-compatible releases such as 'Sweetheart' became extremely important contributors to global sweet cherry production.

5.4 Current Breeding Efforts

Major breeding priorities for sour cherry include suitability for mechanical harvesting and processing, late flowering to avoid spring freeze damage, excellent fruit quality, round pit, resistance to cherry leaf spot, high yielding, self-compatible, and a range of ripening dates. Breeding objectives for sweet cherry vary with the production area. Climatic factors such as the amount and distribution of rainfall, excessive temperatures, occurrence of frosts and freezes and the diurnal temperature fluctuation determine to a large extent the type of cherry that will succeed in a specific area. In general, all sweet cherry breeding programs are interested in developing new cultivars that are high yielding, self-fertile, resistant to fruit cracking with firm large fruit with good shipping quality. Other desirable attributes are resistance to diseases and cold, and early to late ripening. Although precocity in bearing and compact growth habit of scions differs, dramatic increases in precocity and size reduction require rootstocks that confer these attributes to the scion.

The vast majority of sweet and sour cherry breeding programs are in Europe, as these countries represent the ancestral home of cherry (Table 5.1). More recently, sweet cherry breeding programs have been initiated in Asia. The current

Table 5.1 Major governmental or university supported sweet and sour cherry breeding programs

Region	Country	Sweet	Sour	Location
Europe	Belarus	×	×	Research Institute for Fruit Growing – Minsk
	Estonia	×	×	Polli Research Center of Horticulture – Karksi
	Bulgaria	×	×	Fruit Growing Institute, Plovdiv
		×	×	Institute of Agriculture, Kyustendil
	Czech Republic	×		Research and Breeding Institute of Pomology, Holovosy
	Italy	×		Dipartimento Colture Arboree, University of Bologna
		×		Istituto Sperimentale Frutticoltura, Ministry of Agriculture, Rome
		×	×	Istituto Sperimentale Frutticoltura, Verons Province
	France	×	×	INRA, Station de Recherches Fruitieres, Bordeaux
	Germany	×	×	BAZ Institute for Fruit Growing, Dresden

Table 5.1 (continued)

Region	Country	Sweet	Sour	Location
		×		Fruit Research Station, York
	Hungary	×	×	Research Institute for Fruit Growing and Ornamentals, Budapest
	Latvia	×	×	Latvia State Institute of Fruit-Growing, Dobele
	Lithuania	×	×	Lithuanian Institute of Horticulture, Babtai
	Romania	×	×	Research Institute of Fruit Growing, Pitesti
		×	×	Iasi
		×	×	Bistrita
	Russia		×	All-Russian Scientific Research Institute of Horticultural Breeding, Orel
			×	Russian Research Institute of Genetics and Breeding of Fruit Plants, Michurinsk
		×	×	Horticultural Experiment Station, Krymsk
	Serbia	×	×	Fruit and Grape Research Center, Cacak
	Switzerland	×		Swiss Federal Resarch Station for Fruit Growing, Wadensville
	Ukraine	×	×	Institute of Horticulture, Donetsk
		×	×	Institute of Irrigated Horticulture, Melitopol
	United Kingdom	×		East Malling Research Station, East Malling
		×		John Innes Institute, Norwich
America	Canada	×		Summerland, British Columbia
			×	University of Saskatchewan
	U.S.A.		×	Michigan State University, East Lansing
		×		Cornell University, Geneva, New York
		×		Washington State University, Prosser, Wash.
Asia	China	×		Department of Horticulture and Landscape, Hebeo
		×		Department of Pomology, Beijing
	Japan	×		Yamagata Agricultural Research Center Horticultural Experiment Station
		×		Hohhaido Central Agricultural Experiment Station
		×		Kennan Fruit Tree Research Center Aomori Prefectural Agriculture
		×		Yamanashi Prefecture
	Korea	×		National Horticultural Research Institute, RDA

list of breeding programs is impressive, but it gives a false sense of the potential and progress being made in cherry breeding. In a recent survey of 15 sweet cherry breeding programs, these programs reported generating a total of only 5,430–6,280 seedlings per year. On average this works out to only 362–419 seedlings per program per year (Sansavini and Lugli 2005). This extremely low number is due to the difficulty of seed germination in sweet cherry and the field space requirement and number of years necessary before a seedling starts to fruit. The use of marker-assisted selection to pre-screen seedlings prior to field planting and propagating these 'selected' seedlings on precocious rootstocks holds promise to reduce these time and space barriers.

5.5 Genetics of Economically Important Traits

5.5.1 Pest and Disease Resistance

Many diseases and insects attack cherry; however the most costly to the growers are fungal diseases due to the large number of sprays required. In arid areas, controlling powdery mildew on sweet cherry is the major challenge, while in more humid, cold regions, bacterial and *Cytospora* canker are the most problematic. Unfortunately, strains of *P. s. syringae* have been isolated from several orchards in Michigan that are resistant to the copper sprays used to control bacterial canker (Sundin et al. 1989). Another very significant worldwide pathogen is cherry leaf spot (CLS). American brown rot is a major fungal disease on sweet cherries in North America, while European brown rot is an important control issue in Europe. *Phytophthora* root and crown rot and *Armillaria* root rot (syn. Oak root rot) are widespread soil pathogens that severely limit the productivity of infected orchards. Information on the genetics of resistance to cherry pathogens is generally lacking, although powdery mildew resistance is known to be conferred by a single dominant gene termed *Pmr*-1 (Olmstead et al. 2001).

Cultivars and native germplasm have been identified that vary in their tolerance or resistance to most of the diseases listed above, as well as shothole (*Stigmina carpophila*), *Leucostoma* canker (syn. Perennial canker, *Cytospera* canker) and silver leaf (*Chondrostereum purpureum*) (Table 5.2). However, no genetic source of resistance is available to *Armillaria* root rot (Proffer et al. 1988). The most important sour cherry cultivar in the U.S.A., Montmorency, is among those cultivars that are tolerant to European brown rot. In some instances, sweet and sour cherries vary in their levels of susceptibility to pathogens. For example, sour cherry is much more susceptible to European brown rot than sweet cherry, and sweet cherry is much more susceptible to American brown rot than sour cherries.

Some of the most extensive disease screening has been done on cherry leaf spot (CLS). Sour cherry cultivars have been shown to differ greatly in their tolerance to CLS (Table 5.2) with the cultivar NorthStar being the most tolerant (Sjulin et al. 1989). Interspecific hybridization in Russia with the tetraploid species

Table 5.2 Genetics of disease and pest resistance in cherry

Disease	Plant material screened	References
Bacterial		
Bacterial canker *Pseudomonas* ssp.	Sweet cherry cultivars	Fischer (1996), Theiler-Hedrich et al. (1985), Matthews (1979), Matthews (1968)
	Sweet cherry cultivars and interspecific hybrids	Garrett (1977), Kaltschmidt and Fischer (1986)
	Prunus germplasm	Matthews (1973a,b)
Fungal		
Cherry leaf spot *Blumeriella jaapii* (Rehm) Arx	*Prunus* germplasm	Wharton et al. (2003), Sjulin et al. (1989), Kolesnikova et al. (1985), Enikeev et al. (1975), Zhukov et al. (1980), Keitt et al. (1918)
	P. fruticosa	Cummings (1972)
	Sour cherry	Budan et al. (2005), Alderman et al. (1950)
	Sour × sweet cherry hybrids	Arsenijevic et al. (1984), deRavel d'Esclapon (1980), Fourcade (1979)
American Brown rot *Monilinia fructicola (G. Wint.)* Honey	Sour cherry	Grover (1963), Moore et al. (1963)
	Sweet cherry	Biggs and Northover (1989), Brown and Bourne (1988)
European brown rot *Monilinia laxa* (Aderh. & Ruhl.) Honey	Sour cherry	Budan et al. (2005), Tehrani (1984), Filippova (1983), Voronin (1981), Zwintzscher (1979, 1963)
Powdery mildew *Podosphaera clandestina* (Wallr.:Fr.) Lev.	Sweet cherry	Olmstead and Lang (2002), Olmstead et al. (2001), Toyama et al. (1993)
Shothole *Stigmina carpophila*	Sweet cherry	Zwintzscher (1963)
Perennial canker (syn. *Cytospora* canker) *Leucostoma persoonii* and *L. cincta*	Sweet cherry	Rozsnyay and Apostol (2005), Fischer (1996), Kaltschmidt and Fischer (1986), Voronin and Stepanova (1980)
	Sour cherry	Rozsnyay and Apostol (2005)

Table 5.2 (continued)

Disease	Plant material screened	References
Silverleaf *Chondrostereum purpureum*	*Prunus* species	Baumann and Engle (1986), Bennett and Watkins (1976)
Phytophthora root and crown rot	Sour cherry	Mircetich and Matheron (1981)
	P. interspecific hybrids	Cummins et al. (1986)
Armillaria root rot	*Prunus* species	Proffer et al. (1988)
Virus and protoplasma		
PNRSV	Sweet cherry	Uphoff et al. (1988), Kegler et al. (1978)
PNRSV & PDV	*Prunus* species	Lang et al. (1998)
Cherry leaf roll virus	Interspecific hybrid	Cropley (1968)
Little cherry virus	Sweet cherry	Fogle (1975), Posnette et al. (1968)
X-disease	Sweet cherry	Wadley (1970)
	Prunus species	Gilmer and Blodgett (1974)
Insects		
Black cherry aphid *Myzum cerasi*	Sweet cherry	Börner (1943)
	Prunus species	Gruppe (1988)
San Jose scale *Quadraspidiotus perniciosus*	Sour cherry	Jenser and Sheta (1969)
Gall mite *Vasates fockeui*	Sour cherry	Breton (1980), Anonymous (1976)

Source: Iezzoni et al. 1990b

P. maackii resulted in a CLS resistant hybrid cultivar 'Almaz' (Zhukov et al. 1980). A seedling from this hybrid, named R1(1) and the Gisela 6 *P. cerasus* × *P. canescens* hybrid were determined to be resistant to CLS when challenged with a wide range of CLS isolates (Wharton et al. 2003).

There are a number of virus diseases that cause problems in cherry including green ring mottle virus (GRMV), plum pox (Sharka), *Prunus* necrotic ringspot virus (PNRSV), prune dwarf virus (PDV) and tomato ringspot virus (TomRSV). GRMV is a closterovirus which has only been shown to be spread by grafting. Plum pox disease is a leafhopper transmitted disease that is beginning to spread throughout Europe; the cherry strain of the plum pox virus is not known in North America. PNRSV and PDV are ilarviruses that spread through virus-contaminated pollen and seed. These viruses are the most prevalent ones worldwide and are frequently found in older sour cherry orchards in the U.S. TomRSV is a nematode-transmitted nepovirus that is found exclusively in North America. There is also a serious phytoplasma disease of cherries called X-disease which is transported by various leafhopper species. X-disease has the potential to infiltrate pockets of production due to its ability to survive in alternate *Prunus* hosts. An important natural host is wild choke cherry (*P. virginiana*) which is eradicated in many cherry growing regions. Cherry selections have been shown to differ in their tolerance to PNRSV, PDV and X-disease, along with the cherry leaf roll virus and the little cherry virus (Table 5.2).

Among insect pests, fruit flies are a huge economic threat to the industry, as many markets have 'zero tolerance' for even the presence of one larvae in the fruit. In North America three cherry fruit fly species are important pests. The eastern cherry fruit fly [*Rhagoletis cingulata* (Loew)] and the black cherry fruit fly [*R. fausta* (Osten Sacken)] are found in eastern North America. The western cherry fruit fly [*R. indifferens* Curran] occurs in western North America. The European cherry fruit fly [*R. cerasis* L] is a common pest in Europe. The plum curculio (*Conotrachelus nenuphar* (Herbst), which is native to North America, is a serious pest of stone and pome fruits throughout most of the eastern U.S.A. The most common mite species on cherry trees are the European red mite (ERM) [*Panonychus ulmi* (Koch)] and the two-spotted spider mite (TSSM) (*Tetranychus urticae* Koch). Little resistance is available to these very significant pests, although sources of resistance have been described to the more minor insect problems, black cherry aphid (*Myzus cerasi*), San Jose scale (*Quadraspidiotus perniciosus*) and gall mite (*Vasates fockeui*).

5.5.2 Morphological and Physiological Traits

One of the most important barriers to profitable sweet cherry production is the cost of labor, as sweet cherries are all hand harvested for the fresh market. Utilization of precocious dwarfing rootstocks are of intense commercial interest, as such rootstocks significantly reduce production costs and dramatically increase profitability. The most widely used rootstocks are the triploid Gisela hybrids 5 and 6 that resulted from the inter-specific cross between the sour cherry cultivar 'Schattenmorelle' and *P. canescens* (Schmidt 1985).

In sour cherry, the factor most limiting to new cultivar development is finding genotypes with yields comparable to the current high yielding cultivars, such as 'Montmorency' in the U.S.A. and 'Schattenmorelle' in Northern Europe. The vast majority of sour cherry seedlings produce very low yields, due to the poor fruit set associated with meiotic irregularities in the megagametophyte (Schuster and Wolfram 2005).

In sweet cherry, low-temperature damage to flower buds is the most important factor limiting yields in the colder growing regions. Temperatures below −30°C can result in significant injury, although cultivars do vary in their hardiness (Strauch and Gruppe 1985).

Sour cherry is more tolerant to winter cold than sweet cherry, but shows a range in hardiness, with those types more closely resembling sweet cherry being less hardy (Kolesnikova 1975, Mathers 2004a,b). Ground cherry is the most tolerant cherry to cold mid-winter temperatures and in Russia its wood and flower buds can survive winter temperatures of −40°C and −50°C, respectively (Kolesnikova 1975).

A major breeding goal in Italy is the development of small productive trees that can be used for high density plantings by combining a compact tree habit with high spur density. Sansavini et al. (1998) defined three classes to facilitate genetic studies: compact, spur and standard. Compact trees were defined as those having a shoot internode length shorter than 2.3 cm ('Burlat C1' and 'Durone Compactoo di Vignola'). Spur type trees were defined as those with over 25 spurs per meter of length of two three-year-old limbs ('Bing spur', 'Lambert compact' and 'Lapins'). All the other trees were classified as standard. Both compact and spur habit appear to be controlled by multiple recessive genes, as the majority of the progeny from the crosses between classes had the standard growth habit. Even when both parents had the spur habit, only 38% of the progeny exhibited this fruiting habit. When both parents had compact habit, only 8% of the progeny had this growth habit.

5.5.3 Fruit Quality

Excellent fruit size and quality are demanded by the market place for fresh sweet cherries and achieving these parameters are major goals of all breeding programs. Rain induced fruit cracking is one of the most important yearly threats to fruit quality. Currently, breeders use genotypes that are widely reported to be less sensitive to cracking as parents; however, breeding for resistance to rain-induced fruit cracking has been hampered by a lack of knowledge of the mechanism of resistance and the absence of an adequate screening technique (see review by Christensen 1996).

In warm climates such as California, fruit doubling of sweet cherry can result in a significant quantity of unmarketable fruit. This defect is believed to be caused by high summer temperatures at the time of flower bud differentiation. A severe post-harvest disorder is surface pitting. This is the development of sunken depressions on the surface of the cherry that occur after the fruit has been mechanically damaged. Sweet cherry varieties differ in their susceptibility to doubling (Micke et al. 1983) and surface pitting (Patten et al. 1983).

The majority of cherry cultivars have red skin with colored flesh and juice. However, some cultivars have yellow-red skin (e.g. 'blush' fruit), with cream flesh and colorless juice (Fig. 5.1). Within each of these two categories there are grades of red to purple and cream to pink. A few cultivars have bright yellow fruit skin typical of the old German cultivar 'Donissens Gelbe' (syn. 'Gold', 'Drogans Gelbe'). There is close correlation of skin, flesh and juice color, yet types with a mix of traits can be obtained. For example, from certain crosses it is possible to obtain a low percentage of seedlings with red fruit combined with cream flesh and colorless juice (Schmidt 1998). This provides evidence that skin color is under different genetic control than flesh and juice color.

Fogle (1958b) suggested that a major dominant gene A controls skin color, with red/black being dominant to blush. Results of Zwintzscher (1966), Matthews (1973b) and Schmidt (1998) support the presence of a dominant gene for skin color and identify cultivars that are heterozygous for the postulated A locus. However, deviations from this model in some crosses suggest the presence of more than one locus (Fogle 1958). None of the skin color segregants exhibited the bright yellow

Fig. 5.1 (*Left*) Montmorency, (*Right*) Újfehértói Fürtös. Tart cherries are classified into two major groupings, morello and amarelle. Amarelle cherries, such as Montmorency, only have red pigment in the fruit skin while the fruit flesh is clear. Morello cherries, such as Újfehértói Fürtös, have red pigment in the fruit skin and throughout the flesh

Fig. 5.2 Fruit from a breeding population in which progeny are segregating for fruit size

fruit characteristic of 'Donissens Gelbe' suggesting that this skin color is controlled by factors other than the A locus (Schmidt 1998). Cream colored flesh is recessive to red flesh as crosses between two red-fleshed parents can result in progeny individuals with cream flesh (Schmidt 1998).

The red pigments in cherries are anthocyanins and there has been increased interest in these pigments due to their potential as a good source of antioxidants in the human diet. In sour cherry, cyanidin e-glucosylrutinoside is the major anthocyanin, whereas in sweet cherry cyanided 3-rutinoside is the predominant one. Levels and presence/absence of different pigments have been shown to differ among selections (Gao and Mazza 1974, Chandra et al. 1992).

Other fruit quality traits such as fruit size (Fig. 5.2), flesh firmness and flavor are known to segregate in seedling populations according to the expectations of a quantitative traits and parents that contribute more favorable progeny have been identified (Apostol 1998, Brown and Bourne 1988, Hansche et al. 1966, Sansavini and Lugli 1996).

5.6 Crossing and Evaluation Techniques

5.6.1 Breeding Systems

The majority of sweet cherry cultivars are self-incompatible, requiring pollen from a compatible pollinator tree. In commercial orchards, a pollinizer tree of a different incompatibility group must be planted at every third position in every third row for standard size trees. This design provides the minimum number of pollinator trees but ensures that every tree is adjacent to a pollinator. However, for trees on dwarfing rootstocks, the pollinator cultivar can be as few as every seventh tree. Besides providing compatible pollen, the cultivar selected for use as a pollinator should have a bloom time that overlaps with the target cultivar.

Self-incompatibility is controlled by alleles at the S-locus whereby any pollen tube bearing an S-allele in common with either of the two alleles in the stylar parent will be unable to grow down the style. Pollen from the same plant is rejected as well as fertile pollen from any other cultivar sharing the same two S-alleles. The cloning and sequencing of the S-RNase from sweet cherry (Tao et al. 1999) led to the development of PCR methods to distinguish S-allele types (Fig. 5.3). There are numerous PCR primers now available that can be used to genotype the S-RNase (Sonneveld et al. 2005a, Wiersma et al. 2001), and most importantly, the S_4' allele conferring self compatibility (Ikeda et al. 2004). The identification of over 20 S-RNases has resulted in the assignment of 23 incompatibility classes and two SC classes containing the S_4' allele. However, these 23 incompatibility groups only include 10 S-alleles ($S_1 - S_7$, S_9 S_{12}, S_{13}) that have been identified in more than one selection and confirmed in genetic crosses to confer different S-allele specificities (Table 5.3) (Iezzoni et al. 2005, Tobutt et al. 2001).

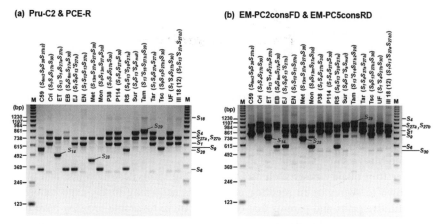

Fig. 5.3 PCR amplification for *S-RNase* alleles of 17 sour cherry selections. (a) Genomic DNA was amplified by PCR with Pru-C2 (Tao et al. 1999) and PCE-R (Yamane et al. 2001) primer set. (b) Genomic DNA was amplified by PCR with EM-PC2consFD and EM-PC5consRD (Sutherland et al. 2004) primer set. PCR products were separated on 2% agarose gels and detected with ethidium bromide staining. The color of black and white is inverted in this image. M: 123 bp DNA ladder (Invitrogen, Carlsbad, California, U.S.A.). Lane Abbreviations are: C59 = 'Cigány 59'; Cri = 'Crisana'; ET = 'Englaise Timpurii'; EB = 'Erdi Botermo'; EJ = 'Erdi Jubileum'; EN = 'Erdi Nagygyumolcsu'; Met = 'Meteor'; Mon = 'Montmorency'; P38 = 'Pandy 38'; P114 = 'Pandy 114'; RS = 'Rheinische Schattenmorelle'; Sur = 'Surefire'; Tam = 'Tamaris'; Tar = 'Tarina'; Tsc = 'Tschernokorka'; UF = 'Újfehértói f rt s'; III 18 (12) = 'MSU III 18 (12)'

Self-compatible sweet cherry cultivars have been developed by utilizing the self-compatible mutant of the S_4 locus, S_4'. Trees with the S_4' are self-compatible and also compatible with all other sweet cherries. Self-compatible cultivars are advantageous since the entire orchard can be planted to one high-value cultivar, thus simplifying orchard operations as well as marketing. In addition, when the weather is not favorable for bee activity, fruit set is higher on self-compatible selections.

Sour cherry cultivars can be either self-incompatible or self-compatible. Because of the economic advantages of self-compatibility, the vast majority of the commercial cultivars are self-compatible. As in sweet cherry, multiple functional S-alleles have been identified (Table 5.4). However, pollen part mutants (identified with a) and stylar part mutants (identified with a_m) have also been identified in genetic crosses. Marker assisted selection for early detection of SC progeny is possible in sweet cherry breeding programs due to the ability to distinguish the pollen-part mutant S_4' from its wild-type allele. This mutant allele has a 4 bp deletion in SFB_4' (Sonneveld et al. 2005, Ushijima et al. 2004) and can be distinguished from its wild-type allele by a dCAPS marker (Ikeda et al. 2004). Therefore, early screening for SC progeny just involves the identification of the S_4'-haplotype and data interpretation is very simple as any individual possessing an S_4'-haplotype is SC. However, unlike sweet cherry, individual sour cherry seedlings would need to be screened for multiple alleles.

Table 5.3 Incompatibility groups and *S*-allele genotype of most widely used sweet cherry cultivars. Nomenclature according to Tobutt et al. (2001). For extensive reviews in sweet cherry *S*-allele genotypes see Iezzoni et al. (2005) and Tobutt et al. (2001, 2004). SC: Self-compatible cultivar. O: Universal donor[a]

Incompatibility Group	S-genotype	Representative Cultivar
I	S_1S_2	Summit
II	S_1S_3	Regina
III	S_3S_4	Bing
IV	S_2S_3	Vega
V	S_4S_5	Late Black Bigarreau
VI	S_3S_6	Ambrunes
VII	S_3S_5	Hedelfingen
VIII	S_2S_5	Vista
IX	S_1S_4	Rainier
X	S_6S_9	Black Tartarian
XII	S_6S_{13}	Noble
XIII	S_2S_4	Sam, Schmidt
XIV	S_1S_5	Valera
XV	S_5S_6	Colney
XVI	S_3S_9	Chelan
XVII	S_4S_6	Larian
XVIII	S_1S_9	Brooks
XIX	S_3S_{13}	Reverchon
XXI	S_4S_9	Inge
XXII	S_3S_{12}	Schneiders
XXIII	S_2S_6	Arcina
SC/O	S_3S_4'	Sweetheart
SC/O	S_1S_4'	Lapins

[a]The *S*-RNase sequence data for S_8, S_{11} and S_{15} suggest that they are synonyms of S_3, S_7 and S_5, respectively (Sonneveld et al. 2001, 2003)

5.6.2 Pollination and Seedling Culture

Cherry breeders select parents based upon their superior traits. If the sweet or sour cherry to be used as the seed parent is self-incompatible, it is not necessary to emasculate the flowers of the seed parent prior to pollen application. However, the branches need to be enclosed to prevent bee pollination using bags or cages that exclude insects. The protective covering should be removed following fruit set.

Table 5.4 *S*-allele genotypes of some major sour cherry cultivars (in preparation)

Cultivar	SC or SI[y]	S-genotype
Érdi Bőtermő	SC	$S_4S_{6m}S_{27a}S_{30}$
Meteor	SC	$S_{13m}S_{27a}S_{27b}S_{28}$
Montmorency	SC	$S_6S_{13m}S_{27a}S_{30}$
Pándy[z]	SI	$S_1S_4S_{27b}S_{30}$
Schattenmorelle	SC	$S_6S_{13}'S_{26}S_{27a}$
Újfehértói fürtös	SC	$S_1'S_4S_{27b}S_{30}$

[y]Self compatible (SC) or self-incompatible (SI)
[z]Syn. = Crşana, Körös, Köröser Weichsel

If the seed parent is self-compatible, all the anthers from the flowers need to be removed prior to pollen shedding. Emasculation is accomplished by tearing away the calyx cup by pinching it with fingernails or notched scissors. The outer portion of the calyx cup to which the stamens are attached, is excised and the pistil is exposed. It is generally considered that bees will not visit the emasculated flowers since the petals have been removed.

Pollen is collected from unopened flower buds, preferably just before the petals separate. Flowers are rubbed across a fine wire screen sieve and the anthers are collected underneath. The anthers are dried for at least 12 h at 22°C. Dried anthers shed pollen readily, but some crushing increases the yield. Cherry pollen can be stored in the freezer with desiccant for many years. Viability is maximized by keeping the pollen cold and dry when taking it to the field. Pollen is applied to the stigma immediately after emasculation or up to two days after emasculation; pollination 1 day after anthesis is considered to be most effective.

Fruits from crosses are harvested at or slightly before normal harvest maturity. The pits are immediately removed and cleaned of all the flesh. The pits containing the seeds are then soaked in a fungicide treatment and/or a 10%–50% chlorine bleach and water for 2–5 min, followed by a rinse in distilled water. Seeds must be stratified in moist cool conditions for 3–6 months prior to germination. After-ripening proceeds best under uniform moisture conditions at 0°C–5°C. Seedlings can be packaged in plastic bags along with vermiculite or sand. Seeds from early maturing cultivars may germinate poorly since the embryos may have aborted.

Even given sufficient stratification, cherry seed germination can be very poor. For early maturing selections the poor germination is attributed to insufficient embryo growth due to a reduced duration of stage II, the seed maturation period of *Prunus* fruit growth. Therefore, embryo culture is used to 'rescue' the immature embryos. In general the procedure not only involves removing the embryo from the stony endocarp (pit) but also removal of the integument. Embryo culture is most successful if the embryos have a length of 3 to 4 mm (Ivanicka and Pretova 1980). Following placement in embryo culture, the best results occur when the cultures subsequently have a 2–4 month chilling period of 5°C (Bassi et al. 1984).

Even for mid and late maturing varieties, seed germination can be very poor, as low as 10%. The reasons for this are not known, but surely involve a complex combination of environmental and genetic factors. To improve seed germination percentage, various strategies can be implemented. Removing the seed from the pit for those seeds that have not germinated after approximately six months of stratification can result in a flush of germination. Gibberellin soaks have been reported to substitute for the latter half of the cold requirement (Fogle 1958b, Fogle and McCrory 1960).

Seedling vigor and survival is maximized if the seedlings are not removed from stratification until the radical reaches between 0.5 and 1 cM in length. As the seedlings grow very weakly, they should be planted shallowly, just covered with a light layer of soil mix or vermiculite. Commercial trays or plant bands that are

~6 cm in diameter and 25 cm deep are ideal for initial planting. This container maximizes the space for root growth and minimizes the possibility that the seedlings will be over watered, as excess moisture can quickly lead to seedling death. This small initial planting container requires that the seeding be repotted or planted into a seedling nursery prior to field planting. The poor seed germination for cherries plus the care required to nurture those seedlings to a size suitable for field planting has resulted in very small population sizes in cherry breeding programs compared to other fruit breeding programs.

5.6.3 Evaluation Techniques

After field planting, sour and sweet cherry seedlings typically begin flowering in years 3–4, respectively, if growing conditions are optimal. Superior seedlings are selected from segregating populations based upon assessments of fruit quality and disease and pest resistance. These selections are then propagated on one or multiple rootstocks to make trees available for field testing at multiple locations. The time from the initial cross to cultivar release takes approximately 20 years. The use of marker assisted selection and precocious rootstocks hold promise for reducing the time for cultivar development.

5.7 Biotechnological Approaches to Genetic Improvement

5.7.1 Genetic Mapping and QTL Analysis

The construction of genetic maps in cherry has lagged behind that of other *Prunus* species, as cherry is of minor importance compared to peach, and linkage map construction has been difficult due to the poor transferability of peach derived SSRs to cherry. Nevertheless, five partial linkage maps of cherry are available. The first cherry map constructed was from a haploid microspore-derived population from the sweet cherry cultivar 'Emperor Francis' using random amplified polymorphic DNA (RAPD) markers (Stockinger et al. 1996). Two allozymes and 89 RAPD markers were mapped to 10 linkage groups totaling 503 cM. Subsequently, isozyme maps were constructed from two interspecific F_1 cherry progenies: 'Emperor Francis' × *P. incisa* E621 and 'Emperor Francis' × *P. nipponica* F1292 (Boskovic and Tobutt 1998). A total of 47 segregating allozymes were scored of which 34 were aligned into seven linkage groups. The group at East Malling is continuing linkage map construction using an interspecific cross between *P. avium* cv. 'Napoleon' and *P. nipponica* and microsatellite markers (D. Sargent, pers.comm.).

Another intraspecific sweet cherry genetic linkage map is being constructed in INRA of Bordeaux (France) from 133 F_1 individuals from the cross between 'Regina' and 'Lapins'. These cultivars were chosen primarily due to their differences in rain cracking resistance, as 'Regina' is resistant and 'Lapins' is susceptible.

However, they differ for other characters including blooming and maturity dates, peduncle length, and fruit color, weight, firmness, titratible acidity and soluble solids. Partial linkage maps of each parent and their comparison with the *Prunus* reference map, 'Texas' × 'Earlygold' (T×E) are described in Dirlewanger et al. (2004). The 'Regina' and 'Lapins' maps have 30 and 28 SSRs markers, respectively, that were used to test for peach-cherry collinearity. Only one non-collinear marker was detected suggesting a high level of synteny between cherry and peach. However, this conclusion of synteny remains to be rigorously tested.

A sweet cherry genetic linkage map is also being constructed at Michigan State University (U.S.A.) from a F_1 progeny from a cross between a wild forest cherry with a small ($\sim$2 g) highly acid dark-red colored fruit (NY54) and a domesticated variety with large ($\sim$6 g), yellow/pink, sub-acid fruit 'Emperor Francis'. The objective of the study is to identify QTLs that control fruit quality traits that have been improved during domestication. The F_1 population is composed of approximately 600 individuals, 180 of them will be used for map construction and initial QTL analysis. The remaining progeny will be used for fine mapping the major QTL identified. The population will also be used to fine map the *S*-locus region. The advantages of this segregating family are its large family size and the absence of skewed segregation ratios that exist in many of the other *Prunus* mapping populations. This cross is fully compatible and progeny segregation for the *S*-locus fits the expected 1:1:1:1 ratio (Ikeda et al. 2005).

In sour cherry, linkage maps were constructed at Michigan State University from 86 individuals from the cross between two cultivars 'Rheinishce Schattenmorelle' (RS) and 'Erdi Botermo' (EB). Since sour cherry is a tetraploid, informative restriction fragment length polymorphism were scored as single dose restriction fragments according to Wu et al. (1992). Due to the limited number of shared markers between the RS × EB map compared to other *Prunus* maps, putative homologous linkage groups could not be identified for the *Prunus* linkage groups 2, 4, 6 and 7. The other linkage groups were arbitarilty numbered from the longest to shortest. The difficulty of identifying SDRFs and eliminating progeny that resulted from non-homologous pairing illustrates the complexity of linkage mapping in a segmental allotetraploid.

The only QTL study published to date on cherry is an analysis of flower and fruit traits in the sour cherry cross 'Rheinische Schattenmorelle' × 'Erdi Botermo' (Wang et al. 1998). Eleven QTL (LOD > 2.4) were identified for six traits (bloom time, ripening time, % pistil death, % pollen germination, fruit weight, and soluble solids concentration) and the percentage of phenotypic variation explained by a single QTL ranged from 12.9% to 25.9%. At the time the 'Rheinische Schattenmorelle' and 'Erdi Botermo' maps were constructed, *Prunus* SSRs were not widely used, but linkage groups 2, 4, 6 and 7 which contain QTL for bloom date, ripening date, fruit weight and soluble solids were suspected to be homologous to the peach and almond linkage groups 2, 4, 6 and 7 using shared markers.

The identification of bloom time QTL in sour cherry is of particular interest, as sour cherry exhibits extreme diversity for bloom time, with many cultivars blooming exceedingly late in the spring. This late blooming character in sour cherry is likely due to its hybridization and continued introgression with the very late blooming ground cherry, *P. fruticosa*. Since there is high heritability for the bloom time trait ($H_{BS} = 0.91$), increased resolution in the linkage map location of bloom time QTL is a high priority for sour cherry genetic studies.

The identification of two QTL for pistil death indicates that there is genetic variation within sour cherry for the ability of the pistils to survive freezing temperatures (Wang et al. 2000). The two QTL were identified in different years which is logical as the freeze events were totally different. The first QTL, *pd1*, was identified in response to a night of $-10°C$ that occurred 21 days before bloom. The second QTL, *pd2*, was identified in response to two freezing events. The first freezing even occurred 12 days before the population started blooming when the temperature declined to $-2.6°C$ for 11 h. The second freezing event was 4 days after the population started blooming when the air temperature was below $-1.5°C$ for 3 h.

5.7.2 Regeneration and Transformation

Regeneration of cherry has been most successful with sour cherry and rootstock germplasm, as opposed to sweet cherry. James et al. (1984) reported shoot regeneration from leaf disks and internode callus of the cherry rootstock 'Colt' (*P. avium* × *P. pseudocerasus*, $2n = 3x = 24$). Ochatt et al. (1987) successfully regenerated whole plants from mesophyll protoplast of 'Colt' and utilized protoplast cultures of 'Colt' to select for salt and drought tolerance (Ochatt and Power 1989). Regeneration systems for sour cherry have also been developed using leaf disks (Dolgov and Firsov 1999) and cotyledons (Mante et al. 1989, Tang et al. 2000). Song and Sink (2006) transformed the sour cherry 'Montmorency' and the rootstock 'GiSela 6' (P. *cersasus* × *P. canescens*) using *Agrobacterium* and transformants with the GUS reporter gene were obtained with a normal phenotype.

References

Alderman WH, Brierley WG, Trantanella S, Weir TS, Wilcox AN, Winter JD, Hanson KW, Snyder LC (1950) Northstar cherry and Lakeland apple. Minnesota Agr Expt Sta Misc Rep 11

Apostol J (1998) Inheritance of some economically important characteristics in sour cherry population. Acta Hort 468:173–180

Anderson MA, Cornish EC, Wau S-L, Williams EG, Hoggart R, Atkinson A, Bönig I, Grego B, Simpson R, Roche PJ, Haley JD, Penschow JD, Miall HD, Tregear GW, Coughlan JP, Crawford RJ, Clarke AE (1986) Cloning of cDNA for a stylar glycoprotein associated with expression of self-incompatibility in *Nicotiana alata*. Nature 321:38–44

Anonymous (1976) Australia, Queensland Department of Primary Industries Annual Report 1975–1976

Baumann G, Engel G (1986) Clonal selection in *Prunus mahaleb* rootstocks. Acta Hortic 180:91–44

Beaver JA, Iezzoni AF (1993) Allozyme inheritance in tetraploid sour cherry (*Prunus cerasus* L.). J Amer Soc Hort Sci 118:873–877

Bennett M, Watkins R (1976) Pears, cherry and plum rootstocks. Annual Report of the East Malling Research Station for 1975. East Malling Research Station, East Malling, pp 99

Biggs AR, Northover J (1989) Association of sweet cherry epidermal characters with resistance to *Monilinia fructicola*. HortScience 24:126–127

Blažek J (1985) Precocity and productivity in some sweet cherry crosses. Acta Hortic 169:105–113

Börner C (1943) On the problem of breeding against the black cherry aphid. Z Pflanzenkrankh (in German) 53:129–141

Bošković RI, Tobutt KR (1996) Correlation of stylar ribonuclear zymograms with incompatibility alleles in sweet cherry. Euphytica 90:245–250

Bošković RI, Wolfram B, Tobutt KR, Cerović R, Sonneveld T (2006) Inheritance and interaction of incompatibility alleles in the tetraploid sour cherry. Theor Appl Genet 112:315–326

Brenton S (1980) The cherry (in French). Centre Technique Interprofessionnal des Fruits et Legumes 21–31

Brown SK, Bourne MC (1988) Assessment of components of fruit firmness in selected sweet cherry genotypes. HortScience 23:902–904

Brown SK, Iezzoni AF, Fogle HF (1996) Cherries, In: Janick J, Moore JN (eds) Fruit breeding, volume I: tree and tropical fruits, John Wiley & Sons, Inc., NewYork, pp 213–255

Budan S, Mutafa I, Stoian IL, Popescu I (2005a) Screening of 200 sour cherry genotypes for *Monilia laxa* field resistance. Acta Hortic 667:145–151

Budan, S, Mutafa I, Stoian IL, Popescu I (2005b) Field Evaluation of cultivar susceptibility to leaf spot at Romania's sour cherry Genebank. Acta Hortic 667:153–157

Chandra A, Nair MG, Iezzoni A. (1992) Evaluation and characterization of anthocyanin pigments in tart cheries (*Prunus cerasus* L.). J Agric Food Chem 40:967–969

Christensen JV (1996) Rain-induced cracking of sweet cherries: Its causes and prevention. In: Webster AD, Looney NE (eds) Crop Physiology, Production and Uses, CAB International, Oxon, UK, pp 297–327

Cropley R (1968) Testing cherry rootstocks for resistance to infection by raspberry ringspot and cherry leaf roll viruses. Ann Rep E Malling Res Sta for 1967, pp 141–143

Cummings JN (1972) Vegetatively propagated selections of *Prunus fruticosa* as dwarfing stocks for cherry. Fruit Var Hortic Dig 26:76–79

Cummings JN, Wilcox WF, Forsline PL (1986) Tolerance of some new cherry rootstocks to December freezing and to *Phytophthora* root rots. Compact Fruit Tree 19:90–93

de Nettancourt D (2001) Incompatibility and incongruity in wild and cultivated plants. Springer, Berlin

Dirlewanger E, Graziano E, Joobeur T, Garriga-Caldere F, Cosson P, Howad W, Arús P (2004) Comparative mapping and marker-assisted selection in Rosaceae fruit crops. Proc Natl Acad Sci 101:9891–9896

Dolgov SV, Firsov AP (1999) Regeneration and *Agrobacterium* transformation of sour cherry leaf discs. Acta Hortic 484:577–580

Entani T, Iwano M, Shiba H, Che F-S, Isogai A, Takayama S (2003) Comparative analysis of the self-incompatibility (*S-*) locus region of *Prunus mume*: Identification of a pollen-expressed F-box gene with allelic diversity. Genes Cells 8:203–213

FAOSTAT (2004) Food and Agricultural Organization of the United Nations. http://faostat.fao.org/site/336/default.aspx

Filippova EL (1983) New sweet cherry cultivars (in Russian). Sadovodstvo 4:27–28

Fischer M (1996) Resistance breeding in sweet cherries. Acta Hortic 410:87–96

Fogle HW (1958a) Inheritance of fruit color in sweet cherries (*P. avium*). J Hered 49:294–298

Fogle HW (1958b) Effects of duration of after-ripening, gibberellin and other pretreatments on sweet cherry germination and seedling growth. Proc Am Soc Hort Sci 72:129–133

Fogle HW (1961) Inheritance of some fruit and tree characteristics in sweet cherry crosses. Proc Amer Soc Hort Sci 78:76–85

Fogle HW (1975) Cherries. In: Janick J and Moore JN (eds.) Advances in fruit breeding. W. Lafayette, Ind., Purdue Univ Press, pp 348–366

Fogle HW, McCrory CS (1960) Effects of cracking, after-ripening and gibberellin on germination of Lambert cherry seeds. Proc Am Soc Hortic Sci 76:134–138

Fourcade M (1979) Maladies cryptogamiques du feuillage du cerisiers. Proc Journees Fruitieres d'Avignon 43–45

Gao L, Mazza G (1995) Characterization, quantitation, and distribution of anthocyanins and colorless phenolics in sweet cherries. J Agric Food Chem 43:343–346

Garrett CME (1977) Selection of resistant rootstocks. Ann Rept E Malling Res Stat for 1976, 128

Gilmer RM, Blodgett EC (1974) X-disease. In: Virus diseases and noninfections disorders of stone fruit in North America. USDA, Agr. Handbook 437

Grover RK (1963) Soluble solid utilization method for determination of natural resistance in sour cherries against brown rot fungi. Indian Phytopathology 16:205–209

Gruppe A (1988) Investigations on the biology, ecology and taxonomy of the Black Cherry Aphid, *Myzus cerasi* Fabricius 1775, and their interactions with the primary host, *Prunus* species and species hybrids. Thesis, Justus Liebig University, Giessen

Hansche PE, Beres V, Brooks RM (1966) Heritability and genetic correlation in the sweet cherry. Proc Am Soc Hortic Sci 88:173–183

Hauck NR, Ikeda K, Tao R, Iezzoni AF (2006a) The mutated S_1-haplotype in sour cherry has an altered *S*-haplotype specific F-box protein gene. J Hered 97:514–520

Hauck NR, Yamane H, Tao R, Iezzoni AF (2006b) Accumulation of non-functional *S*-haplotypes results in the breakdown of gametophytic self-incompatibility in tetraploid *Prunus*. Genetics 172:1191–1198

Hillig KW, Iezzoni AF (1988) Multivariate analysis of a sour cherry germplasm collection. J Am Soc Hortic Sci 113:928–934

Iezzoni, AF, Andersen RL, Schmidt H, Tao R, Tobutt KR, Wiersma PA (2005) Proceedings of the *S*-allele workshop at the 2001 International Cherry Symposium. Acta Hortic 667:25–35

Iezzoni A, Schmidt H, Albertini A (1990a) Cherries (*Prunus* spp). In: Moore JN, Ballington JR (eds) Genetic resources of temperate fruit and nut crops. Int Soc Hortic Sci, Wageningen, Netherlands, pp 110–173

Iezzoni AF, Schmidt H, Albertini A (1990b) Cherry spp. Genetic resources of temperate fruit and nut crops. Acta Hortic 190:110–173

Ikeda K, Igic B, Ushijima K, Yamane H, Hauck NR, Nakano R, Sassa H, Iezzoni AF, Kohn JR, Tao R (2004a) Primary structure futures of the *S* haplotype-specific F-box protein, *SFB*, in *Prunus*. Sex Plant Reprod 16:235–243

Ikeda K, Ushijima K, Yamane H, Tao R, Hauck NR, Sebolt AM, Iezzoni AF (2005) Linkage and physical distances between the *S*-haplotype *S-RNase* and *SFB* genes in sweet cherry. Sex Plant Reprod 17:289–296

Ikeda K, Watari A, Ushijima K, Yamane H, Hauck NR, Iezzoni AF, Tao R (2004b) Molecular markers for the self-compatible $S^{4'}$-haplotype, a pollen-part mutant in sweet cherry (*Prunus avium* L.). J Am Soc Hortic Sci 129:724–728

Ivanicka J, Pretova A (1980) Embryo culture and micropropagation of cherries in vitro. Sci Hortic 12:77–82

James DJ, Passey AJ, Malhotra SB (1984) Organogenesis in callus derived from stem and leaf tissues of apple and cherry rootstocks. Plant Cell Tissue Organ Cult 3:333–341

Jenser G, Sheta IB (1969) Investigation of the resistance of a few Hungarian sour cherry hybrids against the San Jose scale (*Quadraspidiotus perniciosus*). Acta Phytopathol 4:313–315

Kao T-h, Tsukamoto T (2004) The molecular and genetic basis of S-RNase-based self-incompatibility. Plant Cell 16(Suppl):S72–S83

Kegler H, Schimanski HH, Verderevskaja TD, Trifonov D (1978) On the importance of the tolerance of woody plants to virusese and mycoplasmas. Nachrichtenblatt f.d. Pflanzenschutz id DDR 32:248–251

Kolesnikova AF (1975) Breeding and some biological characteristics of sour cherry in central Russia zone of the RSRSR, p 328

Kolesnikova AF, Ossipov Y, Ossipov V, Kolesnikova AI (1985) New hybrid rootstock for cherries. Acta Hortic 169:159–162

Lamb RC (1953) Notes on the inheritance of some characters in the sweet cherry *Prunus* avium. Proc Am Soc Hortic Sci 61:293–298

Lang G, Howell W, Ophardt D (1998) Sweet cherry rootstock/virus interactions. Acta Hortic 468:307–314

Lansari A, Iezzoni A (1990) A preliminary analysis of self-incompatibility in sour cherry. HortScience 25:1636–1638

Lapins K (1970). The Stella cherry. Fruit Var Hortic Dig 24:19–20

Lewis D, Crowe CK (1954) Structure of the incompatibility gene. IV Types of mutations in *Prunus avium* L. Heredity 8:357–363

Livermore JR, Johnstone FE (1940) The effect of chromosome doubling on the crossability of *Solanum chacoense*, *S. jamesii* and *S. bulbocastanum* with *S. tuberosum*. Am Potato J 17:170–173

McClure BA, Gray JE, Anderson MA, Clarke AE (1990) Self-incompatibility in *Nicotiana alata* involves degradation of pollen rRNA. Nature 347:757–760

McClure BA, Haring V, Ebert PR, Anderson MA, Simpson RJ, Sakiyama R, Clarke A (1989) Style self-incompatibility gene products of *Nicotiana alata* are ribonucleases. Nature 342:955–957

McCubbin AG, Kao T-H (2000) Molecular recognition and response in pollen and pistil interactions. Ann Rev Cell Dev Biol 16:333–364

Mante SR, Scorza R, Cordts JM (1989) Plant regeneration from cotyledons of *Prunus persica*, *P. domestica* and *Prunus cerasus*. Plant Cell Tissue Organ Cult 19:1–11

Marshall RE (1954) Cherries and cherry products. In: Economic crops, vol 5. Interscience, New York

Mathers HM (2004a) Supercooling and cold hardiness in sour cherry germplasm: Flower buds. J Am Soc Hortic Sci 129:675–681

Mathers HM (2004b) Supercooling and cold hardiness in sour cherry germplasm: Vegetative tissue. J Am Soc Hortic Sci 129:682–689

Matthews P (1968) Breeding for resistance to bacterial canker in the sweet cherry. Proc Symposium on cherries and cherry growing (Bonn):153–164

Matthews P (1973a) Recent advances in breeding sweet cherries at the John Innes Institute. In: Napier E (ed) Fruit present and future, vol. II. W. Clowes & Son LTD, London

Matthews P (1973b) Some recent advances in sweet cherry genetics and breeding. Proc Eucarpia Fruit Section, Canterbury, pp 84–107

Matthews P (1979) Progress in breeding cherries for resistance to bacterial canker. Proc Eucarpia Fruit Breeding Symposium (Angers) 157–174

Micke WC, Doyle JF, Yeager JT (1983) Doubling potential of sweet cherry cultivars. California Agric., March–April 24–25

Mircetich SM, Matheron ME (1981) Differential resistance of various cherry rootstocks to *Phytophthlogy* species. Phytopathlogy 71:243

Moore JD, Berger RD, Ravenscroft VA (1963) Plant introduction of sour cherry selections – probably sources of brown rot resistance. Phytopathology 53:883

Ochatt SJ, Cocking EC, Power JB (1987) Isolation, culture and plant regeneration in Colt (*Prunus avium* × *pseudocerasus*) cherry protoplasts. Plant Sci 50:139–143

Ochatt SJ, Power JB (1989) Selection for salt and drought tolerance in protoplast and explant-derived tissue cultures of Colt cherry (*Prunus avium* × *pseudocerasus*). Tree Physiol 5:259–266

Olden EJ, Nybom N (1968) On the origin of *Prunus cerasus* L. Hereditas 59:327–345

Olmstead JW (2006) Linkage map construction and analysis of fruit size in sweet (*Prunus avium* L.) and sour cherry (*Prunus cerasus* L.). PhD Thesis, Michigan State University, p 166

Olmstead JW, Lang GA (2002). *Pmr*1, a gene for powdery mildew resistance in sweet cherry. HortScience 37:1098–1099

Olmstead JW, Lang GW, Grove GG (2001) Inheritance of powdery mildew resistance in sweet cherry. HortScience 36:337–340

Patten KD, Patterson ME, Kupferman M (1983) Reducing sweet cherry surface pitting. Coop Ext Bull EB 1219. Tree Fruit Production Series. Washington State University of Pullman

Posnette AF, Cropley R, Swait AAJ (1968) The incidence of virus diseases in English sweet cherry orchards and their effect on yield. Ann Appl Biol 61:352–360

Proffer TJ, Jones AL, Perry RL (1988) Testing of cherry rootstocks for resistance to infection by species of *Armillaria*. Plant Dis 72:488–490

Redalen G (1984) Fertility in sour cherries. Gartenbauwissenschaft 49:212–217

Rozsnyay ZS, Apostol J (2005) Breeding for sweet and sour cherry disease resistance in Hungary. Acta Hortic 667:117–122

Sansavini S, Lugli S (1996) Self-fertility, compact spur habit and fruit quality in sweet cherry: Preliminary findings of the University of Bologna's breeding program. Acta Hortic 410:51–64

Sansavini S, Lugli S (2005) Sweet cherry breeding programs & new varieties released in Europe and Asia. 5th International Cherry Symposium, Bursa, Turkey

Sansavini S, Lugli S, Lugli A, Pancaldi DM (1998) Breeding sweet cherry for self-fertile, compact/spur tree habit and high quality fruits: Trait segregation. Acta Hortic 468:45–51

Sassa H, Hirano H, Ikehashi H. (1992) Self-incompatibility-related RNases in styles of Japanese pear (*Pyrus serotina* Rehd.). Plant Cell Phys 105:751–752

Schmidt H (1985) First results on a trial with new cherry hybrid rootstock candidates at Ahrensburg. Acta Hortic 169:235–243

Schmidt H (1998) On the genetics of fruit colour in sweet cherry. Acta Hortic 468:77–81

Schuster M (2000) Genome investigation of sour cherry, *Prunus cerasus* L. Acta Hortic 538:375–379

Schuster M, Wolfram B (2005) Sour cherry breeding at Dresden-Pillnitz. Acta Hortic 667:127–130

Sjulin TM, Jones AL, Andersen RL (1989) Expression of partial resistance to cherry leaf spot in cultivars of sweet, sour, duke and European ground cherry. Plant Dis 73:56–61

Song G-Q, Sink KC (2006) Transformation of Montmorency sour cherry (*Prunus cerasus* L.) and Gisela 6 (*P. cerasus* × *P. canescens*) cherry rootstock mediated by *Agrobacterium tumefaciens*. Plant Cell Rep 25:117–123

Sonneveld T, Robbins TP, Boskovic R, Tobutt KR (2001) Cloning of six cherry self-incompatibility alleles and development of allele-specific PCR detection. Theor Appl Genet 102:1046–1055

Sonneveld T, Tobutt KR, Robbins TP (2003) Allele-specific PCR detection of sweet cherry self-incompatibility *S*-alleles in cherry and high throughput genotyping by automated sizing of first intron PCR products. Plant Breed 125:305–307

Sonneveld T, Tobutt KR, Robbins TP (2005a) Allele-specific PCR detection of sweet cherry self-incompatibilty (S) alleles S_1 to S_{16} using consensus and allele-specific primers. Theor Appl Genet 107:1059–1070

Sonneveld T, Tobutt KR, Vaughan SP, Robbins TP (2005b) Loss of pollen-*S* function in two self-compatible selections of *Prunus avium* is associated with deletion/mutation of an *S* haplotype-specific F-box gene. Plant Cell 17:37–51

Stockinger EJ, Mulinix CA, Long CM, Brettin TS, Iezzoni AF (1996) A linkage map of sweet cherry based on RAPD analysis of a microspore-derived callus culture population. J Hered 87:214–218

Strauch H, Gruppe W (1985) Results of laboratory tests for winter hardiness of *Prunus avium* cultivars and interspecific cherry hybrids (*Prunus* × spp.). Acta Hortic 169:281–287

Sundin GW, Jones AL, Fulbright DW (1989) Copper resistance in *Pseudomonas syringae* pv. *syringae* from cherry orchards and its associated transfer in vitro and *in planta* with a plasmid. Phytopathology 79:861–865

Tao R, Yamane H, Sugiura A, Murayama H, Sassa H, Mori H (1999) Molecular typing of *S*-alleles through identification, characterization and cDNA cloning for S-RNases in sweet cherry. J Am Soc Hortic Sci 124:224–233

Tang H-R, Ren Z-L, Krczal G (2000) Somatic embryogenesis and organogenesis from immature embryo cotyledons of three sour cherry cultivars (*Prunus cerasus* L.). Scientia Hortic 83:109–126

Tehrani G (1984) 'Viscount' sweet cherry. HortScience 19:451

Tobutt KR, Boskovic R, Cerovic R, Sonneveld T, Ruzic D (2004) Identification of incompatibility alleles in tetraploid species sour cherry. Theor Appl Genet 108:775–785

Tobutt KR, Boskovic R, Sonneveld T (2001) Cherry (in) compatibility genotypes – harmonization of recent results from UK, Canada, Germany, Japan and USA. Eucarpia Fruit Breed Sect Newsletter 5:41–46

Toyama TK, Ophardt DR, Howell WE, Grove GG (1993) New powdery resistant sweet cherry. Fruit Var J 47:234–235

Tukey HB (1933) Artificial culture of sweet cherry embryos. J Hered 24:7

Tsukamoto T, Hauck NR, Tao R, Jiang N, Iezzoni AF (2006) Molecular characterization of three non-functional *S*-haplotypes in sour cherry (*Prunus cerasus*). Plant Mol Biol 62:371–383

Uphoff H, Eppler A, Gruppe W (1988) Reaction patterns of some cherry hybrid rootstock clones towards infection with a PNRSV and a PDV isolate. Proc Intern Symp on Crop Protection, Gent, pp 491–498

Ushijima K, Sassa H, Dandekar AM, Gradziel TM, Tao R, Hirano H (2003) Structural and transcriptional analysis of the self-incompatibility locus of almond: Identification of a pollen-expressed F-box gene with haplotype-specific polymorphism. Plant Cell 15:771–781

Ushijima K, Sassa H, Tao R, Yamane H, Dandekar AM, Gradziel TM, Hirano H (1998) Cloning and characterization of cDNAs encoding *S*-RNases in almond (*Prunus dulcis*): primary structure features and sequence diversity of the *S*-RNases in Rosaceae. Mol Gen Genet 260:261–268

Ushijima K, Yamane H, Watari A, Kakehi E, Ikeda K, Hauck NR, Iezzoni AF, Tao R (2004) The *S*-haplotype-specific F-box protein gene, *SFB*, is defective in self-compatible haplotypes of *Prunus avium* and *P. mume*. Plant J 39:573–586

Vavilov NI (1951) The origin, variation, immunity and breeding of cultivated plants. Ronald, NewYork

Voronin EI (1981) Resistance of sour cherry varieties to *Monilia* in the Crimea (in Russian). S'ezd genetikov I selektsionerov Ukraine, Odessa 4:68–69

Voronin EI, Stepanova AK (1980) Varietal resisatnce of sweet cherries and peaches to Cytospora in the Crimea (in Russian). Byulleten 'Vsesoyuznogo Ordena Lenina Institutat Rastenievodstva imeni N.I. Vavilov 103:33–36

Wadley BN (1970) Developments in disease resistant sweet cherries. Proc Utah Sate Hortic Soc, pp 46–48

Wang D, Karle R, Brettin TS, Iezzoni AF (1998) Genetic linkage map in sour cherry using RFLP markers. Theor Appl Genet 97:1217–1224

Wang D, Karle R, Iezzoni AF (2000) QTL analysis of flower and fruit traits in sour cherry. Theor Appl Genet 100:535–544

Wharton PS, Iezzoni A, Jones AL (2003) Screening of cherry germplasm for resistance to leaf spot. Plant Dis 87:471–477

Wiersma PA, Wu Z, Zhou L, Hampson C, Kappel F (2001). Identification of new self-incompatibility alleles in sweet cherry (*Prunus avium* L.) and clarification of incompatibility groups by PCR and sequencing analysis. Theor Appl Genet 102:700–708

Wünsch A, Hormaza JI (2004) Genetic and molecular analysis in Cristobalina sweet cherry, a spontaneous self-compatible mutant. Sex Plant Reprod 17:203–210

Xue Y, Carpenter R, Dickinson HG, Coen ES (1996) Origin of allelic diversity in *Antirrhinum S* locus RNases. Plant Cell 8:805–814

Yamane H, Ikeda K, Hauck NR, Iezzoni AF, Tao R (2003a) Self-incompatibility (*S*) locus region of the mutated S^6-haplotypes of sour cherry (*Prunus cerasus*) contains a functional pollen *S*-allele and a non-functional pistil *S* allele. J Exp Bot 54:2431–2434

Yamane H, Ikeda K, Ushijima K, Sassa H, Tao R (2003b) A pollen-expressed gene for a novel protein with an F-box motif that is very tightly linked to a gene for S-RNase in two species off cherry, *Prunus cerasus* and *P. avium*. Plant Cell Physiol 44:764–769

Yamane H, Tao R, Sugiura A, Hauck NR, Iezzoni AF (2001) Identification and characterization of
    S-RNases in tetraploid sour cherry (*Prunus cerasus*). J Am Soc Hortic Sci 126:661–667
Zwintzscher M (1963) Sekunda. Der Erwerbsobstbau 5:101–103
Zwintzscher M (1966) Handbuch der Pflanzenzuchtung VI. Steinobst, Kirschen 573–602
Zwintzscher M (1979) Twenty-five years of sour cherry breeding in Germany. A survey. Proceed-
    ings of Eucarpia Fruit Symposium (Angers), pp 223–227

acid (vitamin C). The development of market-assisted

Chapter 6
Currants and Gooseberries

R.M. Brennan

Abstract The cultivation of *Ribes* fruits (black- and redcurrants and gooseberries) is aimed at both fresh and processing markets, with the blackcurrant *R. nigrum* particularly important in the latter. Breeding is increasingly focused on national and regional requirements, and for blackcurrant the specific quality requirements of the processing industry are key objectives in many programmes, alongside agronomic traits such as yield and pest resistance. Durable resistance to foliar pathogens and damaging pests such as gall mite remains a high priority, partly due to increasing interest in integrated crop management systems. Many *Ribes* species, especially the darker-fruited blackcurrant types, contain high concentrations of polyphenolic compounds, notably anthocyanins and flavonols, and these components are of growing importance due to their link to human health, together with high levels of ascorbic acid (vitamin C). The development of marker-assisted breeding strategies in *Ribes* is in progress, to improve breeding efficiency and time to cultivar, and QTLs affecting several important phenological, agronomic and fruit quality traits have been located on the recently-developed linkage map for blackcurrant. Additionally, markers linked to key traits such as gall mite resistance are under development.

6.1 Introduction

The crops within the *Ribes* genus have in recent years been the subject of increased interest, manly due to the perceived health benefits associated with their consumption. The genus comprises over 150 species of shrubs, some with nodal spines, native throughout northern temperate Europe, North America and Asia, and in mountain regions of South America and North Africa. Of the available germplasm within the genus, only about 10–12 species of *Ribes* have been used as the primary genepool for the breeding of cultivars (Hummer 2006).

The cultivated forms of blackcurrant are mainly derived from the species *Ribes nigrum* L. (blackcurrant) and its subspecies, whilst cultivated redcurrants and

R.M. Brennan
Scottish Crop Research Institute, Invergowrie, Dundee DD2 5DA, Scotland, UK
e-mail: rex.brennan@scri.ac.uk

J.F. Hancock (ed.), *Temperate Fruit Crop Breeding*,
© Springer Science+Business Media B.V. 2008

whitecurrants are based on the species *R. sativum* Syme (= *R. vulgare* Jancz.), *R. petraeum* Wulf. and *R. rubrum* L. In the case of gooseberries, the main species are *R. grossularia* L. (= *R. uva-crispa* L.), the European gooseberry, and *R. hirtellum* Michx., the American gooseberry. In all of these crop types, the involvement of other species within the genus is increasingly prevalent in the breeding process, predominantly as donors of important and commercially useful traits.

Production of blackcurrant in Europe is on large plantations with a fully mechanised harvest, and much of the fruit is now used for juice production and other processing applications. However, there is a small but increasing fresh market sector, including some protected cropping in some parts of Europe such as Belgium. The cultivars used for processing are generally different from fresh market types, with widely divergent characteristics and requirements. Poland is the largest world producer of blackcurrant in recent years, followed by Russia, the United Kingdom and Scandinavia, with some production also taking place in New Zealand.

Redcurrant production is generally harvested by hand and the processing market is mainly focused on jams and preserves. The main producers are Poland, Holland, Belgium, Germany and France, with very small quantities now appearing in Chile. In gooseberry, production is mainly for the fresh market with some use as pie fillings etc., and the main producing countries are Poland, Hungary and the Baltic states.

Hybrids between *Ribes nigrum* and *R. grossularia* have limited popularity commercially, mainly as the cultivar 'Josta' from a colchicine-derived amphidiploid between these two species developed by Bauer (1973) as *R. nidigrolaria*. The plants, whilst spine-free, resistant to foliar diseases and vigorous, are mainly used for self-pick outlets.

6.2 Taxonomy, Evolutionary Biology and Germplasm Resources

Ribes is usually regarded as a member of the Saxifragaceae (Engler and Prantl 1891), and as such it has few related crop genera. However, more recent studies by Cronquist (1981) and Sinnott (1985) place the genus in the Grossulariaceae due to its floral morphology. All species are diploids, and considerable debate has been generated over the past century whether the currants and gooseberries form a single genus, *Ribes* (as proposed by Janczewski, 1907) or two, *Ribes* and *Grossularia* (as proposed by Berger, 1924). A single genus is now the most common consensus, based on morphological, crossability and molecular grounds (Senters and Soltis, 2003).

The principal evolutionary pressure in *Ribes* is considered by Sinnott (1985) to be geographical adaptation, with associated cytological uniformity and lack of crossability barriers within subgenera. The main centres of diversity for the currants are found in northern Scandinavia and Russia, into parts of Asia, whilst for the gooseberry species it is North America and parts of continental Europe.

Molecular systematics studies of *Ribes* by Messenger et al. (1999), looking at restriction site variation in two cpDNA regions in *Ribes* subgenera, led to their proposal of two possible evolutionary patterns – long periods of stasis broken by sudden radiation of species or gene flow due to hybridisation leading to diversification (Hummer 2006).

Species of the Eucoreosma subgenus that are considered by Keep (1995) of importance in the past and future evolution of blackcurrants include *R. nigrum* and its subspecies *europaeum*, *scandinavicum* and *sibiricum* (ranging from northern Europe, central and northern Asia to the Himalayas), *R. dikuscha* Fisch. (central to eastern Russia), *R. ussuriense* Jancz. (Manchuria to Korea), *R. bracteosum* Dougl. (Alaska to northern California) and *R. petiolare* Dougl. (North America). These species have been identified as the source of useful genes in regard to resistance and fruit quality traits (Table 6.1).

Redcurrant species of the subgenus Ribesia that are important in the crop's evolution include *R. sativum* (from western Europe), *R. petraeum* (montane regions of Europe, north Africa and Siberia), *R. rubrum* (central and northern Europe and northern Asia), *R. multiflorum* Kit. (southern Europe) and *R. longeracemosum* (Asia) (Keep 1995). The continued use of these species in breeding is mainly aimed at the production of germplasm with longer strigs (floral/fruiting racemes) and larger berries (Table 6.2).

In gooseberry, *R. grossularia* is the main species that has been utilized in cultivar development in Europe, although work in North America has used *R. hirtellum*, *R. divaricatum* Dougl. and *R. oxyacanthoides* L. in conjunction with *R. grossularia* to increase berry size (Table 6.3).

Natural polyploidy occurs only rarely in *Ribes* (Brennan 1996), although experimentally induced polyploids have been utilised by Nilsson (1959 and 1966) and Keep (1962), mainly in the development of allotetraploids through interspecific hybridization.

6.3 History of Improvement

6.3.1 Blackcurrant

Domestication of blackcurrant has occurred within the last 400 years (Keep 1995), and first records in the U.K. are found in seventeenth century herbals, such as Gerard (1636) and Parkinson (1629), reflecting the medicinal properties of the plant (Roach 1985). By 1826, five cultivars of blackcurrant were described by the Royal Horticultural Society in the U.K., and these had increased to a total of 26 by 1920 (Hatton 1920), mainly due to the efforts of private individuals and nurserymen. From this time, the numbers of cultivars have steadily increased throughout Europe, with releases from several active breeding programmes, often in state-funded institutes. The most important cultivar in the U.K. from the late nineteenth century until the 1980s was 'Baldwin', a seedling of unknown provenance valued for its flavour characteristics. After the development and release of a number of cultivars derived from 'Baldwin', there was a move towards the incorporation of

Table 6.1 Sources of key traits in blackcurrant breeding (adapted from Barney and Hummer, 2005)

Trait	Species
Pests	
Resistance to gall mite	*R. grossularia*
	R. glutinosum
	R. pauciflorum
Resistance to leaf curling midge	*R. dikuscha*
Resistance to aphids	*R. glutinosum*
	R. sanguineum
Diseases	
Resistance to leaf spot (*Drepanopeziza ribis*)	*R. dikuscha*
	R. nigrum var.
	sibiricum
Resistance to leaf spot	*R. americanum*
(*Mycosphaerella ribis*)	*R. pauciflorum*
Resistance to powdery mildew	*R. americanum*
	R. carrierei
	R. sanguineum
	R. dikuscha
	R. nigrum var. *scandinavicum*
	R. petiolare
	R. pauciflorum
	R. glutinosum
Resistance to white pine blister rust	*R. ussuriense*
Resistance to BRV*	*R. dikuscha*
	R. fuscescens
	R. nigrum var. *sibiricum*
	R. procumbens
Agronomic and phenological traits	
Winter hardiness	*R. nigrum* var. *sibiricum*
	R. dikuscha
	R. pauciflorum
	R. procumbens
Growth habit	*R. bracteosum*
	R. sanguineum
	Ribesia sp.
Early ripening	*R. dikuscha*
	R. nigrum var. *sibiricum*
Late ripening	*R. bracteosum*
	R. palczewskii
	R. manshuricum
Strig length	*R. bracteosum*
	R. fuscescens
Fruit traits	
Ascorbic acid content	*R. nigrum* var. *sibiricum*
	R. dikuscha
	R. pauciflorum
Resistance to premature fruit shedding ('run off')	*R. hudsonianum*

* BRV = Blackcurrant Reversion Virus

Table 6.2 Sources of key traits in redcurrant breeding (adapted from Barney and Hummer, 2005)

Trait	Species
Diseases	
Resistance to mildew	*R. multiflorum*
	R. petraeum
	R. warsewiczii
	R. longeracemosum
Resistance to leafspot	*R. petraeum*
(*Drepanopeziza ribis*)	*R. rubrum*
	R. pubescens
	R. warsewiczii
	R. longeracemosum
	R. moupinense
	R. multiflorum
Resistance to white pine blister rust	*R. petraeum*
	R. rubrum
Agronomic traits	
Winter hardiness	*R. rubrum*
	R. triste
Late flowering	*R. petraeum*
	R. manschuricum
	R. multiflorum
Yield potential	*R. multiflorum*
Fruit-related traits	
No. of strigs/flowers & strig length	*R. multiflorum*
Fruit size	*R. macrocarpum*

new germplasm especially from northern regions (Tydeman 1938). This can be seen in the use of germplasm from the Nordic countries, where the cultivar 'Brödtorp' (Finland) and the Swedish 'Öjebyn' and 'Sunderbyn II' were selected from wild populations, and led to an increase in late-flowering frost-tolerant cultivars in other parts of Europe, notably in the U.K. Elsewhere in Europe, breeders of blackcurrant in Russia have utilized wild germplasm of *R. nigrum* var. *sibiricum, R. pauciflorum* and *R. dikuscha*, initially leading to cultivar releases such as 'Primoskij Čempion'. Subsequent breeding within state institutions has continued to produce a large number of cultivars with local adaptation, from programmes based across the Russian Federation including St. Petersburg, Orel, Minchurinsk, and Siberia.

In the U.S.A., blackcurrants were probably introduced along with redcurrants in the seventeenth century, but received little attention in terms of domestication (Barney and Hummer 2005). Additionally, in the early 20th century, legislative restrictions on the growing of blackcurrants due to the threats posed by white pine blister rust (see below) limited the development of blackcurrant growing and cultivar production. However, breeding of blackcurrants in Canada in the 1930s, using *R. ussuriense* as a source of rust resistance, led to the release of a series of resistant cultivars including 'Consort'. The introduction of newer resistant cultivars such as 'Titania' from Europe and changes in legislative restrictions has led to increasing interest in blackcurrant in the U.S.A.

Table 6.3 Sources of key traits in gooseberry breeding (adapted from Barney and Hummer, 2005)

Trait	Species
Pests	
Resistance to aphids	*R. alpestre*
	R. leptanthum
	R. watsonianum
Diseases	
Resistance to leafspot (*Drepanopeziza ribis*)	*R. divaricatum*
Resistance to leafspot (*Mycosphaerella ribis*)	*R. divaricatum*
Resistance to powdery mildew	*R. divaricatum*
	R. hirtellum
	R. oxyacanthoides
	R. leptanthum
	R. watsonianum
Resistance to GVBV*	*R. divaricatum*
Agronomic traits	
Winter hardiness	*R. divaricatum*
	R. aciculare
	R. burejense
Growth habit	*R. watsonianum*
	R. leptanthum
Spinelessness	*R. oxyacanthoides*
	R. cynosbati
	R. inerme
	R. robustum
Ease of propagation	*R. hirtellum*
	R. divaricatum
Fruit-related traits	
Late flowering	*R. divaricatum*
Colour (dark fruit)	*R. cynosbati*
	R. niveum
	R. robustum

* GVBV = Gooseberry Vein-Banding Virus

6.3.2 Red- and Whitecurrants

The first reports and descriptions of redcurrants originate from Germany in the early 15th century, from where its occurrence spread to France and the remainder of Europe (Brennan 1996). These first reports are thought to refer to *R. sativum*; *R. petraeum* was not introduced to the U.K. until the early 17th century, and the other main redcurrant species *R. rubrum* was introduced later still. Cultivars derived from *R. rubrum,* such as 'Houghton Castle' and 'Raby Castle' were introduced in the early 19th century, and by the 1920s more than 30 cultivars of redcurrant were identified in the U.K. Subsequent development of redcurrant has been more restricted than for blackcurrants. The cultivar 'Laxton's No. 1' was introduced in the U.K. in the early 20th century, and from the U.S.A. the cultivars 'Red Lake' and 'Fay's Prolific' were produced; all three are still popular today. Much of the more recent development of this crop has been undertaken in Holland, resulting in cultivars such as 'Jonkheer van Tets', 'Rovada' and 'Rondom', the latter including R. *multiflorum* in its parentage from which its late flowering and fruiting is derived.

Whitecurrants were described from Holland in the mid-seventeenth century (Hedrick 1925), but subsequent development has been erratic, with fluctuating interest in both whitecurrants and other colour forms such as pink variants, although several of the latter are listed in 18th and 19th century fruit catalogues (Keep 1995). The most widespread cultivars at present are 'White Versailles', a French seedling of unknown parentage from the late 19th century, and the more recent 'Zitavia' from Holland, although cultivars such as 'White Dutch', first described in 1778, can still be found.

6.3.3 Gooseberry

The first records of gooseberry in the U.K. is from the Middle Ages, when they were included in plants supplied in 1275 to Edward I from France (Roach 1985). By 1778 over 20 cultivars were described by Mawe (1778, quoted by Hedrick 1925), and from this time until the early 20th century a large number of cultivars were developed by amateur 'gooseberry clubs' in the Midlands and north of England – over 1000 cultivars by 1925 (Hedrick 1925). Whilst few of these cultivars survive today, some old cultivars are still popular, including 'Whinham's Industry' and 'Careless', both of which were developed in the nineteenth century.

Gooseberry production went into decline due to the arrival in the U.K. of American gooseberry mildew [*Sphaerotheca mors-uvae* (Schw.) Berk.] in 1905, and the cheaper growing and labour costs associated with the crop in eastern Europe have further reduced its potential in recent times.

In the U.S.A., cultivar development was initially based on R. *hirtellum*, with early cultivars such as 'Poorman' and 'Houghton'. The use of other native species such as R. *oxyacanthoides* produced cultivars such as 'Spinefree' and 'Captivator' with reduced or absent nodal spines. Most of the American cultivars have a small berry size, and subsequent development has been directed towards using European germplasm to increase this.

6.4 Current Breeding Efforts

There are at present around 15 active *Ribes* breeding programmes worldwide, most of them focused on blackcurrant cultivar production. The breeding of redcurrants and gooseberry is now generally located in Eastern Europe and Russia, producing cultivars for local production and market conditions. The Polish programme, at the Research Institute of Pomology in Skierniewice, is active in generating both new gooseberry and blackcurrant cultivars. The breeding programmes are for the most part publicly-funded, although the blackcurrant programme based at the Scottish Crop Research Institute has had commercial funding since 1990 and the New Zealand blackcurrant programme, based at HortResearch near Christchurch, also has an industrial component in the funding.

Ribes breeding in recent years has become more focused on national and regional requirements, and for blackcurrant the specific quality requirements of the processing industry are key objectives in many programmes. The 'Ben' series of cultivars from the Scottish Crop Research Institute began in 1972 with the release of 'Ben Lomond' as a spring frost-escaping alternative to 'Baldwin' and other sensitive types; this was based on the incorporation of late-flowering Scandinavian germplasm into the higher quality central European types. Further releases in the series have proved commercially popular, and the most widely grown cultivars in the U.K. are the partially gall mite-resistant 'Ben Hope' (a complex cross between 'Westra', a mutant form of the older 'Westwick Choice' with a very upright growth habit, and a blackcurrant × gooseberry hybrid) and the reversion virus-resistant 'Ben Gairn' ('Ben Alder' × 'Golubka'). New additions to the 'Ben' series combining high juice quality with resistance to key pests and diseases, notably gall mite, are currently in trials.

Other active blackcurrant breeding programmes are found in Poland, Estonia, Lithuania, Russia and New Zealand. The overproduction of blackcurrants in Europe during the period 2003–5 reduced retail prices for the fruit, with the result that the aim now is for cultivars producing high quality and nutritionally improved fruit, often to specifications set by end-users, grown in low-input systems.

6.5 Genetics of Key Traits

6.5.1 Disease and Pest Resistance

Gall mite (*Cecidophyopsis ribis* Westw.) infestation of blackcurrant, causing the typical 'big bud' symptoms (Fig. 6.1), remains the most serious problem facing commercial blackcurrant growers in most areas where the crop is grown, although no reports of this pest have been noted from Tasmania or from North America. The reduction in available chemical controls in most parts of Europe has placed increasing importance on the development of resistant cultivars of blackcurrant (Brennan 1996), and resistance is available from other *Ribes* species, notably *R. grossularia* (Knight et al. 1974) and some accessions of *R. pauciflorum* Turcz. and *R. procumbens* Pall. (Sabitov et al. 2002). Mite resistance has also been reported from *R. nigrum* var. *sibiricum*, controlled by a single gene designated P, although several of the cultivars containing this gene have not proved to be fully resistant in northern Europe. The resistance in *R. grossularia* is controlled by a single gene Ce, and its introgression into blackcurrant was achieved by the development of resistant F_1 allotetraploids (Knight et al. 1974) using colchicine, followed by an extensive

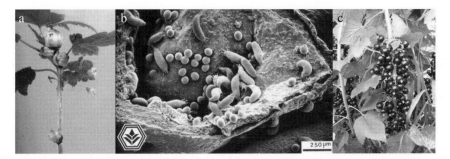

Fig. 6.1 Gall mite (*Cecidophyopsis ribis*) on blackcurrant: (**a**) Stem of susceptible blackcurrant cv. showing typical 'big bud' symptoms, (**b**) SEM of mite-infested blackcurrant bud, and (**c**) Mite-resistant blackcurrant seedling SCRI 8872-1, a BC_6 selection containing the resistance gene *Ce* from *R. grossularia*

backcrossing programme. The continuing development of the backcross progenies to BC_7 and beyond at the Scottish Crop Research Institute is now producing resistant hybrids of commercial quality.

The blackcurrant leaf-curling midge (*Dasyneura tetensi* Rübs.) is becoming more important in commercial plantations due to a reduction in broad-spectrum pesticide applications and increase in integrated management systems. There are clear cultivar differences in sensitivity to this pest, with cultivars 'Ben Alder' and 'Ben Tirran' showing high susceptibility. Within the Eucoreosma, the species *R. americanum* and *R. dikusha* are resistant, together with cultivars such as 'Ben Connan'. Resistance is controlled by a single gene, designated *Dt*, (Keep 1985), and larval antibiosis appears to be the main mechanism for resistance (Crook et al. 2001)

Clearwing (*Synanthedon tipuliformis* Clerck.) affects black-, red- and whitecurrants in many production areas, especially New Zealand, with redcurrants particularly susceptible. There are noted differences in cultivar and species susceptibility (Cone 1967), with the redcurrant species *R. multiflorum* and some of its derivatives found by Hummer and Sabitov (2004) to be resistant. Jermyn (2002) hypothesized that the composition of volatiles may form the basis for resistance, and that this information could be developed into potential semiochemical control strategies and also in the identification of appropriate parental material for the breeding of resistant blackcurrant cultivars.

Mildew (*Sphaerotheca mors-uvae* (Schw.) Berk.) is a common foliar problem in many *Ribes* species and cultivars, although there are now robust sources of resistance. The disease has been one of the main limiting factors to commercial gooseberry production in the 19th and 20th centuries, and in blackcurrant there are several instances, e.g. 'Ben Lomond', where resistance has broken down fairly quickly. A range of resistance genes are available to breeders: the *Sph* series, including Sph_1 from *R. oxyacanthoides* (Keep 1974) and Sph_2 from the Swedish cultivar 'Öjebyn' are all dominant genes, whilst the resistance in the Finnish 'Brödtorp' was thought by Ravkin (1986) to be controlled by two complementary genes. In the breeding of blackcurrant cultivars, *R. americanum* is now being investigated as a potential sources of resistance in interspecific hybrids (Siksnianas et al. 2005), and some of the older Scandinavian types such as 'Sunderbyn II' and their derivatives remain useful parental material for mildew resistance breeding (Anderson 1969), although most recent cultivar releases also show high levels of resistance. Keep (1985) found evidence of a negative linkage between gall mite resistance (controlled by Ce) and mildew resistance controlled by Sph_2.

In redcurrant, mildew resistance was found by Keep (unpublished, reported in Brennan 1996) to be controlled by one or more dominant major genes, with resistance found in *R. multiflorum*, *R. longeracemosum* and derivatives of *R. petraeum*. Resistance in gooseberry species can be found in *R. divaricatum*, *R. oxyacanthoides* and *R. hirtellum* (Keep 1970), and the American cultivar 'Houghton'. In commercial terms, there is need to combine resistance with larger fruit size in these species.

Leafspot (*Drepanopeziza ribis* (Kleb.) von Höhn.) occurs throughout Europe, North America and Australasia, and leads to defoliation followed by reduced yields and crop quality in the subsequent season (Blodgett 1936). Evidence of a race structure in *D. ribis* was provided by Klebahn (1906) and Zakhryapina (1959), and Blodgett described 11 isolates with distinct morphology. Resistance is controlled by two complementary genes designated Pr_1 and Pr_2 by Anderson (1972), with the resistant *R. dikuscha* heterozygous for both genes. Other sources of resistance in species related to *R. nigrum* are *R. pauciflorum*, *R. americanum* and *R. glutinosum*.

In redcurrant, the most resistant cultivars are those derived from *R. petraeum* and *R. multiflorum*, with *R. sativum* and its derivatives showing high levels of susceptibility. Gooseberry suffers less from leafspot than the other *Ribes* crops, but the main sources of resistance are American species and cultivars such as *R. divaricatum* and *R. oxyacanthoides*.

The introduction of white pine blister rust (WPBR, *Cronartium ribicola* Fisch.) to North America on timber stocks imported from Europe in the 1920s brought the emergent *Ribes* industry there to a halt (Barney and Hummer 2005), since many *Ribes* species are alternate hosts for the fungus along with several of the most valuable pine species. *Ribes nigrum* is among the most susceptible species (Fig. 6.2), although several North American native species such as *R. americanum* and *R. sanguineum* are also susceptible. Whilst the effects on blackcurrant and other species are generally limited to premature defoliation and a possible reduction in the subsequent year's crop, the effect on pine species is more serious, often resulting in death of the tree. Eradication programmes and federal legislation were introduced, although the latter was repealed in 1966. State restrictions still apply in a few states, although most now allow *Ribes* production, but quarantine procedures are stringent for *Ribes* germplasm entering the US.

Resistance is controlled by a dominant gene, *Cr*, found in *R. ussuriense*, and blackcurrant cultivars bred in Canada from this species, including 'Kerry' and the later 'Consort', 'Coronet' and 'Crusader' are similarly resistant to the fungus. The Swedish cultivar 'Titania' is also resistant, with 'Consort' in its parentage. Further resistant seedlings are under development in many breeding programmes, notably those in Scotland, in collaboration with the University of Minnesota, and Poland.

Fig. 6.2 Blackcurrant leaf with infection of white pine blister rust (*Cronartium ribicola*)

The *Cr*-based resistance has proven to be fairly robust, and research is in progress to align resistance in *Ribes* with resistance in the *Pinus* spp.

Blackcurrant reversion virus (BRV) is one of the most serious diseases of *Ribes* spp., and is certainly the main viral problem. It occurs in most areas where *Ribes* are found, except North America and Australia (Jones 2002), and can be found in two forms: the European form and a more severe variant found in Finland and Russia. Redcurrants can be infected with BRV, although the symptoms are much less severe, but gooseberry appears to be immune. Both variants' sole means of transmission is through the gall mite *Cecidophyopsis ribis* (see above), and control of BRV is invariably targeted at this vector.

The genetic control of resistance is not entirely clear, but durable resistance to both BRV variants has been accessed by blackcurrant breeders from *R. dikuscha*, leading to the cultivars 'Golubka' from Russia and 'Ben Gairn' (a 'Golubka' derivative) from Scotland. The identification of BRV-resistant germplasm has until recently been dependent on time-consuming grafting to indicator plants, but the development of a PCR-based test for BRV should ensure both that further resistant cultivars are brought forward in the near future and that the genetic control of resistance is elucidated.

6.5.2 Environmental Adaptation

Winter hardiness levels in many *Ribes* species are high, especially within the currants, where species such as *R. dikuscha* and derivatives of *R. nigrum* var. *sibiricum* have proved especially tolerant of severe winter temperatures (Brennan 1996). In the eastern part of Russia, the species *R. pauciflorum*, *R. procumbens* Pall. and *R. fontaneum* have been cited as useful sources of winter hardiness in blackcurrant, and in redcurrants the species *R. palczewskii* (Jancz.) Pojark., *R. pallidiflorum* Pojark. and *R. manschuricum* (Maxim.) Kom. are thought to have similar utility (Sabitov et al. 2002). Most modern blackcurrant cultivars are hardy enough for growing in all but the most extreme environments, although the combination of strong winds and very low temperatures place more stress on the plant's survival mechanisms.

Spring frost tolerance, enabling flowers to survive and produce berries, has been a major objective in many blackcurrant breeding programmes. In the 20th century, blackcurrant production was seriously affected by occasional and unpredictable frost events, causing massive yield losses in susceptible cultivars. Breeding strategies, such as the use of northern *Ribes* germplasm from Scandinavia, have led to a succession of later-flowering, and hence frost-escaping cultivars, such as 'Ben Alder' and 'Ben Tirran' from the Scottish Crop Research Institute (Brennan 1991). Some of the northern germplasm available to breeders has been shown by Dale (1987) and Brennan (1991) to have an increased level of physiological tolerance to spring frost damage at flowering as whole plants, although Carter et al. (1999), using detached flowers from three blackcurrant cultivars, found no significant genotypic differences in response. Further work (Carter et al. 2001) uncovered some evidence of barriers to ice propagation through the floral racemes, and suggested that these may form the basis of the different responses between genotypes. Studies by Dale (1981 and 1987) provided some evidence of an additive genetic control of spring frost tolerance, although common control of the mechanisms for spring frost tolerance and winter hardiness has not been so far demonstrated.

Barney and Hummer (2005) estimated that blackcurrants generally require between 800 and 1600 hours of temperatures below 7°C during the dormant period depending on cultivar before buds will break in the spring, although other unpublished estimates put the chilling requirement of some northern cultivars as high as 2400 hours < 7°C. However, in recent years in Europe, a succession of mild winters has led to problems with a lack of winter chilling in blackcurrant crops in some southern locations, causing erratic budbreak and reductions in crop yields and quality (Fig. 6.3). Lantin (1973 and 1977) identified this as a problem in France, although he recognised the risk that earlier budbreak poses in terms of potential damage by spring frosts. He also identified differences in chilling requirement between cultivars and differences in chilling responses (fluctuating vs. more constant temperatures). Examination of the climate data in England shows a decline in chilling hours below 7.2°C (Atkinson et al. 2005) in the past 50 years of ca. 10%, whilst the incidence of spring frosts over the same periods has declined by up to 50%. The differences in chill requirement between blackcurrant cultivars imply that there is scope for selection of appropriate phenotypes through breeding if the genetic base of the programme is sufficiently wide. Breeding in New Zealand has produced several cultivars adapted to the low-chill environment there (Snelling and Langford 2002), with *ca.* 1300 hours <7°C annually.

Fig. 6.3 Uneven development on a stem of blackcurrant, caused by insufficient winter chilling

Several dehydrin-like genes were isolated by Lanham et al. (2001) from black-currant, using cold-acclimated leaves, and one of these was demonstrated to be chill-induced at 4°C, suggesting possible adaptive mechanisms in blackcurrant germplasm that may be exploitable.

6.5.3 Flowering and Fruiting Habit

Ribes flowers are produced in racemes of 2–70 flowers depending on species, from the second year of growth. Most species are monoecious, but there are dioecious species such as *R. alpinum*. The number of flowers per raceme, or strig, in commercial types of blackcurrant is usually between 6 and 12, while in redcurrant the number of flowers per inflorescence is often higher (from 16 to over 20 in *R. multiflorum* derivatives) and in gooseberry there are usually only 1–4 flowers at each node. Flowers are usually inferior and pentamerous.

Floral initiation commences in the northern hemisphere in late June/early July and is completed by late August (Nasr and Wareing 1961), with photoperiodic effects on initiation noted by Tinklin et al. (1970), whereby a critical daylength of *ca.* 16 hours was proposed for blackcurrant. QTLs for the timing of flowering have been identified by Brennan et al. (2007). The duration of flowering in blackcurrant is usually between three and four weeks, although historical data show that flowering of blackcurrant in the U.K. has in recent years become earlier and more protracted, possibly due to climatic changes (Atkinson et al. 2005).

The overwhelming majority of modern blackcurrant, redcurrant and gooseberry cultivars are self-fertile, enabling large plantations of single genotypes to be grown successfully. However, the available data for *Ribes* species (Brennan 1996) suggests that most wild species are outbreeders, ensuring gene flow between populations.

An important goal in blackcurrant breeding is the development of cultivars with different ripening seasons and with larger berry size, although the latter depends on the end-user requirements – juice processing requires a smaller berry than the fresh market. Early-ripening cultivars are increasingly sought as a means of season extension, since there are already successful very late-fruiting cultivars such as 'Ben Tirran' from the U.K.

Berry size in *Ribes* is considered to be under polygenic control, and QTLs linked to berry size have been identified by Brennan et al. (2007). A linkage between self-fertility and berry size was also reported by Tamás (1963). Many wild accessions of *Ribes*, including North American gooseberry species, have fairly small fruit.

6.5.4 Fruit Quality

The health attributes of blackcurrant are among the key reasons for their continuing and growing popularity. Whilst much of the emphasis is now on the phenolic components contained within the fruit, the high levels of ascorbic acid (vitamin C) are a major factor in the growth of blackcurrant cultivation worldwide. Levels found in blackcurrant cultivars are highly variable, but most contain between 130 and 200 mg/100 ml juice and some breeding lines at the Scottish Crop Research Institute and elsewhere have achieved over 350 mg/100 ml. There are reports (Volunez and Zazulina 1980) that accessions of *R. nigrum* var. *sibiricum* contain even higher levels.

High ascorbate levels are heritable, with strong maternal influences (Franchuk and Manaenkova 1971), but there are also significant variations caused by environmental factors. It has been shown by Viola et al. (2000) that levels of ascorbic acid are genetically determined at an early stage of development. The biosynthetic pathway for ascorbic acid has now been largely described (Smirnoff and Wheeler 2000), and work is in progress to examine allelic variation in key control points in the pathway between high and low phenotypes in order to develop rapid selection methods for breeders. Linkage mapping of traits by Brennan et al. (2007) has identified QTLs linked to ascorbate content in blackcurrant.

Many *Ribes* species, especially the darker-fruited blackcurrant types, contain high concentrations of polyphenolic compounds, notably anthocyanins and flavonols that are increasingly sought for their antioxidant activity (Lister et al. 2002). The main anthocyanins in blackcurrant are cyanidin-3-glucoside, cyanidin-3-rutinoside, delphinidin-3-glucoside and delphinidin-3-rutinoside, with the relative proportions of these compounds varying between cultivars. For juice products, dephinidins have been preferred for their higher stability, and this is reflected in the breeding objectives for this market (Brennan 1996). Blackcurrant also contains a range of other compounds, such as cyanidin- and delphinidin-3-sophoroside plus pelargonidin-3-rutinoside, but these minor compounds are significantly less abundant (*ca.* 5%

of total anthocyanin). Proanthocyanins are also thought to be of significance in the health-related benefits of blackcurrant consumption, and work is currently in progress examining the bioavailability of blackcurrant anthocyanins and phenolic compounds. There is little information at present on the inheritance of high phenolic levels, although significant maternal influences have been identified by breeders.

In recent years, improvement of sensory quality of fruit cultivars, for both fresh market and processing, has increased in importance, and the selection of blackcurrants with superior sensory attributes is now an integral part of some breeding programmes. Blackcurrant germplasm shows considerable variation in sensory profile (Brennan et al. 1997), and these variations are retained throughout processing for juice (Brennan et al. 2003). Some compounds have been associated with specific aroma descriptors, e.g. the link between ketothiols with 'catty' aroma notes (Lewis et al. 1980). Other assessments of the links between chemical composition and sensory quality were made by Latrasse and Lantin (1974), with compounds such as monoterpenes associated with positive sensory traits. For fresh market outlets, there is a requirement for sweeter fruit, with a higher concentration of soluble solids.

Although the exact genetic control of many of the traits described in the sensory vocabulary for blackcurrant is not clear, some species, notably *R. ussuriense* and *R. petiolare*, are known to transmit off-notes to the profile of progenies developed from them (Melekhina et al. 1980).

6.5.5 Mechanical Harvesting

Blackcurrants in commercial production are almost all harvested mechanically, since the berries ripen simultaneously and the main outlet is for processing (Dale et al. 1994). Harvesters are expensive, however, requiring a fairly large cropping area to justify the cost. The increasing use of mechanical harvesting has put additional emphasis on the selection of upright-growing bushes, in preference to the more prostrate forms found in the wild and in older cultivars such as 'Öjebyn'. The upright growth habit shown by the cultivar 'Ben Hope', derived from an induced mutant of 'Westwick Choice', has proved to have a high degree of heritability. However, it is important that improvements in growth habit are not accompanied by changes in vigour. Hybrids between *R. nigrum* × redcurrant, *R. nigrum* × *R. bracteosum* and *R. nigrum* × *R. sanguineum* were considered promising by Jennings et al. (1987) in terms of improved growth habit for machine harvestability.

For the growing fresh market, berries are generally hand-harvested, often as whole racemes. Fresh market outlets are important for redcurrant and gooseberry, since the growth and branch structure of these crops make machine-harvesting difficult to achieve effectively without considerable damage to the plant.

6.6 Crossing and Evaluation Techniques

For the controlled pollination of *Ribes*, individual flowers are emasculated using fine forceps and scalpel up to 5 days prior to anthesis. For redcurrants, individual anthers are easy to remove, but in gooseberry and blackcurrants the entire calyx is removed. To prevent unwanted pollinations, branches bearing emasculated flowers must be securely bagged or the plant must be retained within an insect-proof glasshouse. Pollen is obtained by collecting anthers or flowers from the male parental plants in Petri dishes, where the pollen will remain viable for 2–3 weeks after anthesis. Pollination is effected using a fine paintbrush, and each pollination is usually repeated within 2–3 days.

When the berries are ripe, they are collected and stored at 4°C until the seeds are extracted. This is done in water using a blender, and the viable seeds sink to the bottom and can then be collected after the water is decanted. The seed is dried on filter paper and then stored until required at 4°C. Usually, most seed is used fairly soon after extraction, but the seed can remain viable for several years if stored in good conditions (Brennan 1996).

Seeds are sown on compost and covered with a thin layer of vermiculite and then stratified at 2°C for 13 weeks in the dark, after which they are brought into warm greenhouse conditions (22°C) for germination, which generally takes place within 2–3 weeks.

Seedlings are raised in the glasshouse and then planted in the field at the start of an evaluation process that invariably takes several fruiting seasons to complete. Initial field plantings of seedlings are done at 30 cm spacing, so that vegetative characters, such as foliar disease resistance and growth habit, can be assessed in the first (non-fruiting) year. Selected seedlings are then assessed for their fruiting traits, together with further evaluation of agronomic and vegetative characters; fruit quality characters are measured using small fruit samples (up to 500 g).

In addition to field assessment of breeding populations, there are several laboratory and glasshouse phenotyping procedures now available to breeders, including estimates of winter chilling requirement using glasshouse forcing and the use of PCR-based detection methods for reversion virus resistance studies. Further development of molecular marker-based screens is in progress (Brennan and Gordon 2002). Additionally, field-based infestation plots are used for the identification of gall mite-resistant blackcurrant seedlings, although the hope is that molecular screens will replace these plots in the near future.

6.7 Biotechnological Approaches to Genetic Improvement

6.7.1 Genetic Mapping and QTL Analysis

A number of marker systems have been developed for *Ribes* (Lanham et al. 1995 and 1998, Lanham and Brennan 1999, Lanham et al. 2000, Brennan et al. 2002). Apart from fingerprinting of germplasm, the most suitable and informative for the development of robust genotyping of breeding progeny at the present time are AFLPs

(Lanham and Brennan 1999) and SSRs (Brennan et al. 2003), although the development of SNP markers offers considerable potential for the identification of desirable alleles.

The main targets for marker-assisted selection procedures are pest and disease resistances, notably gall mite, and quality traits. The first linkage map of *Ribes* has been developed by Brennan et al. (2007), using SSRs (genomic and EST-derived), AFLP and SNP markers, and QTLs affecting several important phenological, agronomic and fruit quality traits have been located on the map. The population used for the map was developed from two SCRI breeding lines, one of which carries the *Ce* gall mite resistance gene. The parents were also diverse in their quality attributes. As a result, mapped traits so far include gall mite resistance, and also high ascorbate and anthocyanin levels in the extracted juice. The map can now provide a framework for the development of marker-assisted breeding strategies for blackcurrant, to improve breeding efficiency and time to cultivar. Details of the SSRs used in the mapping can be found at http://www.fruitbreeding.co.uk/ribes_genomics.asp.

One of the most immediate benefits from this mapping work is likely to be a marker linked to gall mite resistance, a trait which currently takes around four years to effectively screen for in field infestation plots (Brennan 1996). An AFLP marker linked to resistance is currently undergoing validation at the Scottish Crop Research Institute, after which it will be deployed in the blackcurrant breeding programme.

6.7.2 Genomic Resources

Ribes americanum is one of the model species used in the Floral Genome project, funded by the US National Science Foundation, due its unique phylogenetic position (http://fgp.bio.psu.edu). cDNA libraries and ESTs from this species have now been developed and are available from the Project.

Work on ripening blackcurrant fruit by Woodhead et al. (1998) using cDNA libraries from different stages of development led to the identification and cloning of a series of genes involved and upregulated during ripening processes. Further work on differential expression of genes linked to dormancy break in *Ribes* is in progress using microarray analysis. Using this technology, it is hoped to monitor the activity of a large number of genes simultaneously and to thereby identify differentially expressed genes encoding dormancy-associated proteins.

References

Anderson MM (1969) Resistance to leaf spot and Amercian gooseberry mildew. Rep Scot Hort Res Inst 1968:39

Anderson MM (1972) Resistance to blackcurrant leaf spot (*Pseudopeziza ribis* Nal.) in crosses between *Ribes dikuscha* and *R. nigrum*. Euphytica 21:510–517

Atkinson C, Sunley R, Jones H, Brennan R, Darby P (2005) Winter Chill in Fruit. UK Government Department of Food and Rural Affairs Report N CTC 0206

Barney DL, Hummer KE (2005) Currants, gooseberries and jostaberries – a guide for growers, marketers and researchers in North America. Haworth Press, Binghamptom, NY

Bauer R (1973) 'True breeding' for combined resistance to leaf, bud and shoot diseases by amphidiploidy in *Ribes*. J Yugosla Pomol 7:17–19

Berger A (1924) A taxonomic review of currants and gooseberries. Bull New York State Agric Exp Sta 109

Blodgett EC (1936) The anthracnose of currant and gooseberry caused by *Pseudopeziza ribis*. Phytopathology 26:115–152

Brennan RM (1991) The effect of simulated frost on black currant (*Ribes nigrum* L.). J Hortic Sci 66:607–612

Brennan RM (1996) Currants and gooseberries. In: Janick J, Moore JN (eds) Fruit breeding, vol II : Small Fruits and vine crops. John Wiley & Sons, NY, pp 191–295

Brennan RM, Gordon SL (2002) Future perspectives in blackcurrant breeding. Acta Hortic 585:39–46

Brennan RM, Hunter EA, Muir DD (1997) Descriptive sensory profiles of blackcurrant juice and genotypic effects on sensory quality. Food Res Intern 30:381–390

Brennan RM, Hunter EA, Muir DD (2003) Relative effects of cultivar, heat-treatment and sucrose content on the sensory properties of blackcurrant. Food Res Intern 36:1015–1020

Brennan R, Jorgensen L, Hackett C, Woodhead M, Gordon SL, Russell J (2007) The development of a genetic linkage map of blackcurrant (*Ribes nigrum* L.) and the identification of regions associated with key fruit quality and agronomic traits. *Euphytica* (in press)

Brennan R, Jorgensen L, Woodhead M, Russell J (2002) Development and characterisation of SSR markers in *Ribes* species. Mol Ecol Notes 2:327–330

Carter JV, Brennan RM, Wisniewski M (1999) Low-temperature tolerance of blackcurrant flowers. HortScience 34:855–859

Carter JV, Brennan RM, Wisniewski M (2001) Patterns of ice formation and movement in *Ribes nigrum* L. HortScience 36:1027–1032

Cone WW (1967) Insecticidal response of currant borer and two-spotted mite on currants in Central Washington. J Econ Entomol 60:436–441

Cronquist A (1981) An integrated system of classification of flowering plants. Columbia University Press, NY

Crook DJ, Cross J, Birch ANE, Brennan RM, Mordue A (2001) Oviposition and larval survival of *Dasineura tetensi* on four *Ribes* cultivars in the UK. Entomologia Experimentalis et Applicata 101:183–190

Dale A (1981) The tolerance of black currant flowers to induced frosts. Ann Appl Biol 99:99–106

Dale A (1987) Some studies in spring frost tolerance in black currant (*Ribes nigrum* L.). Euphytica 36:775–781

Dale A, Hanson EJ, Yarborough DE, McNicol RJ, Stang EJ, Brennan RM, Morris JR, Hergert GB (1994) Mechanical harvesting of berry crops. Hortic Rev 16:255–382

de Janczewski E (1907) Monographie des groseilliers, *Ribes* L. Mémoires de la Société de physique et d'histoire naturelle de Genève 35:199–517

Engler A, and Prantl K (1891) Ribesioideae. Natu Pflanzenfam 3:97–142

Franchuk EP, Manaenkova NI (1971) The inheritance of vitamin C in the hybrid progeny of blackcurrants (in Russian). Sb Nauch Rabo. VNII Sadovodstva 15:196–205

Gerard (1636) The herball or general historie of plantes (revised by Johnson T). Norton and Whittakers, London

Hatton RG (1920) Black currant varieties: a method of classification. J Pomol 1: 65–80, 145–154

Hedrick UP (1925) The small fruits of New York. Rep New York State Agric Exp Sta 33:243–354

Hummer K (2006) *Ribes* genetic resources. http://www.ars.usda.gov/Main/docs.htm?docid=11353

Hummer K, Sabitov A (2004) Genetic resistance to currant borer in *Ribes* cultivars. J Am Pomol Soc 58:215–219

Jennings DL, Anderson MM, Brennan RM (1987) Raspberry and red current breeding. In: Abbott AJ, Atkin RK (eds) Improving vegetatively propagated crops. Academic Press, London, pp 135–147

Jermyn WA (2002) Differential infestation of *Ribes nigrum* cultivars by currant clearwing moth *Synanthedon tipuliformis*. Acta Hortic 585:355–357

Jones AT (2002) Important virus diseases of *Ribes*, their diagnosis, detection and control. Acta Hortic 585:279–285

Keep E (1962) Interspecific hybridisation in *Ribes*. Genetica 33:1–23

Keep E (1970) Response of *Ribes* species to American gooseberry mildew, *Sphaerotheca mors-uvae* (Schw.) Berk. Rept Malling Res Sta for 1973, pp 133–137

Keep E (1974) Breeding for resistance to American gooseberry mildew, *Sphaerotheca mors-uvae*, in the gooseberry (*Ribes grossularia*). Ann Appl Biol 76:131–135

Keep E (1985) The blackcurrant leaf curling midge, *Dasyneura tetensii* Rübs.: its host range and the inheritance of resistance. Euphytica 34:801–809

Keep E (1995) Currants (*Ribes* spp.). In: Smartt J, Simmonds NW (eds) Evolution of Crop Plants 2nd edn., Longman, London pp 235–239.

Klebahn H (1906) Studies on certain Fungi Imperfecti and the associated Ascomycetous forms 3: *Gloeosporium ribis* Mont. et Desm. (in German). Z Pflkrankh 16:65–83

Knight RL, Keep E, Briggs JB, Parker JH (1974) Transference of resistance to blackcurrant gall mite, *Cecidophyopsis ribis*, from gooseberry to blackcurrant. Ann Appl Biol 76:123–130

Lanham PG, Brennan RM (1999) Genetic characterisation of gooseberry (*Ribes* subgenus *Grossularia*) germplasm using RAPD, ISSR and AFLP markers. J Hortic Sci Biotechnol 74:361–366

Lanham PG, Brennan RM, Hackett C, McNicol RJ (1998) RAPD fingerprinting of blackcurrant (*Ribes nigrum* L.) cultivars. Theor Appl Genet 90:166–172

Lanham PG, Brennan RM, McNicol RJ (1995). Fingerprinting of blackcurrant (*Ribes nigrum* L.) cultivars using RAPD analyses. Theor Appl Genet 90:166–172

Lanham PG, Kemp RJ, Jones H, Brennan RM (2001) Expression of dehydrin-like genes in response to chilling in leaves of blackcurrant, *Ribes nigrum* L. J Hortic Sci Biotechnol 76:201–207

Lanham PG, Korycinska A, Brennan RM (2000) Genetic diversity within a secondary gene pool for *Ribes nigrum* L. revealed by RAPD and ISSR markers. J Hortic Sci Biotechnol 75: 371–375

Lantin B (1973) The chilling requirements of the buds of blackcurrant, *Ribes nigrum*, and of some redcurrants, *Ribes* sp. Ann Amèlior Plantes 23:27–44

Lantin B (1977) Estimation of the cold requirements necessary to break dormancy in buds of blackcurrant (*Ribes nigrum* L.) and other currants (in French). Ann Amèlior Plantes 27:435–450

Latrasse A, Lantin B (1974) Varietal differences between the monoterpene hydrocarbons of the essential oil of blackcurrant buds (in French). Ann Technol Agr 23:65–74

Lewis MJ, May HV, Williams AA (1980) Fruit quality: Black currant. Rept Long Ashton Exp Sta 1978:156–157

Lister CE, Wilson PE, Sutton KH, Morrison SC (2002) Understanding the health benefits of blackcurrants. Acta Hortic 585:443–449

Melekhina AA, Yankelevich BB, Eglite MA (1980) Prospects for the uise of *Ribes petiolare* Dougl. in breeding of blackcurrant (in Russian). Latvijas PSR Zinatu Akad 3:116–123

Messenger W, Liston A, Hummer K (1999) *Ribes* (*Grossulariaceae*) phylogeny as indicated by restriction-site polymorphisms of PCR-amplified chloroplast DNA. Plant Syst Evol 217:185–195

Nasr T, Wareing PF (1961) Studies on flowering initiation in blackcurrant 1: some internal factors affecting flowering. J Hortic Sci 36:1–10

Nilsson F (1959) Polyploidy in the genus *Ribes*. Genet Agric 11:225–242

Nilsson F (1966) Cytogenetic studies in *Ribes*. In: Proc. Balsgård fruit breeding Symp., Fjalkestad, pp 197–204

Parkinson J (1629) Paradisi in sole paradises terrestris. Humfrey, Turner and Robert Young, London

Ravkin A (1986) The genetics of resistance to powdery mildew of some blackcurrant species (in Russian). In: Geneticheski Mekhanizmy Selektsii i Evolyutsii, Moscow, pp 74–80.

Roach FA (1985) Cultivated fruits of Britain. Their origin and history. Blackwell, Oxford

Sabitov AS, Vendenskaya IO, Hummer KE (2002) *Ribes* from the Russian Far East: perspectives for breeding. Acta Hortic 585:161–165

Senters AE, Soltis DE (2003) Phylogenetic relationships in *Ribes* (Grossulariaceae) inferred from ITS sequence data. Taxon 52:51–66

Sinnott Q (1985) A revision of *Ribes* L. subg. Grossularia (Mill.) per. Sect. *Grossularia* (Mill.) Nutt. (Grossulariaceae) in North America. Rhodora 87:189–286

Siksnianas T, Stanys V, Staniene G, Sasnauskas A, Rugienius R (2005) American black currant as donor of leaf disease resistance in black currant. Biologija 3:65–68

Smirnoff N, Wheeler GL (2000) Ascorbic acid in plants: biosynthesis and function. Crit Rev Biochem Mol Biol 35:291–314

Snelling C, Langford G (2002) The development of low chill blackcurrants from the New Zealand breeding programme. Acta Hortic 585:167–169

Tamás P (1963) The interactions between fertility and berry size in blackcurrant (in German). Züchter 33:302–306

Tinklin IG, Wilkinson EH, Schwabe WW (1970) Factors affecting flower initiation in the blackcurrant (*Ribes nigrum* L.). J Hortic Sci 45:275–282

Tydeman HM (1938) Some results of experiments in breeding blackcurrants, 2: first crosses between the main varieties. J Pomol 16:224–250

Viola R, Brennan RM, Davies HV, Sommerville L (2000) L-ascorbic acid accumulation in berries of *Ribes nigrum* L. J Hortic Sci Biotechnol 75:409–412

Volunez AG, Zazulina NA (1980) Selection of blackcurrant hybrids with a high content of vitamin C (in Russian). Plodovodstvo 4:60–64

Woodhead M, Taylor MA, Brennan R, McNicol RJ, Davies HV (1998) Cloning and characterisation of the cDNA clones of five genes that are differentially expressed during ripening in the fruit of blackcurrant (*Ribes nigrum* L.). J Plant Physiol 153:381–393

Zakhryapina TD (1959) Differentiation of the pathogen of anthracnose of currant and gooseberry (in Russian). Bot Z 54:836–843

Chapter 7
Grapes

C.L. Owens

Abstract Grape is a major crop worldwide in which production is primarily driven by the ability to grow high-quality fruit. Breeding objectives vary by region and market class of grape, but many programs seek to combine high quality fruit with improved disease resistance and environmental adaptation, or to continue advances in quality attributes. Grapevines are predominantly a grafted crop, making grape rootstocks, and rootstock breeding, vitally important in the growth of the global viticulture industry. There are vast germplasm resources available within the genus *Vitis*, but worldwide production is dominated by cultivars of one species, *V. vinifera*. Species other than *V. vinifera* are of significant interest as useful sources of desirable traits in many modern breeding programs. Little is known concerning the genetic control of most traits in grape beyond the fact that many are quantitatively controlled. Substantial international effort has occurred in the development of molecular genetic and genomic resources for grape. Many tools are now in place to identify the causal genes underlying important traits and to better understand the allelic diversity that exists in important genes.

7.1 Introduction

Grapevine is the most valuable horticultural crop in the world. The majority of the fruit is processed into wine, but significant portions of the worldwide crop are destined for fresh consumption, dried into raisins, processed into non-alcoholic juice, and distilled into spirits. Significant grape acreage exists on all continents of the globe, save for Antarctica. Worldwide estimates are that approximately 8 million hectares are currently planted to grapevine and 60 million metric tons of fruit are produced annually (FAO production statistics). Spain, France, and Italy are the largest grape producers in the world. Many additional European countries, the

C.L. Owens
USDA-ARS, Grape Genetics Research Unit, 308 Sturtervant Hall, Cornell University, Geneva, New York 14456, USA
e-mail: chris.owens@ars.usda.gov

J.F. Hancock (ed.), *Temperate Fruit Crop Breeding*,
© Springer Science+Business Media B.V. 2008

United States, Argentina, Chile, Australia, South Africa, and China are all major producers of grapes. This review emphasizes research conducted since the mid-1990s and readers are referred to excellent earlier reviews on this subject (Alleweldt et al. 1990, Einset and Pratt 1975, Reisch and Pratt 1996).

Grapevines are predominantly grafted in many production regions and as a result significant breeding efforts have been conducted on both grape scions and grape rootstocks over the last 125 years. The importance of grape rootstock breeding in the growth of global viticulture cannot be under-estimated. Many species of *Vitis* have been employed in the breeding of grape rootstocks, primarily for resistance to the root-zone pest, phylloxera. Historically, grape rootstock breeding has also focused on additional soil-borne pathogens and on developing a range of rootstocks with broader adaptation to soil and climatic conditions.

The most widely planted cultivated species of grapevine, both for scion cultivars and for own-rooted vines, is *Vitis vinifera*. *V. vinifera* is the dominant species of grape for wine, raisins, and fresh market table grapes. Despite the existence of thousands of cultivars of *V. vinifera*, only a few dozen cultivars account for the vast majority of world-wide production. The most widely planted grape cultivars in the world include two cultivars planted in Spain at low-planting densities, 'Airen' and 'Grenache', as well as 'Sultanina' (syn. 'Thompson Seedless') the preeminent raisin grape in the world. Other wine grape cultivars that have become widely accepted in many viticultural regions or play a significant role in the bulk wine production of a major wine producing region are: 'Merlot', 'Ugni Blanc', 'Cabernet Sauvignon', 'Carignan', 'Chardonnay' and 'Syrah'.

Numerous interspecific hybrids historically played a significant role in many viticultural regions and continue to be useful in modern breeding efforts. Both *V. labrusca* and associated hybrids, and *V. rotundifolia* have regional importance in the eastern United States as multi-use grapes for wine, juice, and fresh consumption. Many *Vitis* species have been used, both historically and in recent breeding efforts for the development of new fruiting varieties, particularly those adapted to severe temperature or pathogen pressure. However, species other than *V. vinifera* are of only minor importance on a global scale despite evidence that they will be exceptionally useful as sources of desirable traits in future breeding efforts.

Considering the worldwide importance of grapevine, it is not surprising that substantial international effort has occurred in the development of molecular genetic and genomic resources for use by grape scientists, geneticists, and breeders. The manner in which new tools and resources will be utilized for greater understanding of grapevine genetics and breeding is likely to be a major development to follow at the beginning of the 21st century.

7.2 Evolutionary Biology and Germplasm Resources

The genus *Vitis* belongs in the family Vitaceae. Members of the genus *Vitis* are distinguished by the presence of actinomorphic flowers that have fused petals at the tip, axillary buds at every node, dioecy (wild species), and tendrils opposite leaves which appear in an alternate phyllotactic pattern. There are approximately 60 species of *Vitis* in the world, with large centers of diversity existing in North America (approximately 30 species) and east Asia, particularly China (approximately 30 species) (Table 7.1). However, the Asian *Vitis* species are not well described outside of the Chinese literature and the germplasm is often unavailable outside of Asia. Similarly, the *Vitis* species of Mexico, Central America and into the extreme northern portions of South America are poorly characterized, as evidenced by the recent discovery of new Mexican species (Comeaux 1987). The most widely cultivated species, *V. vinifera* is the sole native species in Europe, the Near East and northern Africa.

The genus *Vitis* is divided into two sections: *Euvitis* and *Muscadinia*. These two sections differ in several morphological characteristics, and significantly, in chromosome number: *Euvitis* ($2n = 2x = 38$) and *Muscadinia* ($2n = 2x = 40$). *Euvitis* comprises most of the grape species in the world, *Muscadinia* consisting only of *V. rotundifolia*, *V. rotundifolia* var. *munsoniana*, and *V. popenoi*. The species of the section *Muscadinia* have also been considered within a separate genus, and their placement is one of the many ongoing debates in the taxonomy of *Vitis*. In particular, there are few published taxonomic treatments of the Asian *Vitis* species. The two best sources on *Vitis* taxonomy are currently available only in draft form for Chinese *Vitis* and N. America *Vitis* north of Mexico (Moore unpubl. manuscript, Zhiduan and Wen draft manuscript). Additionally, all of the 38 chromosome species are interfertile, allowing hybrid zones to exist where species are sympatric and bloom time overlaps.

The domestication of *V. vinifera* is thought to have occurred approximately 6,000 to 10,000 years ago (Levadoux 1956, McGovern 2003, Zohary and Hopf 2000), although it is likely that prehistoric hunter-gatherers used wild grapes as a food source (Zohary 1996). There are several morphological and biochemical traits associated with the domestication of *V. vinifera* that were derived from the progenitor species *V. vinifera* subsp. *sylvestris*. The most notable changes are the emergence of perfect flowers, greater uniformity of berry maturity within clusters, higher sugar content, and the selection for a wide range of fruit colors (Levadoux 1956, Olmo 1995, Zohary and Spiegel-Roy 1975). Extant, isolated patches of *V. vinifera* ssp. *sylvestris* can be found from Western Europe to central Asia and North Africa. Habitat loss and the ease with which the wild species can cross with cultivated forms has led to a sharp decline in the number of *V. vinifera* ssp. *silvestris* present and the existence of complexes of feral and wild forms.

Historically, geographical origins and morphological characteristics have been used to sub-divide *V. vinifera* into three morphotypes: *occidentalis*, *pontica* and *orientalis* (Negrul 1938). The *occidentalis* group is characterized by small berries, small clusters, highly fruitful shoots, and is associated with cultivars of Western European origin. The *orientalis* group consists of large berried, loose clustered cultivars from Central Asia. The *pontica* group comprises an intermediate grouping of cultivars from Eastern Europe and the Black Sea Basin. Debate exists concerning the number of domestication events and the location of their occurrence, as *V. vinifera* ssp. *sylvestris* had a wide geographic range and wild populations were likely used as

Table 7.1 Grape species of the world

Species	Major synonyms[a]	Geographic Location
V. acerifolia Raf.	*V. longii*	E. NM and CO, Kansas, OK, N. TX
V. aestivalis Michx.		
V. aestivalis var. *aestivalis*	*V. smalliana* *V. rufotomentosa*	E. U.S.A. from FL to CN and west to MO, TX, NE
V. aestivalis var. *bicolor* Deam	*V. aestivalis* var. *argentifolia*	NE U.S.A. to Northcentral U.S.A.
V. aestivalis var. *lincecumii* (Buckley) Munson	*V. lincecumii*	AR, LA, OK, TX
V. amurensis Rupr.		China
V. arizonica Egelm.	*V. treleasei*	S.W. U.S.A. from AR to W. TX
V. balanseana Planch.		China, S.E. Asia
V. bashanica P.C. He		China (Shanxi)
V. bellula (Rehd.) W.T. Wang		China
V. bellula var. *bellula*		
V. bellula var. *pubigera*		China
V. betulifolia Diels & Gilg		China
V. blancoi Munson		Mexico
V. biformis Rose		Mexico
V. bloodworthiana Comeaux		Mexico
V. bourgaeanna Planch.		Mexico
V. bryoniifolia Bge.		China
V. bryoniifolia var. *Bryoniaefolia*		
V. californica Benth.		Central CA to S. OR
V. × *champinii* Planch.		South Central TX on and adjacent to the Edwards Plateau, natural hybrid between *V. mustagensis* × *V. rupestris* – rare
V. chungii Metcalf		China (Fujian, Guangdong, Guangxi, Jiangxi)
V. chunganensis Hu		China
V. cinerea (Engelm.) Engelm. ex Millardet		
V. cinerea var. *baileyana* (Munson) Comeaux		Interior regions of the SE U.S.A. from AL and GA in the south north to OH, WV and PA.
V. cinerea var. *cinerea*		Mississippi River Basin – from Gulf of Mexico to KS/NE/IA
V. cinerea var. *floridana* Munson	*V. simpsonii*	Coastal regions of SE U.S.A. stretching from LA to MD.
V. cinerea var. *helleri* (Bailey) M.O. Moore	*V. berlandieri*	South Central TX and extreme Northern Mexico, most common on Edwards Plateau but also on Cross Timbers and Prairies and the Blackland Prairies
V. coignetiae Pulliat ex Planch.		Japan, Korea, E. Asia
V. × *doaniana* Munson ex Viala		North Central TX and OK/ natural hybrid between *V. mustagensis* × *V. acerifolia*
V. davidii (Roman. Du Caill.) Föex		China

Table 7.1 (continued)

Species	Major synonyms[a]	Geographic Location
V. erythrophylla W.T. Wang		China (Jiangxi, Zhejiang)
V. fengqinensis C.L. Li		Yunnan
V. flexuosa Thunb.		China
V. girdiana Munson		Southern and Baja CA
V. hancockii Hance		China
V. heyneana Roem. & Schult		China
V. heyneana subsp. heyneana		
V. heyneana subsp. ficifolia		
V. hui Cheng		China (Jiangxi, Zhexi)
V. jacquemontii R. Parker		Central Asia, Pakistan, Afghanistan
V. jaegeriana Comeaux		Mexico
V. jinggangensis W.T. Wang		China
V. labrusca L.		East Coast from Maine to SC, west to OH, MI down to LA, AL
V. lanceolatifoliosa C.L. Li		China (Guangdong, Hunan, Jiangxi)
V. longquanensis P.L. Qiu		China (Fujian, Jiangxi, Zhejiang)
V. luochengensis W.T. Wang		Guangdong, Guangxi
V. menghaiensis C.L. Li		Yunnan
V. mengziensis C.L. Li		Yunnan
V. monticola Buckley		South Central TX, isolated to limestone hills on the Edwards Plateau
V. mustangensis Buckley	*V. candicans*	AL, AK, LA, OK, TX
V. nesbittiana Comeaux		Mexico
V. × *novae-angliae* Fernald		Natural hybrid of *V. labrusca* × *V. riparia*, N.E. U.S.A.
V. palmata Vahl	*V. rubra*	From S.E. U.S.A. (FL, GA, AL) west to LA, north to OK, TN, MO, IN, IL)
V. peninsularis M.E. Jones		Baja California
V. piasezkii Maxim		China
V. piloso-nerva Metclaf		China
V. popenoei J.H. Fennell		Mexico/Central America
V. pseudoreticulata W.T. Wang		China
V. retordii Roman. Du Cail. Ex Planch.		China
V. riparia Michaux		Large range from great plains into Canada through all of northeastern U.S.A., south to northern LA, VA on the coast
V. romanetii Roman du Caill. ex Planch.		China
V. rotundifolia Michx.		S.E. U.S.A.

Table 7.1 (continued)

Species	Major synonyms[a]	Geographic Location
V. rotundifolia var. *munsoniana* (J.Simpson ex Munson) M.O. Moore	*V. munsoniana*	FL, AL, GA
V. rupestris Scheele		Originally Central TX, AK, MS, TN, KY, WV, nw MD, sw PA; now rare mostly southern MO, northern AR
V. ruyuanensis C.L. Li		Guangdong
V. shuttleworthii House		Florida
V. shenxiensis C.L. Li		Shaanxi
V. silvestrii Pamp.		China (W Hubei, S Shaanxi)
V. ×slavinii Rehder		Natural hybrid of *V. argentifolia* × *V. riparia*
V. sinocinerea W.T. Wang		China
V. tiliifolia Humb. & Bonpl. ex Schult.	*V. caribaea*	Mexico, Central America, Carribean
V. tsoii Merr.		China
V. vinifera L.		Western and Central Europe, North Africa, Near East, Caucases
V. vinifera ssp. *sylvestris* (C.C. Gmel.) Hegi	*V. sylvestris*	
V. vulpina L.	*V. cordifolia*	Large range within S.E. U.S.A.
V. wenchowensis C. Ling ex W.T. Wang		Zhejiang
V. wuhanensis C.L. Li		China
V. wilsonae Veitch		China
V. yunnanensis C.L. Li		China (Yunnan)
V. zhejiang-adstricta P.L. Qiu		China (Zhejiang)

[a]Many synonyms for *Vitis* species exist within the literature, particularly species that have been of historical importance for early rootstock and hybrid direct producer breeding. The species names utilized by T.V. Munson and Pierre Galet are indicated here only in those cases where they differ from current usage (Galet 1988, Munson 1909).
Sources: Moore 1987, Moore unpubl. manuscript, Zhiduan and Wen draft manuscript

a food source across much of that range. Recent evidence from the use of chloroplast molecular markers supports the presence of at least two major domestication centers, approximately corresponding with Negrul's *occidentalis* and *orientalis* group (Arroyo-Garcia et al. 2006). Additional attempts at finding genetic relationships between cultivars have provided only weak discrimination among geographic groupings and the presence of secondary domestication centers have been proposed based on evidence from nuclear markers (Aradhya et al. 2003, Grassi et al. 2003).

An important source of genetic variation in *V. vinifera* is the presence of numerous bud sports, or somatic mutations. Stable somatic mutants of a relatively subtle nature are typically grouped under the heading of 'clones'. More substantive mutations, particularly those altering berry pigmentation are often elevated to the state of a new cultivar name (e.g. 'Pinot noir', 'Pinot gris', and 'Pinot blanc' or

'Cabernet Sauvignon', 'Malian', and 'Shalistan' (Walker et al. 2006). Additionally, due to the ease of clonal propagation in grapevine, it is conceivable that a substantial proportion of phenotypically recognized mutants are chimeric in nature (Einset and Lamb 1951, Thompson and Olmo 1963). In fact, molecular markers have been utilized to confirm the presence of chimerism in several cultivar groupings (Franks et al. 2002, Hocquigny et al. 2004, Riaz et al. 2002). Polyploid sports and periclinal chimeras containing tissue layers of differing ploidy level have been reported for grapevine (Einset and Lamb 1951, Einset and Pratt 1954, Sauer and Antcliff 1969). The utilization of naturally occurring mutants of grapevine for the dissection of individual genes controlling important phenotypic traits has recently begun (Boss and Thomas 2002, Fernandez et al. 2006b, Kobayashi et al. 2004).

In many cases, it has proven difficult to distinguish or properly identify clones of a given cultivar. Early attempts at using molecular markers to distinguish clones were unsuccessful (Ye et al. 1998), but as marker technologies advance and are based on known DNA sequence variants, it is becoming easier to identify polymorphic markers that distinguish members within classes of clones (Benjak et al. 2006, Scott et al. 2000a). Despite recent successes in discovering polymorphic markers for some clones, there is still no single reliable way to easily distinguish clones using molecular markers. It remains extremely difficult to discriminate a large set of mixed clones into their appropriate clonal identities.

A combination of the very real difficulties in introducing new wine grape cultivars in many parts of the world and the differences that exist between clones of major cultivars has led to the development of clonal selection and evaluation programs. The long-term and widespread asexual propagation of very old cultivars provides a pool of genetic variants upon which to base selection. Virus infection has also been identified as a non-genetic source of clonal variation. Maintenance of virus-free, clonally identified propagation stock of many cultivars is a major resource for grape growers worldwide.

Artificial polyploidization has been explored as a means to overcome the difficulties in crossing between *Euvitis* and *Muscadinia* (Nesbitt 1962, Olmo 1942a, 1952). Polyploid bud sports have been commonly observed and are typically recognizable by increased berry size (Einset and Pratt 1954). Utilizing polyploid germplasm or manipulating ploidy level to increase fruit size has been conducted in some instances but the autotetraploids often have poor fruitfulness, low vigor, brittle shoots, and decreased cold hardiness (Olmo 1942a, 1952, Ourecky et al. 1967). Most success with the breeding of tetraploid grapes has occurred in Japan where many cultivars have been released that are often derived from *V. vinifera* and *V. labrusca* hybrids (Shirasi et al. 1986). Triploid grapes have been released and the Japanese cultivar 'Takao' is an aneuploid with 75 chromosomes (Ashikawa 1972, Notsuka et al. 2000).

Wild *Vitis* species, including the wild progenitor of *V. vinifera* are dioecious. The appearance of perfect flowers during the domestication and early selection of cultivated *V. vinifera,* and the hybridization of *V. vinifera* with additional species, has lead to the nearly universal presence of perfect flowered cultivated grapevines. Sex expression in *V. vinifera* appears to be controlled primarily by a single locus exhibiting a dominance series in which the expression of staminate flowers is dominant to perfect flowers, which in turn are both dominant to the expression of pistillate flowers (Antcliff 1980, Levadoux 1946). Hermaphroditism has also arisen in *V. rotundifolia*, although the genetic control of this character appears to differ depending on the initial genetic source (Dearing 1917, Detjen 1917, Loomis 1948, Loomis and Williams 1957, Loomis et al. 1954, Reimer and Detjen 1910, Williams 1954).

Grape species of the same chromosome number are highly inter-fertile. Geographical isolation and differences in flowering time appear to be the primary forces in maintaining species identity in natural environments, although interspecific hybrids can be observed when species boundaries do overlap. Selfing of hermaphroditic cultivars is possible, although inbreeding depression is typically observed and can be severe. Crossing between sections has had only limited success due to the differences in chromosome number. However, a small number of viable offspring can be recovered and utilized in breeding programs (Bloodworth et al. 1980, Bouquet 1986, Olmo 1971, Ramming et al. 2000). *V. rotundifolia*, a 40 chromosome member of the section *Muscadinia*, has been identified as a source of dominant resistance to the primary fungal disease of grape worldwide, powdery mildew. Crosses between *V. rotundifolia* and *V. vinifera* have yielded breeding lines and genetic resources that have been useful in determining the nature of this resistance (Bouquet 1986, Doligez et al. 2002, Donald et al. 2002).

The number of existing cultivars of *V. vinifera* has been estimated to be approximately 5,000 (Alleweldt and Dettweiler 1994, This et al. 2006). Due to the ease of asexual propagation, the age of some cultivars, the ease by which desirable cultivars can be transported, and the importance of viticulture in many regions, a situation has arisen in which there are a large number of synonyms and homonyms of cultivar names. Most of the grape-growing countries of the world maintain grape germplasm collections and microsatellite markers have been extensively used to better characterize and inventory those collections (Aradhya et al. 2003, Lopes et al. 1999, Martin et al. 2003, Sefc et al. 2000). A reference set of cultivars and markers has been put forth to ease comparisons amongst locations (This et al. 2004). An international database exists for grape genetic resources (http://www.genres.de/eccdb/vitis) but does not currently include molecular marker data.

Considering the thousands of cultivars of *V. vinifera*, there has been substantial interest in utilizing molecular markers for germplasm management, assessment of genetic diversity, and determination of degrees of relatedness among cultivars and wild accessions (Dangl et al. 2001, Lopes et al. 1999, Thomas et al. 1994). Molecular markers, primarily microsatellites, have been used to identify the parents of many major cultivars of *V. vinifera*, including 'Syrah', 'Cabernet Sauvignon', 'Müller-Thurgau', 'Muscat Hamburg', and 'Petite Sirah' (Bowers and Meredith 1997, Cervera et al. 1998, Crespan 2003, Dettweiler et al. 2000, Lopes et al. 2006, Meredith et al. 1999, Vouillamoz and Grando 2006). Notably, the cultivars 'Pinot' and 'Gouais' have been shown to be the parents of a large number of important European cultivars, including 'Chardonnay', 'Auxerrois', 'Gamay noir', and 'Melon' (Bowers et al. 1999a). Similarly, microsatellites have been employed to trace the geographic origin of cultivars that have been introduced to areas outside the region of initial cultivation (Maletic et al. 2004).

Microsatellite markers have also been developed in additional *Vitis* species, as well as tested for transferability amongst species and interspecific hybrids (Pollefeys and Bousquet 2003, Sefc et al. 1999). The ability to transfer microsatellites broadly across species and interspecific hybrids has useful applications in the molecular fingerprinting of grape rootstocks (Lin and Walker 1998).

Molecular markers have been used to better understand relationships amongst autochthonous cultivars and to identify synonyms and homonyms within numerous collections of cultivars around the world, including: Italy (Labra et al. 2003, 2001, Rossoni et al. 2003), Iran (Fatahi et al. 2003), Spain (Martin et al. 2003), Portugal (Lopes et al. 2006) Albania (Ladoukakis et al. 2005), Turkey (Ergul et al. 2006), Japan (Goto-Yamamoto et al. 2006), and Bulgaria (Hvarleva et al. 2004) as well as groups of ambiguous cultivar names, such as the Pinots (Regner et al. 2000), and Trebbiano (Labra et al. 2001).

7.3 History of Improvement

Archaeological evidence suggests that the early domestication of grapes spread first from the mountainous regions between the Caspian and Black Seas to regions southwards in the Jordan Valley, Egypt, and the western side of the Fertile Crescent by 5000 B.P. (McGovern 2003, McGovern and Michel 1995, Zohary and Hopf 2000). Continued western expansion of viticulture occurred in Crete and both coasts of the Iberian and Italian peninsulas by approximately 2800 B.P. (McGovern 2003). Viticulture and wine production then spread throughout the Mediterranean and were an important part of the cultures of ancient Egypt, Greece, and Rome; although it is possible that separate domestication events took place in other regions within the range of *V. vinifera* ssp. *sylvestris*. Wine and grapes remained an important component of European culture and agriculture through the middle ages and continues to be so.

Controlled crosses of grapevines for cultivar improvement are known to have been conducted prior to the discovery of Mendel's laws and before the spread of North American pest and pathogens around the world. Louis Bouschet de Bernard, and later with his son Henri, is believed to have begun generating hybrids between 'Teinturier du cher' and 'Armaon' in 1824 on his estate, La Calmette, in Mauguio in southern France (Paul 1996). These crosses lead to the intensely pigmented varieties possessing color within the berry flesh as well as the skin. The Bouschet crosses resulted in the successful cultivars 'Petit Bouschet' and 'Alicante Boucschet' and more recently were used to develop 'Rubired', a highly pigmented interspecific hybrid that is widely planted in California.

The birth of modern grape breeding is intimately intertwined with the arrival of North American diseases and insects on European shores. In successive waves in the mid 19th-century the root louse, phylloxera (*Daktulosphaira vitifoliae* Fitch) and a cadre of fungal pathogens, principally powdery mildew (*Uncinula necator* Burr), downy mildew (*Plasmopara viticola* Berl.), and black rot (*Guignardia bidwellii* Ellis) were exported to European vineyards where they caused substantial losses on the highly susceptible *V. vinifera* vines planted there. While phylloxera has received more attention as a primary motivator to develop modern grape breeding, it is important to note that the three fungal pathogens were also crucially important instigators.

Several major advances in viticulture and grape breeding occurred as a result of the epidemics spreading through Europe in the late 19th and early 20th centuries. The first was the advent of rootstock breeding as an effective and immediate means to control phylloxera that allowed European vignerons to continue to cultivate their historically important scion cultivars. Successive waves of wild vines from North America were first imported to be used as rootstocks, principally cuttings of *V. riparia*, and *V. rupestris*, that provided phylloxera resistance. Subsequent importations of *V. cinerea* var. *helleri* (*V. berlandieri*) vines for their combined resistance to phylloxera and adaptation to calcareous soils provided much of the initial genetic material, along with selections of *V. aestivalis* var. *lincecumii*, in the earliest wave of grape rootstock breeding (Campbell 2005). Many significant breeders, nurserymen, and viticulturists working in the late 19th century to early 20th century generated many of the grape rootstocks that have been planted around the world: Federico Paulsen in Sicily – *berlandieri* × *rupestris* hybrids (775, 779, 1103, 1447), Sigmund Teleki in Hungry – *berlandieri* × *riparia* hybrids (5BB, 5C, 8B, 125AA), Richter, Ruggeri, Kober, Alexis Millardet in Bordeaux (101–14, 420A, 41B), and Georges Couderc (C.3309 *riparia* × *rupestris*).

Early development of fungicides, in the form of Bordeaux mix (hydrated lime and copper sulphate) allowed for the limited control of the suite of fungal pathogens that had infiltrated European vineyards. These control measures were far from perfect as application of Bordeaux mix was time consuming, required frequent applications and a high-level of vigilance on the part of the viticulturists tending their vines. Breeding programs to develop cultivars that possessed resistance to phylloxera as well as the fungal pathogens in one vine were begun as early as 1874. Collectively these hybrid vines became known as 'Les hybrids producteurs directes' (the hybrid direct producers, HPDs) (Cahoon 1998). Many of the initial HPDs were imported from the United States and would later be outlawed in France: 'Clinton', 'Noah', 'Herbemont', 'Othello' and others. These cultivars were primarily hybrids of *V. labrusca*, *V. aestivalis*, *V. riparia*, and *V. vinifera*. Due to the unpopularity of flavors associated with *V. labrusca*, breeders in France attempted to produce HPDs without utilizing this species.

The early French breeders primarily relied on *V. rupestris*, *V. riparia*, and *V. aestivalis* var. *lincecumii*. These breeders included Eugene Contassot, Albert Seibel, Georges Couderc, Fernand Gaillard, Francois and Maruice Baco, Bertille Seyve, Eugene Kuhlmann, Pierre Castel and Christian Oberlin. Additional French grape breeders who were active during the 20th century were: Bertille Seyve-Villard, Joannes Seyve, J-F Ravat, Joanny Burdin, Jean-Louis Vidal, Aflred Galibert, Peirre Landot, and Eugen Rudelin.

The hybrid direct producers developed by these grape breeders accounted for substantial grape acerage in Europe for many years in the early 20th century. However, European acreage of hybrids has precipitously decreased over the course of the 20th century, due to improvements in fungal pathogen control and the desires of grape-growers and government officials to discourage the planting of these varieties. Some of the original hybrid grapes continue to be grown in Europe, notably 'Baco blanc' for distillation in the Armagnac region of France.

Many of the hybrid direct producers or French-hybrids had a much longer life as superior wine grape cultivars in the eastern United States and Canada. The intense disease and climatic challenges of growing *V. vinifera* cultivars in eastern North America caused many to cultivate the hybrids for grape and wine production. These cultivars and breeding lines developed in France also have had a major role in providing parental genotypes for many North American breeding programs.

American grape breeding began as early as 1830 when William Robert Prince and Dr. William W. Valk of Flushing, NY heeded the advice of Harvard University's Professor Nuttall who suggested the development of 'hybrids betwixt the European vine and those of the United States which would better answer the variable climates of North America' (Cattell and Miller 1980). Dr. Valk named the first reported cultivar as the result of a cross between a native American variety and *V. vinifera*, 'Ada', in 1852. Other notable grape breeders of the mid-19th century in the United States include E.S. Rogers of Roxbury, Massachusetts, J.H. Ricketts of Newburgh, NY, and Jacob Moore of Brighton NY who developed the important early varieties 'Brighton', 'Diana', 'Hamburg', and 'Diamond'. Hermann Jaeger and Jacob Rommel in Missouri (Rommel produced 'Elvira') also developed many cultivars and had a direct influence on Thomas Volney Munson.

T.V. Munson of Denison, TX became one of the most significant early grape hybridizers and botanists in the United States (McLeRoy and Renfro Jr. 2004) and published the influential book *Foundations of American Grape Culture* (Munson 1909). Munson also had a significant role in providing rootstock material to French breeders and viticulturists looking for parental material for phylloxera resistant rootstocks, particularly those that would be adapted to highly calcareous soils. Some of Munson's more notable cultivars are 'America', 'Bailey', 'Brilliant', 'Headlight', and 'President'.

There have been many notable contributions to the field of grape breeding during the 20th century in North America. The program at the New York State Agricultural Experiment Station in Geneva, NY has generated and provided parental material for many of the earliest U.S. grape breeding programs; early breeding efforts were lead by Richard Wellington and John Einset. Notable introductions include the first seedless grape varieties adapted to the Eastern U.S.: 'Interlaken', 'Himrod', and 'Einset' as well as many high-quality wine grapes: 'Cayuga White', and 'Traminette'.

Loren Stover and John Mortenson, of the University of Florida have developed several hybrids by crossing existing French and American hybrids varieties with material derived from species adapted to the climate and disease pressures of the south eastern United States, such as *V. cinerea* var. *floridana*, *V. aestivalis* var. *aestivalis*, and *V. shuttleworthii* (Okie 1997). The cultivars released by this program are collectively termed 'Florida Bunch Grapes' and provide a novel source of germplasm for hybrid grape production in the southeast United States not dependent on *V. rotundifolia* based cultivars. Major cultivars released from this program include: 'Lake Emerald', 'Blue Lake', 'Conquistador', and 'Blanc du Bois'.

Additional grape breeders working in the Southeastern United States at the end of the 19th century were T.V. Munson, F.C. Reimer and L.R. Detjen (Reimer and Detjen 1914). They made hybridizations of *Euvitis* as well as *Muscadinia* grapes, and hybridizations between the two sections. Additional muscadine breeding continued throughout the 20th century by many individuals, including : B.O. Fry and Ronald P. Lane (Georgia Agricultural Experiment Station), R.G. Goldy and W.B. Nesbitt (North Carolina State University), Carlos F. Williams (USDA-North Carolina St. University), and Robert Dunstan (North Carolina) (Dunstan 1962).

Important grape breeding programs for the colder regions of North America have existed in Summerland, British Columbia, Vineland, Ontario, and the University of Minnesota. A pioneer of grape breeding for cold climates was Elmer Swenson of Osceola, Wisconsin. Swenson released many notable cultivars, including: 'St. Pepin', 'Lacrosse', 'St. Croix', and 'Prairie Star'. Many of these releases are complex hybrids utilizing French and American hybrid cultivars as well as North American species adapted to cold regions, especially *V. riparia*. Active grape breeding continues at the University of Minnesota and at the University of Guelph's Vineland Station in Ontario.

Grape breeding has also been conducted in the Midwestern and Plains of the United States. Key figures included N.E. Hansen of the South Dakota Agricultural Experiment Station and H.C. Barrett of the University of Illinois. A program focusing primarily on the development of improved table grape cultivars for the lower Midwest is still active at the University of Arkansas, a program initiated by J.N. Moore. Introductions from this program include: 'Reliance', 'Mars', 'Jupiter', and 'Neptune'.

Major contributions to 20th century grape breeding were also made by the group of Elmer Snyder, Frank N. Harmon, and J.H. Weinberger with the USDA in Fresno, CA. They developed important rootstocks and table grape varieties. Cultivars that have had major impact on table grape production and breeding include: 'Cardinal' and 'Flame Seedless'. Dr. Harold Olmo at the University of California, Davis released several table grape and wine grape cultivars of major significance including 'Rubired' and 'Ruby Cabernet' wine grapes, and 'Perlette', and 'Redglobe' table grapes. Dr. Olmo also has had extensive influence through his research and publications on many aspects of grape breeding and genetics.

Wine grape production in the 20th century has primarily been dominated by traditionally grown *V. vinifera* cultivars. However, there have been some significant contributions from grape breeders working primarily or exclusively with *V. vinifera* hybridizations. Notably, the cultivar 'Müller-Thurgau', which is the most widely planted grape in Germany, was developed by Dr. Hermann Müller-Thurgau at the Geisenheim research station at the end of the 19th century. Luigi Pirovano, along with his son Alberto introduced over 500 cultivars of table grapes in Italy during the 20th century, including the very important cultivars 'Italia', 'Verona', and 'Sultana Moscato'.

7.4 Current Breeding Efforts

Most significant grape producing countries in the world maintain active grape breeding programs. Seedless table grape breeding is perhaps the most active area, in which new cultivars continue to be produced at a rapid pace. The introduction of new wine grape cultivars has generally been difficult, particularly in regions in which traditional cultivars of *V. vinifera* thrive. However, there are significant portents of future changes. Wine grapes are an important agricultural commodity for many countries, and the genetic vulnerability is high, and there is an interest in growing grapes in many regions of the world in which *V. vinifera* is not well adapted. The recent introduction of interspecific hybrids in Germany, the continued expansion of viticulture in the New World and into non-traditional production regions, potentially threatening disease pressures, and continued advances in the understanding of grapevine genetics and the utilization of that knowledge in grape breeding all suggest that continued development of improved grape cultivars will continue for the foreseeable future.

Publicly funded grape breeding continues in North America at Cornell University in New York, University of Minnesota, University of California-Davis, USDA-ARS Geneva, NY, USDA-ARS Parlier, CA, University of Georgia (muscadines), Florida A&M University (muscadines and florida bunch grapes), the University of Arkansas and the University of Guelph Vineland, Ontario. Additionally, there are several privately run table grape breeding programs in the United States that have made significant contributions to the table grape industry worldwide, notably Sunworld of Bakersfield, California with their introduction of 'Sugraone' (Superior Seedless™) which accounts for significant acreage of seedless table grapes worldwide.

Many programs in Europe are active and continue to release varieties in all of the major grape producing countries. One program of particular note is at Geweilerhof, Germany which has successfully introduced interspecific hybrid wine grape cultivars, such as 'Regent' and 'Sirius', under the rubric of high quality wine grapes traditionally reserved for *V. vinifera* based cultivars only. There are several major table and wine grape breeding programs in Europe including programs in: Spain, Germany, France, Italy, and Hungary.

Outside of Europe and North America, seedless table grape breeding is being conducted in Israel (e.g. 'Prime' and 'Mystery'), Chile, and Australia ('Maroo seedless', 'Millennium Muscat'). Breeding efforts in Australia are also producing new raisin ('Sunmuscat') and wine grape cultivars ('Tyrian', 'Cienna', and 'Rubienne'). Doubtless, additional grape breeding is occurring and producing outstanding locally adapted cultivars in many regions that have yet to gain a large international audience.

7.5 Genetics of Important Traits

Very little experimental information has been generated concerning the inheritance of important traits of grapevine. Grapes have a relatively long generation time, and require a large amount of land and resources to maintain large populations of vines.

Some phenotypic traits have been speculated to be controlled by single loci that segregate in a Mendelian fashion, but the data supporting these conclusions are often scant to non-existent in the scientific literature. Similarly, information on the heritability, and combinatory gene action of quantitative traits is virtually non-existent in grapes. Large numbers of somatic mutants have been identified in which the phenotypic alteration is presumably derived from a lesion in a single gene, but in most cases these resources have yet to submit to precise genetic analysis.

7.5.1 Pest and Disease Resistance

There are many major fungal pathogens of grape. Powdery mildew is the most significant fungal disease of grape as it affects many production regions worldwide. Several sources of powdery mildew resistance have been identified among North American *Vitis* species (Table 7.2). Among the 38 chromosome species of the *Euvitis* section, resistance is quantitatively inherited, in which narrow-sense heritability estimates have been made ranging from 0.31 to 0.51 (Eibach et al. 1989). A single, dominant locus for resistance to powdery mildew, *Run1*, has been identified from the 40 chromosome *V. rotundifolia*. The locus has been introgressed into a *V. vinifera* background in which multiple generations of backcrossing have now occurred (Bouquet 1986). The *Run1* locus has now been mapped, first by identifying candidate genes in the region showing similarity to conserved plant resistance genes (Donald et al. 2002, Pauquet et al. 2001) and subsequently through the fine genetic and physical mapping of this locus (Barker et al. 2005). The map-based cloning of *Run1* and confirmation of resistance through transformation is now taking place.

The analysis and mapping of quantitative trait loci (QTL) in grape has thus far concentrated on the study of resistance loci to several important diseases. QTL for powdery mildew and downy mildew resistance have been identified in multiple interspecific crosses which have utilized *Euvitis* sources of resistance (Dalbo et al. 2001, Fischer et al. 2004). Additional efforts to identify candidate genes that have a high probability of being linked to disease resistance loci has been conducted by identifying resistance gene analogs and resistance gene-like genes from numerous grape species (Di Gaspero and Cipriani 2002, 2003). Recently an ergosterol-induced gene, *VvLTP1* was implicated in the natural protection mechanisms against the pathogen *Botrytis cinerea* (Laquitaine et al. 2006).

Downy mildew resistance has been identified in several North American species and is quantitatively inherited. Resistance to downy mildew has also been reported for several Chinese *Vitis* species despite the causal pathogen not being endemic to Asia (He and Wang 1986). However, many Chinese grape species thrive in regions of high humidity and moisture and may contain broad-scale resistance to numerous pathogens. Narrow-sense heritability of 0.26–0.39 and broad-sense heritability of 0.83 to 0.94 have been estimated (Eibach et al. 1989).

There have been few studies examining the inheritance or genetic control of resistance to additional fungal pathogens. The narrow sense heritability of *Botrytis* resistance, measured as the estimated stilbene production in leaf tissue, was estimated

Table 7.2 Sources of disease resistance in grapes

Disease	Species	References
Fungal		
Powdery Mildew	*V. riparia, V. aestivalis, V. cinerea,* *V. cinerea* var. *helleri, V. rotundifolia*	Alleweldt et al. 1990 Pearson et al. 1988
Downy Mildew	*V. riparia, V. rupestris, V. aestivalis* var. *lincecumii, V. labrusca, V. amurensis,* *V. rotundifolia, V. yenshanensis,* *V. pseudoreticulata, V. piasezkii,* *V. romanetii, V. flexuosa, V. bryoniifolia*	Alleweldt et al. 1990 Eibach et al. 1989 He and Wang 1986
Black rot	*V. riparia, V. mustangensis, V. rotundifolia,* *V. cinerea, V. rupestres*	Alleweldt et al. 1990 Jabco et al. 1985 McGrew 1976
Anthracnose	*V.cinerea* var. *floridana, V. aestivalis* var. *aestivalis, V. shuttleworthii, V. labrusca,* *V. rotundifolia, V. rotundiolia* var. *munsoniana*	Mortenson 1981 Olmo 1986
Botrytis bunch rot	*V. vinifera, V. riparia, V. rupestres*	Alleweldt et al. 1990
Rust	*V. shuttleworthii, V.cinerea* var. *floridana,* *V. rotundifolia, V. tiliifolia*	Fennell 1948
Rotbrenner	*V. vinifera, V. cinerea*	Alleweldt et al. 1990
Bacterial		
Pierce's disease	*V. rotundifolia, V. mustangensis, V. ×* *champinii, V. vulpine, V. shuttleworthii,* *V.cinerea* var. *floridana, V. aestivalis* var. *aestivalis, V. arizonica*	Mortenson 1977 Olmo 1986 Stover 1960 Krivanek et al. 2005
Crown gall	*V. amurensis, V. labrusca*	Alleweldt et al. 1990 Szegedi et al. 1984 Pearson et al. 1988
Flavescence doree	*V. labrusca, V. rupestris*	Pearson et al. 1988
Virus		
Grapevine fanleaf virus	*V. rotundifolia, V. vinifera, V. arizonica,* *V. aesitvalis* var. *aestivalis, V. × slavinii,* *V. mustangensis, V. riparia*	Walker et al. 1985 Walker and Meredith 1990 Bouquet 1981

at 0.82–0.92 (Eibach et al. 1989). Resistance to black rot in fruit was postulated to be controlled by 2 dominant genes (Mortenson 1977), while a contrasting report suggests black rot resistance is quantitatively controlled (Barrett 1955). Anthracnose resistance is thought to be quantitatively inherited (Mortenson 1981).

Pierce's disease has traditionally limited the cultivation of *V. vinifera* in the southeastern United States and in recent years has become a more serious concern in California due to the spread of insect vectors capable of spreading the causal bacterium to wider production regions. Mortenson (1968) suggested that the resistance of Pierce's disease is a dominant trait, and qualitatively controlled by three independent loci. He came to this decision after observing the segregation of resistance for five years in several populations derived from *V. aestivalis* var. *aestivalis, V. cinerea* var. *floridana*, and *V. shuttleworthii* in Florida under field conditions. More recently, the narrow-sense heritability of Pierce's disease resistance was estimated to range from 0.37 to 0.63 for different populations of the pathogen, *Xylella fastidiosa*, in a hybrid population derived from *V. rupestris* × *V. arizonica* (Krivanek et al. 2005). These results indicated the existence of a major gene for Pierce's disease resistance, *PdR1*, which has now been placed on a genetic linkage map of this cross (Krivanek et al. 2006, Riaz et al. 2006). To date, no additional loci for additional bacterial, fungal, or viral pathogens have been placed on molecular maps.

Resistance to additional bacterial pathogens of grape, including crown gall, which is of significant concern, particularly in many cooler viticultural climates, has been identified from several wild sources (Table 7.2), yet no published reports concerning the inheritance of these traits have been produced.

Grapevine fanleaf virus is the most damaging viral pathogen of grape worldwide. This pathogen can be at least partially controlled through the use of rootstocks resistant to the nematode vector, *Xiphinema index*. Resistance to the virus itself has been identified in some wild *V. vinifera* accessions and *V. rotundifolia* (Walker et al. 1985). The *vinifera* source of resistance was suggested to be controlled by at least two genes with resistance being a recessive trait (Walker and Meredith 1990). No other reports on the genetic control of virus existence are known in grapevine.

Resistance to the vector of grapevine fanleaf virus, *Xiphinema index*, has been observed to be controlled by a small number of genes, potentially one dominant and one recessive (Meredith et al. 1982). Numerous sources of resistance for the root-knot nematode *Meloidogyne incognita* have been identified (Table 7.3). Resistance has been described as being primarily due to a single, dominant gene found in *V. champinii*, *V. mustangensis* and the interspecific rootstock 1613C (Lider 1954). Hybrids of *V. vinifera* × *V. rotundifolia* and their derivatives have been used to estimate the heritability of root-knot nematode resistance at 0.391 (Firoozabady and Olmo 1982a). More recently, a series of crosses amongst six pistillate-flowered rootstocks and 4 staminate-flowered rootstocks showed segregation ratios consistent with the presence of a single, dominant allele for root-knot nematode resistance (Cousins and Walker 2002).

Phylloxera is a major insect pest of *V. vinifera* and necessitates the grafting of *V. vinifera* onto resistant rootstocks in most grape production regions around the world. Several sources of resistance to phylloxera have been identified and

Table 7.3 Sources of resistance to insect pests

Insect	Species	References
Root knot nematodes	*V.* × *champinii*, *V. mustangensis*, *V. rotundifolia*, *V. nesbittiana*, *V.* × *slavinii*, *V. aesitvalis* var. *aestivalis*, *V. vulpina*	Lider 1954, Firoozabady and Olmo 1982, 1986, Bloodworth et al. 1980, Cousins and Walker 2002, Boyden 2005, Anwar et al. 2002
Dagger nematodes	*V. aesitvalis* var. *aestivalis*, *V. cinerea*, *V. rutondifolia*	Alleweldt et al. 1990, Meredith et al. 1982, Becker and Sopp 1990
Phylloxera	*V. riparia*, *V. rupestris*, *V. cinerea* var *helleri*, *V. cinerea*, *V.* × *champinii*, *V. rotundifolia*	Alleweldt et al. 1990, Olmo 1986

resistance is thought to be quantitatively controlled (Boubals 1966, Firoozabady and Olmo 1982b). More detailed studies on the inheritance of phylloxera have not been reported, although it is known that resistance can be overcome if rootstocks possessing a significant portion of the *V. vinifera* genome in their background are employed, such as AXR1.

7.5.2 Resistance to Abiotic Stresses

Cultivation of *V. vinifera* has spread globally to many regions that possess a climate amenable to its survival, typically with dry summers and mild winters. *Vitis vinifera* is not-well adapted to climates with severe winter temperature or regions possessing large fluctuations in temperature during the dormant season, and this has led directly to the use of wild germplasm to improve abiotic stress tolerance. *V. vinifera* is a relatively drought tolerant grape species and breeding directly for improvements in drought tolerance have been limited. If water becomes a more limiting factor and the potential problems of increased soil salinity in irrigated regions arise, additional investigation into genetic sources of water deficit and salt stress are likely to occur. Several sources of resistance for abiotic stress: cold, drought, salt, and high pH, have been identified (see Table 7.4), yet to our knowledge no work on the inheritance or genetic control of these traits has ever been reported. Recently, QTL for magnesium-deficiency were identified and placed on a map of 'Welschriesling' × 'Sirius' (Mandl et al. 2006).

Low-temperature stress is a significant concern in many of the more extreme climates in which grape production has been extended, notably in northern North America and northern Europe. A family of low-temperature-induced regulatory genes, the CBFs, which are of key importance in regulating cold acclimation in numerous plant species have recently been identified in both *V. vinifera* and *V. riparia* (Xiao et al. 2006).

Additional work examining the processes leading to dormancy induction and release has been conducted to better understand this critical aspect of winter survival as well as for the manipulation of dormancy in climates with insufficient chilling for dormancy release (Or et al. 2002, Pacey-Miller et al. 2003, Wake and Fennell 2000).

Table 7.4 Sources of adaptation to abiotic stress

Stress	Species	References
Cold Damage	*V. riparia*, *V. labrusca,V. amurensis,V. acerifolia,V. vulpina,V. adstricta*	Alleweldt et al. 1990, He and Lixin 1989, Luby 1991
Drought stress	*V. vinifera, V. rupestris,V. champinii,V. cinerea* var. *helleri*	Alleweldt et al. 1990, During 1986
Iron chlorosis	*V. vinifera, V. cinerea* var. *helleri*	Alleweldt et al. 1990, Pouget 1980
Salinity	*V. cinerea* var. *helleri*, *V.* × *champinii, V. acerifolia*	Alleweldt et al. 1990, Antcliff et al. 1983, Galet 1988

7.5.3 Fruit Quality

Many factors contribute to and ultimately determine grape fruit quality, particularly in the case of wine grapes. Visual attributes, such as color, berry size, and cluster size are critical for acceptance of table grapes and in some cases wine and juice grapes. Considerable variation exists for all these traits (Fig. 7.1). Critical factors contributing to taste are sugar levels/potential alcohol content, organic acid accumulation, tannin levels and how they effect tactile impression of a wine on the palette, and aroma. Several grape secondary metabolites have also been linked to beneficial effects on human health and are of major interest for promoting quality aspects of grapes and grape products.

The genetic control and inheritance of fruit color or anthocyanin production in grapevine is not fully understood despite evidence that the primary determination of anthocyanin production in berries is controlled by a single dominant locus in *V. vinifera,* with white fruit being recessive (Doligez et al. 2002, Riaz et al. 2004). This observation is supported by numerous reports showing that controlled crosses between white fruited vines universally result in white fruited progeny (Barritt and Einset 1969, Hedrick and Anthony 1915, Madero et al. 1986, Snyder and Harmon 1939, 1952, Wellington 1939).

Recently, it has been shown that white-fruited cultivars of *V. vinifera* carry *Gret1,* a Ty3-*gypsy*-type retro-transposon in the promoter region of a *myb*-like regulatory gene with sequence similarity to anthocyanin regulators from maize and other plants (Kobayashi et al. 2004). *Gret1* is recessive, with pigmented cultivars carrying at least one allele without this insertion at the *VvmybA1* locus. *VvmybA1* co-segregates with the berry color locus (Lijavetzky et al. 2006) and mutations in *VvmybA1* are

Fig. 7.1 Diversity found in grape color, size and bunch composition

associated with the vast majority of white fruited *V. vinifera* accessions, as well as many pink and red accessions (Lijavetzky et al. 2006, This et al. 2007). However, there is a large amount of phenotypic diversity in grapevine and it is likely that additional loci and alleles will be identified that further explain the genetic control of variation in flavonoid compounds.

The inheritance of muscat aroma has been investigated and hypothesized to be controlled by five complimentary dominant genes plus a modifier gene (Wagner 1967). The concentration of volatile monoterpenol compounds can vary considerably amongst differing muscat varieties and can easily be influenced by environmental conditions. No heritability estimates for muscat aroma components have been identified, but QTL for muscat flavor and specific monoterpenes have been placed on a linkage map of the cross ('Olivette' × 'Ribol') × 'Muscat Hamburg' (Doligez et al. 2006). Methyl anthranilate, the dominant volatile aroma in *V. labrusca* and its hybrid derivatives, has been hypothesized to be controlled by three dominant, complementary genes (Reynolds et al. 1982).

Many differing and contradictory hypotheses have been put forth over the last 70 years to explain the inheritance of stenospemocarpic seedlessness in grapes. Many of these hypotheses are based on small population sizes and limited numbers of populations. Additionally, the phenotypic definition of stenospermocarpic seedlessness can vary and should take into account such factors as: seed trace size, ovule abortion, hardness of the seed coat, and extent of endosperm development, each of which may be under different genetic control. The most recent hypothesis to explain the inheritance of seedlessness attempts to take into consideration all prior reports on the segregation patterns of this trait and concludes seedlessness is controlled by three independent recessive genes plus one dominant acting regulator gene (Bouquet and Danglot 1996).

The inheritance of seedlessness has been extensively studied for much of the past century, and several markers linked to seedlessness have been identified (Adam-Blondon et al. 2001, Lahogue et al. 1998). Additional fruit quality traits are just beginning to be the subject of molecular genetic analysis and early reports of the mapping of QTL for yield components and muscat aroma and monoterpene content are emerging (Doligez et al. 2006, Fanizza et al. 2005). The major sex expression locus has now been mapped in multiple populations (Dalbo et al. 2000, Riaz et al. 2006).

7.5.4 Inheritance of Additional Traits

Grapes are a high value crop, and as such, extensive cultural manipulations are carried out in the vineyard to maximize fruit quality at the highest sustainable yields for a target quality level. Knowledge of the molecular genetic regulation of factors affecting such areas as yield, vigor, kinetics of fruit ripening, and the interaction between fruit quality and cultural manipulations as well as with the environment are of significant interest to grape growers in order to obtain a better fundamental understanding of how to manipulate fruit quality.

Many of these traits are poorly studied at the genetic level and are thought to be quantitatively inherited. The inheritance of yield components has been studied by several investigators and the broad sense heritability has been estimated as moderate to high for cluster compactness, berry weight, skin texture, and pulp texture (Firoozabady and Olmo 1987, Schneider and Staudt 1979). Recently, QTL for grape yield components have been identified, (berry weight, berries per cluster, clusters per vine, cluster weight) and placed on a genetic map of an 'Italia' (seeded) × 'Big Perlon' (seedless) cross (Fanizza et al. 2005). The heritability of time to ripening has been estimated and shown to be of moderate value (Fanizza and Raddi 1973, Hedrick and Anthony 1915).

Sex expression in *V. vinifera* appears to be primarily controlled by a single locus exhibiting a dominance series in which the expression of staminate flowers is dominant to perfect flowers which in turn are both dominant to the expression of pistillate flowers (Antcliff 1980, Levadoux 1946).

The molecular genetic analysis and characterization of interesting physiological and developmental processes is rapidly expanding in grapevine, particularly, but not exclusively in *V. vinifera*. Considering the importance of many flavonoids for grape, juice, and wine quality, their role in promoting human health, and their role in abiotic and biotic stress responses, significant efforts have been made to further characterize the members of the phenylpropanoid pathway in grapevine (Bais et al. 2000, Bogs et al. 2005, 2006, Boss et al. 1996a,b, Castellarin et al. 2006, Deluc et al. 2006, Downey et al. 2003, 2004, Kobayashi et al. 2001, Sparvoli et al. 1994).

Additional aspects of grape berry development have been examined, notably sugar metabolism (Ageorges et al. 2000, Atanassova et al. 2003, Cakir et al. 2003, Conde et al. 2006, Davies et al. 2006, Fillion et al. 1999, Leterrier et al. 2003, Vignault et al. 2005), potassium transporters (Davies et al. 2006), brassinosteroids (Symons et al. 2005), aquaporins (Picaud et al. 2003), cell-wall modifying enzymes (Nunan et al. 2001), and abscisic acid signaling (Cakir et al. 2003).

In addition to berry development, the regulation of grapevine flowering and inflorescence development is of significant interest. Many of the flowering regulatory genes identified and characterized in model species have been identified in *V. vinifera* and subsequently characterized in greater detail (Boss et al. 2006, Joly et al. 2004, Sreekantan and Thomas 2006, Sreekantan et al. 2006).

7.6 Crossing and Evaluation Techniques

7.6.1 Breeding Systems

Wild *Vitis* species are universally dioecious. The presence of hermaphroditic flowers is restricted to cultivated *V. vinifera* and some cultivated selections of *V. rotundifolia*. The three major types of grape flowers are hermaphroditic, staminate with an un-developed pistil, and pistillate with reflexed stamens. However, intermediate types are observed, and environmental conditions appear to have the capability of altering the sex expression of certain genotypes to a small degree. All of the approx-imately 60 members of the section *Euvitis* are inter-fertile.

Wild grape species are obligate outcrossers and the hermaphroditic cultivars that have been developed are relatively sensitive to inbreeding depression, although it is possible to recover viable offspring vigorous enough to produce fruit. Most scion breeding programs rely on a modified pedigree breeding scheme in which elite parents are chosen for each generation of inter-crossing. Wild species have been frequently used as sources for desirable traits, but often require multiple generations of backcrossing to cultivated types to recover adequate fruit quality. However, the needs of worldwide breeding programs can vary considerably and often have their own singularities.

7.6.2 Pollination and Seedling Culture

Inflorescences are produced in the buds produced during the growth of shoots of the preceding year. Final differentiation of the inflorescence does not occur until the spring of the current year's bloom. The whorls are produced in the following order: sepals, petals and stamens jointly, and finally the pistils. The ovary typically has two locules, each containing two ovules (Pratt 1971). Anthesis is thought to occur either early or late in the day (Pratt 1971). Hermaphroditic cultivars will self very easily and must be emasculated in the generation of controlled crosses. Some suc-cess has been reported in the utilization of plant growth regulators to convert flower types. The greatest success in the conversion of male flowers to hermaphroditic ones has been through the use of prebloom applications of cytokinins (Boyden 2005, Hashizume and Iizuka 1981, Moore 1970, Negi and Olmo 1966, 1970).

Controlled pollination can be conducted most easily by collecting previously bagged clusters as a source of fresh pollen and then tapping them on an emasculated or pistillate cluster. Alternatively, pollen can be collected and applied with a camel's hair brush. Storage of pollen at $-20°C$ has been successful for as long as 12 months (Bamzai and Randhawa 1967, Boyden 2005, Nebel and Ruttle 1936) and in some instances up to four years (Olmo 1942b). Attempts at the cryo-preservation of pollen have also been successful in maintaining adequate levels of pollen viability for up to approximately five years (Akihama and Omura 1986, Ganeshan 1985, Ganeshan and Alexander 1990, Parfitt and Almehdi 1983).

Grape seed germination can be quite variable depending on many factors in-cluding parents, environmental and cultural conditions during fruit set, and seed handling. Seeds can be extracted through manual pressing, or a laboratory blender on low speed. Viable grape seeds will sink when immersed in water, so that copious quantities of water can be used to further extract seeds from the berry flesh and to separate the seeds. Seeds are usually extracted at or close to fruit maturity and then provided a cold stratification period or a combination of chemical treatment and cold stratification. A treatment of a 24 hour soak in water, followed by a 24 hour soak in hydrogen peroxide, a 24 hour soak in a 1% GA_4 solution and 30 days of chilling at

4°C in a moist media have proven reliable and effective in our programs. Following stratification, seeds are germinated on dampened bloater paper in a germination box or enclosed Petri dish in the dark at 30°C until radical emergence. Germinated seed is then transferred to soil-less potting media and kept constantly moist until the emergence in the greenhouse.

7.6.3 Evaluation Techniques

The juvenility period in grapes can range from 2 to 4 years or longer depending on genotype and environmental conditions. Attempts to shorten this period through pruning and growth regulator treatment have provided only moderate successes. Young seedlings were induced to convert tendrils to inflorescences following treatment with cytokinins and auxins (Boyden 2005, Srinivasan and Mullins 1981).

Embryo rescue has been a significant contributor to the breeding of seedless table grapes. The percentage of seedless progeny can be greatly increased by conducting crosses between two seedless parental genotypes (Bouquet et al. 1989, Cain et al. 1983, Emershad and Ramming 1984, Spiegel-Roy et al. 1985). Embryo rescue has also proven to be useful in increasing ovule or seed germination in early-ripening progeny (Ramming 1990, Ramming et al. 1990). Embryo rescue has been useful in transferring the seedless characteristic to hybrids of *V. rotundifolia* (Goldy et al. 1988, 1989, Ramming et al. 2000).

Seedling and selection evaluation techniques vary by program, particularly depending on the type of grapes under selection. Typically, seedlings are planted in the field and will fruit by their 3rd year in which selections can be driven by fruit quality as well as other characteristics, such as disease resistance, plant architecture, and fruitfulness. Microvinification for wine grape selections can also be made. Selections are made, and continued evaluations can occur for a substantial number of years, particularly in the case of wine grapes. A new variety is not typically released in less than 8 to 10 years following the initial hybridization.

7.7 Biotechnological Approaches to Genetic Improvement

The development of molecular genetic, genomic, and biotechnological tools for the study of grapevine has proceeded at a remarkable pace at the dawn of the 21st century. The number of researchers working in this area has increased dramatically over recent years, yet the grape geneticist suffers many of the same challenges confronting all geneticists studying woody, perennial species. Moving from the description of broad genomic regions of interest based on QTL on low-density genetic maps and the generation of large numbers of candidate genes through various '–omics' technologies to the precise and careful analysis of individual alleles affecting plant function will be the next major challenge for the grape genetics community. The manner in which these new tools and knowledge will be utilized for grapevine improvement and grapevine management in the vineyard will be the next story to follow in the history of grape breeding and genetics.

7.7.1 Genetic Mapping and QTL Analysis

Many molecular markers have been developed over the last couple of decades for use in grapevine. These molecular markers have multiple uses in grape breeding and genetics, including: variety identification and germplasm management, mapping of traits of interest, and estimation of genetic diversity. (Bowers et al. 1993, Bowers and Meredith 1996, Dalbo et al. 2000, Ye et al. 1998). The development and utilization of molecular markers has been refined over time, mimicking developments in other crops, beginning with isozymes, RFLPs and RAPDs. Since the early 1990s microsatellites have been one of the most widely used molecular marker systems in grapevine (Thomas et al. 1993, Thomas and Scott 1993) and to a lesser extent AFLPs (Cervera et al. 1998). The number of publicly available microsatellite markers has greatly expanded (Bowers et al. 1999b, 1996), including descriptions of multiplexes (Merdinoglu et al. 2005) reference sets of alleles and accessions (This et al. 2004), the development of microsatellites directly from transcribed sequence of ESTs (Scott et al. 2000b), and linkage maps based on microsatellite markers (Adam-Blondon et al. 2004, Riaz et al. 2004). The possibility of using molecular markers to distinguish the cultivar identities of mixtures of wine would be of considerable value to the wine industry, but to date attempts have only been successful prior to fermentation of the must (Faria et al. 2000, Siret et al. 2000).

Many genetic linkage maps have been constructed for grapevine since the first reported genetic map utilizing DNA based markers in 1995 (Lodhi et al. 1995). These maps represent *V. vinifera* intraspecific crosses (Adam-Blondon et al. 2004, Doligez et al. 2006, Fanizza et al. 2005, Riaz et al. 2004) as well as interspecific crosses utilizing *V. vinifera* (Grando et al. 2003), and more complex interspecific crosses (Doucleff et al. 2004, Fischer et al. 2004, Lodhi et al. 1995, Lowe and Walker 2006, Mandl et al. 2006). Grape species have been observed to have a DNA content similar to that found in rice (*Oryza sativa*), approximately 450 to 500 Mbp (Lodhi and Reisch 1995). More recently progress has been made in the construction of a physical map for *V. vinifera* based on BAC end sequencing, as well as anchored markers from previously reported linkage maps (Adam-Blondon et al. 2005, Lamoureux et al. 2006). Whole genome sequencing is proceeding and the complete genomic sequence of a grape was recently announced (Jaillow et al. 2007).

Linkage disequilibrium (LD) based association mapping is of interest to grape geneticists considering the potentially high resolution and the time savings associated with the utilization of existing germplasm collections (This et al. 2006). Significant haplotypic LD was observed over 30 cM in a *V. vinifera* core collection when estimating LD with 38 microsatellite markers scattered among the 19 *Vitis* linkage groups (Barnaud et al. 2005). Utilizing a subset of the pigmented accessions from the same core collection a much more rapid decay in LD at the single locus level was observed (This et al. 2007). The discrepancy between these results will likely be resolved as a larger number of loci from a diverse pool of germplasm are investigated.

7.7.2 Regeneration and Transformation

Somatic embryogenesis has been documented in grapevine for over thirty years (Hirabayashi et al. 1976, Mullins and Srinivasan 1976). Success has been reported primarily for the use of sporophytic anther tissue (Rajasekaran and Mullins 1979), although leaf, petioles, and stem segments have also been used to establish embryogenic calli (Krul and Worley 1977). Success in regeneration of grapevine has been limited to a relatively small number of cultivars, but the list of successful source material has been steadily increasing (Perrin et al. 2004, Torregrosa 1998). Many modifications and improvements have been reported (Iocco et al. 2001, Perl and Eshdat 1998, Perl et al. 1995, Perrin et al. 2001, Wang et al. 2004).

Early attempts to transform grape using *Agrobacterium tumefaciens* met with difficulty despite the bacterium being a naturally occurring pathogen of the species. The use of high-quality embryogenic suspension cell cultures has allowed the transformation of grape using *Agrobaceterium* to become routine in many laboratories around the world (Perl and Eshdat 1998). Biolistic transformation of a coated microprojectile was reported in grape in the early 1990s (Hebert et al. 1993) for the interspecific hybrid 'Chancellor' (Kikkert et al. 1996) and has been expanded over time to successfully include cultivars of *V. vinifera* (Vidal et al. 2003). Presently, *Agrobacterium*-mediated methods are the predominantly employed protocols for grape transformation worldwide.

Many reports exist concerning the transformation of grapevine for enhanced pathogen resistance, altered fruit quality, such as seedlessness or higher sugar content, and other traits. However, there are very few reports of stably transformed grapevines with altered phenotypes. Perl et al. (2004) has a good review of attempts to transform grapevine for various traits, approaches taken, and recent progress of the projects.

The most notable success in grapevine transformation is the insertion of antimicrobial genes (Vidal et al. 2003) and antifungal genes (Yamamoto et al. 2000) to potentially confer greater bacterial and fungal disease resistance. The insertion of magainins, short peptides with broad-spectrum antimicrobial activity into 'Chardonnay' was tested for the ability to confer resistance to crown gall and powdery mildew (Vidal et al. 2006). Lines expressing magainins appeared to be effective in reducing crown gall symptoms under controlled conditions, and to be somewhat effective in reducing powdery mildew disease symptoms, although strong powdery mildew resistance was not observed. To date, testing of disease resistance of transgenic grapevines under field conditions has not been reported. The advantages of reduced pesticide use and stable, long-term disease resistance are counter-balanced by public concern over the release of transgenic grapevines. It is difficult to foresee when, or if, this promising technology will make it to the commercial vineyard.

Transformation of grapevine has not been restricted to scion varieties but has also been used to transfer useful traits to rootstocks. *Vr-ERE*, a gene encoding a NADPH-dependent aldehyde reductase which coverts eutypine to the alcohol eutpinol has been shown in laboratory tests with both cultured *V. vinifera* cells and whole vines to have some efficacy in de-toxifying the toxin produced by the fungus *Eutypa lata* (Guillen et al. 1998, Legrand et al. 2003). Rootstocks transformed with a chimeric gene containing the alfalfa PR 10 promoter and a *Vitis* stilbene synthase gene (*Vst1*) showed enhanced foliar resistance to *Botrytis cinerea*, primarily a pathogen of fruit (Coutos-Thevenot et al. 2001). Attempts to transform rootstocks with genes potentially capable of conferring resistance to crown gall and multiple viruses has been successful, but no reports on whether or not disease resistance was conferred have been published (Xue et al. 1999). Release of transgenic grapevines is a concern considering the ease with which pollen flow can occur and the existence of wild grape species in many production regions. Assessment of the field safety of transgenic vines containing the coat protein gene from Grapevine fanleaf virus has recently been conducted (Vigne et al. 2004). A small number of nontransgenic GFLV recombinants were detected outside of the field locations, but no evidence was found that the molecular diversity of indigenous GFLV populations was affected by the presence of transgenic grapevines.

7.7.3 Genomic Resources

As of this writing there are over 300,000 publicly available Expressed Sequence Tags (ESTs) from *V. vinifera*. Comprehensive analysis of these ESTs has begun, particularly the examination of gene expression changes during berry ripening, including a comparison of shared gene expression changes between tomato and grapevine (da Silva et al. 2005, Fei et al. 2004, Moser et al. 2005). Additionally, large-scale changes in gene expression of ripening grape berries has been examined utilizing cDNA-AFLP technology (Burger and Botha 2004) and through the use of 50-mer oligonucletoide based arrays designed from a set of 3175 *Vitis* Unigenes (Ageorges et al. 2006, Terrier et al. 2005). An Affymetrix based platform for microarray analysis containing approximately 15,000 unique *Vitis* features is currently available and initial studies utilizing this tool have recently been published (Cramer et al. 2007, Espinoza et al. 2007). Cramer et al. (2007) found that as water deficit progressed, a greater number of affected transcripts were involved in metabolism, transport, and the biogenesis of cellular components than under salinity stress. Salinity resulted in a higher percentage of transcripts involved in transcription, protein synthesis, and protein fate (Fig. 7.2). Espinoza et al. 2007 found numerous genes that were induced or repressed in viral infected grapevines leaves; the primary changes in gene expression involved processes of translation and protein targeting, metabolism, transport, and cell defense.

Many naturally occurring mutants exist within the holdings of research stations and germplasm collections worldwide, particularly for *V. vinifera*, which in some cases has had thousands of years of clonal propagation to accumulate somatic

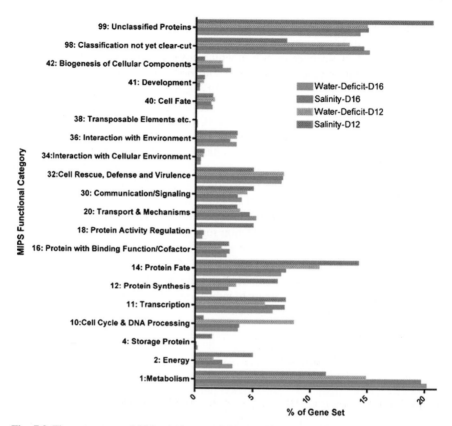

Fig. 7.2 The percentage of *Vitis vinifera* cv. Cabernet Sauvignon genes in different functional classes that were differentially expressed in salt and drought stressed plants relative to controls on day 12 and day 16 (Cramer et al. 2007)

mutations. In only a limited number of examples have mutant phenotypes been associated with a specific lesion in a candidate gene, *VvmybA1* and *VvmybA2* for berry skin color (Kobayashi et al. 2004, Lijavetzky et al. 2006, This et al. 2007, Walker et al. 2006, Yakushiji et al. 2006) and *VvGAI*, an ortholog of the 'green revolution' mutants in agronomic crops, that leads to shortened internodes and precocious conversion of tendrils to inflorescences in grapevine (Boss and Thomas 2002). Considering the highly heterozygous nature of *V.vinifera* and the potentially severe inbreeding depression that occurs upon selfing, it is likely that collections of interesting phenotypic mutants and genetic stocks will become attractive targets for molecular genetic and genomic studies. One example is the *fleshless berry* (*flb*) mutation of the cultivar 'Ugni Blanc' which possess a significantly reduced pericarp compared to wild-type (Fernandez et al. 2006b). The *flb* mutation is now being characterized in greater detail through genetic mapping and genomic technologies (Fernandez et al. 2006a).

References

Adam-Blondon A-F, Bernole A, Faes G, Lamoureux D, Pateyron S, Grando MS, Caboche M, Velasco R, Chalhoub B (2005) Construction and characterization of BAC libraries from major grapevine cultivars. Theor Appl Genet 110:1363–1371

Adam-Blondon A-F, Lahogue-Esnault F, Bouquet A, Boursiquot JM, This P (2001) Usefulness of two SCAR markers for marker-assisted selection of seedless grapevine cultivars. Vitis 40:147–155

Adam-Blondon A-F, Roux C, Claux D, Butterlin G, Merdinoglu D, This P (2004) Mapping 245 SSR markers on the *Vitis vinifera* genome: a tool for grape genetics. Theor Appl Genet 109:1017–1027

Ageorges A, Fernandez L, Vialet S, Merdinoglu D, Terrier N, Romieu C (2006) Four specific isogenes of the anthocyanin metabolic pathway are systematically co-expressed with the red colour of grape berries. Plant Sci 170:372–383

Ageorges A, Issaly R, Picaud S, Delrot S, Romieu C (2000) Identification and functional expression in yeast of a grape berry sucrose carrier. Plant Physiol Biochem 38:177–185

Akihama T, Omura M (1986) Preservation of Fruit Tree Pollen. In: Bajaj YPS (ed) Biotechnology in agriculture and forestry, vol 1. Springer, Berlin, pp 101–112

Alleweldt G, Dettweiler E (1994) The genetic resources of Vitis world list of grapevine collections, 2nd edn, BAZ IRZ Geilweilerhof, Siebeldingen

Alleweldt G, Spiegel-Roy P, Reisch BI (1990) Grapes (*Vitis*). In: Moore JN (ed) Genetic resources of temperate fruit and nut crops, Acta Hortic 290:289–327

Antcliff AJ (1980) Inheritance of sex in *Vitis*. Ann Amelior Plant 30:113–122

Anwar SA, McKenry MV, Ramming DW (2002) A search for more durable grape rootstock resistance to root-knot nematode. Am J Enol Vitic 53:19–23

Aradhya MK, Dangl GS, Prins BH, Boursiquot JM, Walker MA, Meredith CP, Simon CJ (2003) Genetic structure and differentiation in cultivated grape, *Vitis vinifera* L. Genet Res 81: 179–182

Arroyo-Garcia R, Ruiz-García L, Bolling L, Ocete R, López MA, Arnold C, Ergul A, Söylemezo Lu G, Uzun HI, Cabello F, Ibáñez J, Aradhya MK, Atanassov A, Atanassov I, Balint S, Cenis JL, Costantini L, others (2006) Multiple origins of cultivated grapevine (*Vitis vinifera* L. *ssp. sativa*) based on chloroplast DNA polymorphisms. Mol Ecol 15:3707–3714

Ashikawa K (1972) New grape variety 'Takao'. Bull Tokyo Agric Exp Sta 7:1–9

Atanassova R, Leterrier M, Gaillard C, Agasse A, Sagot E, Coutos-Thévenot P, Delrot S (2003) Sugar-regulated expression of a putative hexose transport gene in grape. Plant Physiol 131:326–334

Bais A, Murphy PJ, Dry IB (2000) The molecular regulation of stilbene phytoalexin biosynthesis in *Vitis vinifera* during grape berry development. Aust J Plant Physiol 27:425–433

Bamzai RD, Randhawa GS (1967) Effects of certain growth substances and boric acid on germination, tube growth and storage of grape pollen (*Vitis* spp.). Vitis 6:269–277

Barker CL, Donald T, Pauquet J, Ratnaparkhe MB, Bouquet A, Adam-Blondon A-F, Thomas MR, Dry I (2005) Genetic and physical mapping of the grapevine powdery mildew resistance gene, *Run1*, using a bacterial artificial chromosome library. Theor Appl Genet 111:370–377

Barnaud A, Lacombe T, Doligez A (2005) Linkage disequilibrium in cultivated grapevine. *Vitis vinifera* L. Theor Appl Genet 112:708–716

Barrett HC (1955) Black rot resistance of the foliage on seedlings in selected grape progenies. Proc Am Soc Hortic Sci 66:220–224

Barritt BH, Einset J (1969) Inheritance of 3 Major Fruit Colors in Grapes. J Am Soc Hortic Sci 94:87–89

Becker H, Sopp E (1990) Rootstocks with immunity to phylloxera and nematode resistance. In: Proceedings of the 5th International Symposium on Grape Breeding. Vitis special issue, St. Martin/Pfalz p. 294

Benjak A, Konrad J, Blaich R, Forneck A (2006) Different DNA extraction methods can cause different AFLP profiles in grapevine (*Vitis vinifera* L.). Vitis 45:15–21

Bloodworth PJ, Nesbitt WB, Barker KR (1980) Resistance to root knot nematodes in *Euvitis* × *Muscadinia* hybrids. In: Proceedings of the 3rd International Symposium on Grape Breeding, Davis, CA, pp 275–292

Bogs J, Downey MO, Harvey JS, Ashton AR, Tanner GJ, Robinson SP (2005) Proanthocyanidin Synthesis and Expression of Genes Encoding Leucoanthocyanidin Reductase and Anthocyanidin Reductase in Developing Grape Berries and Grapevine Leaves. Plant Physiol 139:652–663

Bogs J, Ebadi A, McDavid D, Robinson SP (2006) Identification of the Flavonoid Hydroxylases from Grapevine and Their Regulation during Fruit Development. Plant Physiol 140:279–291

Boss PK, Davies C, Robinson SP (1996a) Analysis of the expression of anthocyanin pathway genes in developing *Vitis vinifera* L cv. Shiraz grape berries and the implications for pathway regulation. Plant Physiol 111:1059–1066

Boss PK, Davies C, Robinson SP (1996b) Expression of anthocyanin biosynthesis pathway genes in red and white grapes. Plant Mol Biol 32:565–569

Boss PK, Sreekantan L, Thomas MR (2006) A grapevine TFL1 homologue can delay flowering and alter floral development when overexpressed in heterologous species. Funct Plant Biol 33:31–41

Boss PK, Thomas MR (2002) Association of dwarfism and floral induction with a grape 'green revolution' mutation. Nature 416:847–850

Boubals D (1966) Heredite de la resistance au phylloxera radicole chez la vigne. Ann Amelior Plantes 11:401–500

Bouquet A (1981) Resistance to grape fanleaf virus in Muscadine grape inoculated with *Xiphinema index*. Plant Dis 65:791–793

Bouquet A (1986) Introduction dans l'espece *Vitis vinifera* L. d'un caractere de resistance a l'oidium (*Uncinula necator* Schw. Burr.) issu de l'espece *Muscadinia rotundifolia* (Michx.) Small. Vigne Vini 13, Suppl 12:141–146

Bouquet A, Danglot Y (1996) Inheritance of seedlessness in grapevine (*Vitis vinifera* L). Vitis 35:35–42

Bouquet A, Davis HP, Danglot Y, Rennes C (1989) In-ovulo and in vitro embryo culture for breeding seedless table grapes (*Vitis vinifera* L.). Agron J 9:565–574

Bowers J, Boursiquot J-M, This P, Chu K, Johansson H, Meredith C (1999a) Historical Genetics: the parentage of Chardonnay, Gamay, and other wine grapes of Northeastern France. Science 285:1562–1565

Bowers JE, Bandman EB, Meredith CP (1993) DNA fingerprint characterization of some wine grape cultivars. Am J Enol Vitic 44:266–273

Bowers JE, Dangl GS, Meredith CP (1999b) Development and characterization of additional microsatellite DNA markers for grape. Am J Enol Vitic 50:243–246

Bowers JE, Dangl GS, Vignani R, Meredith CP (1996) Isolation and characterization of new polymorphic simple sequence repeat loci in grape. Genome 39:628–633

Bowers JE, Meredith CP (1996) Genetic similarities among wine grape cultivars revealed by restriction fragment-length polymorphism (RFLP) analysis. J Am Soc Hortic Sci 121:620–624

Bowers JE, Meredith CP (1997) The parentage of a classic wine grape, Cabernet Sauvignon. Nature Genetics 16:84–87

Boyden LE (2005) Allelism of root-knot nematode resistance and genetics of leaf traits in grape rootstocks. Cornell University, Ithaca

Burger AL, Botha FC (2004) Ripening-related gene expression during fruit ripening in *Vitis vinifera* L. cv. Cabernet Sauvignon and Clairette blanche. Vitis 43:59–64

Cahoon GA (1998) French hybrid grapes in North America. In: Ferree DC (ed) A history of fruit varieties. Good Fruit Grower Magazine, Yakima, Washington, pp 152–168

Cain DW, Emershad RL, Tarailo R (1983) In-ovulo embryo culture and seedling development of seeded and seedless grapes (*Vitis vinifera* L.). Vitis 22:9–14

Cakir B, Agasse A, Gaillard C, Saumonneau A, Delrot S, Atanassova R (2003) A grape ASR protein involved in sugar and abscisic acid signaling. Plant Cell 15:2165–2180

Campbell C (2005) The botanist and the vintner: How wine was saved for the world. Algonquin Books of Chapel Hill, Chapel Hill

Castellarin SD, Di Gaspero G, Marconi R, Nonis A, Peterlunger E, Paillard S, Adam-Blondon A-F, Testolin R (2006) Colour variation in red grapevine (*Vitis vinifera* L.): genomic organisation, expression of flavonoid 3'-hydroxylase, flavonoid 3',5'-hydroxylase genes and related metabolite profiling of red cyanidin-/blue delphinidin-based anthocyanins in berry skin. BMC Genomics 7:12

Cattell H, Miller LS (1980) The wines of the East. Vol. III. native American grapes. L& H Photojournalism, Lancaster, PA

Cervera M-T, Cabezas JA, Sancha JC, Martinez de Toda F, Martinez-Zapater JM (1998) Application of AFLPs to the characterization of grapevine *Vitis vinifera* L. genetic resources. A case study with accessions from Rioja (Spain). Theor Appl Genet 97:51–59

Comeaux BL (1987) A new *Vitis* (Vitaceae) species from Veracruz, Mexico. SIDA 12:273–277

Conde C, Agasse A, Glissant D, Tavares R, Geros H, Delrot S (2006) Pathways of glucose regulation of monosaccharide transport in grape cells. Plant Physiol 141:1563–1577

Cousins P, Walker MA (2002) Genetics of resistance *to Meloidogyne incognita* in crosses of grape rootstocks. Theor Appl Genet 105:802–807

Coutos-Thevenot P, Poinssot B, Bonomelli A, Yean H, Breda C, Buffard D, Esnault R, Hain R, Boulay M (2001) In vitro tolerance to *Botrytis cinerea* of grapevine 41B rootstock in transgenic plants expressing the stilbene synthase Vst1 gene under the control of a pathogen-inducible PR 10 promoter. J Exp Bot 52:901–910

Cramer G, Ergül A, Grimplet J, Tillett R, Tattersall E, Bohlman M, Vincent D, Sonderegger J, Evans J, Osborne C, Quilici D, Schlauch K, Schooley D, Cushman J (2007) Water and salinity stress in grapevines: early and late changes in transcript and metabolite profiles. Funct Integr Genomics 7:111–134

Crespan M (2003) The parentage of Muscat of Hamburg. Vitis 42:193–197

Da Silva FG, Iandolino A, Al-Kayal F, Bohlmann MC, Cushman MA, Lim H, Ergul A, Figueroa R, Kabuloglu EK, Osborne C, Rowe J, Tattersall E, Leslie A, Xu J, Baek JM, Cramer GR, Cushman JC, Cook DR (2005) Characterizing the grape transcriptome. Analysis of expressed sequence tags from multiple *Vitis* species and development of a compendium of gene expression during berry development. Plant Physiol 139:574–597

Dalbo MA, Ye GN, Weeden NF, Steinkellner H, Sefc KM, Reisch BI (2000) A gene controlling sex in grapevines placed on a molecular marker-based genetic map. Genome 43:333–340

Dalbo MA, Ye GN, Weeden NF, Wilcox WF, Reisch BI (2001) Marker-assisted selection for powdery mildew resistance in grape. J Am Soc Hortic Sci 126:83–89

Dangl GS, Mendum ML, Prins BH, Walker MA, Meredith CP, Simon CJ (2001) Simple sequence repeat analysis of a clonally propagated species: A tool for managing a grape germplasm collection. Genome 44:432–438

Davies C, Shin R, Liu W, Thomas MR, Schachtman DP (2006) Transporters expressed during grape berry (*Vitis vinifera* L.) development are associated with an increase in berry size and berry potassium accumulation. J Exp Bot 57:3209–3216

Dearing C (1917) The production of self-fertile muscadine grapes. Proc Am Soc Hortic Sci 14:30–34

Deluc L, Barrieu F, Marchive C, Lauvergeat V, Decendit A, Richard T, Carde J-P, Mérillon J-M, Hamdi S (2006) Characterization of a grapevine R2R3-MYB transcription factor that regulates the phenylpropanoid pathway. Plant Physiol 140:499–511

Detjen LR (1917) The inheritance of sex in *Vitis rotundifolia*. N C Agric Exp Stn Tech Bull, 12:1–42

Dettweiler E, Jung A, Zyprian E, Topfer R (2000) Grapevine cultivar Muller-Thurgau and its true to type descent. Vitis 2:63–65

Di Gaspero G, Cipriani G (2002) Resistance gene analogs are candidate markers for disease-resistance genes in grape (*Vitis* spp.). Theor Appl Genet 106:163–172

Di Gaspero G, Cipriani G (2003) Nucleotide binding site/leucine-rich repeats, Pto-like and receptor-like kinases related to disease resistance in grapevine. Mol Genet Genomics 269:612–623

Doligez A, Audiot E, Baumes R, This P (2006) QTLs for muscat flavor and monoterpenic odorant content in grapevine (*Vitis vinifera* L.). Mol Breed 18:109–125

Doligez A, Bouquet A, Danglot Y, Lahogue F, Riaz S, Meredith C, Edwards J, This P (2002) Genetic mapping of grapevine (*Vitis vinifera* L.) applied to the detection of QTLs for seedlessness and berry weight. Theor Appl Genet 105:780–795

Donald TM, Pellerone F, Adam-Blondon AF, Bouquet A, Thomas MR, Dry IB (2002) Identification of resistance gene analogs linked to a powdery mildew resistance locus in grapevine. Theor Appl Genet 104:610–618

Doucleff M, Jin Y, Gao F, Riaz S, Krivanek AF, Walker MA (2004) A genetic linkage map of grape, utilizing *Vitis rupestris* and *Vitis arizonica*. Theor Appl Genet 109:1178–1187

Downey MO, Harvey JS, Robinson SP (2003) Synthesis of flavonols and expression of flavonol synthase genes in the developing grape berries of Shiraz and Chardonnay (*Vitis vinifera* L.). Aust J Grape Wine Res 9:110–121

Downey MO, Harvey JS, Robinson SP (2004) The effect of bunch shading on berry development and flavonoid accumulation in Shiraz grapes. Aust J Grape Wine Res 10:55–73

Dunstan RT (1962) Vinifera-type grapes for the East. Fruit Varieties Hortic Dig 17:6–8

During H (1986) Testing for drought tolerance in grapevine scions. Angewandte Botanik 60:103–111

Eibach R, Diehl H, Alleweldt G (1989) Untersuchungen zur Vererbung von Resistenzeigenschaften bei Reben gegen *Oidium tuckeri*, *Plasmopara viticola* und *Botrytis cinerea*. Vitis 28:209–228

Einset J, Lamb B (1951) Chimeral Sports of Grapes. J Hered 42:158–162.

Einset J, Pratt C (1954) Giant Sports of Grapes. Proc Am Soc Hortic Sci 63:251–256

Einset J, Pratt C (1975) Grapes. In: Janick J, Moore JN (eds) Advances in fruit breeding. Purdue University Press, West Lafayette, IN, pp 130–153

Emershad RL, Ramming DW (1984) In-ovulo embryo culture of *Vitis vinifera* L. cv. 'Thompson Seedless'. Am J Bot 71:873–877

Ergul A, Kazan K, Aras S, Cevik V, Celik H, Soylemezoglu G (2006) AFLP analysis of genetic variation within the two economically important Anatolian grapevine *(Vitis vinifera L.)* varietal groups. Genome 49:467–475

Espinoza C, Vega A, Medina C, Schlauch K, Cramer G, Arce-Johnson P (2007) Gene expression associated with compatible viral diseases in grapevine cultivars. Funct Integr Genomics 7:95–110

Fanizza G, Lamaj F, Costantini L, Chaabane R, Grando MS (2005) QTL analysis for fruit yield components in table grapes (*Vitis vinifera*). Theor Appl Genet 111:658–664

Fanizza G, Raddi P (1973) The heritability of fruit ripening date in *Vitis vinifera* L. Vitis 12:93–96

Faria MA, Magalhaes R, Ferreira MA, Meredith CP, Monteiro FF (2000) *Vitis vinifera* must varietal authentication using microsatellite DNA analysis (SSR). J Agric Food Chem 48:1096–1100

Fatahi R, Ebadi A, Bassil N, Mehlenbacher SA, Zamani Z (2003) Characterization of Iranian grapevine cultivars using microsatellite markers. Vitis 42:185–192

Fei Z, Tang X, Alba RM, White JA, Ronning CM, Martin GB, Tanksley SD, Giovannoni JJ (2004) Comprehensive EST analysis of tomato and comparative genomics of fruit ripening. Plant J 40:47–59

Fennell JL (1948) Inheritance studies with the tropical grape. J Hered 39:54–64

Fernandez L, Doligez A, Lopez G, Thomas MR, Bouquet A, Torregrosa L (2006a) Somatic chimerism, genetic inheritance, and mapping of the fleshless berry (flb) mutation in grapevine (*Vitis vinifera* L.). Genome 49:721–728

Fernandez L, Romieu C, Moing A, Bouquet A, Maucourt M, Thomas MR, Torregrosa L (2006b) The grapevine fleshless berry mutation. A unique genotype to investigate differences between fleshy and nonfleshy fruit. Plant Physiol 140:537–547

Fillion L, Ageorges A, Picaud S, Coutos-Thévenot P, Lemoine R, Romieu C, Delrot S (1999) Cloning and expression of a hexose transporter gene expressed during the ripening of grape berry. Plant Physiol 120:1083–1093

Firoozabady E, Olmo HP (1982a) The heritability of resistance to root-knot nematode (*Meloidogyne incognita* acrita CHIT.) in *Vitis viniera* × *V. rotundifolia* hybrid derivatives. Vitis 21:136–144

Firoozabady E, Olmo HP (1982b) Resistance to grape phylloxera in *Vitis vinifera* × *V. rotundifolia* grape hybrids. Vitis 21:1–4

Firoozabady E, Olmo HP (1987) Heritability and correlation studies of certain quantitative traits in table grapes, *Vitis* spp. Vitis 26:132–146

Fischer BM, Salakhutdinov I, Akkurt M, Eibach R, Edwards K, Töpfer R, Zyprian E (2004) Quantitative trait locus analysis of fungal disease resistance factors on a molecular map of grapevine. Theor Appl Genet 108:501–515

Franks T, Botta R, Thomas MR (2002) Chimerism in grapevines: implications for cultivar identity, ancestry and genetic improvement. Theor Appl Genet 104:192–199

Galet P (1988) Cepages et vignobles de France. Vol. 1 Les Vignes Americaines, 2nd edn. Charles Dehan, Montpellier

Ganeshan S (1985) Cryogenic preservation of grape (*Vitis vinifera* L.) pollen. Vitis 24:169–173

Ganeshan S, Alexander MP (1990) Fertilizing ability of cryopreserved grape (*Vitis vinifera* L.) pollen. Vitis 29:145–150

Goldy RG, Emershad RL, Ramming DW, Chaparro J (1988) Embryo culture as a means of introgressing seedlessness from *Vitis vinifera* to *V. rotundifolia*. Hortscience 23:886–889

Goldy RG, Ramming DW, Emershad RL, Chaparro JX (1989) Increasing production of *Vitis vinifera* × *V. rotundifolia* hybrids through embryo rescue. Hortscience 24:820–822

Goto-Yamamoto N, Mouri H, Azumi M, Edwards KJ (2006) Development of grape microsatellite markers and microsatellite analysis including oriental cultivars. Am J Enol Vitic 57:105–108

Grando MS, Bellin D, Edwards KJ, Pozzi C, Stefanini M, Velasco R (2003) Molecular linkage maps of *Vitis vinifera* L. and *Vitis riparia* Mchx. Theor Appl Genet 106:1213–1224

Grassi F, Labra M, Imazio S, Spada A, Sgorbatti S, Scienza A, Saqla F (2003) Evidence of a secondary grapevine domestication centre detected by SSR analysis. Theor Appl Genet 107:1315–1320

Guillen P, Guis M, Martínez-Reina G, Colrat S, Dalmayrac S, Deswarte C, Bouzayen M, Roustan JP, Fallot J, Pech JC, Latché A (1998) A novel NADPH-dependent aldehyde reductase gene from *Vigna radiata* confers resistance to the grapevine fungal toxin eutypine. Plant J 16:335–343

Hashizume T, Iizuka M (1981) Induction of female organs in male flowers of *Vitis* species by zeatin and dihydrozeatin. Phytochemistry 10:2653–2655

He P, Lixin N (1989) Study of cold hardiness in the wild *Vitis* native to China. Acta Hortic Sinica 16:81–88

He PH, Wang G (1986) Studies on the resistance of wild *Vitis* species native to china to downy mildew, *Plasmopora viticola* (Berk. et Curtis) Berl. et de Toni. Acta Hortic Sinica 13:17–24

Hebert D, Kikkert JR, Smith FD, Reisch BI (1993) Optimization of biolistic transformation of embryogenic grape cell suspensions. Plant Cell Rep 13:405–409

Hedrick UP, Anthony RD (1915) Inheritance of certain characters of grapes. NY State Agric Coll Tech Bull No 45, pp 3–19

Hirabayashi T, Kozaki I, Akihama T (1976) *In vitro* differentiation of shoots from anther callus in *Vitis*. Hortscience 11:511–512

Hocquigny S, Pelsey F, Dumas V, Kindt S, Heloir M-C, Merdinoglu D (2004) Diversification within grapevine cultivars goes through chimeric states. Genome 47:579–589

Hvarleva T, Rusanov K, Lefort F, Tsvetkov I, Atanassov A, Atanassov I (2004) Genotyping of Bulgarian *Vitis vinifera* L. cultivars by microsatellite analysis. Vitis 43:27–34

Iocco P, Franks T, Thomas MR (2001) Genetic Transformation of Major Wine Grape Cultivars of *Vitis vinifera* L. Transgenic Res V10:105–112

Jabco JP, Nesbitt WB, Werner DJ (1985) Resistance of various classes of grapes to the bunch and muscadine grape forms of black rot. J Am Soc Hortic Sci 110:762–765

Joly D, Perrin M, Gertz C, Kronenberger J, Demangeat G, Masson JE (2004) Expression analysis of flowering genes from seedling-stage to vineyard life of grapevine cv. Riesling. Plant Sci 166:1427–1436

Kikkert JR, Hebert-Soule D, Wallace PG, Striem MJ, Reisch BI (1996) Transgenic plantlets of 'Chancellor' grapevine (*Vitis* sp.) from biolistic transformation of embryogenic cell suspensions. Plant Cell Rep 15:311–316

Kobayashi S, Goto-Yamamoto N, Hirochika H (2004) Retrotransposon-induced mutations in grape skin color. Science 304:982

Kobayashi S, Ishimaru M, Ding CK, Yakushiji H, Goto N (2001) Comparison of UDP-glucose: Flavonoid 3-O-glucosyltransferase (UFGT) gene sequences between white grapes (*Vitis vinifera*) and their sports with red skin. Plant Sci 160:543–550

Krivanek AF, Famula TR, Tenscher A, Walker AR (2005) Inheritance of resistance to *Xylella fastidiosa* within a *Vitis rupestris* × *Vitis arizonica* population. Theor Appl Genet 111:110–119

Krivanek AF, Riaz S, Walker MA (2006) Identification and molecular mapping of *PdR1*, a primary resistance gene to Pierce's disease in *Vitis*. Theor Appl Genet 112:1125–1131

Krul WR, Worley JF (1977) Formation of adventitious embryos in callus cultures of 'Seyval', a French hybrid grape. J Am Soc Hortic Sci 102:360–363

Labra M, Imazio S, Grassi F, Rossoni M, Citterio S, Sgorbati S, Scienza A, Failla O (2003) Molecular approach to assess the origin of cv. Marzemino. Vitis 42:137–140

Labra M, Winfield M, Ghiani A, Grassi F, Sala F, Scienza A, Failla O (2001) Genetic studies on Trebbiano and morphologically related varieties by SSR and AFLP markers. Vitis 40:187–190

Ladoukakis ED, Lefort F, Sotiri P, Bacu A, Kongjika E, Roubelakis-Angelakis KA (2005) Genetic characterization of Albanian grapevine cultivars by microsatellite markers. J Int Des Sci De La Vigne Et Du Vin 39:109–119

Lahogue F, This P, Bouquet A (1998) Identification of a codominant scar marker linked to the seedlessness character in grapevine. Theor Appl Genet 97:950–959

Lamoureux D, Bernole A, Le Clainche I, Tual S, Thareau V, Paillard S, Legeai F, Dossat C, Wincker P, Oswald M, Merdinoglu D, Vignault C, Delrot S, Caboche M, Chalhoub B, Adam-Blondon A-F (2006) Anchoring of a large set of markers onto a BAC library for the development of a draft physical map of the grapevine genome. Theor Appl Genet 113:344–356

Laquitaine L, Gomes E, Francois J, Marchive C, Pascal S, Hamdi S, Atanassova R, Delrot S, Coutos-Thevenot P (2006) Molecular basis of ergosterol-induced protection of grape against *Botrytis cinerea*: Induction of type I LTP promoter activity, WRKY, and stilbene synthase gene expression. Mol Plant Microbe Interact 19:1103–1112

Legrand V, Dalmayrac S, Latche A, Pech J-C, Bouzayen M, Fallot J, Torregrosa L, Bouquet A, Roustan J-P (2003) Constitutive expression of *Vr-ERE* gene in transformed grapevines confers enhanced resistance to eutypine, a toxin from Eutypa lata. Plant Sci 164:809–814

Leterrier M, Atanassova R, Laquitaine L, Gaillard C, Coutos-Thevenot P, Delrot S (2003) Expression of a putative grapevine hexose transporter in tobacco alters morphogenesis and assimilate partitioning. J Exp Bot 54:1193–1204

Levadoux L (1946) Study of the flower and sexuality in grapes (in French). Ann Ecol Nat Agri Montpellier NS 27:1–89

Levadoux L (1956) Les populations sauvages et cultivees de *Vitis vinifera* L. Annales de l'amelioration des plantes 6:59–118

Lider LA (1954) Inheritance of resistance to a root-knot nematode (*Meloidogyne incognita* var. *acrita* Chitwood) in *Vitis* spp. Proc Helminthol Soc Wash 21:53–60

Lijavetzky D, Ruiz-García L, Cabezas J, Andrés M, Bravo G, Ibáñez A, Carreño J, Cabello F, Ibáñez J, Martínez-Zapater J (2006) Molecular genetics of berry colour variation in table grape. Mol Genet Genomics 276:427–435

Lin H, Walker MA (1998) Identifying grape rootstocks with simple sequence repeat (SSR) DNA markers. Am J Enol Vitic 49:403–407

Lodhi MA, Daly MJ, Ye GN, Weeden NF, Reisch BI (1995) A Molecular Marker Based Linkage Map of *Vitis*. Genome 38:786–794

Lodhi MA, Reisch BI (1995) Nuclear-DNA Content of *Vitis* Species, Cultivars, Anti Other Genera of the Vitaceae. Theor Appl Genet 90:11–16

Loomis NH (1948) A note on the inheritance of flower type in muscadine grapes. Proc Am Soc Hortic Sci 52:276–278

Loomis NH, Williams CF (1957) A new genetic flower type of the muscadine grape. J Hered 48:294–304

Loomis NH, Williams CF, Murphy MM (1954) Inheritance of flower types in muscadine grapes. Proc Am Soc Hortic Sci 64:279–283

Lopes MS, dos Santos MR, Dias JEE, Mendonca D, da Camara Machado A (2006) Discrimination of Portuguese grapevines based on microsatellite markers. J Biotechnol 127:34–44

Lopes MS, Sefc KM, Eiras Dias E, Steinkellner H, da Camara Machado LM, da Camara Machado A (1999) The use of microsatellites for germplasm management in a Portuguese germplasm grapevine collection. Theor Appl Genet 99:733–739

Lowe KM, Walker MA (2006) Genetic linkage map of the interspecific grape rootstock cross Ramsey (Vitis champinii) × Riparia Gloire (Vitis riparia). Theor Appl Genet 112:1582–1592

Madero E, Boubals D, Truel P (1986) Transmission hereditaire des principaux caracters des cepages Cabernet Franc, Cabernet Sauvignon et Merlot (V. vinifera L.). Vigne Vini 13, Suppl 12:209–219

Maletic E, Pejic I, Kontic JK, Piljac J, Dangl GS, Vokurka A, Lacombe T, Mirosevic N, Meredith CP (2004) Zinfandel, Dobricic, and Plavac mali: The genetic relationship among three cultivars of the Dalmatian Coast of Croatia. Am J Enol Vitic 55:174–180

Mandl K, Santiago JL, Hack R, Fardossi A, Regner F (2006) A genetic map of Welschriesling × Sirius for the identification of magnesium-deficiency by QTL analysis. Euphytica 149:133–144

Martin JP, Borrego J, Cabello F, Ortiz JM (2003) Characterization of Spanish grapevine cultivar diversity using sequence-tagged microsatellite markers. Genome 46:10–18

McGovern PE (2003) Ancient wine: the search for the origins of viticulture. Princeton University Press, Princeton

McGovern PE, Michel RH (1995) The analytical and archaeological challenge of detecting ancient wine: two case studies from the ancient Near East. In: McGovern PE, Fleming SJ, Katz SH (eds) The origins and ancient history of wine. Gordon and Breach, Amsterdam, pp 57–67

McGrew JR (1976) Screening grape seedlings for black rot resistance. Fruit Varieties J 30:31–32

McLeRoy SS, Renfro Jr. RE (2004) Grape Man of Texas. Eaking Press, Austin, TX

Merdinoglu D, Butterlin G, Bevilacqua L, Chiquet V, Adam-Blondon A-F, Decroocq S (2005) Development and characterization of a large set of microsatellite markers in grapevine (Vitis vinifera L.) suitable for multiplex PCR. Mol Breed 15:349–366

Meredith CP, Bowers JE, Riaz S, Handley V, Bandman EB, Dangl GS (1999) The identity and parentage of the variety known in California as Petite Syrah. Am J Enol Vitic 50:236–242

Meredith CP, Lider LA, Raski DJ, Ferrari NL (1982) Inheritance of Tolerance to Xiphinema-Index in Vitis Species. Am J Enol Vitic 33:154–158

Moore JJ (1970) Cytokinin-induced sex conversion in male clones of Vitis species. J Am Soc Hortic Sci 95:387–393

Moore MO (1987) A study of selected taxa of Vitis (Vitaceae) in the southeastern United States. Rhodora 89:75–91

Moore MO (unpubl. manuscript) Vitaceae. In: Flora of North America editorial committee (eds) Flora of North America North of Mexico. Cambridge, New York

Mortenson JA (1968) The inheritance of resistance to Pierce's disease in Vitis. J Am Soc Hortic Sci 92:331–337

Mortenson JA (1977) Segregation for resistance to black rot in selfed grape seedlings. Fruit Varieties Journal 31:59–60

Mortenson JA (1981) Sources and inheritance of resistance to anthracnose in Vitis. J Hered 72:423–426

Moser C, Segala C, Fontana P, Salakhudtinov I, Gatto P, Pindo M, Zyprian E, Toepfer R, Grando MS, Velasco R (2005) Comparative analysis of expressed sequence tags from different organs of Vitis vinifera L. Funct Integr Genomics 5:208–217

Mullins MG, Srinivasan C (1976) Somatic embryos and plantlets from an ancient clone of the grapevine (cultivar Cabernet Sauvignon) by apomixis in vitro. J Exp Bot 27:1022–1030

Munson T (1909) Foundations of American Grape Culture. T.V. Munson & Son, Denison, Texas

Nebel BR, Ruttle ML (1936) Storage experiments with pollen of cultivated fruit trees. J Pomol Hort Sci 14:347–359

Negi SS, Olmo HP (1966) Sex conversion in a male *Vitis vinifera* L. by a kinen. Science 152:1624–1625

Negi SS, Olmo HP (1970) Studies on sex conversion in male *Vitis vinifera* L. (*sylevstris*). Vitis 9:89–96

Negrul AM (1938) Evolucija kuljturnyx from vinograda. Dokl Akad Nauk SSSR 8:585–588

Nesbitt WB (1962) Polyploidy and interspecific hybridzation of *Vitis*. Department of Horticulture. North Carolina State University, Raleigh, NC

Notsuka K, Tsuru T, Shiraishi M (2000) Induced polyploid grapes via in vitro chromosome doubling. J Jpn Soc Hortic Sci 69:543–551

Nunan KJ, Davies C, Robinson SP, Fincher GB (2001) Expression patterns of cell wall-modifying enzymes during grape berry development. Planta 214:257–264

Okie WR (1997) Register of new fruit and nut varieties: Brooks and Olmo list 38. HortScience 32:785–805

Olmo HP (1942a) Breeding of new tetraploid grape varieties. Proc Am Soc Hortic Sci 41:219–224

Olmo HP (1942b) Storage of grape pollen. Proc Am Soc Hortic Sci 41:219–224

Olmo HP (1952) Breeding tetraploid grapes. Proc Am Soc Hortic Sci 59:285–290

Olmo HP (1971) Vinifera rotundifolia hybrids as wine grapes. Am J Enol Vitic 22:87–91

Olmo HP (1986) The potential role of (*vinifera* × *rotundifolia*) hybrids in grape variety improvement. Experientia 42:921–926

Olmo HP (1995) The origin and domestication of the *Vinifera* grape. In: McGovern PE (ed) The origins and ancient history of wine. Gordon and Breach, Amsterdam, pp 31–43

Or E, Vilozny I, Fennell A, Eyal Y, Ogrodovitch A (2002) Dormancy in grape buds: isolation and characterization of catalase cDNA and analysis of its expression following chemical induction of bud dormancy release. Plant Sci 162:121–130

Ourecky DK, Pratt C, Einset J (1967) Fruiting behavior of large-berried and large-clustered sports of grapes. Proc Am Soc Hortic Sci 91:217–223

Pacey-Miller T, Scott K, Ablett E, Tingey S, Ching A, Henry R (2003) Genes associated with the end of dormancy in grapes. Funct Integr Genomics 3:144–152

Parfitt DE, Almehdi AA (1983) Cryogenic storage of grape pollen. Am J Enol Vitic 34:227–228

Paul HW (1996) Science, vine, and wine in modern France. Cambridge University Press, Cambrdige

Pauquet J, Bouquet A, This P, Adam-Blondon AF (2001) Establishment of a local map of AFLP markers around the powdery mildew resistance gene Run1 in grapevine and assessment of their usefulness for marker assisted selection. Theor Appl Genet 103:1201–1210

Pearson RC, Goheen AC (1988) Compendium of grape diseases. APS Press, St. Paul, MN

Perl A, Colova-Tsolova V, Eshdat Y (2004) *Agrobacterium*-mediated transformation of grape embryogenic calli. In: Curtis IS (ed) Transgenic crops of the world: Essential protocols. Kluwer Academic Publishers, pp 229–242

Perl A, Eshdat Y (1998) DNA transfer and gene expression in transgenic grapes. Biotechnol Genet Eng Rev 15:365–386

Perl A, Saad S, Sahar N, Holland D (1995) Establishment of long-term embryogenic cultures of seedless *Vitis vinifera* cultivars – a synergistic effect of auxins and the role of abscisic acid. Plant Sci 104:193–200

Perrin M, Gertz C, Masson JE (2004) High efficiency initiation of regenerable embryogenic callus from anther filaments of 19-grapevine genotypes grown worldwide. Plant Sci 167: 1343–1349

Perrin M, Martin D, Joly D, Demangeat G, This P, Masson JE (2001) Medium-dependent response of grapevine somatic embryogenic cells. Plant Sci 161:107–116

Picaud S, Becq F, Dedaldechamp F, Ageorges A, Delrot S (2003) Cloning and expression of two plasma membrane aquaporins expressed during the ripening of grape berry. Funct Plant Biol 30:621–630

Pollefeys P, Bousquet J (2003) Molecular genetic diversity of the French-American grapevine hybrids cultivated in North America. Genome 46:1037–1048

Pouget R (1980) Breeding grapevine rootstocks for resistance to iron chlorosis. Proceedings of the 3rd International Symposium on Grape Breeding, Davis, University of California, California, pp 191–197

Pratt C (1971) Reproductive Anatomy in Cultivated Grapes – Review. Am J Enol Vitic 22:92–109

Rajasekaran K, Mullins MG (1979) Embryos and plantlets from cultured anthers of hybrid grapevines. J Exp Bot 30:399–407

Ramming DW (1990) The use of embryo culture in fruit breeding. Hortscience 25:393–398

Ramming DW, Emershad RL, Spiegel-Roy P, Sahar N, Baron I (1990) Embryo culture of early ripening seeded grape (*Vitis vinifera*) genotypes. Hortscience 25:339–342

Ramming DW, Emershad RL, Tarailo R (2000) A stenospermocarpic, seedless *Vitis vinifera* × *Vitis rotundifolia* hybrid developed by embryo rescue. Hortscience 35:732–734

Regner F, Stadlbauer A, Eisenheld C, Kaserer H (2000) Genetic relationships among Pinots and related cultivars. Am J Enol Vitic 51:7–14

Reimer FC, Detjen LR (1910) Self-sterility of the scuppernong and other muscadine grapes. N C Agric Exp Sta Bull, 209:1–23

Reimer FC, Detjen LR (1914) Breeding rotundifolia grapes: a study of transmission of character. North Carolina Agricultural Experiment Station Technical Bulletin, 10:1–47

Reisch BI, Pratt C (1996) Grapes. In: Janick J, Moore JN (eds) Fruit breeding, vol. II: Vine and small fruits. John Wiley & Sons, Inc, New York

Reynolds AG, Fuleki T, Evans WD (1982) Inheritance of methyl anthranilate and total volatile esters in *Vitis* spp. Am J Enol Vitic 33:14–19

Riaz S, Dangl GS, Edwards KJ, Meredith CP (2004) A microsatellite marker based framework linkage map of Vitis vinifera L. Theor Appl Genet 108:864–872

Riaz S, Garrison KE, Dangl GS, Boursiquot J-M, Meredith CP (2002) Genetic divergence and chimerism within ancient asexually propagated winegrape cultivars. J Am Soc Hortic Sci 127:508–514

Riaz S, Krivanek AF, Xu K, Walker MA (2006) Refined mapping of the Pierce's disease resistance locus, *PdR1*, and *Sex* on an extended genetic map of *Vitis rupestris* × *V. arizonica*. Theor Appl Genet 113:1317–1329

Rossoni M, Labra M, Imazio S, Grassi F, Scienza A, Sala F (2003) Genetic relationships among grapevine cultivars grown in Oltrepo Pavese (Italy). Vitis 42:31–34

Sauer W, Antcliff AJ (1969) Polyploid mutants of grapes. Hortscience 4:226–227

Schneider W, Staudt G (1979) Estimation of broad sense heritability of some characters of *Vitis vinifera* (in German with English summary). Vitis 18:238–243

Scott KD, Ablett EM, Lee LS, Henry RJ (2000a) AFLP markers distinguishing an early mutant of Flame Seedless grape. Euphytica 113:245–249

Scott KD, Eggler P, Seaton G, Rossetto M, Ablett EM, Lee LS, Henry RJ (2000b) Analysis of SSRs derived from grape ESTs. Theor Appl Genet 100:723–726

Sefc KM, Lopes MS, Lefort F, Botta R, Roubelakis-Angelakis KA, Ibáñez J, Pejif I, Wagner HW, Glössl J, Steinkellner H (2000) Microsatellite variability in grapevine cultivars from different European regions and evaluation of assignment testing to assess the geographic origin of cultivars. Theor Appl Genet 100:498–505

Sefc KM, Regner F, Turetschek E, Glossl J, Steinkellner H (1999) Identification of microsatellite sequences in *Vitis riparia* and their adaptability for genotyping of different *Vitis* species. Genome 42:367–373

Shirasi S-I, Watanabe Y, Okubo H, Uemoto S (1986) Anthocyanin pigments of black-purple grapes related to variety 'Kyoho' (*Vitis vinifera* L. × *V. labrusca* L.). J Jpn Soc Hortic Sci 55:123–129

Siret R, Boursiquot J-M, Merle MH, Cabanis JC, This P (2000) Toward the authentication of varietal wines by the analysis of grape (*Vitis vinifera* L.) residual DNA in must and wine using microsatellite markers. J Agric Food Chem 48:5035–5040

Snyder E, Harmon FN (1939) Grape progenies of self-pollinated vinifera varieties. Proc Am Soc Hortic Sci 37:625–626

Snyder E, Harmon FN (1952) Grape Breeding Summary 1923–1951. Proc Am Soc Hortic Sci 60:243–246

Sparvoli F, Martin C, Scienza A, Gavazzi G, Tonelli C (1994) Cloning and molecular analysis of structural genes involved in flavonoid and stilbene synthesis in grape (*Vitis vinifera* L.). Plant Mol Biol 24:743–755

Spiegel-Roy P, Sahar N, Baron J, Lavi U (1985) In vitro culture and plant formation from grape cultivars with abortive ovules and seeds. J Am Soc Hortic Sci 110:109–112

Sreekantan L, Thomas MR (2006) VvFT and VvMADS8, the grapevine homologues of the floral integrators FT and SOC1, have unique expression patterns in grapevine and hasten flowering in Arabidopsis. Funct Plant Biol 33:1129–1139

Sreekantan L, Torregrosa L, Fernandez L, Thomas MR (2006) VvMADS9, a class B MADS-box gene involved in grapevine flowering, shows different expression patterns in mutants with abnormal petal and stamen structures. Funct Plant Biol 33:877–886

Srinivasan C, Mullins MG (1981) Induction of precocious flowering in grapevine seedlings by growth regulators. Agronomie 1:1–5

Stover LH (1960) Progress in the development of grape varieties for Florida. Proc Fla State Hort Soc 73:320–323

Symons GM, Davies C, Shavrukov Y, Dry IB, Reid JB, Thomas MR (2005) Grapes on steroids. Brassinosteroids are involved in grape berry ripening. Plant Physiol 140:150–158

Szegedi E, Korbuly J, Koleda I (1984) Crown gall resistance in East-Asian *Vitis* species and in their *V. vinifera* hybrids. Vitis 23:21–26

Terrier N, Glissant D, Grimplet J, Barrieu F, Abbal P, Couture C, Ageorges A, Atanassova R, Leon C, Renaudin JP, Dedaldechamp F, Romieu C, Delrot S, Hamdi S (2005) Isogene specific oligo arrays reveal multifaceted changes in gene expression during grape berry (*Vitis vinifera* L.) development. Planta 222:832–847

The French-Italian Public Consortion for Grapevine Genome Characterization. (2007). The Grapevine Genome Sequence Suggests Ancestral Hexaplodization in Major Angiosperm Phyla. Nature 449: 463–467.

This P, Jung A, Boccacci P, Borrego J, Botta R, Costantini L, Crespan M, Dangl G, Eisenheld C, Ferreira-Monteiro F, Grando S, Ibáñez J, Lacombe T, Laucou V, Magalhães R, Meredith C, Milani N, Peterlunger E, Regner F, Zulini L, Maul E (2004) Development of a standard set of microsatellite reference alleles for identification of grape cultivars. Theor Appl Genet 109:1448–1458

This P, Lacombe T, Cadle-Davidson M, Owens CL (2007) Wine grape (*Vitis vinifera* L.) color associates with allelic variation in the domestication gene *VvmybA1*. Theor Appl Genet 114:723–730

This P, Lacombe T, Thomas MR (2006) Historical origins and genetic diversity of wine grapes. Trends Genet 22:511–519

Thomas MR, Cain P, Scott NS (1994) DNA Typing of Grapevines – a Universal Methodology and Database for Describing Cultivars and Evaluating Genetic Relatedness. Plant Mol Biol 25:939–949

Thomas MR, Matsumoto S, Cain P, Scott NS (1993) Repetitive DNA of grapevine: classes present and sequences suitable for cultivar identification. Theor Appl Genet 86:173–180

Thomas MR, Scott NS (1993) Microsatellite Repeats in Grapevine Reveal DNA Polymorphisms When Analyzed as Sequence-Tagged Sites (STSs). Theor Appl Genet 86:985–990

Thompson MM, Olmo HP (1963) Cytohistological studies of cytochimeric and tetraploid grapes. Am J Bot 50:901–906

Torregrosa L (1998) A simple and efficient method to obtain stable embryogenic cultures from anthers of *Vitis vinifera* L. Vitis 37:91–92

Vidal JR, Kikkert JR, Malnoy MA, Wallace PG, Barnard J, Reisch BI (2006) Evaluation of trans-genic 'Chardonnay' (*Vitis vinifera*) containing magainin genes for resistance to crown gall and powdery mildew. Transgenic Res 15:69–82

Vidal JR, Kikkert JR, Wallace PG, Reisch BI (2003) High-efficiency biolistic co-transformation and regeneration of 'Chardonnay' (*Vitis vinifera* L.) containing npt-II and antimicrobial peptide genes. Plant Cell Rep 22:252–260

Vignault C, Vachaud M, Cakir B, Glissant D, Dédaldéchamp F, Büttner M, Atanassova R, Fleurat-Lessard P, Lemoine R, Delrot S (2005) VvHT1 encodes a monosaccharide transporter expressed in the conducting complex of the grape berry phloem. J Exp Bot 56:1409–1418

Vigne E, Komar V, Fuchs M (2004) Field safety assessment of recombination in transgenic grapevines expressing the coat protein gene of *Grapevine fanleaf virus*. Transgenic Res 13:165–179

Vouillamoz JF, Grando MS (2006) Genealogy of wine grape cultivars: 'Pinot' is related to 'Syrah'. Heredity 97:102–110

Wagner R (1967) Etude de quelques disjonctions dans des descendances de Chasselas, Muscat Ottonel et Muscat a petits grains. Vitis 6:353–363

Wake CMF, Fennell A (2000) Morphological, physiological and dormancy responses of three *Vitis* genotypes to short photoperiod. Physiol Plant 109:203–210

Walker AR, Lee E, Robinson SP (2006) Two new grape cultivars, bud sports of Cabernet Sauvignon bearing pale-coloured berries, are the result of deletion of two regulatory genes of the berry colour locus. Plant Mol Biol 62:623–635

Walker MA, Meredith CP (1990) The genetics of resistance to grapevine fanleaf virus in *Vitis vinifera*. In: Alleweldt G (ed) Proceeding of the 5th International Symposium on Grape Breeding. St. Martin, Pfalz, Germany, Vitis special Issue, pp 228–238

Walker MA, Meredith CP, Goheen AC (1985) Sources of resistance to grapevine fanleaf virus (GFV) in *Vitis* species. Vitis 24:218–228

Wang Q, Mawassi M, Sahar N, Li P, Violeta C-T, Gafny R, Sela I, Tanne E, Perl A (2004) Cryopreservation of Grapevine (*Vitis* spp.) Embryogenic Cell Suspensions by Encapsulation & Vitrification. Plant Cell, Tissue and Organ Cult 77:267–275

Wellington R (1939) The Ontario grape and its seedlings as parents. Proc Am Soc Hortic Sci 37:630–634

Williams CF (1954) Breeding perfect-flowered muscadine grapes. Proc Am Soc Hortic Sci 64:274–278

Xiao H, Siddiqua M, Braybrook S, Nassuth A (2006) Three grape CBF/DREB1 genes respond to low-temperature, drought and abscisic acid. Plant, Cell and Environ 29:1410–1421

Xue B, Ling KS, Reid CL, Krastanova S, Sekiya M, Momol EA, Sule S, Mozsar J, Gonsalves D, Burr TJ (1999) Transformation of five grape rootstocks with plant virus genes and a *virE2* gene from *Agrobacterium tumefaciens*. In Vitro Cell Dev Biol Plant 35:226–231

Yakushiji H, Kobayashi S, Goto-Yamamoto N, Tae Jeong S, Sueta T, Mitani N, Azuma A (2006) A skin-color mutation of grapevine, from black-skinned Pinot Noir to white-skinned Pinot Blanc, is caused by deletion of the functional *VvmybA1* allele. Biosci Biotechnol Biochem 70:1506–1508

Yamamoto T, Iketani H, Ieki H, Nishizawa Y, Notsuka K, Hibi T, Hayashi T, Matsuta N (2000) Transgenic grapevine plants expressing a rice chitinase with enhanced resistance to fungal pathogens. Plant Cell Rep V19:639–646

Ye GN, Soylemezoglu G, Weeden NF, Lamboy WF, Pool RM, Reisch BI (1998) Analysis of the relationship between grapevine cultivars, sports and clones via DNA fingerprinting. Vitis 37:33–38

Zhiduan C, Wen J (draft manuscript) *Vitaceae*. In: Committee FoCE (ed) Flora of China, vol. 12 (Hippocastanaceae through Pentaphylaceae). Science Press, Beijing, and Missouri Botanical Garden Press, St. Louis

Zohary D (1996) In: Harris DR (ed) The Origins and Spread of Agriculture and Pastoralism in Eurasia. University College London Press, London, pp 142–147

Zohary D, Hopf M (2000) Domestication of Plants in the Old World, 3rd edn. Oxford University Press, London

Zohary D, Spiegel-Roy P (1975) Beginnings of fruit growing in the old world. Science 187:319–327

Chapter 8
Kiwifruit

A.R. Ferguson and A.G. Seal

Abstract Kiwifruit are still a relatively minor crop making up perhaps 0.2% of total world annual production of fruit. The kiwifruit of commerce are large-fruited selections of two closely related species *Actinidia chinensis* and *A. deliciosa*. Most current kiwifruit cultivars are selections from the wild or from seedling populations and only a few result from planned hybridizations. The main emphasis in the breeding programs underway is on fruit novelty, flavor, size, time of harvest, flesh color, length of storage life, environmental adaptation and vine productivity. Until recently, nearly all the kiwifruit grown commercially outside China were of one green-fruited cultivar of *A. deliciosa*; now yellow-fleshed, sweeter flavored kiwifruit are becoming important in international trade. To take advantage of the considerable diversity within the genus requires good germplasm resources and a better knowledge of the reproductive biology of kiwifruit. The main constraints to breeding include dioecy, the long generation time and the complexity of some key traits as well as the need for support structures, the exuberant vegetative growth and the need to control growth to ensure fruiting. Many of the traits associated with fruit quality are quantitatively inherited. Use of molecular biological and biotechnological techniques should facilitate improvement programs.

8.1 Introduction

The kiwifruit of international commerce are selections of two closely related species *Actinidia chinensis* Planch. and *A. deliciosa* (A. Chev.) C.F.Liang et A.R.Ferguson. Currently about 1.5–1.6 million t of kiwifruit are produced each year. Italy (25% of present world production), New Zealand (20%) and Chile (7.5%) are the three most important producing countries in terms of international trade. Italy exports c. 66% of

A.R. Ferguson
The Horticulture and Food Research Institute of New Zealand Ltd (HortResearch), Mt Albert Research Centre, Private Bag 92 169, Auckland, New Zealand
e-mail: rferguson@hortresearch.co.nz

J.F. Hancock (ed.), *Temperate Fruit Crop Breeding*,
© Springer Science+Business Media B.V. 2008

its kiwifruit, New Zealand at least 94% and Chile c. 88%. The recently established plantings and rapidly increasing production in China (340,000 t in 2002) have yet to have a major impact. About two thirds of current kiwifruit plantings are in the northern hemisphere and one third in the southern hemisphere; there is an increasing tendency towards complementary marketing of fruit from the two hemispheres to ensure year-round supply in the marketplace.

Kiwifruit are large, long-lived vines that are demanding in their growth requirements. They require strong and expensive support structures. They need very well-drained soils but also an adequate supply of moisture, since their large leaves transpire 80–100 L of water per day. Throughout the vegetative period they can transpire in total the equivalent of at least 700 mm rainfall. They need a long frost-free period of about 270–300 days from budburst to commercial harvest and they are therefore susceptible to late spring or early autumn frosts. Although they cannot withstand winter temperatures much below 0°C, they do require a period of winter chilling to break dormancy and to ensure adequate flowering. Their young shoots are easily blown out by strong winds in spring and the fruit are susceptible to windrub throughout the whole season. High summer temperatures can result in extensive premature fruit drop and leaf fall. Kiwifruit can therefore be grown successfully in only a relatively narrow band between about 35–45° north or south of the equator. The two main species grown differ in their requirements: *A. chinensis* breaks dormancy earlier than *A. deliciosa* and is more likely to be damaged by spring frosts but it is more tolerant of hot summer conditions. These differences are probably due to the natural distribution of the species; *A. chinensis* is found at warmer altitudes to the east of China and *A. deliciosa* mainly higher in the inland mountains under cooler, wetter conditions.

Kiwifruit are amongst the most recently domesticated of all fruit crops (Ferguson and Bollard 1990, Ferguson and Huang 2007). *Actinidia deliciosa* was first cultivated outside China at the beginning of the 20th century. The first commercial orchards were established in New Zealand by about 1930, but it was not until the successful export of fruit from there in the 1970s and 1980s that serious attempts were made to grow kiwifruit commercially in other countries. The domestication of *A. chinensis* is even more recent. The first known systematic cultivation was in China in 1961 and significant quantities of fruit became traded internationally only at the end of the 20th century. Today, about 85% of the kiwifruit produced commercially are of *A. deliciosa*, 15% of *A. chinensis*.

The most widely planted kiwifruit cultivar is *A. deliciosa* 'Hayward' (Fig. 8.1), selected in New Zealand in about 1925. 'Hayward' and its associated pollenizer males account for about half of kiwifruit plantings throughout the world and 'Hayward' fruit represent about 90–95% of the kiwifruit traded internationally. The second most widely planted cultivar is *A. deliciosa* 'Qinmei' which is grown commercially only in China; this cultivar and its males account for c. 30% of Chinese kiwifruit plantings or c. 15% of the world total. The most widely planted cultivar of *A. chinensis* is 'Hort16A', the fruit of which are marketed as ZESPRI[TM] GOLD Kiwifruit. In 2006, 'Hort16A' accounted for about 20% of the kiwifruit plantings in

Fig. 8.1 *A. deliciosa* 'Hayward' kiwifruit close to harvest maturity. Fruit are produced on the current season's growth

New Zealand and it is also being planted under license in other countries, a current total of about 3000 ha throughout the world.

Other *Actinidia* species are of very minor commercial importance. There are very small plantings of *A. arguta* (Sieb. et Zucc.) Planch. ex Miq., perhaps 100 ha worldwide, and even more limited plantings of *A. eriantha* Benth., *A. kolomikta* (Maxim. et Rupr.) Maxim. and *A. polygama* (Sieb. et Zucc.) Maxim.

8.2 Evolutionary Biology and Germplasm Resources

Actinidia species (family Actinidiaceae) share a number of characteristics (Ferguson 1990b): (1) all are climbing or straggling plants; (2) all species are dioecious, that is, staminate and pistillate flowers occur on different plants and although both types of flowers produce pollen, the pollen of female plants is sterile; (3) the ovary in the staminate flower is rudimentary and does not contain ovules. The syncarpic ovary of the pistillate flower is formed by the fusion of many carpels and has a whorl of free, radiating styles; (4) the fruit are technically berries, having many seed embedded in a juicy flesh. The genus is widespread in Asia, occurring between the equator and about latitude 50° North, but most species are restricted to the Yangzi River Valley and southern China, the apparent centre of evolution of the genus. Currently 50–60 *Actinidia* species are recognized in China, with a few species being found in neighboring countries.

Although the genus has been revised several times in recent years (Liang 1984, Li et al. 2007), there is still debate as to the infrageneric subdivisions of the genus and the evolutionary relationships between species (Ferguson and Huang 2007). Many *Actinidia* taxa are morphologically variable and are rather ill-defined, lacking clear taxonomic boundaries. Molecular analyses indicate that many taxa are polyphyletic, having undergone recurrent hybridization. The very broken topography of much of southern China, coupled with the resulting heterogeneous environments, spatial separation, mutation and recurrent hybridization between geographically overlapping species and polyploidization has resulted in polyphyletic groups of species. These groups assort naturally according to geographic distributions corresponding to north China, the Yangzi River Valley, southeastern China, southern China and southwestern China. *Actinidia chinensis* and *A. deliciosa* belong to a grouping mainly centered on the Yangzi River Valley (Huang et al. 2002a). In nature, there is probably considerable gene flow between species, one reason for the intermingling of morphological and molecular characters and hence the difficulties in delimiting taxa (Zhang et al. 2007).

The relationship between *A. chinensis* and *A. deliciosa* is still unresolved (Ferguson 1990c, Ferguson and Huang 2007). There is no doubt that these two species, and a third, *A. setosa* (Li) C.F.Liang et A.R.Ferguson, are closely related, but they have been variously treated as distinct species or as varieties of the one species. In this discussion, we retain them as distinct species but if they are combined they would correctly be *A. chinensis* var. *chinensis*, *A. chinensis* var. *deliciosa* (A.Chev.) A.Chev. and *A. chinensis* var. *setosai* (Li). A number of other species, mainly of limited distributions, appear morphologically to be closely related and may be natural hybrids with *A. chinensis* or *A. deliciosa* or may have contributed to the genome of *A. deliciosa*. Further work is required to examine the relationships between these taxa.

All *Actinidia* species are climbing or straggling plants (Li 1952, Liang 1984). Some species are naturally weaker growing or at their altitudinal or latitudinal limits may be reduced to inter-twined thickets. Other species, such as *A. chinensis*, *A. deliciosa* and *A. eriantha*, are vigorous, indeed rampant growers, and can climb to the tops of tall trees and from there cascade down in great masses of vegetation, flowering and fruiting when exposed to the light.

Polyploidy in *Actinidia* is common with a structured reticulate pattern of diploids, tetraploids, hexaploids and octoploids in diminishing frequency (Ferguson and Huang 2007). The basic chromosome number, $x = 29$, is unusually high and there may have been chromosome duplication early in the evolution of the genus followed by rediploidization (McNeilage and Considine 1989, He et al. 2005). Ploidy races have been detected in 15 taxa but, as only a few chromosome counts have been made for most taxa, intrataxal variation in ploidy may be much more general within the genus than so far observed. Of the commercially important kiwifruit, all *A. deliciosa* cultivars are hexaploid ($6x$), *A. chinensis* cultivars are either diploid ($2x$), e.g., 'Hort16A', or tetraploid ($4x$), e.g., Jintao®, and *A. arguta* cultivars are mainly tetraploid.

Although ploidy races are sometimes separated geographically within taxa, they cannot usually be readily distinguished morphologically, e.g., there are no obvious morphological differences between the diploid and tetraploid races of *A. chinensis*. Counting of *Actinidia* chromosomes is a slow and tedious process because of the high basic chromosome number, the smallness of the chromosomes and the frequency of polyploidy. Fortunately, flow cytometry can be used to determine ploidy races within taxa or the ploidy of interspecific crosses (Ferguson and Huang 2007). The ploidy of the offspring from interspecific crosses (particularly interploidy crosses) is not always what might have been predicted.

Polyploidy in *Actinidia* has probably evolved mainly by recurrent sexual polyploidization through the production of numerically unreduced gametes as has been observed in *A. chinensis* (Yan et al. 1997). Bud mutations of diploid *A. chinensis* 'Hort16A' in commercial orchards occasionally produce tetraploid shoots which are apparently cytologically stable and carry larger fruit. This might also occur in nature.

All *Actinidia* species appear to be dioecious, although functional dioecy has been confirmed only in *A. deliciosa* and *A. polygama* (McNeilage 1991, Kawagoe and Suzuki 2004). Gender determination in *Actinidia* appears to be of the $X_n X/X_n Y$ type with the male being heterogametic (Testolin et al. 1995a, Harvey et al. 1997b). Two genes are probably involved in sex expression, one to suppress pistil development in staminate flowers, the other to stop pollen development in pistillate flowers (Harvey et al. 1997a). It is likely that in the evolution of *Actinidia*, dioecy preceded polyploidy and speciation (Harvey et al. 1997a) and has been maintained through sexual polyploidization (Yan et al. 1997, Testolin et al. 1999).

Dioecy means that *Actinidia* plants are obligatorily outcrossing. This helps to maintain the extensive heterozygosity observed in morphological characters, even in closely related plants (Beatson 1992, Blanchet and Chartier 1992, Testolin et al. 1995b, Zhu et al. 2002, Cheng et al. 2004) and in phenological attributes such as timing and duration of flowering (Zhu et al. 2002, Cheng et al. 2006b). Heterozygosity is also encouraged by polyploidy and the high chromosome number probably promotes frequent recombination. Molecular studies confirm remarkably high levels of heterozygosity in *Actinidia*, higher than in most other woody plants and higher in hexaploid *A. deliciosa* than in tetraploid genotypes of *A. chinensis* which are, in turn, more heterozygous than diploid genotypes of *A. chinensis* (Zhen et al. 2004). The high rates of polyallelism in *Actinidia* may be due, at least in part, to the diploid species being rediploidized palaeopolyploids, i.e., cryptic polyploids (Huang et al. 1998).

Some cultivars of commercial fruiting plants, such as grapes and strawberries, are now strictly hermaphrodite whereas their ancestors were dioecious. Fully hermaphrodite kiwifruit cultivars that are self-fertile and self-setting are realistic goals for breeding programs because gender inconstancy has been observed in *A. arguta*, *A. chinensis*, *A. deliciosa* and *A. eriantha* and probably occurs in other *Actinidia* species as well (Messina et al. 1990, Testolin et al. 1999, Mizugami et al. 2007). Six different types of phenotypic gender expression have been identified in *A. deliciosa*: (1) male, (2) fruiting (inconstant or andromonoecious) male, (3) neuter, (4) inconstant female, (5) female and (6) hermaphrodite. Gender changes are not uncommon: a bud mutation in a mature male vine caused a gender change from male to female (Testolin et al. 2004) and many of the plants regenerated from protoplasts isolated from a female plant proved to be male when mature (He et al. 1995).

Self-fertile, completely hermaphrodite plants of *A. deliciosa* have been produced from crosses involving inconstant males or hermaphrodites (McNeilage and Steinhagen 1998, McNeilage et al. 2007). Hermaphroditism in *A. deliciosa* is stable and heritable. The fruit of the best hermaphrodites can be large, up to 100 g on average, and have many of the qualities required in a commercial cultivar. Such plants demonstrate that hermaphrodite cultivars of *A. deliciosa* worth cultivation are achievable and that good hermaphrodite cultivars of other *Actinidia* species should be possible.

There are many characteristics within other *Actinidia* species that could, with advantage, be incorporated into cultivated kiwifruit (Ferguson 1990a, Testolin and Costa 1994, Huang et al. 2004, Ferguson and Huang 2007) (Table 8.1). *Actinidia* species are particularly diverse in fruit characteristics such as size, shape, skin hairiness, flesh color, flavor, nutrient content, time of maturity and storage life, and by taking advantage of this diversity, it should be possible to develop many new and novel types of kiwifruit (Ferguson 2007). Good examples of what has already been achieved are the *A. chinensis* cultivars with yellow flesh such as 'Hort16A' (Ferguson et al. 1999) and Jintao® (Huang et al. 2002b, Cipriani and Testolin 2007), or the selections of *A. arguta* with their small, smooth-skinned fruit (Williams et al. 2003). It should also be possible to introduce new vine growth characteristics such as a reduced requirement for winter chilling or greater cold hardiness.

Many interspecific *Actinidia* crosses have been described (e.g., Fairchild 1927, Pringle 1986, Wang et al. 1989, 1994, Ke et al. 1992, Testolin and Costa 1994, An et al. 1995) and even crosses between species considered to be only distantly related often produce viable seed (Hirsch et al. 2001). Some crosses are very successful with thousands of hybrids plants being raised, e.g., *A. chinensis* × *A. eriantha* (Wang et al. 2000). Many other crosses, however, especially interploidy ones, set only a few fruit containing only a small number of viable seed. Although some F$_1$ hybrids are considered to have immediate commercial potential, e.g., 'Kosui', probably a hybrid between *A. rufa* and *A. chinensis* (Kataoka et al. 2003), further crossing or backcrossing is usually required and for this the F$_1$ hybrids must be fertile. Furthermore, introgression of characters depends on good chromosome pairing between parental genomes and generally, pairing in *Actinidia* hybrids is better the more closely related the parents are. What appears to be high levels of chromosome pairing in wide species hybrids is often due to pairing within rather than between parental genomes (Datson et al. 2006). Knowledge of the type of chromosome pairing (allosyndetic v. autosyndetic) in hybrids helps in making decisions as to whether the production of F$_2$ hybrids is the better breeding strategy as opposed to backcrossing to one or other of the parents.

Actinidia germplasm repositories provide the starting material for programs of kiwifruit improvement but even large germplasm collections can conserve only a very small part of the wild diversity, especially as *Actinidia* appears to be such a variable genus (Fig. 8.2). Many of the genotypes held in various germplasm collections have been obtained by exchange, making the sampling of the wild diversity more limited than might be first thought.

Table 8.1 Some *Actinidia* species and their important horticultural traits. Some of the traits listed are restricted to only some genotypes

Taxon	Ploidy	Important traits
A. deliciosa (A. Chev.) C.F.Liang et A.R.Ferguson	6*x*	'Traditional' kiwifruit, large fruit, up to c. 120 g, long stiff hairs on brown skin, green fruit flesh, occasionally red around core, good flavor, good vitamin C content, very good storage life
A. chinensis Planch.	2*x*, 4*x*	Other main commercial kiwifruit, large fruit, up to c. 100 g, soft hairs on green or brown skin, green or yellow fruit flesh, sometimes red around core, sweeter flavor, reasonable storage life, high vitamin C content
A. arguta (Sieb. et Zucc.) Planch. ex Miq.	4*x*, 6*x*	Vines cold tolerant, small fruit, c. 10 g, hairless, edible skin, green fruit flesh, sometimes turning light red to purplish red on ripening, often very good sweet flavor, uneven fruit maturation, limited storage life
A. arguta var. *purpurea* (Rehder) C.F.Liang	4*x*, 8*x*	Late flowering, small fruit, hairless, edible skin, external color and green fruit flesh become completely dark purple when ripe, late maturing, usually poor flavor, limited storage life
A. eriantha Benth.	2*x*	Medium-sized fruit, usually 15–40 g, very hairy skins, somewhat peelable, dark green fruit flesh, sometimes fig-like flavor, very high vitamin C content, up to 1% fresh weight
A. henanensis C.F.Liang	-	Small fruit, c. 15 g, hairless, edible skins, fruit skin and core turn red when ripe
A. kolomikta (Maxim. et Rupr.) Maxim.	2*x*, 4*x*	Compact growth habit, short growing season, vines very cold hardy, small fruit, <5 g, hairless edible skins, very sweet, high vitamin C, limited storage life
A. latifolia (Gardn. et Champ.) Merr.	2*x*	Late flowering, very small fruit, <5 g, in large inflorescences, tough brown, slightly tomentose skins, green flesh, very high vitamin C content, up to 2% fresh weight
A. macrosperma C.F.Liang	4*x*	Less vigorous vine, small fruit, 10–15 g, hairless, smooth but tough skins, peppery flavor when unripe, external color and green fruit flesh and core become orange on ripening, very large seed
A. melanandra Franch.	2*x*, 4*x*	Small fruit, hairless edible skins, external color and green fruit flesh become red on ripening, limited storage life
A. polygama (Sieb. et Zucc.) Maxim.	2*x*, 4*x*	Vines cold tolerant, small fruit, 5–10 g, hairless, edible skins, peppery flavor when unripe, external color and fruit flesh and core turn yellow on ripening, limited storage life
A. rufa (Sieb. et Zucc.) Planch. ex Miq.	2*x*	Very productive, small fruit, c. 15 g, brown skins with fine pubescence, green fruit flesh
A. setosa (Li) C.F.Liang et A.R.Ferguson	2*x*	Medium-sized fruit, 30–40 g, very hairy, tough greenish skins, green flesh, maturing very early

Table 8.1 (continued)

Taxon	Ploidy	Important traits
A. tetramera Maxim.	2*x*	Small fruit, <5 g, hairless, smooth skins, turn yellow on ripening
A. valvata Dunn	4*x*, 6*x*	Vines cold tolerant, small fruit, c. 10 g, hairless, smooth skins, external color and green fruit flesh become orange on ripening, peppery flavor when unripe

Sources – Ferguson 1990a, Huang et al. 2004, Ferguson and Huang 2007

The repositories in China are exceptional in terms of the number of taxa represented. The national germplasm repository for *Actinidia* developed since 1978 at the Wuhan Institute of Botany, Chinese Academy of Sciences has as its goal the conservation of natural resources of *Actinidia* in China and the development of superior cultivars (Huang 2003). Nearly 60 taxa of more than 40 species, comprising more than 200 accessions from the wild or from exchange, are conserved as a living collection in the Wuhan Botanic Garden, Moshan, Wuhan, Hubei (Wang et al. 2003b). The characteristics of the fruit have been summarized in Huang et al. (2003 and 2004). Most of the plants come from Hubei, Guangxi, Hunan, Jiangxi and Fujian and there are 63 named cultivars or selections. Some species that come from colder areas, such as *A. kolomikta* or *A. arguta*, do not adapt well to the very hot summers

Fig. 8.2 Fruit of *Actinidia* species vary in size, shape, skin hairiness and external color as well as in internal characteristics such as flavour and chemical composition. Lower centre left, fruit of *A. deliciosa* 'Hayward', lower centre right, *A. chinensis* 'Hort16A'

of Wuhan and other species that were deciduous in their native habitat remain evergreen at Wuhan. Complementary collections are therefore held under cooler conditions at Lushan, Jiangxi or the moister, more subtropical conditions of the Guilin
Botanic Garden, Guangxi. The collection at the Guangxi Institute of Botany, Guilin
has been developed over 25 years and is important for its association with Liang
Choufen who revised the genus in 1984. The collection occupies 0.3 ha and contains
representatives of 74 taxa and 69 clones or cultivars (Li et al. 2000). Phenological
and fruit characteristics of plants in the collection have been described in a series
of papers (e.g., Huang et al. 1983, Li et al. 1985, Li et al. 1996). Another important
collection is that of the Sichuan Provincial Natural Resources Research Institute,
Chengdu, which has probably the best assembly anywhere of wild *A. deliciosa* and
A. chinensis accessions from three provinces, as well as cultivars and examples of
numerous other *Actinidia* taxa (Li and Lowe 2007).

The largest *Actinidia* collection outside China is that of HortResearch,
New Zealand, occupying 6.2 ha at three research orchards at Kerikeri, Te Puke and
Riwaka. This has been developed since 1955 and currently contains about 3500
genotypes from 310 accessions of 24 taxa, with about 1000 of these genotypes
having been selected for long-term retention (Ferguson 2007). In addition, there
is a collection of 80 cultivars or named selections, many of them originating outside New Zealand. The HortResearch collection is particularly strong in the two
commercially important kiwifruit species, but the sampling of wild diversity is limited in that most of the accessions of *A. deliciosa* come from two provinces of
China, whereas the species occurs naturally in seven other provinces. A few taxa
are represented by only a single genotype or genotypes of one gender so in total,
18 species are fruiting. Several of the more cold-hardy species grow vigorously at
all three orchards but seem to fruit more reliably at Riwaka, probably because their
winter-chilling requirements are better met by the colder winters.

The most important collection in Europe is that of the Dipartimento di Agrarie
Scienze e Ambientali, University of Udine, Italy. This occupies 0.8 ha and has about
175 accessions of 28 *Actinidia* taxa (mainly separate species) from introductions of
budwood or seed and named cultivars or selections (R. Testolin, pers. comm.). The
collection is particularly strong in *A. chinensis* cultivars (directly or indirectly from
China), *A. deliciosa* cultivars and selections mainly of Italian or New Zealand origin
and some interesting species seldom found in other collections. Cold-hardy species,
such as *A. arguta*, grow well at this location but the severe winters at Udine prevent
the more subtropical species from surviving. European plant nurseries, especially
those in the United Kingdom, also offer for sale a number of *Actinidia* cultivars and
species, including some of wild origin, and these can be a useful source of additional
material.

The United States Department of Agriculture National Clonal Germplasm Repository for *Actinidia* is mainly at Corvallis, Oregon, with some additional material
held at Davis, California. The collection is very well catalogued and contains nearly
300 distinct genotypes, most of which are named selections or cultivars. It has a
particularly good collection of cold-hardy *Actinidia* including 103 genotypes of
A. arguta and *A. arguta* var. *purpurea* and 48 of *A. kolomikta*; these are mainly

named selections that are winter-hardy in much of the United States. Most other taxa are represented by only a few genotypes, although there are 32 of *A. chinensis* and 15 of *A. deliciosa*, again mostly cultivars or selections. There is also a good collection of cold-hardy *Actinidia* at the University of Minnesota (Start et al. 2007).

There are smaller but still important germplasm collections in other countries. In Korea, the Sung Kyun Kwan University in Suwon has collected 20 *Actinidia* species and 18 cultivars (Shim and Ha 1999) and the Subtropical Fruits Experimental Station in Haenam and the Forest Genetics Research Institute in Suwon have collected wild germplasm of *A. arguta* as well as representatives of other taxa. They have also carried out numerous interspecific crosses. In Japan, the Kagawa Prefectural Experiment Station has a collection of cultivars of *A. chinensis*, *A. deliciosa*, *A. rufa*, *A. arguta* and interspecific hybrids (Kokudo et al. 2003, Mizugami et al. 2007).

8.3 History of Improvement

Actinidia species (*mihoutao*) have long been harvested from the wild by Chinese peasants and brought down from the mountains to be sold in markets (Ferguson and Huang 2007). Even today, fruit of *A. chinensis*, *A. deliciosa* and other species such as *A. arguta* and *A. eriantha* are collected from the wild in China, although such harvests are becoming less important. The potential of kiwifruit as a commercial crop was not recognized in China until kiwifruit industries developed successfully in other countries and as recently as 30 years ago, there was less than one hectare of cultivated kiwifruit in China (Cui 1981). *Actinidia* species were not cultivated for fruit in the other countries where they occur naturally, such as Japan and Korea.

Widespread cultivation of *Actinidia* species began in the mid-19th century with the introduction of the cold-hardy species *A. arguta*, *A. kolomikta* and *A. polygama* to Europe and North America (Ferguson and Huang 2007). Initially, these were valued primarily as ornamentals but, after initial confusion caused by the misidentification of individual plants, it was recognized that *A. arguta* and *A. kolomikta* had fruit which, although small, could be very sweet and have a fine flavor. Many selections of *A. arguta* and *A. kolomikta* have been made over the years, but the origins of most of these are now obscure. Michurin (1949) was the first to make selections from the wild in eastern Siberia and from seedlings arising from several cycles of controlled crosses within *A. arguta* or crosses between *A. arguta* and *A. kolomikta* (Evreinoff 1949). Later selections were probably chance seedlings recognized for the quality of their fruit. Fruit of *A. kolomikta* are probably too small ever to justify commercial cultivation, even in those areas that are too cold for other kiwifruit species, but *A. arguta* has more obvious economic potential.

The two commercially important *Actinidia* species, *A. chinensis* and *A. deliciosa*, occur naturally only in China. Plants of these two species were being cultivated outside China during the first years of the 20th century but it seems that only those of *A. deliciosa* survived; some of these eventually became the basis of the worldwide kiwifruit industry (Ferguson and Bollard 1990, Ferguson 2004). Domestication of *A. chinensis* was to follow much later.

Actinidia deliciosa was initially introduced at about the same time to the United Kingdom, the United States and New Zealand, but it was the introduction to New Zealand that was ultimately critical. All the important New Zealand cultivars of *A. deliciosa* can be traced back to a single introduction in 1904 of seed collected from the wild in China. These cultivars are all seedling selections, but at least one generation removed from the original introduction of seed to New Zealand (Ferguson and Bollard 1990, Ferguson 1997). Initially many different selections were grown but eventually, the best strains were identified and characterized, their origins determined, and they were given names commemorating some of the industry pioneers (Mouat 1958). 'Hayward', named for Hayward Wright who had selected it, became the cultivar of choice because it had large fruit of good flavor and an exceptional storage life. It was an extraordinarily lucky selection as it came from a row of possibly 40 seedlings (Ferguson and Bollard 1990). As other countries started growing kiwifruit, they adopted New Zealand cultivars of *A. deliciosa* because of their perceived superiority (Ferguson and Bollard 1990, Ferguson 1990a). When New Zealand marketers decided to export only 'Hayward' kiwifruit because these appealed most to customers, other countries followed suit so that 'Hayward' quickly became the cultivar of choice everywhere. The New Zealand kiwifruit cultivars were even sent back to China when commercial kiwifruit orchards were first established there and still account for about 20% of all Chinese kiwifruit plantings. That kiwifruit developed so rapidly as a crop is largely due to the excellence of 'Hayward' and the long storage life of its fruit that allowed export by sea.

Modern cultivation of kiwifruit in China began at the Beijing Botanic Gardens, Chinese Academy of Sciences in 1957 with introduction of *A. deliciosa* seed from the Qinling Mountains, Shaanxi (Zhang et al. 1983). Four years later, the Beijing Gardens started growing *A. chinensis* using seed from fruit collected near Neixiang in the Funiu Mountains, Henan (Zhang et al. 1983, Huang 2003, 2004), the first recorded successful cultivation of this species. In late 1977 and early 1978, evaluation of the natural resources of *Actinidia* became more systematic throughout China with the initiation of surveys of wild germplasm of all *Actinidia* species and the selection of superior genotypes for cultivar development. More than 1400 promising individuals were selected for further study (Qian and Yu 1992) and several hundred were named on evaluation after grafting and replicated trials across regions.

Current kiwifruit plantings in China are striking for their diversity. In most countries, only *A. deliciosa* 'Hayward' and its pollenizers are grown, but in China the most widely planted cultivar, *A. deliciosa* 'Qinmei', accounts for only about one third of the total area in kiwifruit. There are another 15–20 important fruiting cultivars of *A. chinensis* and *A. deliciosa*, even if some of these are largely restricted to a single province. Nearly all Chinese cultivars are selections from the wild with budwood having been originally collected from plants that were recognized as having superior fruit (Ferguson and Huang 2007). 'Moshan No. 4', probably the most widely planted pollenizer for *A. chinensis* in China, was likewise selected from the wild. Some selections from the wild are now being cultivated commercially outside China, e.g., *A. chinensis* Jintao® which is being grown in Europe and South America (Huang et al. 2002b).

A few Chinese cultivars were not selected directly from the wild but originated as wild-collected seeds or seedlings, e.g., *A. chinensis* 'Hongyang' was selected from more than 3000 seedlings raised from seed collected in the wild (Wang et al. 2003a). 'Hongyang' is therefore only one generation removed from the wild, whereas the New Zealand cultivars are probably at least two generations removed. Other cultivars grown outside China that are selections from wild-collected seed are *A. deliciosa* 'California Male' ('Chico Male'), from seed sent to the United States in 1908 by E.H. Wilson (Ferguson 1997), and *A. chinensis* ChinaBelle®, recently selected in France from seed collected in China (Blanchet and Chartier 1998).

Some kiwifruit cultivars have also originated as selections of open-pollinated seedlings of named cultivars or selections (Ferguson 1997). 'Jinkui', the third most widely planted *A. deliciosa* cultivar in China, is a seedling from open-pollination of a cultivar itself selected from the wild (Chen 2003). 'Koryoku' is a seedling from open-pollination of 'Hayward', itself a seedling selection. Other cultivars are clonal selections, thought to have some advantages over existing cultivars, e.g., the Kramer strain of 'Hayward', or are budsports of an existing cultivar, e.g., 'Wilkins Super', a budsport of 'Hayward', TopStar®, a budsport of 'Hayward' with hairless fruit, and Green Light®, a sport of 'Hayward', whose fruit are claimed to mature at least a month earlier than those of 'Hayward'.

Most kiwifruit cultivars are thus selections made directly or indirectly from the wild, are chance seedlings or are budsports. Selection has required recognition of plants with superior fruit having useful commercial attributes but has not involved planned, controlled crosses with carefully chosen parents of established breeding value. There are a few exceptions, however: *Actinidia deliciosa* 'Tomua' was produced by crossing 'Hayward' with an early flowering male, the intention being to produce fruit that could be harvested well before those of 'Hayward' (Muggleston et al. 1998). 'Tomua' has large fruit, similar in many respects to those of 'Hayward', but they matures two to four weeks earlier. Unfortunately, they did not respond well to the fruit handling procedures used and 'Tomua' is no longer grown commercially. Some other *A. deliciosa* selections resulting from controlled crosses, such as 'Katiuscia', 'Silvia', and 'Stefania' from Italy (Valmori 1991) and 'Skelton' from New Zealand (US Plant Patent PP 8,334), have not stimulated much interest. Two *A. deliciosa* Summerkiwi[TM] selections from Italy that are attracting attention arose from pollination of 'Hayward' by a fruiting male. They meet the desire for an earlier harvest to avoid frost (Testolin 2005). If the fruit quality proves adequate, the early harvest would be a great advantage particularly in those areas where the climate is marginal for kiwifruit. In 2006, total production of Summerkiwi® fruit was about 4000 t. Only one commercially important cultivar of *A. chinensis* has so far resulted from a controlled hybridization. 'Hort16A' came from the cross of a female, the product of open pollination of seedlings from a seed accession from the wild, and a male raised from seed collected in the wild (Muggleston et al. 1998, Ferguson et al. 1999). The fruit of 'Hort16A' now account for about 25% of the earnings of the New Zealand kiwifruit export industry.

8.4 Current Breeding Efforts

Most breeding programs have broadly similar breeding objectives. In *A. chinensis* and *A. deliciosa*, the main emphasis is on fruit flavor, size, time of harvest, flesh color, length of storage life, and vine productivity. Storage life is very important, as most of the main kiwifruit producers rely heavily on fruit exports. Frequently, there is also a demand for fruit that mature and can be harvested earlier (to avoid the risk of frost), but in China the preference is for cultivars which can be harvested later in the season when temperatures are cooler and the fruit contain less field heat (Huang and Ferguson 2001). In other species and interspecific hybrids, the emphasis tends to be on novelty (e.g., edible or peelable skins, skin and flesh color, nutritional content), flavor and environmental adaptation. One major advantage in kiwifruit breeding is that once desirable genotypes are identified they can be immediately fixed by clonal propagation. This enables all the genetic variance (additive, dominant and epistatic) to be exploited (Zhu et al. 2002).

'Hort16A', now marketed as ZESPRI™ GOLD Kiwifruit (Fig. 8.3), was the first new kiwifruit cultivar in international trade since countries other than New Zealand started growing 'Hayward'. The release and successful commercialization of 'Hort16A' has stimulated increased interest and investment in kiwifruit breeding and selection. In China, the emphasis has begun to shift from selecting cultivars directly from the wild to systematic breeding programs. HortResearch in New Zealand still has the largest breeding program, but programs at the Wuhan Institute of Botany and Hunan Horticultural Research Institute, Changsha in China have increased in scale and significance. There are also programs elsewhere including those at Haenam in Korea, Kagawa in Japan, Minnesota in the United States, Rome and Udine in Italy and Bucharest in Romania as well those of private breeders. Even so, internationally there is only limited effort on breeding new kiwifruit (Seal 2003).

Several new yellow-fleshed cultivars of *A. chinensis* have been released outside China since 'Hort16A' including Jintao® (Huang et al. 2002b, Cipriani and Testolin 2007) originally selected at the Wuhan Institute of Botany, but

Fig. 8.3 Left, fruit of *Actinidia chinensis* 'Hort16A', sold commercially as ZESPRI™ GOLD Kiwifruit, with almost hairless skin and bright gold flesh; right, fruit of *A. deliciosa* 'Hayward', the most important of all green-fleshed kiwifruit cultivars

commercialized internationally by the Italian consortium, Consorzio Kiwigold, and more recently 'Sanuki Gold' (Fukuda et al. 2007) from Japan. Jintao® is a selection from the wild, but 'Sanuki Gold' is the result of a planned cross. Other breeding programs are about to release new selections.

Red-centred cultivars of *A. chinensis* are creating great interest. The first red kiwifruit to be released onto international markets were those of *A. chinensis* 'Hongyang' (Wang et al. 2003a). A similar red-fleshed cultivar, 'Chuhong', has been selected in Changsha, China from the wild (Wang et al. 2004, Zhong et al. 2007). A red-centred cultivar of *A. deliciosa*, 'Hongmei', has also been selected from the mountains of northern Sichuan (Wang et al. 2005). In New Zealand, good progress is being made in selecting superior parents primarily to increase red coloration and fruit size in *A. chinensis* (Cheng et al. 2007).

So far, no new green-fleshed *A. deliciosa* cultivars seem likely to challenge the global standard 'Hayward', although the Summerkiwi™ selections released in Italy might provide an alternative for earlier harvests.

Cultivars from other *Actinidia* species are also starting to emerge, including several of *A. arguta* from New Zealand (Williams et al. 2003), Romania (Stănică and Zuccherelli 2007) and Korea (Jo et al. 2007b), and *A. eriantha* 'Bidan' from Korea (Jo et al. 2007a). There are active programs of interspecific hybridization and selection in New Zealand (Beatson et al. 2007), Korea (Cho et al. 2007), U.S.A. (Guthrie et al. 2007) and Japan (Kataoka et al. 2003).

8.5 Genetics of Important Traits

There have been only a few detailed studies of the inheritance of important traits in *Actinidia* (Table 8.2) but some general trends are emerging. Most important traits appear to be polygenically controlled. Only gender (Testolin et al. 1995a, Harvey et al. 1997a,b, Testolin et al. 1999, 2004) and petal color (A.G. Seal, unpublished) have been shown to be influenced by major genes showing simple Mendelian segregation. Studies have consistently revealed the high levels of genetic variation expected in newly domesticated, outcrossing crop species. While heritabilities often vary according to species, population (Marsh et al. 1999) and method of estimation (Beatson 1992), heritabilities for some key traits including fruit weight, shape and vitamin C content have tended to be consistently high. A small negative genetic correlation between fruit weight and dry matter content (a common measure of kiwifruit quality) might hinder progress in improving these key traits simultaneously but, overall, the enormous diversity in both fruit and vine traits offers plentiful opportunities for improvement by traditional breeding and selection methods.

8.5.1 Disease and Pest Resistance

There is very little published information on the genetics of disease and pest resistance in *Actinidia*. Hill et al. (2007) found that the heritabilities for the incidence of

Table 8.2 Genetics of adaptation, productivity, and fruit quality in *Actinidia*. Note that narrow sense heritabilies vary amongst different breeding populations of *A. chinensis* and *A. deliciosa*

Attribute	Observations and sources
Adaptations	
Date of budburst	High heritability (Testolin et al. 1995b)
Flowering date	High heritability sometimes observed (Testolin et al. 1995b, Marsh et al. 1999, Cheng et al. 2006b)
Flowering duration	Low heritability and strongly affected by environment, although influenced by gender in that males flower for significantly longer (Marsh et al. 1999, Zhu et al. 2002, Cheng et al. 2006b)
Harvest date	High heritability (Beatson 1992)
Scale resistance	Moderate heritability (Hill et al. 2007)
Productivity	
Percentage of floral shoots	High heritability in female plants (Zhu et al. 2002)
Fruit quality	
Size	High heritability (Beatson 1992, Marsh et al. 1999 and 2003a, Zhu et al. 2002, Cheng et al. 2004 and 2006a)
Shape, dimensions	High heritability (Beatson 1992, Zhu et al. 2002, Marsh et al. 2003a)
Core dimensions	Heritability depends on populations (Marsh et al. 2003a)
Vitamin C	High heritability (Beatson 1992, Cheng et al. 2005 and 2006a)
Soluble solids	Heritability depends on populations (Beatson 1992, Marsh et al. 1999, Cheng et al. 2005 and 2006a)
Titratable acids	Heritability depends on populations (Cheng et al. 2005 and 2006a)
Dry matter	Heritability depends on populations (Cheng et al. 2005 and 2006a)

scale insects on leaves and on fruit of diploid *A. chinensis* were moderate. Only low to modest positive genetic correlations were observed between incidence of scale and a range of taste and other fruit traits. This suggests that breeding and selection for lower levels of scale infestation should not impede improvement in key quality traits. Furthermore, glucose, dry matter and soluble solids contents had moderate to high negative genetic correlations with scale levels on leaves and on fruit. Selection for improvement of these fruit quality traits is therefore likely to produce a favorable correlated response in the incidence of scale.

8.5.2 Environmental Adaptation

Response to environmental variation is well known for many traits in *Actinidia*. Phenological and physiological traits such as flowering time and the timing of fruit maturity vary when an individual genotype is grown in different seasons, regions or countries, or at different altitudes in the same region. The red-flesh color of the *A. chinensis* cultivar 'Chuhong' is more pronounced when the vines are grown at higher altitudes (Zhong et al. 2007). Vitamin C content of fruit of 'Hayward' fruit can be markedly affected by growing conditions (e.g., Snelgar et al. 2005). While such environmental variation is likely to affect a wide range of traits, little is yet known about the prevalence of genotype × environment (GxE) interactions. More information on the sensitivity of key traits to G×E interactions would help optimization of breeding strategies and cultivar deployment. Environmental adaptation will probably become more important in breeding programs if global climate change follows current projections. Already, winter chilling is inadequate in some important growing regions.

8.5.3 Flowering and Fruiting Habit

Testolin et al. (1995b) found high heritability for date of budburst but low heritability for a number of other traits in a population of *A. deliciosa* seedlings. Zhu et al. (2002) also estimated heritabilities based on variance components for a range of flower and fruit traits in a seedling population of *A. deliciosa*. Pedicel length had high heritability in both male and female vines. The number of terminal flowers per shoot, number of lateral flowers per shoot and total shoot number had low heritabilities. Date of first bloom and flowering duration had moderate heritability among male vines but low heritability among female vines. The percentage of floral shoots showed the opposite trend, having moderate heritability only among female vines. Marsh et al. (1999) found median flowering date and 'early Brix' (mean soluble solids concentration at a single pre-harvest date–an indication of fruit maturity) to be moderately to highly heritable in two populations of *A. deliciosa*. The duration of flowering and the juvenile period (years from field planting to flowering) had moderate to low heritabilities in these populations. In a recent study of a large population of diploid *A. chinensis* seedlings over two seasons, Cheng et al. (2006b) confirmed that flowering time ('time to reach 50% flowering') was highly heritable whereas flowering duration had low heritability. In contrast to Zhu et al. (2002), the results suggested that the flowering times of male vines were less heritable than those of female vines. However, breeding aimed at taking advantage of significant specific combining ability (SCA) effects of particular male and female parental combinations could be successful in changing the timing of flowering.

Overall, it appears that the timing of flowering might be more amenable to selection than the duration of flowering. The timing of flowering is particularly important in selecting pollenizers to match female cultivars and in selecting female cultivars for deployment in areas prone to spring frosts. Duration of flowering is markedly affected by environment and an extended flowering period could lead to mixed maturities at harvest.

8.5.4 Fruit Quality

Significant genetic variation was found in a population of *A. chinensis* seedlings for quality traits including individual sugar and acid levels, juice pH, titratable acidity, vitamin C content, dry matter content, soluble solids content and weight of mature fruit (Cheng et al. 2004). Narrow sense heritabilities were estimated using variance components: vitamin C content had very high heritability followed by titratable acidity, soluble solids content, fruit weight and dry matter content suggesting that, individually, these traits will be responsive to selection. Fortunately, levels of the sugars fructose, glucose and sucrose were moderately to highly genotypically correlated with each other and all were highly correlated with vitamin C, dry matter and soluble solids contents. However, fruit weight was negatively correlated at a moderate level with sugar content (except myo-inositol), dry matter and soluble solids content. This is likely to hinder progress in breeding programs attempting to improve these traits simultaneously. Glucose content, quinic acid content and fruit pH had low heritabilities.

In a population of tetraploid *A. chinensis* seedlings, Cheng et al. (2007) also found that narrow-sense heritabilities were relatively high for titratable acidity, vitamin C and fruit weight. However, levels of individual sugars, dry matter content and soluble solids content had low heritabilities. Soluble solids content and dry matter content were again genetically correlated with sugar levels. Lower heritabilities in some traits might reflect the narrower genetic base of this tetraploid material, as all known tetraploid *A. chinensis* come from a restricted area in China. Certainly, the analyses indicate that conclusions cannot necessarily be transferred from one population to another.

Soluble solids content was found to be moderately to highly heritable in different populations of *A. deliciosa* (Beatson 1992, Marsh et al. 1999, 2003a). Heritability estimates for flavor and texture based on sensory scores (Marsh et al. 2003a) were low but this might partly reflect the difficulties inherent in using assessment methods based on human perceptions.

Other important aspects of fruit 'quality' include fruit size and shape. Individual mean fruit weight and most components of fruit shape (e.g., maximum and minimum fruit and core diameters and fruit length) showed high heritabilities in populations of *A. deliciosa* (Beatson 1992, Zhu et al. 2002, Marsh et al. 2003a) and *A. chinensis* (Cheng et al. 2004, 2007).

8.5.5 Yield

Total yield and individual fruit weight are obviously key traits in kiwifruit production and breeding. While fruit weight can be expected to respond readily to individual or mass selection, yield is likely to be less responsive. As in many other crops, yield in kiwifruit generally has only low or moderate heritability. In a population of *A. deliciosa* seedlings, Zhu et al. (2002) found moderate heritability for total yield, slightly higher heritability for individual mean fruit weight but low heritability for fruit number and 'vine efficiency' (total fruit weight divided by cross-sectional area of the trunk at a height of 40 cm in spring). In two different populations of *A. deliciosa* seedlings across three seasons, Marsh et al. (1999) found that total yield and fruit number had moderate to low heritability, whereas fruit weight was highly heritable.

8.6 Crossing and Evaluation Techniques

8.6.1 Breeding Systems

The main constraints to progress in kiwifruit breeding are dioecy, the long generation time and the complexity of some key traits as well as the need for support
structures, the exuberant vegetative growth and the need to control growth to ensure
fruiting.

Most breeding programs employ some kind of recurrent selection strategy. Controlled pollination between elite male and female parents, usually in some kind
of factorial mating design, is often used but mass selection using seed resulting
from open pollination might be successful in improving traits with high heritability
(Beatson 1992). Kiwifruit are obligatorily outcrossing, being normally dioecious
and only in rare exceptions is selfing possible (McNeilage et al. 2007). Although
kiwifruit are unusually heterozygous, selfing of hermaphrodites has revealed some
inbreeding effects with fruit malformation or a reduction in vigor. Care must therefore be taken to avoid excessive inbreeding which can lead to a reduction in vigor
and fruit quality.

Fruiting characteristics are not expressed in the male parent and only female
plants can be phenotypically selected (McNeilage et al. 2007). Males have to be
selected at random or progeny tested, a lengthy and expensive process in which
males are chosen for fruit characteristics by assessing the fruit of their daughters
(Marsh et al. 1999). Nevertheless, the choice of male parents is important because it
has been shown that they as well as the female parent contribute significantly to fruit
characteristics in the offspring (Cheng et al. 2004, 2006a). Indirect selection of male
parents based on the performance of their female sibs (sib selection) is an option
but it is likely to be very risky since males from the same family can often have
very different breeding values. Two females with desirable fruit qualities cannot
be directly crossed. As the pollenizers and the fruiting cultivars must coincide in
flowering time, selection of a new fruiting cultivar usually requires selection of new
pollenizers.

A preponderance of male plants has sometimes been reported in wild populations
but a 1:1 sex ratio is usually found when seedling populations are raised. Seedlings
of most *Actinidia* species do not flower until they are 3–4 years old, sometimes
even later. Male plants are usually more precocious than female plants, so the
sex ratio approaches 1:1 as more vines flower each consecutive season (Testolin
et al. 1995b, Marsh et al. 2003a). The main aim of most breeding programs is the
development of new fruiting cultivars and the cultivation of many unwanted male
plants is a major limitation. Use of a DNA sex marker for determining the gender
of young seedlings would greatly increase the efficiency of breeding programs. The
development of hermaphrodite breeding lines offers perhaps the best long-term solution, since hermaphrodites with known fruit quality characteristics will produce
only hermaphrodite and female progeny when used as pollen parents (McNeilage
et al. 2007).

Under New Zealand conditions, it usually takes from two to five years for seedling vines to flower for the first time. It has so far proved difficult to shorten this time significantly under natural conditions. In some other crops, selected clonal rootstocks are used to advance the flowering and fruiting of grafted seedlings but such rootstocks are not yet available in kiwifruit.

Variation in ploidy between taxa and within taxa can create obvious difficulties when trying to introgress characters from the different *Actinidia* species into the current commercially important species. Ploidy can, however, be manipulated by, for example, induced doubling by antimitotic agents such as colchicine or by induced parthenogenesis using lethally irradiated pollen. Generally, however, it is more realistic to choose parents at compatible ploidies, particularly if a taxon contains ploidy races.

8.6.2 Pollination and Seedling Culture

The pistillate kiwifruit flower has an ovary consisting of many carpels fused together, each carpel having two rows of ovules and a free style. After fertilization, the ovary develops into a fleshy fruit or berry which can contain more than a thousand seeds. Every pistillate flower that is pollinated with compatible pollen is usually capable of setting a fruit and very large numbers of seed can thus be obtained from relatively few crosses. With interspecific or interploidy crosses, however, many of the developing seed may abort and the fruit produced may contain only a few viable seed. Even very low rates of pollen contamination can confound results.

The methods for bagging of pistillate flowers, pollen collection from staminate flowers, pollination, seed extraction and germination are summarized in Ferguson et al. (1996). Flower buds that are at full calyx split ('popcorn' stage) are collected from selected male parent vines. Anthers are removed and dried for 24 hours at 35°C to encourage dehiscence. The anthers and shed pollen are then collected into a glass vial and stored in a refrigerator at 4°C until needed for pollination. On selected female parent vines, shoots with unopened flower buds at the 'popcorn' stage are enclosed in a paper bag, secured so that bees and other potential pollinators are excluded. When the pistillate flowers open, the bag is temporarily removed and the stored pollen is applied using a dry brush. The bag is then replaced and resealed. Bags can be removed about 10 days later, after fruit set. Emasculation of flowers is not normally necessary but if hermaphrodite flowers are being used for the seed parent, then the stamens are removed about 3 days before anthesis at the early popcorn stage.

Fruit resulting from crosses are harvested when mature and allowed to ripen (ripening can be hastened using ethylene). Ripe fruit are cut up and immersed in pectinase solution (Rohm Rohapect D5L at 1 ml:200 ml water) for 2–3 days. The pulp is then washed through a sieve to separate the seed. The seed are dried and stored in a refridgerator prior to sowing. Kiwifruit seed usually require a period of cold treatment (stratification) to break dormancy prior to germination. Dried seed are placed on moist filter paper and kept at 1°C–4°C for 3½ to 4 weeks followed by another 10 days at alternating temperatures (5°C for 16 hours, 24°C for 8 hours). Seed can then be sown into punnets of seed raising mix and kept at 20°C–25°C. Germination should begin after about 10 days.

8.6.3 Evaluation

Seedlings are planted in rows in a design appropriate to the population structure and objectives of the trial. Unless DNA markers are available, gender cannot usually be determined until flowering and so male and female vines are effectively distributed at random. In New Zealand, a small proportion of seedlings (mostly male vines) will usually flower after two years in the orchard and most vines will have flowered after four years. A minimum of two fruiting seasons of evaluation is usually required, so the minimum generation time is usually 5–6 years. Selected seedlings can be readily grafted into clonal trials which will typically take at least another 4–5 years to complete. Further propagation, grower trials and test marketing are required prior to commercialization of a new cultivar. Thus, it can easily take at least 15 years from crossing to significant commercial production – much longer if several generations of breeding and selection are required.

Any new fruiting cultivar of kiwifruit must have at least an acceptable eating quality. Eating quality has many components including taste, texture and aroma. Taste is largely determined by the concentration of sugars and acids (Marsh et al. 2003b), and aroma by the composition and concentration of volatile compounds (Paterson et al. 1991). Many of these components can be measured instrumentally. However, it is difficult to predict precisely how consumers will respond to different combinations of these components. There is often significant variation between fruit samples from the same genotype. Maturity of fruit at harvest, the length and conditions of storage and ripeness when tasted can be critical. Perhaps most challenging of all is the variation found among consumers in their perceptions of eating quality (Harker et al. 2007). Assessments by large consumer panels are impractical for testing large numbers of seedlings, each bearing usually no more than 100–200 fruit. In the future, a better understanding of the components of eating quality and their interactions might provide instrumental methods that better predict consumer responses. Until then, evaluation of eating quality at the seedling stage is more about eliminating poor quality seedlings than about identifying the very best.

8.7 Biotechnological Approaches to Genetic Improvement

Applications of biotechnology to kiwifruit breeding have recently been reviewed (Sharma and Shirkot 2004, Oliveira and Fraser 2005, Atkinson and MacRae 2007, MacRae 2007); so far, few are being routinely used because of the difficulties in working with kiwifruit. We can expect some of these techniques, especially the use of DNA markers, to have an increasing impact on kiwifruit breeding in the future.

8.7.1 Genetic Maps and QTL Analysis

Various molecular fingerprinting systems: isozymes (Messina et al. 1991), RAPDs (Randomly Amplified Polymorphic DNA markers) (Cipriani et al. 1996), AFLPs (Amplified Fragment-Length Polymorphism markers) (Prado et al. 2007) and microsatellites (SSR, Simple Sequence Repeats) (Palombi and Damiano 2002, Zhen et al. 2004) have been tested on kiwifruit. The different procedures vary in their ability to discriminate between closely related genotypes and AFLPs and microsatellites seem particularly powerful since they detect higher levels of polymorphism. For example, microsatellite markers are able to discriminate between very closely related genotypes not separable by RAPD markers (Palombi and Damiano 2002). Testolin et al. (2001), using microsatellites derived from genomic libraries, found limited cross-species amplification between two taxonomically distant *Actinidia* species. EST-derived microsatellites should be more conserved because of the selection pressure operating on functional and regulatory genes: all such markers that were fully informative in an *A. chinensis* mapping family showed at least some level of cross amplification over 26 different *Actinidia* taxa (Fraser et al. 2004, 2005). Such EST-derived microsatellites are now being used to prepare a map using a population resulting from an intraspecific cross between two diploid *A. chinensis* plants chosen because they came from distant regions of China and showed wide diversity in fruit characteristics (Fraser et al. 2007). Work on this genetic map, based on frequencies of recombination between markers during crossing over, is being accompanied by construction of BAC (Bacterial Artificial Chromosome) contigs (sets of overlapping segments of DNA) of the *A. chinensis* genome (Hilario et al. 2007). This will eventually allow the physical distances between molecular markers anchored to the genetic map to be determined.

Protein patterns (Khukhunaishvili and Dzhokhadze 2006), isozyme polymorphism (Hirsch et al. 1997, Shirkot et al. 2001) and DNA markers (Gill et al. 1998, Shirkot et al. 2002, Xiao et al. 2003) have been used to screen for gender. The most successful so far are the two RAPD markers from *A. chinensis* which were converted into SCARs (Sequence-Characterized Amplified Regions) for large scale screening. These work very reliably in most, although not all, *A. chinensis* populations and in the *A. deliciosa* accessions tested, but are not useful in other *Actinidia* species (Gill et al. 1998, Xiao et al. 2003). These PCR markers are now being used to screen for gender in many thousands of *A. chinensis* seedlings each year in the HortResearch (New Zealand) kiwifruit breeding programs.

8.7.2 Regeneration and Transformation

Actinidia species are amenable to tissue culture techniques for regeneration, micropropagation, transformation and ploidy manipulation. The extensive literature has been recently reviewed in Xiao (1999), Xu et al. (2003), Sharma and Shirkot (2004) and Oliveira and Fraser (2005).

8.7.2.1 Micropropagation

Kiwifruit are easily propagated from seed but the plants produced are highly variable. Commercial cultivars were usually vegetatively propagated by grafting, budding, softwood cuttings or hardwood cuttings but in vitro propagation has proved both easy and reliable and capable of the very rapid production of the enormous numbers of plants sometimes required. Apparently little or no variability is induced when shoot tips or stem explants with axillary buds are cultured and the adventitious shoots that differentiate are rooted to form plants. A variety of other explants have also been used as starting material (for reviews, see Ferguson et al. 1996, Xiao 1999, Xu et al. 2003, Sharma and Shirkot 2004, Oliveira and Fraser 2005). Micropropagated plants are sometimes reported as being more difficult to establish in orchards than cutting-grown or grafted plants (Díaz Hernandez et al. 1997, Piccotino et al. 1997), but a long-term comparison of kiwifruit plants propagated by different methods indicated that plants obtained by in vitro propagation were as productive as those from cuttings or from grafting (Monastra and Chiariotti 1997). Furthermore, in this comparison, no somaclonal variation was revealed in the micropropagated plants over 11 years. Micropropagation has been successful in producing many of the plants grown by the Italian kiwifruit industry.

Many different tissues from various *Actinidia* species have been successfully used to form callus from which plants have been regenerated (Ferguson et al. 1996, Xiao 1999, Sharma and Shirkot 2004, Oliveira and Fraser 2005). Plants have also been regenerated from protoplasts induced to callus and differentiate (Xiao 1999, Gan et al. 2003). Somatic embryogenesis has so far been most successful in callus cultures of *A. chinensis*. However, regeneration of plants from callus does mean a risk of somaclonal variation (Rugini and Gutierrez-Pesce 2003, Palombi and Damiano 2002, Prado et al. 2007) and this seems particularly common in protoplast-derived kiwifruit plants, with variation in chromosome number often being reported (He et al. 1995, Zhang et al. 1997, Liu et al. 2003), even if aneuploid plants tend to revert to euploidy as they develop (L.G.Fraser, pers. comm.). There are also reports of regenerated plants of *Actinidia* that are mixoploid or have multinucleate cells (Zhang et al. 1998). Somaclonal variation, if stable, could be a useful source of genetic variation, particularly if selected for during in vitro regeneration (Marino et al. 1998, Muelo et al. 2003). In practice, however, somaclonal variation might limit use of some tissue culture techniques. Since aneuploidy has frequently been observed in *Actinidia* plants regenerated from callus or protoplasts, it seems a sensible precaution to assume that, until established otherwise, such regenerated plants might well vary somaclonally, even if this is not obviously expressed. The calli produced during regeneration following transformation are usually chimeras with transformed and non-transformed tissues (Li et al. 2003a).

8.7.2.2 Embryo Rescue

Many interspecific crosses in *Actinidia* fail, especially if the parents have different numbers of chromosomes and ploidy levels in the embryo and endosperm diverge from the normal ratio of 2:3. This is a severe limitation because of the variation in ploidy within *Actinidia* and the current commercial kiwifruit cultivars being variously diploid, tetraploid or hexaploid. Often embryos are produced by interploidy crosses but their development fails at various stages. If seed abortion is due to the endosperm failing to develop properly, in vitro culture can save the embryos if they are isolated early in development. They can then be grown up, preferably directly by embryogenesis or through a callus stage from which plants can be regenerated (Mu et al. 1990). Hirsch et al. (2001) developed culture media that allowed immature embryos to be rescued from a wide range of crosses and thereby recovered plants from some interspecific *Actinidia* crosses that would otherwise have failed.

8.7.2.3 Ploidy Manipulation

Homozygous breeding lines are usually not possible in woody plants but they can be produced through haploidy (Germanà 2006). Haploid plants have not so far been reported in kiwifruit but trihaploid plants of *A. deliciosa* (normally hexaploid) have been successfully produced by induced parthenogenesis by pollination with lethally-irradiated pollen (Pandey et al. 1990, Chalak and Legave 1997) or by pollen from a male that differs from the female in ploidy (A.G. Seal, unpublished). In vitro embryo rescue was not required. Such trihaploid plants can double in chromosome number spontaneously or after treatment with antimitotic agents (Chalak and Legave 1996): however, they are unlikely to be homozygous for many genes, because of duplicate loci arising through polyploidy (MacRae and Atkinson 2003).

Some triploid plants have been produced by tissue culture of *A. chinensis* endosperm but most of them were aneuploid (Gui et al. 1993). A higher proportion of triploid and tetraploid plants can be produced by crossing a diploid with a tetraploid or a diploid with a hexaploid, but these plants are not necessarily fertile and their offspring might be aneuploid.

8.7.2.4 Somatic Hybridization

Protoplasts are readily prepared from *Actinidia*. Successful somatic fusion and re-generation has been reported of protoplasts from *A. chinensis* and *A. deliciosa*, and of *A. chinensis* and *A. kolomikta* (Xiao and Han 1997, Xiao et al. 2004). Chomosome counts, ploidy measurements and DNA analyses confirmed that at least one of the plants raised was a somatic hybrid between *A. chinensis* and *A. kolomikta*, apparently with chilling tolerance similar to that of *A. kolomikta* (Xiao et al. 2004).

8.7.2.5 Transformation

Compared with other fruit crops, relatively high rates of transformation can be achieved in *Actinidia* (Li et al. 2003b). *Agrobacterium*-mediated transformation has been reported for *A. arguta*, *A. chinensis*, *A. deliciosa* and *A. eriantha*, and transgenic plants of *A. deliciosa* and *A. eriantha* have been grown to flowering and fruiting maturity to demonstrate transmission of transgenes to progeny (Rugini et al. 1997, Fung et al. 1998, Wang et al. 2006). Molecular analysis of transformed *A. eriantha* plants showed that the transferred genes were incorporated into the genome, and expressed in leaf, stem, root, flower and fruit tissues (Wang et al. 2006). *A. eriantha* is a particularly rewarding plant for transformation studies because flowering and fruiting can be achieved under containment conditions less than two years after inoculation with *Agrobacterium*. At present, however, many marketers of kiwifruit are unwilling to handle fruit of genetically modified cultivars; transformation is being used experimentally but not for breeding or commercial kiwifruit improvement.

8.7.3 Genomic Resources

A large database of more than 130,000 expressed sequence tags (ESTs) from kiwifruit has been developed at HortResearch, New Zealand (Atkinson and MacRae 2007, MacRae 2007). About 80% of these ESTs came from over 50 cDNA libraries from *A. chinensis* and *A. deliciosa*; 43% from shoot buds, 29% from fruit, 12% from leaves and 12% from petals (Atkinson and MacRae 2007). These are being used in the preparation of a genetic map, the karyotyping of diploid *A. chinensis* (He et al. 2005, P.M. Datson, pers. comm.), and the cloning, ordering and sequencing of genomic DNA libraries (Hilario et al. 2007). A microarray of >17,000 kiwifruit genes has also been developed to monitor patterns of gene expression.

Acknowledgments We thank L.G. Fraser, F.A. Gunson and J.-H. Wu for reading parts of the manuscript and R. Testolin for providing information on the *Actinidia* germplasm collection at Udine, Italy.

References

An H-X, Cai D-R, Mu X-J, Zheng B, Shen Q-G (1995) New germplasm of interspecific hybridis-ation in *Actinidia*. Acta Hortic Sin 22:133–137

Atkinson RG, MacRae EA (2007) I.13 Kiwifruit. In: Pua EC, Davey MR (eds) Biotechnology in agriculture and forestry, vol 60. Transgenic crops V, Springer-Verlag, Berlin Heidelberg, pp 329–346

Beatson RA (1992) Inheritance of fruit characters in *Actinidia deliciosa*. Acta Hortic 297:79–86

Beatson RA, Datson PM, Harris-Virgin PM, Graham LT (2007) Progress in the breeding of novel interspecific *Actinidia* hybrids. Acta Hortic 753:147–151

Blanchet P, Chartier J (1992) Genetic variability among the progeny of 'Hayward' kiwifruit. Acta Hortic 297:87–92

Blanchet P, Chartier J (1998) Sélection de kiwis chinois pour les zones chaudes ChinaBelle® et PolliChina®. Arb Fruit 513:37–40

Chalak L, Legave JM (1996) Oryzalin combined with adventitious regeneration for an efficient chromosome doubling of trihaploid kiwifruit. Plant Cell Rep 16:97–100

Chalak L, Legave JM (1997) Effects of pollination by irradiated pollen in Hayward kiwifruit and spontaneous doubling of induced parthenogenetic trihaploids. Scientia Hortic 68:83–93

Chen S-D (2003) Germplasm resources and high quality varieties of *Actinidia chinensis* Planch. in China. Acta Hortic 626:451–455

Cheng CH, Seal AG, Boldingh HL, Marsh KB, MacRae EA, Murphy SJ, Ferguson AR (2004) Inheritance of taste characters and fruit size and number in a diploid *Actinidia chinensis* (kiwifruit) population. Euphytica 138:185–195

Cheng CH, Seal A, Boldingh H, Marsh K, MacRae E, Murphy S, Ferguson R (2006a) Inheritance of taste characters and fruit weight in a tetraploid *Actinidia chinensis* (kiwifruit) population. In: Mercer CF (ed) Breeding for success: diversity in action. Proceedings of 13th Australasian Plant Breeding Conference, Christchurch, New Zealand, pp 150–154

Cheng CH, Seal AG, Murphy SJ, Lowe RG (2006b) Variability and inheritance of flowering time and duration in *Actinidia chinensis* (kiwifruit). Euphytica 147:395–402

Cheng CH, Seal AG, Murphy SJ, Lowe RG (2007) Red-fleshed kiwifruit (*Actinidia chinensis*) breeding in New Zealand. Acta Hortic 753:139–146

Cho HS, Jo YS, Liu IS, Ahn CS (2007) Characteristics of *Actinidia deliciosa* × *A. arguta* and *A. arguta* × *A. deliciosa* hybrids. Acta Hortic 753:205–209

Cipriani G, Di Bella R, Testolin R (1996) Screening RAPD primers for molecular taxonomy and cultivar fingerprinting in the genus *Actinidia*. Euphytica 90:169–174

Cipriani G, Testolin R (2007) 'Jintao': A Chinese kiwifruit selection grown in Italy. Acta Hortic. 753:247–251

Cui Z-X (1981) Cultivation of mihoutao in China. In: Qu Z-Z (ed) Mihoutao de zaipei he liyong. Nongye Chubanshe, Beijing, pp 95–104

Datson P, Beatson R, Harris-Virgin T (2006) Meiotic chromosome behavior in interspecific *Actinidia* hybrids. In: Mercer CF (ed) Breeding for success: diversity in action. Proceedings of 13th Australasian Plant Breeding Conference, Christchurch, New Zealand, pp 869–874

Díaz Hernandez MB, Ciordia Ara M, Garcia Rubio JC, Garcia Berrios J (1997) Performance of kiwifruit plant material propagated by different methods. Acta Hortic 444:155–160

Evreinoff VA (1949) Notes sur les variétés d'*Actindia*. Rev Hortic 121:155–158

Fairchild D (1927) The fascination of making a plant hybrid; being a detailed account of the hybridization of *Actinidia arguta* and *Actinidia chinensis*. J Hered 18:49–62

Ferguson AR (1990a) Kiwifruit (*Actinidia*). Acta Hortic 290:603–653

Ferguson AR (1990b) The genus *Actinidia*. In: Warrington IJ, Weston GC (eds) Kiwifruit: science and management. Ray Richards Publisher and NZ Soc Hortic Sci, Auckland, pp 15–35

Ferguson AR (1990c) Botanical nomenclature: *Actinidia chinensis, Actinidia deliciosa*, and *Actinidia setosa*. In: Warrington IJ, Weston GC (eds) Kiwifruit: science and management. Ray Richards Publisher and NZ Soc Hortic Sci, Auckland, pp 36–57

Ferguson AR (1997). Kiwifruit (Chinese gooseberry). In: The Brooks and Olmo register of fruit & nut varieties. 3rd edn. ASHS Press, Alexandria, VA, pp 319–323

Ferguson AR (2004) 1904 – the year that kiwifruit (*Actinidia deliciosa*) came to New Zealand. NZ J Crop Hortic Sci 32:3–27

Ferguson AR (2007) The need for characterisation and evaluation of germplasm: Kiwifruit as an example. Euphytica 154:371–382

Ferguson AR, Bollard EG (1990) Domestication of the kiwifruit. In: Warrington IJ, Weston GC (eds) Kiwifruit: science and management. Ray Richards Publisher and NZ Soc Hortic Sci, Auckland, pp 165–246

Ferguson AR, Huang H-W (2007) Genetic resources of kiwifruit: domestication and breeding. Hortic Rev 33:1–121

Ferguson R, Lowe R, McNeilage M, Marsh H (1999) 'Hort16A': Un nuova kiwi a polpa gialla dalla Nuova Zelanda. Riv Frutticolt Ortofloricolt 61(12):24–29

Ferguson AR, Seal AG, McNeilage MA, Fraser LG, Harvey CF, Beatson RA (1996) Kiwifruit. In: Janick J, Moore JN (eds) Fruit breeding. vol 2. Vine and small crops. John Wiley & Sons, NewYork, pp 371–417

Fraser LG, Harvey CF, Crowhurst RN, De Silva HN (2004) EST-derived microsatellites from *Actinidia* species and their potential for mapping. Theor Appl Genet 108: 1010–1016

Fraser LG, McNeilage MA, Tsang GK, De Silva HN, MacRae EA (2007) The use of EST-derived microsatellites as markers in the development of a genetic map in kiwifruit. Acta Hortic 753:169–175

Fraser LG, McNeilage MA, Tsang GK, Harvey CF, De Silva HN (2005) Cross-species amplification of microsatellite loci within the dioecious, polyploid genus *Actinidia* (Actinidiaceae). Theor Appl Genet 112:149–157

Fukuda T, Suezawa K, Katagiri T (2007) New kiwifruit cultivar 'Sanuki Gold'. Acta Hortic 753:243–246

Fung RWM, Janssen BJ, Morris BA, Gardner RC (1998) Inheritance and expression of transgenes in kiwifruit. NZ J Crop Hortic Sci 26:169–179

Gan L, Xiong X, Wang R, Power JB, Davey MR (2003) Plant regeneration from cell suspension protoplasts of *Actinidia deliciosa*. Acta Hortic 610:197–202

Germanà MA (2006) Doubled haploid production in fruit crops. Plant Cell Tissue Organ Cult 86:131–146

Gill GP, Harvey CF, Gardner RC, Fraser LG (1998) Development of sex-linked PCR markers for gender identification in *Actinidia*. Theor Appl Genet 97:439–445

Gui Y, Hong S, Ke S, Skirvin RM (1993) Fruit and vegetative characteristics of endosperm-derived kiwifruit (*Actinidia chinensis* F) plants. Euphytica 71:57–72

Guthrie RS, Luby JJ, Bedford DS, McNamara ST (2007) Partial dominance in *Actinidia kolomikta* hybrids. Acta Hortic 753:211–218

Harker FR, Jaeger SR, Lau K, Rossiter K (2007) Consumer perceptions and preferences for kiwifruit: A review. Acta Hortic 753:81–88

Harvey CF, Gill GP, Fraser LG (1997a) Sex determination in *Actinidia*. Acta Hortic 444:85–88

Harvey CF, Gill GP, Fraser LG, McNeilage MA (1997b) Sex determination in *Actinidia*. 1. Sex-linked markers and progeny sex ratio in diploid *A. chinensis*. Sex Plant Reprod 10:149–154

He Z-C, Cai Q-G, Ke S-Q, Qian Y-Q, Xu L-M (1995) Cytogenetic studies on regenerated plants derived from protoplasts of *Actinidia deliciosa*. I. Variation of chromosome number of somatic cells. J Wuhan Bot Res 3:97–101

He Z-C, Li J-Q, Cai Q, Wang Q (2005) The cytology of *Actinidia*, *Saurauia* and *Clematoclethra* (Actinidiaceae). Bot J Linn Soc 147:369–374

Hilario E, Bennell T, Crowhurst RN, Fraser LG, McNeilage MA, Rikkerink E, MacRae EA (2007) Construction of kiwifruit BAC contig maps by Overgo hybridisation and their use for mapping the sex locus. Acta Hortic 753:185–189

Hill MG, Mauchline NA, Cheng CH, Connolly PG (2007) Measuring the resistance of *Actinidia chinensis* to armoured scale insects. Acta Hortic 753:685–692

Hirsch AM, Fortune D, Chalak L, Legave JM, Chat J, Monet R (1997) Peroxidase test: a test for sex screening in kiwifruit seedlings. Acta Hortic 444:899–95

Hirsch A-M, Testolin R, Brown S, Chat J, Fortune D, Bureau JM, De Nay D (2001) Embryo rescue from interspecific crosses in the genus *Actinidia* (kiwifruit). Plant Cell Rep 20:508–516

Huang HW (2003) Integrating both nuclear and cytoplasmic genetic information into a high genome coverage conservation approach for *Actinidia*. Acta Hortic 610:87–93

Huang H, Ferguson AR (2001) Kiwifruit in China. NZ J Crop Hortic Sci 29:1–14

Huang H-W, Li Z-Z, Li J-Q, Kubisiak TL, Layne DR (2002a) Phylogenetic relationships in *Actinidia* as revealed by RAPD analysis. J Am Soc Hortic Sci 127:759–766

Huang H-W, Wang S-M, Huang R-H, Jiang Z-W, Zhang Z-H (2002b) 'Jintao', a novel, hairless, yellow-fleshed kiwifruit. HortScience 37:1134–1136

Huang H-W, Wang S, Jiang Z, Zhang Z, Gong J (2003) Exploration of *Actinidia* genetic resources and development of kiwifruit industry in China. Acta Hortic 610:29–43

Huang H-W, Wang Y, Zhang Z-H, Jiang Z-W, Wang S-M (2004) *Actinidia* germplasm resources and kiwifruit industry in China. HortScience 39:1165–1172

Huang WG, Cipriani G, Morgante M, Testolin R (1998) Microsatellite DNA in *Actinidia chinensis*: isolation, characterisation, and homology in related species. Theor Appl Genet 97:1269–1278

Huang Z-F, Liang M-Y, Huang C-G, Li R-G (1983) A preliminary study on the character and nutritive composition of *Actinidia* fruits. Guihai 3:53–56, 66

Jo YS, Cho HS, Park MY, Bang GP (2007a) Selection of a sweet *Actinidia eriantha*, 'Bidan'. Acta Hortic 753:253–257

Jo YS, Ma KC, Cho HS, Park JO, Kim SC, Kim WS (2007b) 'Chiak', a new selection of *Actinidia arguta*. Acta Hortic 753:259–262

Kataoka I, Kokudo K, Beppu K, Fukuda T, Mabuchi S, Suezawa K (2003) Evaluation of characteristics of *Actinidia* interspecific hybrid 'Kosui'. Acta Hortic 610:103–108

Kawagoe T, Suzuki N (2004) Cryptic dioecy in *Actinidia polygama*: a test of the pollinator attraction hypothesis. Can J Bot 82:214–218

Ke S-Q, Huang R-H, Wang S-M, Xiong Z-T, Wu Z-W (1992) Studies on interspecific hybrids of *Actinidia*. Acta Hortic 297:133–139

Khukhunaishvili RG, Dzhokhadze DI (2006) Electrophoretic study of the proteins from *Actinidia* leaves and sex identification. Appl Biochem Microbiol 42:107–110

Kokudo K, Beppu K, Kataoka I, Fukuda T, Mabuchi S, Suezawa K (2003) Phylogenetic classification of introduced and indigenous *Actinidia* in Japan and identification of interspecific hybrids using RAPD analysis. Acta Hortic 610:351–356

Li H-L (1952) A taxonomic review of the genus *Actinidia*. J Arnold Arboretum 24:362–374

Li J-Q, Li X-W, Soejarto DD (2007) A revision of the genus *Actinidia* from China. Acta Hortic 753:69–71

Li M, Huang Z-G, Han L-X, Zhao G-R, Li Y-H (2003a) Gene transfer strategy for kiwifruit to obtain pure transgenic plant through inducing adventitious roots in leaf explants with *Agrobacterium tumefaciens*. Acta Hortic 610:495–499

Li M, Huang Z-G, Han L-X, Zhao G-R, Li Y-H, Yao J-L (2003b) A high efficient *Agrobacterium tumefaciens*-mediated transformation system for kiwifruit. Acta Hortic 610:501–507

Li M-Z, Lowe RG (2007) Survey of wild *Actinidia* from Yangtze River region in China. Acta Hortic 753:69–71

Li R-G, Huang C-G, Liang M-Y, Huang Z-F (1985) Investigation of germplasm resources of *Actinidia* in Guangxi. Guihaia 5:253–267

Li R-G, Liang M-Y, Li J-W (2000) The collection and conservation of germplasm of genus *Actinidia*. In: Huang H-W (ed) Advances in *Actinidia* research. Science Press, Beijing, pp 87–89

Li R-G, Liang M-Y, Li J-W, Mao S-H (1996) Studies on the biological characteristics of genus *Actinidia*. Guihaia 16:265–272

Liang C-F (1984) *Actinidia*. In: Feng K-M (ed) Flora Reipublicae Popularis Sinicae, 49/2. Science Press, Beijing, pp 196–268, 309–324

Liu J-H, Xu X-Y, Deng X-X (2003) Protoplast isolation, culture and application to genetic improvement of woody plants. Food Agri Environ 1:112–120

McNeilage MA (1991) Gender variation in *Actinidia deliciosa*, the kiwifruit. Sex Plant Reprod 4:267–273

McNeilage MA, Considine JA (1989) Chromosome studies in some *Actinidia* taxa and implications for breeding. NZ J Bot 27:71–81

McNeilage MA, Duffy AM, Fraser LG, Marsh HD, Hofstee BJ (2007) All together now: the development and use of hermaphrodite breeding lines in *Actinidia deliciosa*. Acta Hortic 753:191–197

McNeilage MA, Steinhagen S (1998) Flower and fruit characters in a kiwifruit hermaphrodite. Euphytica 101:69–72

MacRae EA (2007) Can biotechnology help kiwifruit breeders? Acta Hortic 753:129–138

MacRae E, Atkinson R (2003) Multiple gene copies: carrying out biochemical and molecular studies in kiwifruit. Acta Hortic 610:457–466

Marino G, Certazza G, Buscaroli C (1998) In vivo growth and tolerance to lime-induced iron chlorosis of leaf-derived cv. Tomuri and Hayward kiwifruit (*Actinidia deliciosa*) somaclones. J Hortic Sci Biotechnol 73:670–675

Marsh H, McNeilage M, Gordon I (1999) Breeding value of parents in kiwifruit (*Actinidia deliciosa*). Acta Hortic 498:85–92

Marsh HD, Paterson T, Seal AG, McNeilage MA (2003a) Heritability estimates in kiwifruit. Acta Hortic 622:221–229

Marsh K, Rossiter K, Lau K, Walker S, Gunson A, MacRae E (2003b) The use of fruit pulps to explore flavour in kiwifruit. Acta Hortic 610:229–237

Messina R, Testolin R, Morgante M (1991) Isozymes for cultivar identification in kiwifruit. HortScience 26:899–902

Messina R, Vischi M, Marchetti S, Testolin R, Milani N (1990) Observations on subdioeciousness and fertilisation in a kiwifruit breeding program. Acta Hortic 282:377–386

Michurin IV (1949) Selected works. Foreign Languages Publishing House, Moscow

Mizugami T, Lim JG, Beppu K, Kataoka I, Fukuda T (2007) Observation of parthenocarpy in *Actinidia arguta* selection 'Issai'. Acta Hortic 753:199–203

Monastra F, Chiariotti A (1997) Effect of different propagation methods on vegetative-productive behaviour of kiwifruit plants. J Southern African Soc Hortic Sci 7:1–3

Mouat HM (1958) New Zealand varieties of yang-tao or Chinese gooseberry. NZ J Agric 97:161, 163, 165

Mu SK, Fraser LG, Harvey CF (1990) Rescue of hybrid embryos of *Actinidia* species. Scientia Hortic 44:97–106

Muelo R, Iacona C, Leva AR, Loretti F, Boscherini G, Bernardini M, Buitta M (2003) Variazione somaclonale e miglioramento genetico dell'actinidia. In: Costa G (coord.), Actinidia, la novità frutticolo del XX secolo. Convegno Nazionale del Società Orticolo Italiana, 21 November 2003. Camera di Commercio Industria Artigianato e Agricoltura di Verona, Verona, pp 79–84

Muggleston S, McNeilage M, Lowe R, Marsh H (1998) Breeding new kiwifruit cultivars: the creation of Hort16A and Tomua. Orchardist NZ 71(8):38–40

Oliveira MM, Fraser LG (2005) *Actinidia* spp. Kiwifruit. In: Litz RE (ed) Biotechnology of fruit and nut crops. (Biotechnology in agriculture No. 29) CABI Publ., Wallingford, UK, pp 2–27

Palombi MA, Damiano C (2002) Comparison between RAPD and SSR molecular markers in detecting genetic variation in kiwifruit (*Actinidia deliciosa* A. Chev). Plant Cell Rep 20:1061–1066

Pandey KK, Przywara L, Sanders PM (1990) Induced parthenogenesis in kiwifruit (*Actinidia deliciosa*) through the use of lethally irradiated pollen. Euphytica 51:1–9

Paterson VJ, MacRae EA, Young H (1991) Relationships between sensory properties and chemical composition of kiwifruit (*Actinidia deliciosa*). J Sci Food Agric 57:235–251

Piccotino D, Massai R, Dichio B, Nuzzo V (1997) Morphological and anatomical modifications induced by in vitro propagation of kiwifruit plants. Acta Hortic 444:127–132

Prado MJ, Gonzalez MV, Romo S, Herrera MT (2007) Adventitious plant regeneration on leaf explants from adult male kiwifruit and AFLP analysis of genetic variation. Plant Cell Tiss Organ Cult 88:1–10

Pringle GJ (1986) Potential for hybridization in the genus *Actinidia*. Plant Breeding Symposium, DSIR. Agron Soc NZ Spec Publ 5:365–368

Qian Y-Q, Yu D-P (1992) Advances in *Actinidia* research in China. Acta Hortic 297:51–55

Rugini E, Caricato G, Muganu M, Taratufolo C, Camilli M, Cammilli C (1997) Genetic stability and agronomic evaluation of six-year-old transgenic kiwi plants for rol ABC and rol B genes. Acta Hortic 447:609–610

Rugini E, Gutierrez-pesce (2003) Micropropagation of kiwifruit (*Actinidia* spp.). In: Jain SM, Ishii (ed) Micropropagation of woody trees and fruits. Kluwer, Dordrecht, pp 647–669

Seal AG (2003) The plant breeding challenges to making kiwifruit a worldwide mainstream fresh fruit. Acta Hortic 610:76–80

Sharma DR, Shirkot P (2004) Biotechnological interventions for genetic amelioration of *Actinidia deliciosa* var. *deliciosa* (kiwifruit) plant. Indian J Biotechnol 3:249–257

Shim K-K, Ha Y-M (1999) Kiwifruit production and research in Korea. Acta Hortic 498:127–131

Shirkot P, Sharma DR, Mohapatra T (2002) Molecular identification of sex in *Actinidia deliciosa* var. *deliciosa* by RAPD markers. Scientia Hortic 94:33–39

Shirkot P, Sharma DR, Shirkot CK (2001) Use of isozyme polymorphism for gender evaluation in kiwifruit (*Actinidia deliciosa* var. *deliciosa*). Indian J Plant Genet Resour 14:14–17

Snelgar WP, Hall AJ, Ferguson AR, Blattmann P (2005) Temperature influences growth and maturation of fruit on 'Hayward' kiwifruit vines. Funct Plant Biol 32:631–642

Stănică F, Zuccherelli G (2007) New selections of *Actinidia arguta* from the Romanian breeding program. Acta Hortic 753:263–267

Start MA, Luby J, Filler D, Riera-Lizarazu O, Guthrie R (2007) Ploidy levels of cold-hardy *Actinidia* accessions in the United States determined by flow cytometry. Acta Hortic 753:161–168

Testolin R (2005) Quale spazio produttivo e commerciale per le nuove varietà di actinidia? Riv Frutticolt Ortofloricolt 67(9):18, 20–23

Testolin R, Cipriani G, Costa G (1995a) Sex segregation ratio and gender expression in the genus *Actinidia*. Sex Plant Reprod 8:129–132

Testolin R, Cipriani G, Costa G (1995b) Estimate of variance components and heritability of characters in kiwifruit. Acta Hortic 403:182–188

Testolin R, Cipriani G, Messina R (1999) Sex control in *Actinidia* is monofactorial and remains so in polyploids. In: Ainsworth CC (ed) Sex determination in plants. BIOS Scientific Publishers Ltd, Oxford, pp 173–181

Testolin R, Costa G (1994) Il miglioramento genetico dell'actinidia. Riv Frutticolt Ortofloricolt 56(1):31–42

Testolin R, Huang WG, Lain O, Messina R, Vecchione A, Cipriani G (2001) A kiwifruit (*Actinidia* spp.) linkage map based on microsatellites and integrated with AFLP markers. Theor Appl Genet 103:30–36

Testolin R, Messina R, Lain O, Cipriani G (2004) A natural sex mutant in kiwifruit (*Actinidia deliciosa*). NZ J Crop Hortic Sci 32:179–183

Valmori I (1991) Nuove varietà in frutticoltura. Edizioni Agricole, Bologna

Wang M, Li M, Meng A (2003a) Selection of a new red-fleshed kiwifruit cultivar 'Hongyang'. Acta Hortic 610:115–117

Wang M-Z, Li X-D, Yu Z-S, Li M-Z, Song S-S (2005) Breeding report on the red-fleshed cultivar 'Hongmei'. China Fruits 4:7–9

Wang S-M, Huang H-W, Jiang Z-W, Zhang Z-H, Zhang S-R, Huang H-Q (2000) Studies on *Actinidia* breeding by species hybridization between *A. chinensis* and *A. eriantha* and their hybrids' progenies. In: Huang H-W (ed) Advances in *Actinidia* research. Science Press, Beijing, pp 123–127

Wang S-M, Huang R-H, Wu X-W, Ning K (1994) Studies on *Actinidia* breeding by species hybridization. J Fruit Sci 11:23–26

Wang S, Jiang Z, Huang H, Zhang Z, Ke J (2003b) Conservation and utilization of germplasm resources of the genus *Actinidia*. Acta Hortic 610:365–371

Wang S-M, Wu X-W, Huang R-H, Xiong Z-T, Ke S-Q (1989) Preliminary report on fluctuation of interspecific crosses of Chinese gooseberry. J Wuhan Bot Res 7:399–402

Wang T-C, Ran Y-D, Atkinson RG, Gleave AP, Cohen D (2006) Transformation of *Actinidia eriantha*: a potential species for functional genomic studies in *Actinidia*. Plant Cell Rep 25:425–431

Wang Z-Y, Zhong C-H, Bu F-W (2004) A new red-fleshed kiwifruit cultivar 'Chuhong'. Agr Sci Technol (China) 5:23–24

Williams MH, Boyd LM, McNeilage MA, MacRae EA, Ferguson AR, Beatson RA, Martin PJ (2003) Development and commercialization of 'baby kiwi' (*Actinidia arguta* Planch.). Acta Hortic 610:81–86

Xiao X-G (1999) Advances in *Actinidia* biotechnology and molecular biology. Acta Hortic 498:53–64

Xiao X-G, Zhang L-S, Li S-H, Testolin R, Cipriani G (2003) Molecular markers for early sex determination in *Actinidia*. Acta Hortic 610:533–538

Xiao Z-A, Han B-W (1997) Interspecific somatic hybrids in *Actinidia*. Acta Bot Sin 39:1110–1117

Xiao Z-A, Wan L-C, Han B-W (2004) An interspecific somatic hybrid between *Actinidia chinensis* and *Actinidia kolomikta* and its chilling tolerance. Plant Cell Tissue Organ 79:299–306

Xu X-B, Yao X-H, Chen H (2003) Application of modern biotechnology on kiwifruit. Acta Hortic 610:525–531

Yan G-J, Ferguson AR, McNeilage MA, Murray BG (1997) Numerically unreduced (2*n*) gametes and sexual polyploidization in *Actinidia*. Euphytica 96:267–272

Zhang J, Wang J-R, Cai D-R, An H-X (1983) Studies on the introduction and selection of Chinese gooseberries (*Actinidia chinensis* Planch.). Acta Hortic Sin 10:93–98

Zhang T, Li Z-Z, Liu Y-L, Jiang Z-W, Huang H-W (2007) Genetic diversity, gene introgression and homoplasy in sympatric populations of the genus *Actinidia* as revealed by chloroplast microsatellite markers. Biodiversity Sci 15:1–22

Zhang Y-J, Qian Y-Q, Cai Q-G, Mu X-J, Wei X-P, Zhou Y-L (1997) Somaclonal variation in chromosome number and nuclei transfer of regenerated plants from protoplasts of *Actinidia eriantha*. Acta Bot Sin 39:102–105

Zhang Y-J, Qian Y-Q, Mu X-J, Cai Q-C, Zhou Y-L, Wei X-P (1998) Plant regeneration from in vitro-cultured seedling leaf protoplasts of *Actinidia eriantha* Benth. Plant Cell Rep 17:819–821

Zhen Y-Q, Li Z-Z, Huang H-W, Wang Y (2004) Molecular characterization of kiwifruit (*Actinidia*) cultivars and selections using SSR markers. J Am Soc Hortic Sci 129:374–382

Zhong C-H, Wang Z-Y, Peng D-F, Bu F-W (2007) Selection of a new red-fleshed kiwifruit cultivar 'Chuhong'. Acta Hortic 753:235–241

Zhu D-Y, Lawes GS, Gordon IL (2002) Estimates of genetic variability and heritability for vegetative and reproductive characters of kiwifruit (*Actinidia deliciosa*). Euphytica 124:93–98

Chapter 9
Peaches

J.F. Hancock, R. Scorza and G.A. Lobos

Abstract Common goals of peach breeders are: (1) extending the harvest season, (2) improving flavor and aroma, (3) lengthening self life, (4) controlling tree size, (5) broadening the adaptive range, and (6) developing resistance to sharka (PPV), powdery mildew, brown rot, leaf curl, *Xanthomonas* spp. and the green aphid (the vector of PPV). A number of single genes have been identified that reduce tree size and modify plant shape, and regulate firmness, mealiness, melting flesh, browning, flesh color and the freestone trait. Fruit maturity has been shown to be quantitatively regulated with a very high heritability. A growing number of molecular linkage maps have been developed of peach and its relatives; map coverage ranges from 396 to 1300 cM, with 8 to 23 linkage groups being identified. QTL have been identified for numerous horticulturally important traits including bloom and ripening time, fruit quality, storage life, freestone trait, internode length and pest resistance. Several bacterial artificial chromosome (BAC) libraries have been developed for peach and over 85,000 *Prunus* ESTs have been sequenced and deposited in the NCBI dbEST database. Peaches have been regenerated utilizing several systems, but there are only two reports of stable peach plant transformation.

9.1 Introduction

The peach, and its smooth skinned mutant, the nectarine, are primarily grown in temperate zones, between latitudes 30 and 45 N and S. The peach flower bud is hardy to about $-23°$ C to $-26°$ C which limits its cultivation at higher latitudes. Most peach cultivars require from 100–1000 hours of chilling below $7°$ C and they are highly susceptible to early spring frosts.

The fruits of peach cultivars vary widely across the world and even within regions. Fruit shapes vary from beaked, round to flat, colors vary from yellow, white to red, the flesh can be melting or non-melting and they can be clingstone or freestone.

J.F. Hancock
Department of Horticulture, 342C Plant and Soil Sciences Building, Michigan State University, East Lansing, Michigan 48824, USA
e-mail: hancock@msu.edu

J.F. Hancock (ed.), *Temperate Fruit Crop Breeding*,
© Springer Science+Business Media B.V. 2008

Peaches are eaten fresh, canned or dried and are excellent sources of fiber, vitamins and antioxidants (http://riley.nal.usda.gov/NDL/cgi-bin/list_nut_edit.pl). The highest quality peaches are produced in regions with warm to hot summers.

Worldwide production of peaches is now in excess of 15,000,000 tonnes, with almost half of the production coming from Asia (mostly China) (Sansavini et al. 2006). Among the deciduous fruits, peaches rank second to only apples in tonnage. Europe accounts for about 30% of the peach crop, while North America contributes 11%, South America 6% and Africa 5%. The major producers in Europe are Italy, Greece and Spain; in North America the greatest concentration of production is found on the western and eastern seaboards, and along the Great Lakes. The peach industry in Asia has grown dramatically over the last decade, while peach production in the rest of the world has shown only moderate to little change. The peach industry in South America is still limited, but increasing in Chile and Brazil.

9.2 Evolutionary Biology and Germplasm Resources

The peach [*Prunus persica* (L.) Batsch] is the most widely grown species in a very important genus containing the European plum (*P. domestica* L.), Japanese plum (*P. salicina* Lindl.), apricot [*P. armeniaca* (L.) Kostina], almond (*P. amygdalus* Batsch), sweet cherry (*P. avium* L.), and sour cherry (*P. cerasus* L.). Peach belongs to the family Rosaceae and the subgenus *Amygdalus*. Unusual in its subgenus, the peach is largely self fertile. There are at least 77 wild species of *Prunus* and most of them are found in central Asia. While polyploidy is common in the genus *Prunus*, the cultivated peach is diploid and has a chromosome number of $2n = 2x = 16$.

Five species that can be termed 'peach' are generally recognized: *P. persica*, *P. davidiana* (Carr.) Franch, *P. mira* Koehne, *P. kansuensis* Rehd. and *P. ferganensis* (Kost. & Rjab) Kov. & Kost. All are found in China (Table 9.1). The domesticated

Table 9.1 Native peach species

Species	Common name	Chromosome number ($2n$)	Distribution
P. davidiana (Carr.) Franch	Mountain peach, Shan tao	16	N. China
P. ferganensis (Kost. & Rjab) Kov. & Kost.	Xinjiang tao	16	N.E. China
P. kansuensis Rehd.	Wild peach, Kansu tao	16	N.W. China
P. mira Koehne	Tibetan peach, Xizang-tao	16	W. China & Himalayas
P. persica (L.) Batsch	Peach, Maotao	16	China

Adapted from Scorza and Okie 1990

Table 9.2 *Prunus* species that have been hybridized with *P. persica* that form mostly sterile hybrids

Species	Common name	Origin
P. americana Marsh.	American plum	U.S.A.
P. armeniaca L.	Apricot	Asia
P. besseyi Bailey	Western sand cherry	N. U.S.A., Canada
P. brigantine Vill.	Briancon apricot	France
P. cerasifera Ehrh.	Myrobolan plum	W. Asia
P. cerasus L.	Sour Cherry	W. Asia, S.E. Europe
P. domestica L.	European plum	W. Asia, Europe
P. hortulana Bailey	Wild plum	Central U.S.A.
P. japonica Thunb	Chinese or Korean bush cherry	China
P. munsoniana Wight & Hedr.	Wild goose plum	Central U.S.A.
P. nigra Ait	Canadian plum	N. U.S.A., Canada
P. pumila L.	Eastern sandcherry	N. U.S.A.
P. salicina Lindl.	Japanese plum	China
P. simmonii Carr.	Simon's plum	N. China
P. spinosa L.	Sloe	Europe, W. Asia, N. Africa
P. tenella (=*nana*) Batsch	Siberian almond	S.E. Europe, W. Asia
P. tomentosa Thumb.	Chinese bush cherry	N. & W. China, Japan
P. virginiana L.	Choke cherry	N. U.S.A., Canada

Adapted from Scorza and Okie 1991

peach can be readily hybridized with native populations of *P. persica* and all the other wild species of peach. Successful hybrids have also been produced between peach and almond, apricot, plum and sour cherry (Table 9.2). In most cases, these wide hybrids are largely sterile, although F_1s of almond and peach can be highly fertile (Armstrong 1957) and can be employed as rootstocks for both peach and almond.

9.3 History of Improvement

Peach cultivation probably originated in western China from wild populations of *P. persica* (Hedrick 1917, Scorza and Okie 1991). The peach is mentioned in 4,000 year old Chinese writings, and most of the known variation in cultivated peaches is found in Chinese land races. Peaches arrived in Greece through Persia about 2,500 B. P. and in Rome 500 years later. The Romans spread the peach throughout their empire. The peach came to Florida, Mexico and South America in the mid 1500s via Spanish and Portuguese explorers. It became feral in the southeastern United States and Mexico, and was further spread throughout North America by the Native Americans.

A rapid expansion in fruit culture arose in Europe during the Industrial Revolution of the 16th century, as a growing class of people acquired substantial wealth and began to garden. Numerous cultivars were released during this period by active fruit tree breeders such as John Rivers. Many of these cultivars were released as clones, although many may also have been distributed from seed. Peach breeding began about 100 years ago in the North American colonies, utilizing two major sources of germplasm – naturalized seedlings from the southeastern U.S.A. and Mexico, and cultivars originated in England. Until the American Revolution, peaches were mostly produced in seedling stands of very low quality. The first budded trees were offered for sale by Robert Prince on Long Island just before the Revolutionary War and by John Kenrick of Massachusetts in the 1790s (Hedrick 1950).

A number of cultivars of unknown origin were released in the first half of the 1800s including 'Early Crawford', 'Late Crawford' and 'Oldmixon Cling'. In 1850, Charles Downing introduced 'Chinese Cling' from China to the United States via England, and it was originally planted in South Carolina by Henry Lyons (Scorza and Sherman 1996). After the Civil War, Samuel Rumph planted 'Chinese Cling' in Marshallville, Georgia and released two important cultivars from that field, 'Belle of Georgia' ('Belle') and 'Elberta', which likely had 'Chinese Cling' as a parent. Other important, early cultivars were 'Hiley' (a seedling of 'Belle') and 'J.H. Hale' (a seedling of 'Elberta'). This small group of cultivars formed the foundation of most subsequent breeding activity (Scorza et al. 1985). Cullinan (1937) has provided a list of the most significant cultivars that were released between 1850 and 1900.

Peach breeding began in earnest at a number of State Experiment Stations in the late 1890s and early 1900s. Among the earliest large programs were in California, New Jersey and the United States Department of Agriculture. Stanley Johnston in Michigan began his landmark program in 1924 and developed the 'Redhaven' peach, which dominated peach cultivation in the eastern U.S.A. for decades (Iezzoni 1987). Other early, large public programs in the U.S.A. were at Arkansas, North Carolina, Louisiana, Texas, Florida and South Carolina (Childers and Sherman 1988, Cullinan 1937, Okie et al. 1985). Vineland in Canada has had a breeding program since 1914, along with Harrow since 1960. Significant peach breeding efforts have also been undertaken in Argentina, Australia, Brazil, China, France, Italy, Japan, Mexico and South Africa (Childers and Sherman 1988, Li 1984, Okie et al. 1985, Wang and Lu 1992, Yoshida 1988). In the middle of the century, several major private breeding efforts emerged in the US including Grant Merrill, F. W. Anderson and Armstrong Nursery Company. More recent public companies are Zaiger Genetics, Metzler and Sons, Bradford and Bradford, Paul Friday, and Fruit Acres (A. and R. Bjorge).

9.4 Current Breeding Efforts

Worldwide breeding activity has been very high over the last decade, with likely over a thousand new varieties being released. Sansavini et al. (2006) has called the 20th century, the 'Golden Age of Peach Breeding'. The private sector is responsible for most of the new peach releases, although the new non-melting flesh clingstone varieties for canning have come from the public sector. Over half of the releases (55%) have come from the U.S.A. and 30% from Europe, with France and Italy leading the way (Table 9.3). Most of the cultivar releases are yellow-fleshed peaches and nectarines, although a number of white fleshed cultivars have been developed in France, China, Japan and South Korea.

Among the most important advances are 'a notable enhancement of such fruit quality traits as increased fruit size, fuller and more extensive blush, better skin ground color, increased flesh-to-pit ratio, etc.' (Sansavini et al. 2006). The harvest calendar has been dramatically increased from two-three months to four-six months, and chilling requirements have been substantially lowered to allow expansion into more subtropical climates.

There are a number of traits that are being targeted by breeders as high priorities. Expanding the environmental ranges of peach is a common goal, in some cases to reduce chilling requirements to further expand into the subtropical climates of Spain, France, Italy, U.S.A. and China, but also to increase frost tolerance through

Table 9.3 Peach and nectarines released worldwide by country 1999–2001

Region	Peach		Nectarine		Clingstone		Total
	Yellow	White	Yellow	White	Yellow	White	
Africa							
Egypt	0	3	0	0	0	0	3
South Africa	7	1	8	0	5	0	21
Asia							
China	4	29	11	10	2	1	57
Japan	7	30	0	0	3	0	40
South Korea	1	10	2	0	0	0	13
Taiwan	1	0	0	0	0	0	1
Europe							
Czech Republic	8	0	1	0	0	0	9
France	35	42	24	33	6	0	140
Italy	51	32	37	14	3	4	141
Moldavia	7	0	0	0	0	0	7
Poland	2	0	0	0	0	0	2
Romania	4	0	6	0	0	0	10
Spain	6	3	7	2	3	1	22
Ukraine	10	0	1	0	5	0	16
Oceana							
Australia	1	1	2	0	8	0	12
North Zealand	1	7	0	1	0	0	9
North America							
Canada	2	0	1	0	4	0	7
Mexico	0	0	0	0	18	0	18
U.S.A.	219	107	160	77	21	0	584
South America							
Brazil	0	3	3	1	5	0	12

Source: Sansavini et al. 2006

bloom delay in the colder climates of Canada, Poland and Russia. Considerable effort is also being undertaken to develop a broad range of very early and late ripening types to expand production windows. Strong efforts are being made to increase fruit quality by enhancing appearance, along with improved flavor and aroma. Many European programs are committed to recovering the sensory traits of old cultivars, and Chinese programs are particularly interested in low-acid types (Sansavani et al., 2006). Improving self life by developing firmer fruit is also an important goal of most programs, with the added benefit of reduced damage during handling. The reduction of postharvest disorders related to long-distance shipping of peaches, especially between the northern and southern hemispheres, is an important goal for programs in countries such as Chile, South Africa, New Zealand, and the U.S.A. Control of tree size and vigor is an important goal of most programs, to facilitate mechanization and reduce the costs of pruning, thinning and harvesting (Scorza et al. 2000).

The most widespread disease and pest problems that are being pursued are sharka (PPV), powdery mildew, brown rot, leaf curl, *Xanthomonas* spp. and green aphid (the vector of PPV). Other significant breeding efforts are focusing on nematode (China and the U.S.A.) and phytoplasma resistance (Romania).

Rootstock breeding also remains a high priority at many locations, with most of the research being targeted towards tree vigor management, ease of clonal propagation, soil adaptability (drought and lime), nematode resistance and resistance to bacterial and virus diseases (*Xanthomonas*, *Pseudomonas*, PPV and ACLR) (Layne 1987).

Peach breeding world-wide is a productive endeavor that supplies a large number of improved cultivars each year allowing growers an ample choice of material ripening over a long season, filling a wide range of ecological conditions and satisfying a range of consumer demands. Nevertheless, there are serious needs that remain to be addressed and these needs will become more critical with time. The critical issues that can be at least partially if not fully addressed through breeding include climactic change which may significantly alter biotic and abiotic stress factors, global marketing of fruits increasing competition between peach growing regions and between peach and a vast array of other fruits, and the changing eating habits of populations, especially in developed countries, with emphasis on nutrition and convenience. To meet these challenges will require an even greater commitment to peach breeding that will include exploration of new germplasm, and the application of complementary genomic breeding technologies such as molecular marker assisted breeding and genetic engineering. The development and application of these technologies for the production of new cultivars with improved quality, nutrition, pest and abiotic stress resistance, and market novelty will require additional resources supplied over extended periods of time. Intra and inter institutional collaborations will be necessary in order to utilize diverse genetic improvement technologies. Training the next generation of breeders and the development of fruit improvement teams that span laboratory and field will play critical roles in the continued success of peach as an important crop that sustains grower investment and adds to the health and well being of consumers.

9.5 Genetics of Economically Important Traits

9.5.1 Pest and Disease Resistance

Some of the most widespread disease problems that concern peach breeders are bacterial canker (*Pseudomonas syringae*), bacterial spot (*Xanthmonas campestris*), brown rot (*Monilinia fruticola*), fungal gummosis (*Botrysphaeria dothidea*), leaf curl (*Taphrina deformans*), Leucostoma (Cytospora), canker (*Leucostoma persoonii*), powdery mildew (*Sphaerotheca pannosa*) and sharka (PPV) (Table 9.4). Among the most important pests receiving breeder attention are peach tree borers (*Synanthedon exitiosa*) and the green aphid (*Myzus persicae*) which is a vector of PPV.

Table 9.4 Genetics of disease resistance in peach

Disease	Observations and source
Bacterial	
Bacterial canker *Pseudomonas syringae*	Most cultivars are susceptible; but sources of resistance exist (Gardan et al. 1971, Weaver et al. 1979)
Bacterial spot *Xanthomonas campestris*	Dominant genes may regulate resistance (Sherman and Lyrene 1981); highly resistant cultivars identified (Keil and Fogle 1974, Simeone 1985, Werner et al. 1986)
Fungi	
Brown rot *Monilinia fructicola*	Little resistance in most cultivars, but sources of resistance may exist (Scorza and Okie 1991, Feliciano et al. 1987)
Cytospora canker *Leucotoma persoonii*	Little resistance in most cultivars, but sources of resistance exist (Gairola and Powell 1970, Hampson and Sinclair 1973, Scorza and Pusey 1984)
Fungal gummosis *Botryosphaeria dothidea*	Most cultivars are susceptible, but sources of resistance exist (Daniell and Chandler 1982, Okie and Reilly 1983)
Leaf curl *Taphrina deformans*	Resistance is moderately heritable and polygenic (Monet 1985, Ritchie and Werner 1981); highly resistant cultivars identified (Ackerman 1953, Simeone 1985)

Powdery mildew
  Sphaerotheca pannosa
  Podosphaera clandestina
	Resistance controlled by two loci, with one locus for high resistance (D'Bov 1983); resistance is dominant (Pukanova et al. 1980), few cultivars are highly resistant (Scorza and Okie 1989)

Virus	
Plum pox (PPV)	Little resistance in most cultivars, but sources of resistance exist (Rankovic and Sutic 1980, Surgiannides and Mainou 1985)
Nematode	
Root-knot *Meloidogyne* ssp.	Two dominant resistance genes identified to *M. javanica* (*Mj1* and *Mj2*) (Sharp et al. 1970); a single, dominant resistance gene identified to *M. incognita* (*Mi*) (Weinberger et al. 1943)
Root lesion *Pratylenchus* sp.	Little resistance in most cultivars, but tolerance has been reported (Potter et al. 1984)
Insect	
Green aphid *Myzus persicae*	A single dominant resistance gene identified (*Rm1*) (Monet and Massonie 1994)
Peach tree borer *Synanthedon exitiosa*	Little resistance in most cultivars, but modest resistance has been reported (Chaplin and Schneider 1975, Weaver and Boyce 1965)

Bacterial spot causes severe defoliation and blemishing of fruit, particularly in areas with high rainfall, strong winds, high humidity and sandy soil. There is considerable variation in disease incidence from year to year, and under favorable conditions for infection all cultivars show at least some symptoms, although highly resistant cultivars have been identified (Keil and Fogle 1974, Simeone 1985, Werner et al. 1986). Cultivars in the eastern U.S.A. tend to be more resistant than those in the west; the breeding program in North Carolina has been particularly successful in developing resistant cultivars. Sherman and Lyrene (1981) suggest that resistance is regulated by dominant genes. The PR defense genes, β-1,3-glucanases have been shown to be induced by inoculation with *Xanthomonas campestries* pv. *pruni* (Thimmapuram et al. 2001).

Peach leaf curl is a problem in many peach growing regions. Resistant cultivars have been identified, but immunity has not been reported (Ritchie and Werner 1981, Simeone 1985). Leaf curl resistance in peach is moderately heritable and likely under polygenic control (Monet 1985, Ritchie and Werner 1981). Tolerance to the disease is in a large part dependent upon whether the genotypes begin to leaf-out when conditions are optimal for infection (Ackerman 1953), although there are resistant genotypes that leaf-out under conditions favorable to infection (Ritchie and Werner 1981, Scorza 1992). Eglandular leaf genotypes appear to be more resistant than glandular ones, and nectarines are less susceptible than peaches.

Powdery mildew frequently attacks leaves, young shots and fruits. Mildew resistance appears to be regulated by two loci with one providing strong resistance, and another conditioning intermediate to low resistance (D'Bov 1983, Pukanova et al. 1980). The high resistance found at the first locus is epistatic to moderate and low resistance at the other locus. The allele for moderate resistance is dominant to low resistance. While strong resistance exists in *P. persica*, high levels have only been incorporated into a few cultivars (Scorza and Okie 1990).

Leucostoma or peach canker is a particularly serious disease in northern production areas, where tissue death during the winter serves as an entry point for the pathogen. This canker kills scaffold limbs and ultimately the whole tree. High levels of resistance have not been found among North American cultivars, but resistance does exist in Chinese and Russian germplasm (Gairola and Powell 1970, Hampson and Sinclair 1973, Scorza and Pusey 1984). Resistance to canker appears to be strongly correlated with cold tolerance (Chang et al. 1989) and how well water transport is maintained through the canker zone (Chang et al. 1991).

Fungal gummosis causes severe problems in Australia, China, Japan and the southeastern U.S.A. Most cultivars are highly susceptible, but a few have been identified that are highly resistant (Daniell and Chandler 1982, Okie and Reilly 1983). The genetics of resistance is unknown.

Bacterial canker has been associated with the short life syndrome of peach in the southern U.S.A. (Scorza and Okie 1989). Strong resistance to this disease has not been identified, but moderately resistant cultivars have been found with unspecified genetics (Gardan et al. 1971, Weaver et al. 1979).

Brown rot is a serious disease wherever peaches are grown. Little resistance has been described, although feral peaches in Central Mexico and perhaps the Brazilian cultivar 'Bolinha' have some degree of resistance (Feliciano et al. 1987, Scorza and Okie 1989). 'Bolinha' may not be a useful source, as its resistance is limited to the epidermis and it carries several negative characteristics that are readily transmitted such as a tendency for pre-harvest drop, yellow green epidermis and a susceptibility to bruising (Gradziel 1994, Gradziel and Wang 1993).

A number of serious virus diseases and phytoplasm attack peach including plum pox (PPV), prune dwarf, peach yellows, X-disease, *Prunus* necrotic ringspot, tomato ringspot, peach stunt, willow twig, stubby twig, and peach rosette mosaic. No immunity has been reported to any of these diseases, although large differences in resistance to PPV among genotypes have been found (Rankovic and Sutic 1980, Surgiannides and Mainou 1985).

Peachtree borer is a widespread problem and a tree (particularly young trees) can be girdled and killed in a single season. A few cultivars have been identified that are less susceptible than others to infestation, but no strong resistance has been identified (Chaplin and Schneider 1975, Weaver and Boyce 1965).

Myzus persicae is an aphid species that commonly attacks peach. They damage new growth through their feeding, but more importantly, they are vectors of PPV which causes substantial crop loss. Resistant cultivars and genotypes have been identified (Massonie et al. 1982), and Monet (1985) showed that the resistance is controlled by a single dominant gene. Seedlings carrying this gene are resistant to *Myzus persicae* and *M. varians*, but not *Hyalopterus amygdale* (Massonie et al. 1982). Since PPV is transmitted by aphid probing and not feeding, it is not clear if aphid resistance would affect PPV infection and the spread of the disease.

Several nematodes are commonly associated with peaches across the world and can cause replant problems including *Pratylenchus* ssp. (root lesion nematode), *Xiphinema* spp. (dagger nematode), *Meloidogyne incognita* (root knot nematode) and *Criconemella* spp (ring nematode). Tolerance has been reported to *Pratylenchus*, but not immunity (Potter et al. 1984). Multiple resistance genes to *Meloidogyne incognita* have been identified in peach (Gillen and Bliss 2005). Resistance to *M. javanica* has also been described that may be regulated by duplicate, independent dominant factors (Sharp et al. 1970).

Peach tree short life (PTSL) syndrome is a nematode-related disease syndrome of peach caused by a complex of biotic, abiotic and climatic factors. It affects more than 70% of the peach acreage in the southeastern US. It appears to be due to the extreme physiological stress associated with very high densities of ring nematodes, which results in wilting and a sudden collapse of new growth. Tolerance to this disease was unknown until the recent release of the rootstock 'Guardian' (BY520-9). The genetics of tolerance appears to be complex, as 38 AFLP markers have been associated with the PTSL syndrome, on five peach linkage groups (Blenda et al. 2007).

9.5.2 Morphological and Physiological Traits

Trees that have a thrifty growth habit which can be easily picked and pruned in high density orchards are an important goal of most breeding programs. A number of single, recessive genes have been identified that cause extreme size reduction – dwarf (*dw*, *dw2*, *dw3*), semi-dwarf (*n*), compact (*ct*) and bushy (*bu1* and *bu2*) (Table 9.5), but few commercial cultivars have been developed from these to date, due to poor

Table 9.5 Genetics of adaptation, productivity, plant habit and fruit quality in peach

Attribute	Observations and source
Adaptation	
Chilling requirement	Generally quantitatively inherited, although a few major genes may exist (Lesley 1944, Sharp 1961); a single, recessive gene for evergreen has been identified (*evg*) (Rodriguez et al. 1994); chilling requirements of buds and seed germination are correlated (Rodriquez and Sherman 1985)
Cold hardiness	Quantitatively inherited, largely additive (Mowry 1964); tissues vary in their hardiness (Cain and Anderson 1980); extremely cold hardy germplasm has been identified (Layne 1992, Myers and Okie 1986, Young 1987)
Season of flowering	Considerable variability exists among genotypes, but genetics is complex and quite subject to environmental interactions (Scorza and Sherman 1996)
Harvest date	Quantitatively inherited, with many major genes (Bailey and Hough 1959, Hansche et al. 1972, Vileila-Morales et al. 1981); a gene has been identified (*sr*), that greatly slows ripening (Ramming 1991)
Flower traits	
Flowers per bud	Single genes have been identified for single/double (*Sh/sh*) (Lammerts 1945)
Flower buds per node	Germplasm with high flower density has been identified (Okie and Werner 1990, Werner et al. 1988)
Petal color	Single genes have been identified for colored/white (*W/w*), anthocyanins/anthocyaninless (*AN/an*), dark pink/light pink (*P/p*) and pink/red (*R/r*) (Lammerts 1945, Monet 1967)
Petal number	Single genes have been identified for single/double (*Di/di*) and fewer extra petals/more extra petals (*Dm1/dm1* and independent *Dm2/dm2*) (Lammerts 1945, Yamazaki et al. 1987)
Petal size	Single genes have been identified for nonshowy/showy (*Sh/sh*) and large showy flowers/small showy flowers (*Sh/sh*) (Lammerts 1945)
Pollen fertility	Single genes located for pollen fertile/pollen sterile (*Ps/ps* and *Ps2/ps2*) (Scott and Weinberger 1944, Werner and Creller 1997)
Leaf traits	
Color	Single gene identified for red leaf/green leaf (*Gr/gr*) (Blake 1937); dominance is incomplete (Chaparro et al. 1995)
Foliar glands	Single genes identified for glandular foliage/eglandular foliage (*E/e*) (Conners 1922)
Shape	Single genes identified for smooth leaf margin/wavy leaf margin (*Wa/wa*) and normal/willow leaf (*Wa2/wa2*) (Chaparro et al. 1994, Scott and Cullinan 1942,)

Table 9.5 (continued)

Attribute	Observations and source
Plant habit	
Shape	Several single, recessive genes have been identified that influence plant shape – weeping, *pl* (Monet et al. 1988); compact, *ct* (Mehlenbacher and Scorza 1986); pillar, *br* (Scorza et al. 2002); bushy, *bu1* and *bu2* (Lammerts 1945)
Tree Height	Several single, recessive genes have been identified that influence plant height – dwarf, *dw* (Monet et al. 1988), dw_2 (Hansche 1988), *dw3* (Chaparro et al. 1994); semi-dwarf, *n* (Monet and Salesses 1975)
Fruit quality	
Acidity	Quantitatively inherited (Hansche et al. 1972); a QTL has been found for a single locus (D/d) that regulates low vs. high malic acid (Dirlewanger et al. 2004)
Flesh texture	Three single genes regulate melting flesh/ non-melting flesh (F/f), soft melting flesh/firm melting flesh (M/m) (Bailey and French 1941 and 1949) and melting flesh/ stonyhard flesh (*Hd/hd*) (Yoshida 1970); known dominance relationships are $ST > M > m$, F/f and M/m are on the same linkage group (Dirlewanger et al. 2004); candidate gene (endopolygalacturonase) identified for melting vs. non-melting trait (Peace et al. 2005b)
Pit adherence	Single gene regulating the freestone/clingstone trait (F/f) (Bailey and French 1941 and 1949); QTL identified located on same linkage group as the flesh texture genes *M/m and St/st* (Dirlewanger et al. 2004)
Internal breakdown (IB)	High heritability exists for all the traits associated with IB including mealiness; flesh browning and flesh bleeding; QTL have been found for all of these characteristics (Peace et al. 2005, 2006); the pectic enzyme polygalacturonase (PG) is strongly associated with the melting flesh characteristic and IB (Lester et al. 1996, Peace et al. 2006, Pressey and Avantes 1978)
Pubescence	Single genes regulating pubescent skin/glabrous (G/g) (Blake 1932) and normal pubescence/rough surface (Okie and Prince 1982); level of pubescence is quantitatively inherited (Blake 1940, Weinberger 1944)
Color	A number of single genes regulating color have been identified including *Y* which results in white fruit (Conners 1922), *h* which suppresses red color (Beckman et al. 2005) and *fr* which regulates full red color (Beckman and Sherman 2003); *bf* (blood flesh) is regulated by a single gene (Werner et al. 1998); degree of red skin color is likely regulated quantitatively; red color around the pit is dominant (Blake 1932)
Overall fruit quality	Browning, soluble solids, sweetness and overall flavor are quantitatively inherited (Hansche et al. 1972, Hansche 1986, Hansche and Boynton 1986)
Shape	Mostly quantitatively inherited, but a single, dominant gene has been identified for saucer vs. non-saucer shape (Lesley 1939) that is lethal in the homozygous state (Guo et al. 2002)
Size/weight	Quantitatively inherited with mostly additive genes (Hansche et al. 1972, Weinberger 1955)

fruit quality and issues associated with short internodes and large numbers of spurs (Loreti and Massai 2002). Greater success in developing cultivars for high density plantings has come from the use of genes that modify plant shape. The pillar gene (*br*), which forms a columnar growth habit, has been successfully used in the U.S.A. and Italy to produce a narrower tree that is easier to prune (Fig. 9.1). The weeping gene (*pl*), is also being utilized by the French to develop more efficiently pruned trees, although specific orchard systems will need to be developed to exploit this habit. A potentially useful 'arching' phenotype with a distinctive curvature of the one-year-old shoots has been described in Brbr/plpl genotypes (Werner and Chaparro 2005).

The environmental adaptations that have received the greatest amount of attention from peach breeders are winter cold hardiness, spring frost hardiness and chilling requirement. Cold hardiness is an issue in the cold temperate zones where peaches have been traditionally grown, and reducing the chilling requirement has become very important in expanding the range of peach cultivation into warmer climates. Frost tolerance has been an issue in both warm and cold climates. Winter cold tolerance is influenced by when cold tolerance is initiated, the rate of development of cold tolerance, the maximum cold tolerance that can be achieved, when cold tolerance is lost, the rate of loss of tolerance, and whether cold tolerance can be regained (Stushnoff 1972). The avoidance of spring frost damage can be achieved by developing cultivars with late blooming dates and multiple flowers per node. Later blooming types are less likely to suffer spring frosts and those with higher

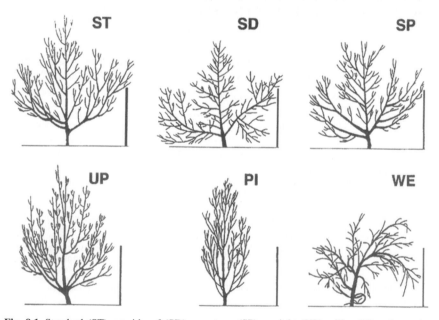

Fig. 9.1 Standard (ST), semidwarf (SD), spur-type (SP), upright (UP), pillar (PI) and weeping (WE) peach tree growth habits from Bassi et al. 1994

flower numbers are more likely to have sufficient numbers of flowers remaining after frosts. Cultivars that receive inadequate chilling commonly display sporadic foliation, irregular flower formation and floral abscission. Most peach cultivars have chilling requirements (hours below 7° C) of 650–1000 hrs, but germplasm has been utilized to produce cultivars with chilling requirements as low as 150 hours.

Most of the information available on cold tolerance has come from natural freezes during test winters, although methods of conducting controlled freezes have been developed (Layne 1989, Quamme 1991, Wisniewski and Arora 1991). Genotypes with high chilling requirements tend to have the least bud death due to winter cold. In general, a range in bud damage is apparent in segregating populations, suggesting quantitative inheritance; however, some segregating populations are skewed and have greater average resistance to cold injury than would be predicted by examining the parents (Mowry 1964). 'Redskin' stood out as a genotype with only modest hardiness that produced many progeny with good bud tolerance to cold. Few studies have sought to isolate the genes associated with cold tolerance in peach, although transcripts of the stress-induced dehydrin gene (*ppdhn1*) have been found to accumulate more in cold-tolerant peach tree cambium than the low cold tolerant 'Evergreen' cultivar (Artlip et al. 1997, Wisniewski et al. 1999).

Many of the genotypes most resistant to mid-winter cold originated from northern China such as 'Chui Lum Tao', 'Hui Han Tao', 'Tzim Pee Tao' and 'Siberian C'. Most of these hardy types have early bloom dates and poor fruit quality which take 3 or 4 generations of backcrossing to breed out, with the subsequent loss of some winter hardiness (Scorza and Sherman 1996). Unusually cold tolerant naturalized North American hybrids with late bloom have also been identified such as 'Reliance' (Cain and Anderson 1980, Layne 1984).

Considerable variability has been observed in numbers of flower buds per node that is stable across years and locations (Okie and Werner 1990) and is highly heritable (Hansche et al. 1972) (Fig. 9.2). Those cultivars developed for the colder climates tend to have higher numbers of buds per node than those developed in warmer climates (Werner et al. 1988). The number of flowers and fruits on 2-year old seedlings has also been shown to be heritable at the $h^2 = 0.16$ and $h^2 = 0.33$ level, respectively (Hansche 1986). While large numbers of flowers are of value in years of frost damage, in the absence of such damage, excessive flowering requires increased thinning and can negatively affect fruit size.

Little work has been conducted to determine the genetics of chilling hour requirements, although segregation patterns suggest that it is a quantitative trait, with a few major genes having important effects (Lammerts 1945, Lesley 1944, Sharp 1961). The inheritance appears to be largely additive, with little dominance effects. The genes regulating a low chilling requirement have come predominately from peaches from south China (Sharp 1974). Lammerts (1945) identified a recessive gene for 'evergreen' that held most of its foliage during mild, frostless winters. More recent work has shown that the wild type gene is incompletely dominant with heterozygotes being intermediate (Rodriquez et al. 1994). This gene now referred to as *Evergrowing* has been mapped (Wang et al. 2002a) and shown to be a result of a deletion in a MADS-box transcription factor sequence(s) (Bielenberg et al. 2004).

Fig. 9.2 High (*top*) and low (*bottom*) flower bud density in peach seedlings

The genetics of bloom date is largely unknown, although Hansche et al. (1972) did show that this trait was moderately heritable at $h^2 = 0.39$. While considerable variability has been described, it has proven difficult to partition the relative effects of chilling requirement, rate of bloom development and environment. There is likely an interaction between cold and heat requirements and the conditioning of other genes appears important. Regardless, cultivars do maintain 'a rather ordered progression of bloom at any given locality' (Scorza and Sherman 1996), making local selection possible.

Much more is known about time of fruit maturity. Considerable variability is found in this trait and it is quantitatively regulated with a very high heritability (Hansche 1986, Hansche et al. 1972). Bailey and Hough (1959) presented a model that involved 9 major or dominant genes and 10 modifying genes. Vileila-Morales et al. (1981) found that early fruiting is regulated by three major genes.

A number of simply inherited foliar and flower traits have been described. Among the foliar traits are red leaf/green leaf (*Gr/gr*), smooth leaf margin/wavy leaf margin (*Wa/wa*), Willow-leaf (*Wa2/wa2*) and glandular foliage/eglandular foliage (*E/e*). *E* has been located on Linkage group 7 (Dirlewanger et al. 2004). Among the flower traits are pollen fertile/pollen sterile (Ps/ps and Ps_2/ps_2), nonshowy/showy (Sh/sh), large showy flowers/small showy flowers (L/l), colored/white (W/w), with anthocyanins/anthocyaninless (AN/an), dark pink/light pink (p/p), pink/red (R/r), single/double (Di/di) and fewer extra petals/more extra petals ($Dm1/dm1$ and independent $Dm2/dm2$) (Fig. 9.3). Two pairs of these loci have been shown to segregate independently, E/e – Ps/ps and Sh/sh – An/an (Monet and Bastard 1983, Monet et al. 1985).

9.5.3 Fruit Quality

Numerous traits related to fruit quality are of importance to peach breeders. In the fresh market, consumers desire a large, well shaped fruit that is flavorful with a high sugar content and low to moderate acidity. For the processed market, several characteristics are appreciated including firm flesh, absence of a tip on the pit, no pit cracking, attractive color and non-browning of the flesh.

A number of single genes have been described that regulate important fruit characteristics (Table 9.5). Bailey and French (1941 and 1949) identified genes for freestone/clingstone (F/f), melting flesh/non-melting flesh (M/m) and soft melting flesh/firm melting (St/st) which are all found on the same chromosome. The dominance relationships between the genes regulating flesh texture are $ST > M > m$. F appears to be epistatic to mm allowing for only St or M expression, although F_ mm could be lethal (Scorza and Sherman 1996). Only a single freestone, non-melting individual has been reported and it has been lost (Blake 1937). Yoshida (1970) described genes for melting flesh/'stonyhard' flesh (Hd/hd); these plants produce little ethylene and remain firm throughout storage (Goffreda 1992, Haji et al. 2001).

A significant recent effort has been undertaken at the University of California, Davis to describe the genetics of a number of traits associated with internal breakdown (IB) of fruit or chilling injury (Peace et al. 2005, 2006). Using a combination of conventional and QTL mapping approaches, they have found high heritability for all the traits associated with IB including mealiness, flesh browning and flesh bleeding and found major QTL for all of these characteristics. The observed segregation patterns suggested that only a few major genes control each of the IB symptoms. Mealiness and browning were positively correlated, and both were negatively associated with bleeding (red coloration). Mealiness and bleeding were positively correlated with flowering date, while browning was positively associated with harvest date. The flesh color locus Y did not have a significant effect on IB.

The expression of a number of genes has been associated with the ripening and softening of peach fruits. Several cell hydrolases that cause cell wall-loosening have been implicated in fruit softening including glucanases, cellulases and pectic enzymes (Bonghi et al. 1998, Callahan et al. 1991, Scorza 2001). Three forms of the pectic enzyme polygalacturonase (PG) have been found in peach fruits, two being

Fig. 9.3 A sample of peach flower types: showy single (*upper left*), non-showy (*upper right*), double showy (*middle right*), double showy extra petals (*middle left*), 'chrysanthemum' petals (*lower left*), variegated petals (*lower right*). Photos by D. Hu and R. Scorza

exo-PG and one being endo-PG. The exo-PG activity is ripening regulated (Downs et al. 1992) and high activity in this enzyme is strongly associated with the melting flesh characteristic (Pressey and Avantes 1978). Lester et al. (1996) found an RFLP for an endo-PG that co-segregated with the melting flesh trait, and they discovered that there was a deletion of endo-PG-related sequences in the nonmelting flesh variety, Fla. 9-26C. Peace et al. (2005) concluded that a single locus with at least one gene for endopolygalacturanase controls the freestone and melting traits with at least three alleles.

When Peace et al. (2006) used a candidate gene approach to identifying specific genes associated with IB, they discovered that a gene encoding endopolygalacturonase co-segregates with the freestone and melting flesh traits and they found a large QTL for mealiness. Endo-β-1,4-glucanases (ppEG1) have been shown to accumulate during fruit abscission and share 76% homology with ripening-related avocado glucanase (Trainotti et al. 1997).

The expression of several genes has been associated with the ethylene climacteric in peach. 1-Aminocyclopropane-1-carboxylic acid (ACC) synthase and ACC oxidase activity have been shown to increase during fruit ripening (Callahan et al. 1993a,b). Two ethylene receptor genes, Pp-$ETR1$ and Pp-$ERS1$, have been isolated from peach that are homologous to $ETR1$ and $ESR1$ in $Arabidopisis$ (Bonghi et al. 2002). The level of expression of Pp-$ETR1$ was unchanged during ripening, while Pp-$ERS1$ expression increased in conjunction with the ethylene climacteric. Application of the ethylene inhibitor 1-methyl-cyclopropane reduced expression of both genes, along with ethylene biosynthesis. Ruperti et al. 2001 found two ACC oxidases to be differentially expressed in flowers, fruits and leaves; one of the genes (PP-$ACO2$) was expressed only in fruit and was not affected by propylene, while the other gene (PP-$ACO1$), was highly expressed in senescing leaves, abscising fruit and ripe mesocarp and was positively regulated by propylene. The transcripts from three genes, $PpAz8$, $PpAz44$ and $PpAz152$ have been isolated from cells of fruit and leaf abscission zones that show homology to PR thaumatin-like proteins and plant and fungal β-D-oxylosidases (Ruperti et al. 2002).

In other genetic work on the biochemical components associated with fruit ripening and taste, Monet (1979) described a gene pair (D/d) that determines low malic acid vs. normal. Initial studies suggested that low fruit acidity was dominant to high acidity, but subsequent work has shown a continuous range of variability. Hansche et al. (1972) found a modest level of heritability for fruit acidity ($h^2 = 0.19$), while heritability for fruit soluble solids was only 0.01. Fruit browning was shown to have a heritability of 0.35 in another study of peach (Hansche and Boynton 1986). Ramming (1991) identified a gene, sr, that slows down fruit ripening. Genotypes that are homozygous for this gene ripen very slowly or not at all, have reduced CO_2 and C_2H_4 production and fail to abscise. Hansche (1986) found low to medium heritability for soluble solids, sweetness, firmness and flavor in peach and nectarine populations dwarfed by the dw gene. Etienne et al. (2002) cloned six peach genes associated with organic acid metabolism and storage during fruit development (Mitochondrial citrate synthase, cytosolic NAD-dependent malate dehydrogenase, vacuolar proton translocating pumps, vacuolar H+-ATPase, and two vacuolar H+-pyrophophatases).

Several compounds have been found in peach that can cause food allergies, including a family of 9 kDa lipid transfer proteins (LTPs) (Malet et al. 1988, Pastorello et al. 1999). These compounds cause type I allergic reactions in humans by binding to immunoglobin E. Transcripts of two LTP genes, *pp-LTP1* and *pp-LTP2*, are found in peach with *pp-LTP1* being expressed in the skin of ripe fruit, while *pp-LTP2* expresses in the ovary (Botton et al. 2002).

Blake (1932) described a gene pair regulating pubescent skin/glabrous (G/g). Heavy pubescence was initially reported as being dominant to light pubescence (Blake and Connors 1936), although the amount of pubescence appeared to be quantitatively inherited in later studies (Blake 1940, Weinberger 1944). Most recently, Okie and Prince (1988) have reported on a gene regulating normal pubescence vs. a rough surface (*Rs/ss*) that also causes glabrous flower buds. Interestingly, it is not expressed in *gg* genotypes.

Conners (1922) and Blake (1934 and 1940) originally suggested that small fruit size was dominant to large fruit size, but later work has indicated that fruit size is controlled by predominantly additive genes with little dominance involved (Hansche et al. 1972, Weinberger 1955). Scorza and Sherman (1996) suggested that 'unimproved genotypes could express a few genes that have major effects on fruit size'. Hansche (1986) found moderate to high heritability for fruit weight in peach and nectarine populations dwarfed by the *dw* gene.

A single, dominant gene regulating saucer vs. non-saucer shape (S/s) has been identified (Lesley 1939) that is lethal in the homozygous state (Guo et al. 2002), although shape in general appears to be quantitatively regulated (Scorza and Sherman 1996). Oval has been described as dominant to round, but other studies suggested a much more complex inheritance (Blake 1940). The S locus is found on Linkage group 6, along with Dwarf (*Dw*), Redleaf (*Gr*) and male sterility (*ps*) (Fig. 9.2). The D/d locus regulating acid level may also be in this linkage group as Monet et al. (1985) found them to be linked by 30 cM to S/s; however, the D locus was found on Linkage groups 2 and 5 in the composite map (Dirlewanger et al. 2004).

A few single genes have been associated with fruit color. An allele (Y) has been described that produces white fleshed fruit (Connors 1920) and another, highlighter (*h*), suppresses red color (Beckman et al. 2005). The relationship between these two alleles has not been explored, although highlighter is known to be independent from the petal coloration alleles anthocyaninless (*An*) and white flower (*W*). The full red color phenotype is regulated by a recessive gene *fr* (Beckman and Sherman 2003), with the degree of red skin color likely regulated by multiple genes with complex environmental interactions. The blood-flesh trait (red-violet mesocarp) is regulated by a single gene, *bf* (Werner et al. 1998). A red surface blush had a heritability of $0.19 + / - 0.04$ in a segregating population of dwarf peaches (Hansche 1986). The degree of red color around the pit varies greatly and is likely polygenic; however, the presence of red color has been reported to be dominant (Blake 1932). Pillar (*Br*), double flowering and the flesh color locus are linked (Rajapakse et al. 1995).

French (1951) studied the segregation of several traits in hybrid peach populations including pubescence, flesh stringiness, coarseness, stone size, juiciness, skin thickness and toughness. French's populations varied greatly between years making conclusions difficult, but he did suggest that stringiness of the flesh and flesh coarseness were mostly recessive to their counterparts. Expression of the juiciness trait and stone size was very dependent on which parents were crossed; some parents appeared to pass the trait in a dominant fashion, although inheritance generally appeared to be quantitative. The thickness of the skin of progeny populations was dependent on the parents. A statistical analysis that estimates both environmental and genetic components of variability needs to be made to better elucidate the genetics of these traits.

9.6 Crossing and Evaluation Techniques

9.6.1 Pollination and Seedling Culture

Pollen is generally collected from well advanced flowers that are not quite open ('balloon stage'). The flowers are usually collected in paper bags, and the anthers are extracted within a few hours of collection by rubbing them over a wire mesh screen with a 4–6 mm mesh. When the flowers must be stored for longer periods of time, they can be held in the collection bags at $2°$ C–$4°$ C for a couple of days. The anthers are most often sifted onto absorbent paper for drying and allowed to dehisce for 12–24 hours at ambient room temperature. After drying, the pollen is commonly placed into glass shell vials and can be held at ambient temperature for a season. For longer storage times, the pollen is generally frozen at $-18°$C (Griggs et al. 1953) or held at $0°$C–$2°$C at 25% relative humidity (King and Hesse 1938). Pollen frozen in liquid nitrogen will retain its viability for many years.

Stamens of peach flowers are attached distally in a ring at the base of the corolla and can be easily removed by pulling the flowers apart using the finger nails. Emasculation is done when the flowers approach anthesis but are not yet open or shedding pollen. Branches are emasculated from the top down, to avoid accidental wind pollination and checked every few days for 7–10 days after pollination to remove any new flowers.

Pollination is accomplished using a camel's hair brush, the rubber tip of a pencil, a finger or a glass rod. A simple touch of the stigmatic surface is all that is necessary. After pollination, 70% alcohol is used to kill any pollen left on the applicator. Pollinators are generally not attracted to petaless flowers, so branches are not generally covered for cultivar development crosses. For genetic crosses, chance pollination is prevented by covering the branches with paper bags or cheese cloth. If wet weather is expected, the paper bags can be protected with polyethylene bags, but they need to be well ventilated by punching holes in them. To protect against frost damage during and after pollination, plastic houses or parachute covers with heat sources have proven effective (Werner and Cain 1985).

Seed are collected from ripe fruit soon after harvest, but before they begin to rot or ferment. Seed are commonly allowed to dry after removal, but the percentage of germination can sometimes be increased by stratifying them before they dry. Stratification is often accomplished by placing a single row of seeds (removed from the endocarp) on the bottom of 250 ml Erlenmeyer flasks and covering them with water containing a fungicide. The next day enough water is removed to uncover the seeds and the flask is stoppered with cotton, film or foil and held at 2° C–4° C (with occasional watering). Seeds are also sometimes stratified in moist perlite in plastic bags with a fungicide. Germination normally begins after 90–120 days, when the rest requirement of the seeds has been met (Hartmann and Kester 1959). Non-germinated seeds can be placed back in cold stratification. When the radicals are 0.5–1 cm long, the seeds are ready for planting. They can be set directly in the field by placing the radical at 5 cm depth, or they can be grown in a greenhouse to get better emergence and early growth. When this is done, the seedlings are generally moved to the field when convenient.

Peach breeders commonly use embryo culture to germinate seed from early-maturing genotypes, particularly in subtropical areas where short development periods are a major goal. Commonly, the flesh of the early ripening types matures before the embryo is fully developed.

Almost all cultivars ripening 70–75 days from full bloom can be successfully cultured, but the culture of younger embryos is dependent on genotype and growth conditions. An index called PF_1 (embryo length/seed length) was proposed by Hesse and Kester (1955) to measure comparative embryo development. In their work, embryos with a PF_1 lower than 70 were difficult to culture, although Ramming (1990) was able to culture embryos at PF_1 as low as 25.

For embryo culture, the fruit are generally surface sterilized with 0.25–1% sodium hypochlorite and the seed is removed from the endocarp. The embryo is then excised from the seed and cultured on 0.6–0.7% agar containing 2–4% sucrose and nutrients (Ramming 1985, Tukey 1934).

9.6.2 Evaluation Techniques

Most commonly, seedlings are planted at 1–2 m within rows and 3.0–4.5 m between rows in the spring following hybridization. Seedlings begin to fruit 1–2 years after planting. High density plantings have also been developed in Florida where seedlings are set at 13 cm apart in rows 1 m apart in August or September in the same year as hybridization (Sherman et al. 1973). This system allows for many more seedlings to be evaluated in small areas of field space, but only the most easily scored traits such as chilling requirement, fruit development period and fruit quality can be successfully evaluated (Rodriquez et al. 1986). The less dense plantings are typically evaluated for four or five years with little yearly rouging of undesirable genotypes, while the high density plantings are evaluated for three years with thinning in the second year.

Selections that appear to have potential are then second tested under commercial field conditions against standard cultivars. The most promising ones are distributed after 2–4 crops to a number of test locations within the expected adaptation zone, including grower cooperators and Agriculture Experiment Stations. When a selection survives these tests by showing high commercial potential, it is released. A minimum of 10 years, and often many more, are required between the initial cross and a genotypes release to the industry.

9.7 Biotechnological Approaches to Genetic Improvement

9.7.1 Regeneration and Transformation

Peaches have been regenerated utilizing several systems including in vitro leaves (Gentile et al. 2002), mature cotyledons (Pooler and Scorza 1995), embryo-derived callus (Scorza et al. 1990) and immature zygotic embryos (Hammerschlag et al. 1985). However, there are only two reports of stable peach plant transformation. Smigocki and Hammerschlag (1991) generated transgenic peach plants from embryogenic cultures of 'Redhaven' using the sooty mutant strain of *A. tumefaciens, tms*:328::Tn5. This strain carries an octopine type Ti plasmid with a functional cytokinin gene and a mutated auxin gene. The transgenic plants with the cytokinin gene were dwarf, produced unusually high numbers of branches and had delayed leaf senescence (Hammerschlag et al. 1997, Hammerschlag and Smigocki 1998). Perez-Clemente et al. (2004) produced transformants using embryo explants from stored seeds, utilizing two strains of *A. tumefaciens* containing the binary plasmid pBIN19 with the CaMV35spor-sGFP-CaMV35ster cassette as a reporter gene. Their highest efficiency rate of transgenic plant production was 3.6%, utilizing *A. tumefaciens* strain C58 and embryo sections. Between these two reports of preach transformation it appears that a total of four transgenic peach plants have been produced. To date there have been no reports replicating these results. An efficient, repeatable peach transformation methodology awaits development.

Efforts are underway to improve peach transformation protocols. For example, Padilla et al. (2006) conducted a large multivariate experiment to determine the optimal conditions for *Agrobacterium*-mediated transformation of peach explants. The GUS (*uidA*) marker gene was tested using two *A. tumifaciens* strains, three plasmids and four promoters, while GFP was evaluated in six *A. tumefaciens* strains, one plasmid and the doubleCaMV35s (dCAMV35s) promoter. The highest rates of transformation were produced with the combination of *A. tumifaciens* EHA105, plasmid pBIN19 and the CaMV35s promoter utilizing peach epicotyl internodes (56.8%), cotyledons (52.7%) and embryotic axes (46.7%). While these studies have enhanced transformation protocols in peach, transformation rates remain rather low and when combined with low regeneration rates the development of transgenic peaches remains problematic.

9.7.2 Genetic Mapping and QTL Analysis

A growing number of molecular linkage maps have emerged of peach and its relatives (Table 9.6); five maps are available of pure *Prunus persica*, two of almond
× *P. persica*, two of *P. persica* × *P. davidiana*, and one each of *P. persica* × nectarine, *P. persica* × *P. ferganensis* and myrobalan plum × an almond – *P. persica*
hybrid. Map coverage ranges from 396–1300 cM, with 8–23 linkage groups being
identified. Molecular markers have also been used to distinguish between peach

Table 9.6 Published genetic linkage maps of peach

Parents	No. Loci	Linkage groups	Size (cM)	Reference
Peaches 'NC174RL' × 'Pillar'	83	15	396	Chaparro et al. 1994
Peaches 'New Jersey Pillar' × 'KV77119'	79	13	540	Abbott et al. 1998, Rajapakse et al. 1995, Sosinski et al. 2000
Peaches 'Suncrest' × 'Bailey'	145	23	926	Abbott et al. 1998, Sosinski et al. 2000
Peaches 'Lovell' × 'Nemared'	153	15	1300	Abbott et al. 1998, Lu et al. 1998, Sosinski et al. 2000
Peaches 'Harrow Blood' × 'Okinawa'	76	10		Gillen and Bliss 2005
Peaches 'Akame' × 'Jueitou'	178	8	571	Shimada et al. 2000 Yamamoto et al. 2002
Peach 'Ferjalou Jalousia' × Nectarine 'Fantasia'	249	11	712	Dirlewanger et al. 2004, 2006
Peach 'Guardian' × 'Nemaguard' (*P. persica* × *P. davidiana*) F$_2$	171	8	737	Blenda et al. 2007
Almond 'Texas' × peach 'Earlygold' F$_2$	562	8	519	Aranzana et al. 2002, Dirlewanger et al. 2004, Joobeur et al. 1998
Almond 'Padre' × peach 54P455 F$_2$	161	8	1144	Bliss et al. 2002, Foolad et al. 1995
Peach 'Summergrand' × *P. davidiana* clone 1908	23, 97[1]	3, 9	159 471	Dirlewanger et al. 1996, Viruel et al. 1998
Peach IF7310828 ('J.H. Hale' × 'Bonanza') × selection of *P. ferganensis* BC$_1$	216	8	665	Dettori et al. 2001, Quarta et al. 2000, Verde et al. 2005
Myrobalan plum P.2175 × almond – peach hybrid GN22	93, 166[1]	8, 7	525, 716	Dirlewanger et al. 2004

[1] Separate maps were generated for each of the parents

cultivars, measure their relatedness and determine their origins (Aranzana et al. 2004, Dirlewanger et al. 2002, Testolin et al. 2000, Xu et al. 2006).

Linkage relationships with molecular markers have been described for 23 monogenic morphological traits associated with adaptation, flower color, fertility, leaf shape and color, plant habit, fruit quality and pest resistance (Table 9.7). QTL have also been identified for 23 horticulturally important traits including bloom and

Table 9.7 Monogenic traits associated with molecular markers in peach

Trait	Linkage group	References
Adaptation		
Evergrowing (*evg*)	1	Dirlewanger et al. 2004, Wang et al. 2002
Flower traits		
Double flower (*Dl*)	2	Dirlewanger et al. 2004, Sosinski et al. 2000
Flower color (*Fc*)	3	Dirlewanger et al. 2004, Yamamoto et al. 2001
Male sterility (*Ps*)	6	Dirlewanger et al. 1999, 2004, 2006
Leaf traits		
Leaf color (*Gr*)	5	Chaparro et al. 1994, Dirlewanger et al. 2004, Yamamoto et al. 2001
Leaf glands (*E*)	7	Dettori et al. 2001, Quarta et al. 2000
Leaf shape (*Nl*)	6	Dirlewanger et al. 2004
Plant habit		
Dwarf plant (*Dw*)	6	Dirlewanger et al. 2004
Pillar growth habit (*Br*)	2	Dirlewanger et al. 2004
Fruit quality		
Blood flesh (*bf*)	4	Gillen and Bliss 2005
Flat fruit (*S*)	6	Dirlewanger et al. 1999, 2004, 2006
Flesh adhesion (*F*)	4	Abbott et al. 1998, Dettori et al. 2001, Dirlewanger et al. 2004, Quarta et al. 2000, Yamamoto et al. 2001
Flesh color (*Y*)	1	Abbott et al. 1998, Bliss et al. 2002, Dirlewanger et al. 2004, Warburton et al. 1996
Flesh color around stone (*Cs*)	3	Dirlewanger et al. 2004, Yamamoto et al. 2005
Non acid fruit (*D*)	2,5	Bliss et al. 2002, Dirlewanger et al. 1999, Dirlewanger et al. 2004
Polycarpel (*Pcp*)	3	Bliss et al. 2002, 2004, 2006
Skin color (*Sc*)	6	Dirlewanger et al. 2004, Yamamoto et al. 2001
Skin hairiness (*G*)	5	Bliss et al. 2002, Dirlewanger et al. 1999, 2004, 2006
Pest resistance		
Leaf curl resistance	3,6	Viruel et al. 1998
Nematode resistance (*Mij*)	2	Abbott et al. 1998, Dirlewanger et al. 2004, Gillen and Bliss 2005, Lu et al. 1998, Lu et al. 1999, Lu et al. 2004, Wang et al. 2002, Yamamoto et al. 2001
Nematode resistance (*Mja*)	7	Blenda et al. 2002, Dirlewanger et al. 2004, Yamamoto et al. 2001
Powdery mildew resistance	7,8	Quarta et al. 2000, Verde et al. 2002
Resistance gene analogs	Many	Gillen and Bliss 2005, Lalli et al. 2005

Table 9.8 QTL[1] associated with major traits of peach

Trait	Linkage group	References
Adaptation		
Flowering time	4	Dirlewanger et al. 1999, Quarta et al. 2000, Verde et al. 2002
Fruit development period	4	Abbott et al. 1998, Etienne et al. 2002, Verde et al. 2002
Internode length	1	Verde et al. 2002
Maturity date	3, 4	Dirlewanger et al. 1999, Etienne et al. 2002
Productivity	6,9	Dirlewanger et al. 1999
Ripening time	2, 6	Dirlewanger et al. 1999, Quarta et al. 2000, Verde et al. 2002
Short life syndrome	1, 2, 4, 5, 6	Blenda et al. 2007
Fruit quality		
Bleeding	1,4	Peace et al. 2006
Browning	5	Peace et al. 2006
Fruit diameter	2	Abbott et al. 1998
Fruit skin color	2,6	Quarta et al. 2000, Verde et al. 2002
Fruit weight	5, 6	Abbott et al. 1998, Dirlewanger et al. 1999, Etienne et al. 2002
Mealiness	4	Peace et al. 2006
pH	5	Abbott et al. 1998, Etienne et al. 2002
Titratable acidity	5, 6	Bliss et al. 2002, Dirlewanger et al. 1999, Etienne et al. 2002
Malic acid content	5, 6	Dirlewanger et al. 1999, Etienne et al. 2002
Citric acid content	5, 6	Dirlewanger et al. 1999, Etienne et al. 2002
Quinic acid	8	Etienne et al. 2002
Soluble solids	2, 4,6	Abbott et al. 1998, Dirlewanger et al. 1999, Etienne et al. 2002, Quarta et al. 2000, Verde et al. 2002
Fructose content	4	Abbott et al. 1998, Etienne et al. 2002
Glucose content	4	Abbott et al. 1998, Dirlewanger et al. 1999, Etienne et al. 2002
Sorbitol	6	Dirlewanger et al. 1999
Sucrose content	5	Dirlewanger et al. 1999, Etienne et al. 2002

[1]QTL that were identified in more than one year

ripening time, fruit quality, storage life, freestone trait, internode length and pest resistance (Table 9.8).

Considerable synteny has been observed among the maps of the various *Prunus* species, allowing for the development of a *Prunus* consensus map [Cmap in the Genome Database for Rosaceae (GDR) at http://www.rosaceae.org]. When the positions of RFLP, SSR and isozyme anchor markers are compared among the individual genetic maps, the genomes of the diploid species of almond, apricot, cherry, *P. davidiana*, *P. cerasifera* and *P. ferganensis* are mostly collinear (Dirlewanger et al. 2004). Only one large chromosomal rearrangement has been found, a reciprocal translocation in the almond ('Garfi') × peach ('Nemared') cross (Jauregui et al. 2001) and the peach F_2 'Akame' × 'Juseitou' (Yamamoto et al. 2001). A high level of synteny also appears to exist between *Prunus* and *Malus*,

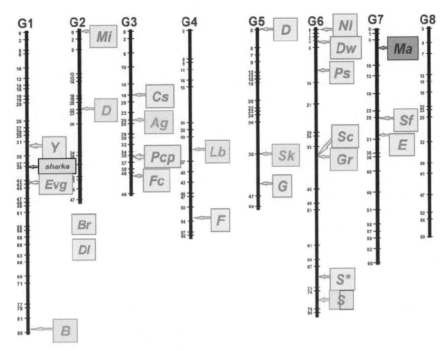

Fig. 9.4 Approximate position of 28 major *Prunus* genes mapped in populations of apricot (blue background), peach (orange background), almond or almond × peach (yellow background), and Myrobalan plum (green background) (Dirlewanger et al. 2004). The gene abbreviations are: *Y*, peach flesh color; *B*, almond/peach petal color; *sharka*, plum pox virus resistance; *B*, flower color in almond × peach; *Mi*, nematode resistance from peach; *D*, almond shell hardness; *Br*, broomy plant habit; *Dl*, double flower; *Cs*, flesh color around the stone; *Ag*, anther color; *Pcp*, polycarpel; *Fc*, flower color; *Lb*, blooming date; *F*, flesh adherence to stone; *D*, non-acid fruit in peach, *Sk*, bitter kernel; *G*, fruit skin pubescence; *Nl*, leaf shape; *Dw*, dwarf plant; *Ps*, male sterility; *Sc*, fruit skin color; *Gr*, leaf color; *S**, fruit shape; *S*, self-incompatibility (almond and apricot); *Ma*, nematode resistance from Myrobalan plum; *E*, leaf gland shape; *Sf*, resistance to powdery mildew. Genes *Dl* and *Br* are located on an unknown position of G2

although only limited numbers of loci have been compared. Dirlewanger et al. 2004 was able to generate a map for all of *Prunus* on which 28 major genes were mapped in populations of apricot, peach, almond and Myrobalan plum (Fig. 9.4).

9.7.3 Genomic Resources

Several bacterial artificial chromosome (BAC) libraries have been developed for peach (Genome Database for Rosaceae (GDR) at http://www.rosaceae.org). Two of the largest are those of Georgi et al. (2002) which was generated from fruit mesocarp of the peach rootstock 'Nemared' and Wang et al. (2001) which was produced from leaves of the traditional cultivar Jingyu. The libraries of Georgi et al. (2002) and Wang et al. (2001) contain 44,160 and 20,736 clones, respectively.

Over 85,000 *Prunus* ESTs have been sequenced and deposited in the NCBI dbEST database (http://www.genome.clemson.edu/gdr/). A high proportion of the ESTs have been found to contain SSRs in transcribed regions, allowing for the placement of known genes on linkage maps (Georgi et al. 2002, Jung et al. 2005, Wang et al. 2002). The EST-derived SSRs are less polymorphic than those from intergenic regions, but are more easily transferred among species, as the transcribed sequences are often more highly conserved. Most recently,18 EST-SSR markers have been developed from a mesocarp cDNA library of the peach cultivar 'Yumyeong', whose primers gave successful amplification in six other *Prunus* species (almond, apricot, sweet cherry, Japanese plum, European plum and *Prunus ferganensis*) (Vendramin et al. 2007).

Horn et al. 2005 used probes of core markers (141) from the 'Texas' × 'Earlygold' peach map to screen the BAC library to provide the framework for a physical and transcript map. When they hybridized 1,236 ESTs from the unigene set and an additional 68 peach cDNA colonies to genetically anchored BACs, they were able to place 11.2% of the ESTs and cDNAs on the peach genetic map. One cluster of 32 ESTs were of special note as most of them were not homologous to sequences in the NCBI data base. It was suggested by Horn et al. (2005) that these 'ESTs might be unique to fruit trees or rapidly evolved from a common ancestor to fulfill new functions in fruit trees'.

Resistance gene analogs (RGAs) representing NBS-LRR, kinase, transmembrane domain classes, pathogen response (PR) proteins and resistance-associated transcription factors have also been hybridized to the peach BAC library to develop a resistance map for *Prunus* (Lalli et al. 2005). Using the peach physical map data base of the Genome Database for Rosaceae (GDR), 42 map locations were identified with possible resistance regions across 7 of the 8 linkage groups of peach.

References

Abbott AG, Rajapakse S, Sosinski B, Lu Z, Sossey-Alaoui K, Gannavarapu M, Reighard G, Ballard RE, Baird WV, Scorza R, Callahan A (1998) Construction of saturated linkage maps of peach crosses segregating for characters controlling fruit quality, tree architecture and pest resistance. Acta Hortic 465:41–49

Ackerman WL (1953) The evaluation of peach leaf curl in foreign and domestic peaches and nectarines grown at the U.S. Plant Introduction Garden, Chico, California. Div. Plant Exploration and Introduction, Bul Plant Ind Soils Agricul Eng, USDA (Mimeo) pp 1–31

Aranzana MJ, Pineda AM, Cosson P, Dirlewanger E, Ascasibar K, Cipriani G, Ryder CD, Testolin R, Abbott AG, King GJ, Iezzoni AF, Arus P (2002) A set of simple sequence repeats (SSR) markers covering the *Prunus* genome Theor Appl Genet 106:819–825

Armstrong DL (1957) Cytogenetic study of some derivatives of the F₁ hybrid *Prunus amygdalus* × *P. persica.* In: Genetics. University of California, Davis

Artlip TS, Callahan AM, Bassett CL, Wisniewski ME (1997) Seasonal expression of a dehydrin gene in sibling deciduous and evergreen genotypes of peach [*Prunus persica* (L.) Batsch]. Plant Mol Biol 33:61–70

Bailey CH, French AP (1949) The inheritance of certain fruit and foliage characteristics in peach. Massachusetts Agric Exp Sta Bull 452

Bailey CH, Hough LF (1959) An hypothesis for the inheritance of season of ripening in progenies from certain early ripening peach varieties and selections. Proc Am Soc Hortic Sci 73: 125–133

Bailey JS, French AP (1941) The genetic composition of peaches. Massachusetts Agric Exp Sta Bull 378 (Annual Report 1940): 91

Beckman TG, Rodriguez AJ, Sherman WB, Werner DJ (2005) Evidence for qualitative suppression of red skin color in peach. HortScience 40:523–524

Beckman TG, Sherman WB (2003) Probable quantitative inheritance of full red skin color in peach. HortScience 38:1184–1185

Bielenberg DG, Wang Y, Fan S, Reighard GL, Scorza R, Abbott AG (2004) A deletion affecting several gene candidates is present in the *Evergrowing* peach mutant. J Hered 95:436–444

Blake MA (1932) The J.H. Hale as a parent in peach crosses. Proc Am Soc Hortic Sci 29:131–136

Blake MA (1934) Relative hardiness of 157 varieties of peaches and nectarines in 1933 and 14 varieties in 1934 at New Brunswick, N.J. New Jersey Agric Exp Sta Circ 303

Blake MA (1937) Progress in peach breeding. Proc Am Soc Hortic Sci 35:49–53

Blake MA (1940) Some results of crosses of early ripening varieties of peaches. Proc Am Soc Hortic Sci 37:232–241

Blake MA, Connors CH (1936) Early results of peach breeding in New Jersey. New Jersey Agric Exp Sta Bull 599

Blenda AV, Reighard GL, Baird WV, Georgi LL, Abbott AG (2002) Molecular markers and candidate resistance genes: a genetic study of tolerance to ring nematode in peach. In: Plant, Animal & Microbe Genomes X Conference, San Diego, USA

Blenda AV, Verde I, Georgi LL, Reighard G, Forrest S, Muñoz-Torres M, Baird WV, Abbott AG (2007) Construction of a genetic linkage map and identification of molecular markers in peach rootstocks for response to peach tree short life syndrome. Tree Genet Genomics 81:281–288

Bliss FA, Arulsekar S, Foolad MR, Becerra V, Gillen A, Warburton ML, Dandekar AM, Kocsine GM, Mydin KK (2002) An expanded genetic linkage map of *Prunus* based on an interspecific cross between almond and peach. Genome 45:520–529

Bonghi C, Ferrarese L, Ruperti B, Tonutti P, Ramina A (1998) Endo-beta-1, 4-glucanases are involved in peach fruit growth and ripening and regulated by ethylene. Physiol Plant 102: 346–352

Bonghi C, Rasori A, Ziliotto F, Ramina A, Tonutti P (2002) Characterization and expression of two genes encoding ethylene receptors in peach fruit. Acta Hortic 592:583–588

Botton A, Begheldo M, Rasori A, Bonghi C, Tonutti P (2002) Differential expression of two lipid transfer protein genes in reproductive organs of peach [*Prunus persica* (L.) Batsch]. Plant Sci 163:993–1000

Cain DW, Anderson RL (1980) Inheritance of wood hardiness among hybrids of commercial and wild Asian peach genotypes. J Am Soc Hortic Sci 105:349–354

Callahan A, Cohen R, Dunn L, Morgens P (1993a) Isolation of genes affecting peach fruit ripening. Acta Hortic 336:47–51

Callahan A, Morgens P, Cohen R (1993b) Isolation and initial characterization of cDNAs for mRNAs regulated during peach fruit development. J Am Soc Hortic Sci 118:531–537

Callahan A, Scorza R, Morgens P, Mante S, Cordts J, Cohen R (1991) Breeding for cold hardiness: searching for genes to improve fruit quality in cold hardy peach germplasm. HortScience 26:522–526

Chang LS, Iezzoni AF, Adams GC, Ewers FW (1989) *Leucostoma persoonii* tolerance and cold hardiness among diverse peach genotypes. J Am Soc Hortic Sci 114:482–485

Chang LS, Iezzoni AF, Adams GC, Ewers FW (1991) Hydrolic conductance in susceptible versus tolerant peach seedlings infected with *Leucostoma persoonii*. J Am Soc Hortic Sci 116:831–834

Chaparro JX, Werner DJ, O'Mally DO, Sederoff RR (1994) Targeted mapping and linkage analysis of morphological, isozyme and RAPD markers in peach. Theor Appl Genet 87:805–815

Chaparro JX, Werner DJ, Whetten RW, O'Malley DM (1995) Inheritance, genetic interaction and biochemical characterization of anthocyanin phenotypes in peach. J Hered 86:32–38

Chaplin CE, Schneider GW (1975) Resistance to the common peach tree borer (*Sanninoidea exitiosa* Say) in seedlings of 'Rutgers Redleaf' peach. HortScience 10:400

Childers NF, Sherman WB (1988) The Peach. Horticultural Publ., Gainesville, FL

Conners CH (1922) Peach breeding: a summary of results. Proc Am Soc Hortic Sci 19:108–115

Connors CH (1920) Some notes on the inheritance of unit characters in the peach. Proc Am Soc Hortic Sci 16:24–36

Cullinan FP (1937) Improvement of stone fruits. pp 605–702. In: USDA yearbook of agriculture. Washington D.C.

Daniell JW, Chandler WA (1982) Field resistance of peach cultivars to gummosis disease. HortScience 17:375–376

D'Bov S (1983) Inheritance of peach resistance to powdery mildew, *Sphaerotheca pannosa* var *persicae*, III: leaf resistance in F₁ of cultivars 'J. H. Hale' × 'Nectarine-Ferganensis-2' (In Russian). Genet Sel 16:146–150

Dettori MT, Quarta R, Verde I (2001) A peach linkage map integrating RFLPs, SSRs, RAPDs and morphological markers. Genome 44:783–790

Dirlewanger E, Cosson P, Boudehri K, Renaud C, Capdeville G, Tauzin Y, Laigret F, Moing A (2006) Development of a second-generation genetic linkage map for peach [*Prunus persica* (L.) Batsch] and characterization of morphological traits affecting flower and fruit. Tree Genet Genom 3:1–13

Dirlewanger E, Cosson P, Howad W, Capdeville G, Bosselut N, Claverie M, Voisin R, Poizat C, Lafargue B, Baron O, Laigret F, Kleinhentz M, Arus P, Esmenjaud D (2004) Microsatellite genetic linkage maps of myrobalan plum and an almond-peach hybrid – location of root-knot nematode resistance genes. Theor Appl Genet 109:827–838

Dirlewanger E, Cosson P, Tavaud M, Aranzana MJ, Poizat C, Zanetto A, Arús P, Laigret F (2002) Development of microsatellite markers in peach [*Prunus persica* (L.) Batsch] and their use in genetic diversity analysis in peach and sweet cherry. Theor Appl Genet 105:127–138

Dirlewanger E, Graziano E, Joobeur T, Garriga-Caldere F, Cosson P, Howad W, Arús P (2004) Comparative mapping and marker-assisted selection in Rosaceae fruit crops. Proc Nat Acad Sci (USA) 101:9891–9896

Dirlewanger E, Moing A, Rothan C, Svanella L, Pronier V, Guye A, Plomion C, Monet R (1999) Mapping QTLs controlling fruit quality in peach [*Prunus persica* (L.) Batsch]. Theor Appl Genet 98:18–31

Dirlewanger E, Pascal T, Zuger C, Kervella J (1996) Analysis of molecular markers associated with powdery mildew resistance in peach [*Prunus persica* (L.) Batsch] × *Prunus davidiana* hybrids. Theor Appl Genet 93:909–919

Dirlewanger E, Pronier V, Parvery C, Rothan C, Guye A, Monet R (1998) Genetic linkage map of peach [*Prunus persica* (L.) Batsch] using morphological and molecular markers. Theor Appl Genet 97:888–895

Downs CG, Brady CJ, Gooley A (1992) Exopolygalacturonase protein accumulates late in peach fruit ripening. Physiol Plantarum 85:133–140

Etienne C, Dirlewanger E, Cosson P, Svanella-Dumas L, Monet R, Moing A, Rothan C (2002) QTLs and genes controlling peach fruit quality. Acta Hortic 592:253–258

Etienne C, Moing A, Dirlewanger E, Raymond P, Monet R, Rothan C (2002) Isolation and characterization of six peach cDNAs encoding key proteins in organic acid metabolism and solute accumulation: Involvement in regulating peach fruit quality. Physiol Plantarum 114:259–270

Etienne C, Rothan C, Moing A, Plomion C, Bodenes C, Svanella-Dumas L, Cosson P, Pronier V, Monet R, Dirlewanger E (2002) Candidate genes and QTLs for sugar and organic content in peach [*Prunus persica* (L.) Batsch]. Theor Appl Genet 105:145–159

Feliciano A, Feliciano AJ, Ogawa JM (1987) *Monilinia fructicola* resistance in the peach cultivar Bolinha. Phytopathology 77:776–780

Foolad MR, Arulsekar S, Becerra V, Bliss FA (1995) A genetic map of *Prunus* based on an inter-specific cross between peach and almond. Theor Appl Genet 91:262–269

French AP (1951) The peach: Inheritance of time of ripening and other economic characteristics. Massachusetts Agric Exp Sta Bull 462

Gairola C, Powell D (1970) *Cytospora* peach canker in Illinois. Plant Dis Rep 54:832–835

Gardan CR, Luisetti J, Prunier JP (1971) Preliminary results on varietal susceptibility of the peach to bacterial canker (in French). Compt Rend Seances Acad Agr 57:1090–1094

Gentile A, Monticelli S, Damiano C (2002) Adventitious shoot regeneration in peach [*Prunus persica* (L.) Batsch]. Plant Cell Rep 20:1011–1016

Georgi LL, Wang Y, Yvergniaux D, Ormsbee T, Inigo M, Reighard G, Abbott AG (2002) Construction of a BAC library and its application to the identification of simple sequence repeats in peach [*Prunus persica* (L.) Batsch]. Theor Appl Genet 105:1151–1158

Gillen AM, Bliss FA (2005) Identification and mapping of markers linked to the *Mi* gene for root-knot nematode resistance in peach. J Am Soc Hortic Sci 130:24–33

Goffreda JC (1992) Stony hard gene of peach alters ethylene biosynthesis, respiration and other ripening related characteristics. HortScience 27:610

Gradziel TM (1994) Changes in susceptibility to brown rot with ripening in three clingstone peach genotypes. J Am Soc Hortic Sci 119:101–105

Gradziel TM, Wang D (1993) Evaluation of brown rot resistance and its relation to enzymatic browning in clingstone peach germplasm. J Am Soc Hortic Sci 118:675–679

Griggs WH, Vansell GH, Iwakiri BT (1953) The storage of hand collected and bee collected pollen in a home freezer. Proc Am Soc Hortic Sci 62:304–305

Guo J, Jang Q, Zhang K, Zhao J, Yang Y (2002) Screening for molecular marker linked to saucer gene of peach fruit shape. Acta Hortic 592:267–271

Haji T, Yaegaki H, Yamaguchi M (2001) Changes in ethylene production and flesh firmness of melting and nonmelting and stony hard peaches after harvest. J Jap Soc Hortic Sci 70:458–459

Hammerschlag FA, Bauchan FA, Scorza R (1985) Regeneration of peach plants from callus derived from immature embryos. Theor Appl Genet 70:248–251

Hammerschlag FA, McCanna IJ, Smigocki AC (1997) Characterization of transgenic peach plants containing a cytokinin biosynthesis gene. Acta Hortic 447:569–574

Hammerschlag FA, Smigocki AC (1998) Growth and in vitro propagation of peach plants transformed with the shooty mutant strain of *Agrobacterium tumefaciens*. HortScience 33:897–899

Hampson MC, Sinclair WA (1973) Xylem dysfuction in peach caused by *Cytospora leucostoma*. Phytopath 63:676–681

Hansche PE (1986) Heritability of fruit quality traits in peach and nectarine breeding stocks dwarfed by the *dw* gene. HortScience 21:1193–1195

Hansche PE (1988) Two genes induce brachytic dwarfism in peach. HortScience 23:604–606

Hansche PE, Boynton B (1986) Heritability of enzymatic browning in peaches. HortScience 21:1195–1197

Hansche PE, Hesse CO, Beres V (1972) Estimates of genetic and environmental effects of several traits in peach. J Am Soc Hortic Sci 97:76–79

Hartmann HT, Kester DE (1959) Plant Propagation. Prentice-Hall, Englewood Cliffs, NY

Hedrick UP (1917) The peaches of New York. New York Experiment Station, Ithaca

Hedrick UP (1950) A history of horticulture in America to 1860. Oxford Univ Press, New York

Hesse CO, Kester DE (1955) Germination of embryos of *Prunus* related to degree of embryo development. Proc Am Soc Hortic Sci 65:251–264

Horn R, Lecouls A-C, Callahan A, Dandekar AM, Garay L, McCord P, Howad W, Jung S, Georgi LL, Forrest S, Mook J, Zhebentysyeva TN, Yu Y, Kim HR, Jesudurai C, Sosinski B, Arús P, Aranzana MJ, Baird WV, Parfitt DE, Reighard G, Scorza R, Tomkins J, Wing R, Abbott AG (2005) Candidate gene database and transcript map for peach, a model species for fruit trees. Theor Appl Genet 110:1419–1428

Iezzoni AF (1987) The 'Redhaven' peach. Fruit Varieties J 41:50–52

Jauregui B, De Vicente MC, Messequer R, Felipe A, Bonnet G, Salesses G, Arus P (2001) A reciprocal translocation between 'Garfi' almond and 'Nemared' peach. Theor Appl Genet 102:1169–1176

Joobeur T, Viruel MA, de Vicente MC, Jauregui B, Ballister J, Dettori MT, Verde I, Truco MJ, Messequer R, Battle I, Quarta R, Dirlewanger E, Arus P (1998) Construction of a saturated map for *Prunus* using an almond × peach F$_2$ progeny. Theor Appl Genet 97:1034–1041

Jung S, Abbott A, Jusudurai C, Tomkins J, Main D (2005) Frequency, type, distribution of simple sequence repeats in Rosaceae ESTs. Funct Integ Genomics 5:136–143

Keil HL, Fogle HW (1974) Orchard susceptibility of some aprocot, peach and plum cultivars and selections to *Xanthomonas pruni*. Fruit Varities J 28:16–19

King JR, Hesse CO (1938) Pollen longevity studies with deciduous fruits. Proc Am Soc Hort Sci 36:310–313

Lalli DA, Decroocq V, Blenda AV, Schurdi-Levraud V, Garay L, Le Gall O, Damsteegt V, Reighard GL, Abbott AG (2005) Identification and mapping of resistance gene analogs (RGAs) in *Prunus*: a resistance map for *Prunus*. Theor Appl Genet 111:1504–1513

Lammerts WE (1945) The breeding of ornamental edible peaches for mild climates, I. Inheritance of tree and flower characteristics. Am J Bot 32:53–61

Layne REC (1984) Breeding peaches in North America for cold hardiness and perennial canker (*Leucostoma* ssp.) resistance: review and outlook. Fruit Varieties J 38:130–136

Layne REC (1987) Peach rootstocks. In: Rom, RC, Carlson RF (eds) Rootstocks for fruit crops. John Wiley & Sons, NewYork

Layne REC (1989) Breeding cold hardy peaches for Canada. Acta Hortic 254:73–78

Layne REC (1992) Breeding cold hardy peaches and nectarines. In: Janick J (ed) Plant Breeding Reviews, vol 10. Wiley, NewYork

Lesley JW (1939) A genetic study of saucer fruit shape and other characteristics in the peach. Proc Am Soc Hortic Sci 38:218–222

Lesley JW (1944) Peach breeding in relation to winter chilling requirements. Proc Am Soc Hortic Sci 70:243–250

Lester DR, Sherman WB, Atwell BJ (1996) Endopolygalacturonase and the *Melting Flesh* (*M*) locus in peach. J Am Soc Hortic Sci 121:231–235

Li ZL (1984) Peach germplasm and breeding in China. HortScience 19:348–351

Loreti F, Massai R (2002) The high density peach planting system: present status and perspectives. Acta Hortic 592:377–390

Lu Z, Sossey-Alaoui K, Reighard G, Baird WV, Abbott AG (1999) Development and characterization of a codominant marker linked to root-knot nematode resistance and its application in peach breeding. Theor Appl Genet 99:115–122

Lu ZX, Sosinski B, Reighard G, Baird WV, Abbott AG (1998) Construction of a genetic linkage map and identification of AFLP markers for resistance to root-knot nematodes in peach rootstocks. Genome 41:199–207

Malet A, Sanosa J, Garcia-Calderon PA (1988) Diagnosis of allergy to peach. A comparative study of in vivo and in vitro techniques. Allergological Immunopathology 16:181–184

Massonie G, Maison P, Monet R, Grasselly C (1982) Resistance to the green peach aphid *Myzus persicae* Sulzer (Homoptera: Aphididae) in *Prunus persica* (L.) Batsch and other *Prunus* species (in French). Agronomie 2:63–69

Mehlenbacher SA, Scorza R (1986) Inheritance of growth habit in progenies of 'Compact Redhaven' trees. HortScience 21:124–126

Monet R (1967) A contribution to the genetics of peaches (in French). Ann Amelior Plant 17:5–11

Monet R (1979) Genetic transformation of the 'fruit sweetness' character-incidence on selection for quality (in French). In: Eucarpia Fruit Section Symposium, Tree Fruit Breeding, Angers, pp 273–276

Monet R (1985) Heredity of the resistance to leaf curl (*Taphrina deformans*) and green aphid (*Myzus persicae*) in the peach. Acta Hortic 173:21–23

Monet R, Bastard Y (1983) New cases of independent segregation for Mendelian characters in peach (in French). Agronomie 3:387–390

Monet R, Bastard Y, Gibault B. (1985) Genetic studies on the breeding of flat peaches (In French) Agronomie 5:727–731

Monet R, Bastard Y, Gibault B (1988) Genetic study of the weeping habit in peach (in French). Agronomie 8:127–132

Monet R, Massonie G (1994) Determination of the genetics of resistance to the green aphid (*Myzus percicae*) (in French). Agronomie 14:177–182

Monet R, Salesses G (1975) A new mutant for dwarfing in peach (in French). Ann Amelior Plant 25:353–359

Mowry JB (1964) Inheritance of cold hardiness of dormant peach flower buds. Proc Am Soc Hortic Sci 85:128–133

Myers SC, Okie WR (1986) Low midwinter temperature injury to peach flower buds in Georgia. Fruit Var J 40:136–139

Okie WR, Prince VE (1982) Surface features of a novel peach x nectarine hybrid. HortScience 17:66–67

Okie WR, Ramming DW, Scorza R (1985) Peach, nectarine and other stone fruit breeding by the USDA in the last two decades. HortScience 20:633–641

Okie WR, Reilly CC (1983) Reaction of peach and nectarine cultivars and selections to infection by *Botryosphaeria dothidea*. J Am Soc Hort Sci 108:176–179

Okie WR, Werner DJ (1990) Effects of genotype and environment on fruit bud density in peach and nectarine. HortScience 25:1069

Padilla IMG, Golis A, Gentile A, Damiano C, Scorza R (2006) Evaluation of transformation in peach *Prunus persica* explants using green florescent protein (GFP) and beta-glucuranidase (GUS) reporter genes. Plant Cell, Tissue Organ Cult 84:309–314

Pastorello EA, Farioli L, Pravettoni V, Ortolani C, Ispano M, Monaz M, Baroglio C, Scibola E, Ansaloni R, Incorvaia C, Conti A (1999) The major allergen in peach (*Prunus persica*) is a lipid transfer protein. J Allergy Clin Immunol 103:520–526

Peace CP, Ahmad R, Gradziel TM, Dandekar AM, Crisosto CH (2005a) The use of molecular genetics to improve peach and nectarine post-storage quality. Acta Hortic 682:403–409

Peace CP, Crisosto CH, Garner DT, Dandekar AM, Gradziel TM, Bliss FA (2006) Genetic control of internal breakdown in peach. Acta Hortic 713:489–496

Peace CP, Crisosto CH, Gradziel TM (2005b) Endopolygalacturonase: a candidate gene for freestone and melting flesh in peach. Mol Breed 16:21–31

Perez-Clemente R, Perez-Sanjuan, Garcia-Ferriz L, Beltran J-P, Canas L (2004) Transgenic peach plants (*Prunus persica*) produced by genetic transformation of embryo sections using green fluorescent protein (GFP) as an in vitro marker. Mol Breed 14:419–427

Pooler MR, Scorza R (1995) Regeneration of peach [*Prunus persica* (L.) Batsch] rootstock cultivars from cotyledons of mature stored seed. HortScience 30:355–356

Potter JW, Dirks VA, Johnson PW, Olthof THA, Layne REC, McDonnell MM (1984) Response of peach seedlings to infection by the root lesion nematode *Pratylenchus penetrans* under controlled conditions. J Nematol 16:317–322

Pressey R, Avantes JK (1978) Differences in polygalactronase composition of clingstone and freestone peaches. J Food Sci 43:1415–1423

Pukanova ZG, Gatina ES, Sokolova SA (1980) Inheritance of resistance to powdery mildew in the F_2 of peach (in Russian). Ref Zhurnal 137:108–115

Quamme HA (1991) Application of thermal analysis to breeding fruit crops for increased fruit hardiness. HortScience 26:513–517

Quarta R, Dettori MT, Sartori A, Verde I (2000) Genetic linkage map and QTL analysis in peach. Acta Hortic 521:233–241

Rajapakse S, Belthoff LE, He G, Estanger AE, Scorza R, Verde I, Ballard RE, Baird WV, Callahan A, Monet R, Abbott AG (1995) Genetic linkage mapping in peach using morphological, RFLP and RAPD markers. Theor Appl Genet 90:503–510

Ramming DW (1985) In ovule embryo culture of early maturing *Prunus*. HortScience 20:419–420

Ramming DW (1990) The use of embryo culture in fruit breeding. HortScience 25:393–398

Ramming DW (1991) Genetic control of a slow-ripening fruit trait in nectarine. Can J Plant Sci 71:601–603

Rankovic M, Sutic D (1980) Investigation of peach as host of sharka (plum pox) virus. Acta Phytopathol Acad Sci Hungaricae 15:201–205

Ritchie DF, Werner DJ (1981) Susceptibility and inheritance of susceptibility to peach leaf curl in peach and nectarine cultivars. Plant Dis 65:731–734

Rodriquez AJ, Sherman WB. 1985. Relationships between parental, seed and seedling chilling requirements in peach and nectarine, *Prunus persica* (L.) Batsch. J Amer Soc Hort Sci 110:627–630

Rodriguez-A J, Sherman WB, Lyrene PM (1986) High-density nursery system for breeding peach and nectarine: a 10-year analysis. J Am Soc Hortic Sci 111:311–315

Rodriguez-A J, Sherman WB, Scorza R, Wisniewski M (1994) 'Evergreen' peach, its inheritance and dormant behavior. J Am Soc Hortic Sci 119:789–792

Ruperti B, Bonghi C, Rasori A, Ramina A, Tonutti P (2001) Characterization and expression of two members of peach 1-aminocyclopropane-1-carboxylate oxidases gene family. Physiol Plantarum 111:623–628

Ruperti B, Pagni S, Ramina A, Cattiveli L (2002) Ethylene responsive genes are differentially regulated during peach fruitlet abscission. Acta Hort 592:623–628. King JR,

Sansavini S, Gamberini A, Bassi D (2006) Peach breeding, genetics and new cultivar trends. Acta Hortic 713:23–48

Scorza R (1992) Evaluation of foreign peach and nectarine introductions in the U.S. for resistance to leaf curl [*Taphrina deformans* (Berk.) Tul.]. Fruit Varieties J 46:141–145

Scorza R (2001) Progress in tree fruit improvement through molecular genetics. HortScience 34:1129–1130

Scorza R, Bassi D, Dima A, Rizzo M (2000) Developing new peach tree growth habits for higher density plantings. The Compact Fruit Tree. 33:19–21.

Scorza R, Cordts JM, Mante S (1990) Long term somatic embryo production and plant regeneration from embryo derived peach callus. Acta Hortic 280:183–190

Scorza R, Mehlenbacher SA, Lightner GW (1985) Inbreeding and coancestry of freestone peach cultivars of the eastern United Sates and implications for peach germplasm improvement. J Am Soc Hortic Sci 110:547–552

Scorza R, Melnicenco L, Dang P, Abbott AG (2002) Testing a microsatellite marker for selection of columnar growth habit in peach [*Prunus persica* (L.) Batsch]. Acta Hortic 465:285–290

Scorza R, Okie WR (1990) Peaches (*Prunus*). In: Moore JN, Ballington, JR (eds). Genetic resources of temperate fruit and nut crops. Acta Hort 290, ISHS, Wageningen, pp 177–231

Scorza R, Pusey PL (1984) A wound-freeing inoculation technique for evaluating resistance to *Cytospora leucostoma* in young peach trees. Phytopath 74:569–572

Scorza R, Sherman WB (1996) Peaches. In: Janick J, Moore JN (eds) Fruit breeding, vol 1.: Tree and tropical fruits. John Wiley & Sons, NewYork, pp 325–440

Scott DH, Weinberger JH (1944) Inheritance of pollen sterility in some peach varieties. Proc Am Soc Hort Sci 45:229–232

Sharp RH (1961) Developing new peach varieties for Florida. Proc Fla State Hortic Soc 74:348–352

Sharp RH (1974) Breeding peach rootstocks for the southern United States. HortScience 9:362–363

Sharp RH, Hesse CO, Lownsberry BF, Perry VG, Hansen CJ (1970) Breeding peaches for root-knot nematode resistance. J Am Soc Hortic Sci 94:209–212

Sherman WB, Lyrene PM (1981) Bacterial leaf spot susceptibility in low chilling peaches. Fruit Var J 35:74–77

Sherman WB, Sharpe RH, Janick J (1973) The fruiting nursery: ultrahigh density for evaluation of blueberry and peach seedlings. HortScience 8:170–172

Shimada T, Yamamoto T, Hayama H, Yamaguchi M, Hayashi T (2000) A genetic linkage map constructed by using an interspecific cross between peach cultivars grown in Japan. J Jap Soc Hortic Sci 69:536–542

Simeone AM (1985) Study on peach and nectarine cultivars susceptibility to the main fungus and bacteria. Acta Hortic 173:541–551

Smigocki AC, Hammerschlag FA (1991) Regeneration of plants from peach embryo cells infected with a shooty mutant strain of *Agrobacterium*. J Am Soc Hortic Sci 116:1092–1097

Sosinski B, Gannavarapu M, Hager LD, Beck LE, King GJ, Ryder CD, Rajapakse S, Baird WV, Ballard RE, Abbott AG (2000) Characterization of microsatellite markers in peach [*Prunus persica* (L.) Batsch]. Theor Appl Genet 97:1034–1041

Stushnoff C (1972) Breeding and selection methods for cold hardiness in deciduous crops. HortScience 7:10–13

Surgiannides GD, Mainou AH (1985) Study of the susceptibility to plum pox virus of 33 varieties of peach and nectarine (in Greek). Georgike Ereuna 9:207–214

Testolin R, Marrazzo T, Cipriani G, Quarta R, Verde I, Dettori MT, Pancaldi M, Sansavini S (2000) Microsatellite DNA in peach (*Prunus persica* L. Batsch) and its use in fingerprinting and testing the genetic origin of cultivars. Genome 43:512–520

Thimmapuram J, Ko TS, Korban SS (2001) Characterization and expression of beta-1,3-glucanase genes in peach. Molec. Genet. Genomics 265:469–479

Trainotti L, Spolaore S, Ferrarese L, Casadoro G (1997) Characterization of ppEG1, a member of a multigene family which encodes endo-beta-1,4-glucanase in peach. Plant Mol Biol 34:791–802

Tukey HB (1934) Artificial culture methods for isolated embryos of deciduous fruits. Proc Am Soc Hortic Sci 32:313–322

Vendramin M, Dettori MT, Giovinazzi J, Micali S, Quarta R, Verde I (2007) A set of EST-SSRs isolated from peach fruit transcriptome and their transportability across *Prunus* species. Mol Ecol Notes 7:307–310

Verde I, Lauria M, Dettori MT, Vendramin M, Balconi C, Micali S, Wang Y, Marrazzo MT, Cipriani G, Hartings H, Testolin R, Abbott AG, Motto M, Quarta R (2005) Microsatellite and AFLP markers in the *Prunus persica* [L. (Batsch)] × *P. ferganensis* BC$_1$ linkage map: saturation and coverage improvement. Theor Appl Genet 111:1013–1021

Verde I, Quarta R, Cedrola C, Dettori MT (2002) QTL analysis of agronomic traits in a BC$_1$ peach population. Acta Hortic 592:291–297

Vileila-Morales EA, Sherman WB, Wilcox CJ, Andrews CP (1981) Inheritance of short fruit development period in peach. J Am Soc Hortic Sci 106:399–401

Viruel MA, Madur D, Dirlewanger E, Pascal T, Kervella J (1998) Mapping quantitative trait loci controlling peach leaf curl resistance. Acta Hortic 465:79–87

Wang Q, Zhang K, Qu X, Jia J, Shi J, Jin D, Wang B (2001) Construction and characterization of a bacterial artificial chromosome library of peach. Theor Appl Genet 103:1174–1179

Wang Y, Georgi LL, Reighard G, Scorza R, Abbott AG (2002a) Genetic mapping of the evergrowing gene in peach [*Prunus persica* (L.) Batsch]. J Hered 93:352–358

Wang Y, Georgi LL, Zhebentysyeva TN, Reighard G, Scorza R, Abbott AG (2002b) High-throughput targeted SSR marker development in peach (*Prunus persica*). Genome 45:319–328

Wang Z, Lu Z (1992) Advances of fruit breeding in China. HortScience 27:729–732

Warburton ML, Becerra-Velasquez VL, Goffreda JC, Bliss FA (1996) Utility of RAPD markers in identifying genetic linkages to genes of economic interest in peach. Theor Appl Genet 93:920–925

Weaver DJ, Doud SL, Wehunt EJ (1979) Evaluation of peach seedling rootstocks for susceptibility to bacterial canker, caused by *Pseudomonas syringae*. Plant Dis Rep 63:364–367

Weaver GM, Boyce HR (1965) Preliminary evidence of host resistance to the peach tree borer *Sanninoidea exitiosa*. Can J Plant Sci 45:293–294

Weinberger JH (1944) Characteristics of the progeny of certain peach varieties. Proc Am Soc Hortic Sci 45:233–238

Weinberger JH (1955) Peaches, apricots and almonds. Handbook der Pflanzenzrechtung 6:624–636

Weinberger JH, Marth PC, Scott DH (1943) Inheritance study of root-knot nematode resistance in certain peach varieties. Proc Am Soc Hortic Sci 42:321–325

Werner DJ, Cain DW (1985) Cages for protection of fruit tree hybridizations. HortScience 20:450–451

Werner DJ, Chaparro JX (2005) Genetic interaction of pillar and weeping peach genotypes. HortScience 40:18–20

Werner DJ, Creller MA (1997) Genetic studies in peach: Inheritance of sweet kernal and male sterility. J Am Soc Hort Sci 122:215–217

Werner DJ, Creller MA, Chaparro JX (1998) Inheritance of blood flesh in peach. HortScience 33:1243–1246

Werner DJ, Mowrey BD, Chaparro JX (1988) Variability in flower bud number among peach and nectarine cultivars. HortScience 23:578–580

Werner DJ, Ritchie DF, Cain DW, Zehr EI (1986) Susceptibility of peaches and nectarines, plant introductions, and other *Prunus* species to bacterial spot. HortScience 21:127–130

Wisniewski M, Arora R (1991) Adaptation and response of fruit trees to freezing temperatures. In: Biggs A (ed) Histology and cytology of fruit tree disorders. CRS Press, Boca Raton, FL, pp 299–320

Wisniewski M, Webb R, Balsamo R, Close TJ, Yu XM, Griffith M (1999) Purification, immunolo- calization, cryoprotective, and antifreeze activity of PCA60: A dehydin from peach (*Prunus persica*). Physiol Plantarum 105:600–608

Xu DH, Wahyuni S, Sato Y, Yamaguchi M, Tsunematsu H, Ban T (2006) Genetic diversity and relationships of Japanese peach (*Prunus persica* L.) cultivars revealed by AFLP and pedigree tracing. Genet Res Crop Evol 53:883–889

Yamamoto T, Mochida K, Imai T, Shi IZ, Ogiwara I, Hayashi T (2002) Microsatellite markers in peach [*Prunus persica* (L.) Batsch] derived from enriched genomic and cDNA libraries. Mol Ecol Notes 2:298–302

Yamamoto T, Shimada T, Imai T, Yaegaki T, Haji T, Matsuta N, Yamaguchi M, Hayashi T (2001) Characterization of morphological traits based on a genetic linkage map in peach. Breeding Science 51:271–278

Yamazaki K, Okabe M, Takahashi E (1987) Inheritance of some characteristics and breeding new hybrids in flowering peach. Bull Kanagawa Hortic Exp Sta 34:46–53

Yoshida M (1970) Genetical studies on the fruit quality of peach varieties, 1: acidity (in Japanese). Bull Hort Res Stat (Hiratsuka) 9:15

Yoshida M (1988) Peach cultivars, breeding and statistics in Japan. In: Childers NF, Sherman WB (eds) The Peach. Horticultural Publ., Gainesville, FL

Young RS (1987) West Virginia peach and nectarine fruit and vegetative bud injury and crop rating resulting from −28°C, January temperatures. Fruit Varieties J 41:68–72

Chapter 10
Pears

J.F. Hancock and G.A. Lobos

Abstract The most important commercial pear species grown are *Pyrus communis* and *P. pyrifolia*, although there are significant acreages of several other species. Climatic adaptation is a concern of all pear breeders, as well as fruit quality, season extension, compatibility with major pollenizer cultivars and disease resistance. Increasing fire blight resistance is an important goal in the eastern and southern parts of North America, and many regions of Europe. In the breeding of pear rootstocks, the common goal is to develop rootstocks that induce size control and precocity in the scion cultivar. Resistance to fire blight is quantitatively inherited in an additive fashion, with a few major genes playing an important role. Genes have also been identified for semidwarf or compact cultivars, short-internode dwarfs and short internode compact pears. Varying amounts are known about the genetics of fruit development and quality. Molecular studies have been conducted on the expression patterns of genes during fruit ripening and storage. Several genetic maps have been developed of pear and DNA markers have been linked to a number of resistance genes including black spot disease, scab and fire blight resistance. Transformation strategies have been employed to generate herbicide and disease resistant plants.

10.1 Introduction

Pears are grown in all temperate regions of the world. Culivars of the European pear, *Pyrus communis*, predominate in Europe, North America, South America, Africa and Australia, and the sand or Japanese pear, *P. pyrifolia,* is the main cultivated species in southern and central China, Japan and Southeast Asia. Other pears grown widely in Asia include *P. ussuriensis* (Ussuri pear) and hybrids of *P. pyrifolia* and *P. ussuriensis*. Interest in Asian pears, primarily Japanese cultivars, continues to increase in Western Europe, North America, New Zealand and Australia, but the European pear has made little impact in Asia, except in Northern Japan, where most

J.F. Hancock
Department of Horticulture, 342C Plant and Soil Sciences Building, Michigan State University, East Lansing, Michigan 48824, USA
e-mail: hancock@msu.edu

J.F. Hancock (ed.), *Temperate Fruit Crop Breeding*,
© Springer Science+Business Media B.V. 2008

Asian pears do not have sufficient winter hardiness. Pears are eaten fresh, cooked, dried or made into a fermented cider-like beverage called 'perry'. They are also processed as canned halves, diced pieces for fruit cocktail and as puree for baby food.

The world production of pears is second only to apples among the deciduous tree fruits. About 19.2 million metric tons of pears were produced in 2005 (FAOSTAT, 2007), up over 2 million tons from 2004. Asia produced the most pears (13.5 million t), followed by Europe (3.2 million t), South America (0.8 million t), North America (0.8 million t), Africa (0.7 million t), and Oceania (0.2 million t). In Asia, China was the largest producer with 60% of the world volume, followed by Japan (2.1%) and Republic of Korea (2.0%). The major producers in Europe were Italy (4.9%), Spain (3.4%) and France (1.2%). In South America, the highest producers were Argentina (2.7%) and Chile (1.1%). In North America, the leading producer by far was the U.S.A. (3.9%), where the main state is Washington with 50% of the countries volume (California and Oregon complete the other half). In Africa, the major producer was the Republic of South Africa (1.8%). Production in Oceania was highest in Australia (0.8%).

Pyrus is genetically quite diverse, with considerable variability in morphology and physiological adaptations (Fig. 10.1) (Knight 1963, Westwood 1982, Lombard and Westwood 1987, Bell et al. 1996). European and Asian pear breeders have utilized this variability to develop high quality cultivars with large size and attractive appearance that are well adapted to local conditions. The European pears are distinguished by their juiciness, delicate flavor and aroma, while the Oriental (Asian or nashi) pears are known for their crispness and sweet flavor. North American breeders have had to focus more on disease resistance and cold hardiness than the Europeans,

Fig. 10.1 Diversity in fruit of *Pyrus* species (Picture by Joseph Postman, USDA-ARS National Clonal Repository in Corvallis, Oregon

although the spread of the bacterial disease fire blight [*Erwinia amylovora* (Burrill) Winslow et al.] throughout Europe is forcing the breeders there to concentrate more on disease resistance (Bell et al. 1996).

10.2 Evolutionary Biology and Germplasm Resources

The genus *Pyrus* is in the subfamily Pomoideae of the Rosaceae. All species of *Pyrus* are diploid and interfertile (Westwood and Bjornstad 1971, Bell and Hough 1986). The genus contains 24 primary species (Table 10.1), up to six natural interspecific hybrids and at least three artificial hybrids (Bell et al. 1996). The species of *Pyrus* are located in Europe, temperate Asia and the mountains of North Africa. The species boundaries are blurred in many instances, resulting in different species designations by some authorities.

Pomoideae are unique in the Rosaceae by having a basic chromosome number of 17 compared to 7–9 for the other subfamilies. The origin of the subfamily may have occurred when two primitive forms of Rosaceae successfully hybridized, one having a basic chromosome number of 8 and the other 9 (Sax 1931, Zielinski and Thompson 1967). These could have been members of the Prunoideae and Spiraeoideae. All the species of *Pyrus* have a chromosome number of $2x = 34$, except for a few higher polyploid cultivars of *P. communis*. The genus *Pyrus* probably arose during the Tertiary period in the mountains of western China and the early taxa likely dispersed east and west through the mountain chains. Speciation may have been associated with geographical isolation of populations in the mountain ranges (Rubtsov 1944, Zeven and Zhukovsky 1975).

A number of pear species are grown commercially (Bell et al. 1996). As mentioned previously, *Pyrus communis* (common pear) is the most important pear in Europe, North America, South America, Africa, and Australia. *Pyrus nivalis*, the snow pear, is grown locally in Europe to make perry and hybrids of *P. communis* and *P. pyrifolia* are grown in North America for processing. The sand pear, *P. pyrifolia*, is the primary pear species cultivated in southern and central China and in Japan, with hectarage of the Ussuri pear, *P. ussuriensis*, hybrids of *P. pyrifolia* and *P. ussuriensis*, and the Chinese white pear (*P. ×bretschneideri*) also being found in northern China and Japan. *Pyrus pashia* (Pashia pear) is grown in southern China and northern India. Several species are used for rootstocks including *P.betulifolia*, *P. calleryana*, *P. pyrifolia*, *P. ussuriensis*, and *P. communis* in Europe, North America and eastern Asia, and *P. pyraster*, *P. amygdaliformis* and *P. elaeagrifolia* in Asia Minor and central Asia. The small fruited species, *P. calleryana*, *P. fauriei*, *P. betulifolia*, *P. salicifolia*, and *P. kawakamii* are grown as ornamentals.

The fruit of the Asian pears are notable in that they can be eaten right after harvest, unlike the European pears. They are sweet and juicy and tend to have crisp-textured flesh, but do not have the smooth, buttery texture of European pears, tend to have less aroma, and some have abundant stone cells (Bell et al. 1996). The Chinese or Japanese sand pear (*P. pyrifolia*) is well adapted to warm climates and is very

Table 10.1 *Pyrus* species of the world

Scientific name	Synonyms	Distribution
European		
P. communis L.	*P. asiae-mediae* Popov	West to southeast Europe,
	P. balansae Decne.	Turkey
	P. boissieriana Buhse	
	P. caucasica Fed	
	P. elata Rubtzov	
	P. medvedevii Rubtzov	
P. korshinskyi Litv.	*P. pyraster* (L.) Burgsd.	South central Asia,
		Afghanistan
P. nivalis Jacq.		West, Central and Southern
		Europe
P.cordata Desv.		S.W. England, W. France,
		Spain, Portugal
P. ×*salviifolia* DC		Europe, Crimea
Circum-Mediterranean		
P. amygdaliformis Vill	*P. sinaica* Dum. Cours.	Mediterranean Europe, Asia
		minor
P. *complexa* Rubtzov		Caucasus
P. elaeagrifolia Pall.	*P. kotschyana* Boiss. ex Decne.	S.E. Europe, Russia, Turkey
P. syriaca Boiss.		Tunisia
P. longipes Coss. & Dur.		Algeria
P. gharbiana Trab.		Morocco, W. Algeria
P. mamorensis Trab.		Morocco
Mid-Asian		
P. glabra Boiss.		Iran
P. salicifolia Pall.		N.W. Iran, N.E. Turkey,
		South Russia
P. regelii Rehd.	*P. heterophylla* Regal &	South Central Asia
	Schmalh.	
P. pashia Buch.-Ham.	*P. kumaoni* Decne.	Pakistan, India, Nepal
ex D. Don.	*P. variolosa* Wall. ex G. Don.	
	P. wilhelmii C.K. Schneid.	
East Asian		
P. ×*bretschneideri*		Northern China
Rehd.		
P. ×*phaeocarpa* Rehd.		Northern China
P. pyrifolia (Burm.)Nak	*P. serotina* Rehd.	China, Japan, Korea
P. ×*serrulata* Rehd.		Central China
P. pseudopashia T.T. Yu	*P. kansuensis* Batalin	N.W. China
P. ussuriensis Maxim	*P. lindleyi* Rehd.	Siberia, Manchuria,
	P. ovoidea Rehd.	N. China, Korea
	P. sinensis Lindley	
P. calleryana Decne.		Central & S. China,
		Vietnam
P. betulifolia Bunge		Central & N. China,
		S. Manchuria
P. fauriei C.K. Scheid.		Korea
P. hondoensis Kik. &		Japan
Nak		
P. dimorphophylla Mak.		Japan
P. kawakamii Hayata	*P. koehnei* C.K. Schneid.	Taiwan, S.E. China

Adapted from Bell et al. 1996

widespread with hundreds of cultivars grown. The fruit vary in size from very small to very large and they are sweet, juicy and gritty, with a russeted skin. *Pyrus pashia* is well adapted to very hot, humid climates, but has only mediocre fruit quality compared to the other species.

The Ussurian pear, *P. ussuriensis,* is the most cold hardy Asian pear and is grown in Northern Japan where *P. pyrifolia* is susceptible to winter injury. It has small, globose medium-sized fruit with persistent calyxes. They are bland flavored but not too gritty and some have melting flesh (Shen 1980). The Chinese white pear, *P. ×bretschneideri* is the second most hardy Asiatic type. Their fruits are of medium size, range in shape from pyriform to obovate and have the most pleasing texture and flavor of the Oriental pears.

It is likely that interspecific hybridization played an important role in pear domestication. A number of species have been implicated as being in the background *P. communis* including wild populations of *P. communis* var. *pyraster, P. caucasica* and *P. nivalis* (Challice and Westwood 1973). Rubtsov (1944) felt that modern cultivars of *P. communis* had characteristics derived from at least three species, *P. elaeagrifolia*, *P. salicifolia* and *P. syriaca.* Cultivars grown in northern China may belong to a hybrid complex involving *P. ussuriensis* and *P. pyrifolia* (Bell et al. 1996). The origin of *P. ×bretschneideri* likely involved the hybridization of *P. betulnefolia* and *P. ussuriensis* or *P. pyrifolia* (Kikuchi 1946). The most recent molecular data suggests that *P.×bretschneideri* is closely related to *P. pyrifolia* and *P. ussuriensis,* and is likely a variety or subspecies of *P. pyrifolia* (Yamamoto et al. 2002a, Bao et al. 2007).

Vavilov et al. (1951) identified three centers of diversity for cultivated pears: (1) A Chinese center, where *P. pyrifolia* and *P. ussuriensis* are found. The primitive species, *P. calleryana*, is located in this center (Challice and Westwood 1973, Pu et al. 1986). (2) A central Asiatic center (northwest India, Afghanistan, Tadjikistan and Uzbekistan), where *P. communis* and hybrids of *P. communis* and *P.×bretschneideri* are found (Yu and Zhang 1979). (3) A Near Eastern center (Caucasus Mountains and Asia Minor), where *P. communis* is also grown. It is in the Near Eastern center that *P. communis* may have been first domesticated.

The USDA-ARS National Plant Germplasm System (NPGS) clonal repository at Corvallis, maintains an extensive and diverse pear germplasm collection that includes historic varieties, landrace varieties, and wild species (more than 2,200 different individuals in total). This collection includes genotypes with desirable traits important for crop improvement, such as resistance to fire blight, pear psylla, Fabraea leaf spot, and pear scab.

10.3 History of Improvement

Originating in the Caucasus region (Southeastern Europe between Black and Caspian seas), pears have been cultivated for at least 3,000 years in Asia (Kikuchi 1946). Indo-European tribes spread the pear as they migrated into Europe and

Northern India. The first mention of pears in the written record was made by Homer in about 1000 BC, when he referred to them as one of the 'gifts of the Gods' (Hedrick et al. 1921). According to Theophrastus (371–286 BC), pear culture was common in ancient Greece, where cultivars were propagated by grafting and cuttings. The Roman, Cato (235–150 BC), described pear cultural methods that are very similar to the techniques practiced today. By the time of Pliny the Elder (23–79 BC), at least 35 cultivars of pear were being grown in Rome, while only three types of apples were noted. The range in fruit characters of these ancient cultivars were similar to those grown today. By the end of the Sung Dynasty (China, AD 1279), over one hundred varieties of pear existed.

In the Middle Ages, pears were grown widely in central and western Europe (Bell et al. 1996). Cordus (1515–1544) described pear cultivars in Germany that had all the fruit characters possessed by modern cultivars, with the exception of buttery texture. France was the leading pear-producing country in the sixteenth and seventeenth centuries, and were most active in the development of new cultivars. In 1628, the amateur fruit collector Le Lectier had 254 pear cultivars in his garden and by the early 1800s there were over 900 cultivars of pears growing in France. Most of these were crisp-fleshed.

In the eighteenth century, Belgium became the center of pear culture and improvement, primarily through the efforts of Nicolas Hardenpont and Jean Baptiste Van Mons (Bell et al. 1996). The Belgian breeders developed the first cultivars that had a melting buttery flesh, and some of their cultivars are still important today, including 'Beurre Bosc', 'Beurre d'Anjou' and 'Winter Nelis'.

The pear was not native to England, but commercial culture probably arose there by 1200. It is not known when the pear was introduced into England, but it is likely to have been before the Roman conquest (Bell et al. 1996). The English became expert pear breeders in the 1800s, evidenced by an 1826 catalog of the Royal Society of London listing 622 cultivars. The most important cultivar now grown in the world, 'Williams Bon Chretien' ('Bartlett'), was identified around 1796 (Fig. 10.2) (Hooker 1818). Another 19th century selection named 'Conference', is also still planted.

Only two species were bred in Europe, *Pyrus communis* (the common European pear), and *P. nivalis* (the perry pear). Most of the improved types were derived from open-pollinated seedlings of existing cultivars, although Knight in England was using controlled hybridization around 1800 (Bell et al. 1996). The father of modern genetics, Gregor Mendel, was also hybridizing pears at Brno in the late nineteenth century to develop late-ripening cultivars with superior flesh quality (Vavra and Orel 1971).

Domestication of the Chinese pear species began about 3300 years ago (Kikuchi 1946) and commercial orchards have existed in China for more than 2000 years (Pieniazek 1966). However, pear breeding did not begin in China until 1956. Pears were cultivated as early as the eighth century in Japan (Kajiura 1966), with large plantings not appearing until 1868. Kikuchi began the first breeding program in Japan in 1915, which is still active (Kanato et al. 1982).

Fig. 10.2 The 'Bartlett' pear (watercolor published in The Fruits of New York, U.P. Hedrick et al., 1921)

Pears were introduced to North America by the early French and English settlers; the first reference to pear culture was made in New England in 1629 (Hedrick et al. 1921). Pears were introduced into the west Coast by the Franciscan monks, and by 1800, pear cultivation extended from British Columbia to southern California. In the 17th and 18th centuries, only *P. communis* was grown in North America and it was not until the early 1800s that *P. pyrifolia* arrived at the west coast of the United States via Chinese immigrants. The first hybrid of these two species, 'Le Conte', appeared in 1846, followed by 'Kieffer' (1873) and 'Garber' (1880) (Hedrick et al. 1921). The fruit of these hybrids was not of as high quality as the existing European cultivars, but the trees were more resistant to fire blight. These early hybrids were likely accidental, but controlled crosses were soon being made in hopes of combining high fruit quality with blight resistance.

Cold hardy cultivars of *P. communis* were introduced to North America from Russia in 1879. These trees had poor fruit quality and were susceptible to fire blight, but they proved to be an excellent source of cold hardiness (Magness 1937). Additional sources of cold hardiness came with the introduction of *P. ussuriensis* into Iowa by Patten around 1867. Chance hybrids of *P. ussuriensis* × *P. communis* proved to be very cold hardy. Reimer (1925) introduced a number of European and Oriental cultivars to North America in the early 1900s, and some of the Oriental introductions had superior blight resistance. This stimulated the U.S. Department

of Agriculture and several State Agricultural experiment stations to begin actively
breeding for fire blight resistance (Bell et al. 1996).

In general, pear breeding is only a recent activity in South America, Australia,
New Zealand and Africa. However, 'Packham's Triumph', was introduced commer-
cially in Australia around 1900 and is still important there and in New Zealand
(Lombard and Westwood 1987).

10.4 Current Breeding Efforts

Climatic adaptation is a concern of all pear breeders. Cold hardiness is of paramount
importance in the more temperate and most northern production regions, while in
the warmer regions the emphasis is on drought and heat tolerance. A goal of most
pear breeders is to generate a series of cultivars that produce a continuous supply of
quality fruit throughout the season. Compatibility with major pollenizer cultivars is
important everywhere.

Disease resistance is also very important to most pear breeders. Increasing fire
blight resistance is a major goal in the eastern and southern parts of North America,
and many regions of Europe. Resistance is also sought in many areas to pear psylla
[*Cacopsylla pyricola* (Foerster)], powdery mildew [*Podosphaera leucotricha* (Ellis
& Everh.) E.S. Salmon], leaf spot and scab (*Venturia pirina* Aderh.). In Japan, resistance to black spot of Asian pear (*Alternaria alternata* (Fr.) Keissler) is particularly
important, as well as resistance to the Asian pear scab (*Venturia nashicola* Tan. &
Yan.) and pear rust (*Gymnosporangium asiaticum* Miyabe ex Yamada).

All pear breeders seek high fruit quality, although what constitutes good quality
varies by species and location, as was previously mentioned. A soft, buttery flesh is
desired in European pears along with aromatic flavor and low fiber. A crisp, breaking
flesh is preferred in Oriental pears, as well as a sweet flavor and lack of grittiness.
In European pears, the ideal fruit is considered large (7 cm long and 6 cm wide)
with a pyriform shape. The skin can be a wide range of colors from golden yellow
with or without a red blush, green or greenish-yellow and even bright red. Russet
free skin is generally preferred, although russeting that is uniform and smooth is
acceptable. Fruit are sought that are resistant to bruising, white fleshed and contain limited quantities of stone cells. In Asian pears, an even larger size is desired
(up to 10 cm wide), along with a regular, round shape (Kanato et al. 1982). The
skin can be yellow or a light green, glossy and uniformly covered with a golden,
smooth russet. In both European and Oriental pears, uniformity of ripening, a long
postharvest storage life and a low susceptibility to physiological disorders are all
important.

A trend that is of increasing importance is the ability of a cultivar to be harvested
mechanically for both fresh markets and processing. Pear cultivars adapted to this
approach need to produce fruit of uniform size and maturity that separates easily on
shaking, and has a tough skin that resists bruising (Bell et al. 1996). It is likely that

trees adapted to mechanical harvesting will also need to be adapted to trellises or other support systems, and have trunks and limbs that can withstand the vibrations associated with mechanical shaking.

In the breeding of pear rootstocks, the common goal is to develop rootstocks that induce size control and precocity in the scion cultivar. They should also be compatible, winter hardy, and disease resistant. Adaptation to the specific climatic conditions of each production region is also very important. A reduction in root suckering contributes to decreased risk of infections and herbicide damage. Because seedlings rootstocks are derived from parent cultivars which are highly heterozygous and self-incompatible, the rootstocks are genetically not identical, although they are usually uniform in all important characteristics. Japanese and Chinese breeding programs have focused on reducing the physiological disorder 'black end'. In Europe, a strong emphasis has been placed on improved cold hardiness in *Pyrus* and *Cydonia*, along with adaptation to iron chlorosis in quince (Chevreau and Bell 2004). Several disease problems of rootstocks receive primary attention in Europe and North America including fire blight and the pear decline phytoplasma.

Much effort was devoted in the past on improving quince rootstocks, but more energy is now being devoted to developing dwarfing *Pyrus* rootstocks. There is less incompatibility encountered between intrageneric than intergeneric grafts, and some of the *Pyrus* species are more winter hardy, disease resistant, drought tolerant and have better anchorage than the quince rootstocks (Bell et al. 1996). Only a modest amount of root stock breeding is being conducted in the United States, but active programs exist in England, France, Italy, Sweden, the Soviet Union and Romania.

Numerous pear breeding programs are found across the world. Some of the largest programs in North America are those of the U.S. Department of Agriculture at Kearneysville, West Virginia and Agriculture and Agri-Food Canada at Harrow, Ontario. At least three private companies are also producing pear cultivars in the U.S.A. Some of the most important programs in Europe are at the Institute National de la Recherche Agronomique at Angers and Dax, France, the Instituto Sperimentale per la Frutticoltura in Rome and Forli, Italy, and the East Malling Research Station in England. Other pear breeding programs are found in Norway, Sweden and Romania.

In Asia, major programs exist in China at Xingcheng (Liaoning Province), concentrating on *P.×bretschneideri*, and Zhengzhou (Henan Province) and Hangzhou (Zhejiang Province), focusing on *P. pyrifolia* (Wang 1990). Breeding is also being done in Korea at the National Horticultural Research Institution (Kim and Ko 1991, 1992) and in Taiwan (Hsu and Lin 1987). In Japan, the largest breeding program is at Kikuchi (Kanato et al. 1982).

In the southern hemisphere, only a few breeding programs exist. A major program was established in 1983 at HortResearch in New Zealand that focuses on fresh fruit for export to the Northern Hemisphere. In the 1990s, breeding work was also begun in South Africa (Human 2005) and Brazil (Barbosa et al. 2007).

10.5 Genetics of Economically Important Traits

10.5.1 Pest and Disease Resistance

Fire blight is the most important disease of pears in North America and Western Europe. In fact, commercial pear growing has been largely abandoned in the warm humid regions of the southeastern, southern, and central United States because of the severity of this problem. The most widely grown cultivars are generally susceptible, but several new varieties have been released that have improved resistance (Quamme and Spearman 1983, Bell et al. 2002, Hunter et al. 2002a,b). 'Seckel' and 'Old Home' were the original sources of resistance. Resistance is quantitatively inherited in an additive fashion, with a few major genes playing an important role (Decourtye 1967, Quamme et al. 1990). The highest levels of resistance are found in the Asian species *P. calleryana* and *P. ussuriensis* (Table 10.2).

Identifying resistance to fire blight has been problematic due to a number of factors, including (1) age, vigor, succulence, and kind of tissue infected; (2) temperature and humidity relationships during the pre- and post infection period; (3) inoculum purity and concentration; (4) method of inoculation, and (5) virulence of isolates (Bell et al. 1996). A useful rating system for estimating blight resistance in the field was devised by van der Zwet et al. (1970).

Leaf blight caused by the fungus *Fabraea maculata* Atk. is a widespread problem in pears, along with Pseudomonas blight (also called bacterial canker and blossom blast) caused by *Pseudomonas syringae* pv. *syringae* van Hall. Leaf blight resistant genotypes are relatively common in a number of *Pyrus* species, while resistance to Pseudomonas blight is limited in European and Asian pears (Bell et al. 1996). Red-skinned mutants of 'Beurre d'Anjou' have been reported to be more resistant to Pseudomonas blight than green-skinned ones (Whitesides and Spotts 1991).

Two species of *Venturia* cause widespread scab diseases in pears: (1) *V. pirina* Aderh, which is particularly important in Europe, but is found everywhere *P. communis* is grown, and (2) *V. nashicola* Tanaka & Yamamoto which infects Asian pears throughout their range. Resistant genotypes have been found to *V. pirina* in *P. communis*, *P. pyrifolia*, *P. ussuriensis* and *P. nivalis* (Kovalev 1963, Westwood 1982, Bell 1991) and resistant cultivars have been released (Fisher and Mildenberger 2000, Bellini et al. 2000). However, resistance can be variable regionally due to the existence of multiple fungal biotypes (Shabi et al. 1973, Vondracek 1982). Few major Asian pear cultivars are resistant to *V. nashicola*, although a number of minor ones carry resistance (Kanato et al. 1982).

Other locally important diseases that effect pears are black spot [Alternaria alternata (Fr.) Keissler], Monilia fruit rot [*Monilinia fructicola* (Wint.) Honey], Fabraea leaf spot (*Fabraea maculata* Atk.), powdery mildew [*Podosphaera leucotricha* (Ell. & Ev.) Salm.], Asiatic pear rust [*Gymnosporangium haraeanum* Syd.], along with several virus diseases including pear bud drop, stony pit and ring spot mosaic. Black spot resistant cultivars have been released (Kanato et al. 1982, Kajiura 1992) and resistance sources have been identified to all of the other diseases listed above (Bell et al. 1996, Bell and van der Zwet 2005, Serdani et al. 2006).

Table 10.2 Useful horticultural traits carried by native pear species (from Bell 1991)

Scientific name	Useful characteristics
European	
P. communis L.	Resistance to fire blight, white spot, pear scab, pear decline, pear psylla; cold hardy
P. nivalis Jacq.	Resistance to pear psylla; cold hardy
P. cordata Desv.	Resistance to codling moth; adaptation to warm winters, high pH, drought
Circum-Mediterranean	
P. amygdaliformis Vill	Resistance to codling moth; adaptation to warm winters, high pH, drought and clay soils
P. elaeagrifolia Pall.	Resistance to codling moth; cold hardy
P. syriaca Boiss.	Adaptation to drought
P. longipes Coss. & Dur.	Cold hardy; adaptation to wet and dry soils
P. gharbiana Trab.	Adaptation to drought
P. mamorensis Trab.	Adaptation to low pH
Mid-Asian	
P. glabra Boiss.	
P. salicifolia Pall.	Adaptation to drought
P. regelii Rehd.	Resistance to pear psylla
P. pashia Buch.-Ham. ex D. Don.	Adaptation to warm winters, low pH, clay soils
East Asian	
P. pyrifolia (Burn.) Nak	Resistance to pear scab
P. pseudopashia T.T. Yu	Resistance to codling moth; winter hardy
P. ussuriensis Maxim	Resistance to fire blight, pear psylla; winter hardy
P. calleryana Decne.	Resistance to fire blight, Fabraea leaf spot, pear psylla, codling moth; adaptation to warm winters, low pH, wet, dry and clay soils
P. betulifolia Bunge	Resistance to pear psylla, codling moth; adaptation to warm winters, low pH, wet, dry and clay soils
P. fauriei C.K. Scheid.	Resistance to fire blight, pear psylla, codling moth; adaptation to warm and cold winters, low pH, wet and clay soils
P. hondoensis Kik. & Nak	Resistance to fire blight
P. dimorphophylla Mak.	Resistance to powdery mildew, pear psylla, codling moth; adaptation to low pH and wet soils
P. kawakamii Hayata	Resistance to codling moth; adaptation to warm winters

The phytoplasma induced disease pear decline is a very widespread problem. It is vectored by the pear psylla, *Cacopsylla* spp. (Hibino and Schneider 1970), with *C. pyricola* Förster being the causative species in North America, and *C. pyri* L. and *C. pyrisuga* Förster being the vectors in Europe. Pear species vary widely in their resistance to pear psylla, with the east Asiatic species being more resistant than those from Asia Minor and Europe (Bell et al. 1996). Improved resistance to pear psylla has been developed in Romania (Braniste 2002).

A number of insect pests cause problems in pear orchards (Table 10.3). The woolly pear aphid (*Eriosoma pyricola* Baker & Davidson) is a common pest of pear trees in the nursery and young orchards, particularly in Oregon. Resistance to this aphid is widespread in *Pyrus*, with at least eight species carrying

Table 10.3 Genetics of pest and disease resistance in pear

Pest or disease	Observations and sources
Bacterial	
Fire blight	Resistance in *P. communis* is rare but exists (van der Zwet and Keil 1979,
Erwinia amylovora	Quamme and Spearman 1983, Thibault et al. 1987, van der Zwet and
	Bell 1990, Hevesi et al. 2004), High levels of resistance are present in
	P. calleryana and *P. ussuriensis*, but not immunity (Hartman 1957),
	Resistance is quantitatively inherited, although major genes have been
	implicated (Layne 1968, Thompson et al. 1975, Bell et al. 1977),
	Additive effects predominate (Quamme et al. 1990)
Fungi	
Black spot	Many sources of resistance exist (Hiroe et al. 1958, Kanato et al. 1982),
Alternaria alternata	Susceptibility is controlled by a single dominant gene (Kozaki 1973)
Leaf blight	
Pseudomonas syringae	Resistance has been identified in most *Pyrus* species (Bell et al. 1996)
Pear scab	
Venturia pirina	Resistant genotypes have been found in *P. communis*, *P. pyrifolia*,
	P. ussuriensis, *P. nivalis* (Kovalev 1963, Westwood 1982, Bell 1991),
	Resistance varies regionally due to the existence of multiple fungal
	biotypes (Shabi et al. 1973, Vondracek 1982)
Venturia nashicola	The major cultivars are not resistant, but many minor cultivars of many
	species carry resistance (Kanato et al. 1982)
Pear rust	
Gymnosporangium	Resistance has been identified (Kanato et al. 1982)
haraeanum	
Powdery mildew	Resistance has been identified (Fisher 1922, Kanato et al. 1982,
Podosphaera	Westwood 1982, Serdani et al. 2006)
leucotricha	
Pseudomonas blight	Most European and Asian pears are susceptible, red-skinned mutants
Pseudomonas syringae	of 'Beurre d'Anjou' are less susceptible than green-skinned ones
	(Whitesides and Spotts 1991)
Phytoplasma	
Pear decline	*Pyrus communis* is generally tolerant, *P. betulifolia* is very tolerant,
	while *P. pyrifolia*, *P. ussuriensis* and *P. calleryana* are susceptible
	(Lombard and Westwood 1987), Resistance is inherited additively
	(Westwood 1976)
Insects	
Pear psylla	Resistance is common in *P. betulifolia*, *P. calleryana*, *P. fauriei*,
Cacopsylla spp.	*P. ussuriensis*, *P.×bretschneideri* and *P. communis × P. ussuriensis*
	hybrids (Westigard et al. 1970, Quamme 1984, Bell 1991), resistant
	genotypes have been identified in *P. nivalis* and *P. communis* (Bell and
	Stuart 1990, Bell 1992), resistance was positively correlated with large
	fruit size in progeny of *P. communis × P. ussuriensis* (Harris and
	Lamb 1973)
Pear sawfly	Variation in resistance exists (Shaw et al. 2004)
Caliroa cerasi	
Wooly pear aphid	Most pear species are resistant or have variable resistance (Westwood and
Eriosoma pyricola	Westigard 1969), Immunity has been found in two Indian rootstocks of
	P. pashia (Khan 1955)
Nematodes	
Root-knot	Resistant genotypes of *P. communis* have been reported (Tufts and
Meloidogyne spp.	Day 1934)

resistance (Westwood and Westigard 1969). Other insect pests that periodically cause problems are the pear leaf blister mite (*Eriophyes pyri* Pagenstecher), pear sawfly (*Caliroa cerasi* L.) and the root-knot nematode (*Meloidogyne* spp.). There are no reports of resistance to these pests, except pear sawfly (Shaw et al. 2003, 2004).

10.5.2 Morphological and Physiological Traits

Improving winter cold hardiness has played a key role in expanding the range of pear growing into northern regions of North America, Europe, and Asia. The very cold hardy species *P. ussuriensis* from northern Russia has proven extremely useful in the development of cold hardy cultivars for the northern U.S.A. and Canada (Ronald and Temmerson 1982, Stushnoff and Garley 1982, Luby et al. 1987, Peterson and Waples 1988), as well as Europe (Ludin 1942, Zavoronkov 1960, Sansavini 1967). These cold hardy hybrid types do not have as high quality as the cultivars grown in the main pear-growing regions, but they can withstand winter temperatures as low as −30 to −40°C and have much higher fruit quality than their *P. ussuriensis* parents. Inheritance of cold hardiness has not been studied in pear, but it is likely to be similar to apple where inheritance is quantitative and mostly additive (Bell et al. 1996).

Resistance of blossoms to spring frost is also an important goal in cold climates. Two approaches have been taken to solving this problem: (1) breeding for a late blooming periods, and (2) identifying genotypes with direct resistance of blossoms to frost. Bloom date has been shown to be highly heritable (Anjou 1954) and late flowering does not appear to be strongly correlated with late ripening (Baldini 1949). Date of leaf break may be used to predict flowering date (Bell et al. 1996). A large variation in frost tolerance of blossoms has been identified, but parental and progeny assessment of frost resistance has proven to be difficult due to yearly microclimate differences, variations in the impact of fruit set reductions (Perraudin 1955) and cultivar differences in parthenocarpic fruit set (Simovski et al. 1968).

Reducing the chilling requirement of *P. communis* is a goal of some southern U.S.A. breeding programs to expand the range of cultivation. A number of interspecific *P. communis* × *P. pyrifolia* hybrids have been released with low chilling requirement by the University of Florida. Efforts have also been taken to reduce the chilling requirement of Asian pears at numerous locations in S.E. Asia, China and India. Detailed information about the chilling hour requirements of pear varieties can be found in Spiegel-Roy and Alston (1979) and Ghariani and Stebbins (1994). To our knowledge, no studies have been conducted on the genetics of chilling requirement in pear.

Drought resistance of scions and rootstocks can be critical when pears are grown in arid to semiarid conditions. Pruss and Eremeev (1969) found considerable variability in the drought tolerance of 99 pear cultivars of Russian, European,

Australian, and American origin. Pears grafted onto *P. salicifolia* performed much better in dry soils than those on *P. communis* rootstocks and had higher tolerance to extreme temperature changes and alkaline soils. They were also more resistant to pear scab and woolly aphid (Kuznetzov 1941).

Dwarf fruit trees are now favored to standard types, as they are easier to manage and yield more per area due to more efficient light interception (Tukey 1964). However, there are no completely dwarf pear cultivars, and potential sources of small size are much harder to find in pears than in apples. Quince rootstocks are sometimes used to control pear plant size, but the adaptability of these rootstocks is too limited for widespread use (Bell et al. 1996).

There are several possible approaches to reducing pear tree size (Table 10.4). Semidwarf or compact cultivars of *P. communis* have been identified (Tuz 1972) and when they are crossed with standard types, there appears to be dominance for the compact form. Decourtye (1967) identified a short-internode dwarf, 'La Nain Vert', that is under single dominant gene control. This gene has been used in U.S.A. and Italian breeding programs, but the dwarf parent has poor fruit quality and a very slow growth rate. Jingxian et al. (1988) identified short internode compact pears from seedlings of 'Jin-xiang' that are also under single gene control. Few genetically dwarfed scion cultivars have been released to date (Bellini et al. 2000), but active efforts continue in Italy and France to reduce stature (Rivalta et al. 2002, Chevreau and Bell 2006).

Precocity is very important in pears, both to the breeder who is anxious to evaluate the fruit and the producer who wants a return on his planting investment as soon as possible. The length of the juvenile period is heritable and under additive genetic control (Zielinski 1963, Visser 1967, 1976, Zimmerman 1976, Bell and Zimmerman 1990). Li et al. (1981) found that the amount of linear growth to the first flower ('juvenile span') was positively correlated with the length of the juvenile period and was heritable. Bell and Zimmerman (1990) found that in a population of inter-specific hybrid origin, the juvenile period was more dependent on the individual genotype than species pedigree. Seedlings of *P. pyrifolia* are more precocious than *P. ×bretschneideri* and *P. communis* (Zhejiang Agricultural University 1978).

A number of characters may be useful in the early identification of precocity. Bell et al. (1996) suggested that juvenile seedlings can be identified by the presence of thorns, irregular leaf margins and wide-angled branches. Seedlings with short juvenile periods lose these features quickly. Vigor, measured as stem diameter, is not an accurate predictor of the length of the juvenile period (Zimmerman 1977, Shen et al. 1982).

While there is a need for pear cultivars that ripen all across the season, early ones are especially important for the fresh market. In *P. communis,* heritability for date of ripening was calculated to be 0.49 (Thibault et al. 1988), while in *P. pyrifolia* a value of 0.33 was estimated (Zhejiang Agricultural University 1977). Bell et al. (1996) suggested that to breed most effectively for earliness or lateness, both parents should be either early or late.

Table 10.4 Genetics of adaptation, productivity, plant habit and fruit quality in pear

Attribute	Observations and sources
Adaptation	
Cold hardiness	Quantitatively inherited, largely additive; *P. ussurienis* is extremely hardy (Bell et al. 1996)
Season of flowering	Quantitatively inherited (Anjou 1954, Bell et al. 1996); late flowering only weakly correlated with late ripening (Baldini 1949)
Harvest date	Quantitatively inherited with high heritability observed in *P. communis* (Thibault et al. 1988) and moderate heritability in *P. pyrifolia* (Zhejiang Agricultural University 1977), although some studies have shown parents to be poor predictors of progeny performance (Crane and Lewis 1949)
Productivity	
Juvenile phase length	Quantitatively inherited under additive control (Zielinski 1963, Visser 1967 and 1976, Zimmerman 1976, Bell and Zimmerman 1990)
Incompatibility	Numerous S-alleles exist, but there are compatible combinations (Sanzol and Herrero 2002, Kim et al. 2004, Moriya et al. 2007, Wu et al. 2007)
Plant habit	
Dwarfing	Several sources of dwarfing exist that are regulated by single genes (Decourtye 1967, Jingxian et al. 1988)
Fruit quality	
Firmness	Quantitatively inherited and highly heritable (Machida and Kozaki 1976, Kajiura and Sato 1990)
Flavor	Quantitatively inherited but with low heritability in *P. communis* (Bell and Janick 1990); poor flavor of *P. ussuriensis* is inherited quantitatively in interspecific crosses with some dominance (Lantz 1929); Aroma of *P. ussuriensis* is highly heritable in interspecific crosses (Pu et al. 1963)
Juiciness	Regulated by a limited number of dominant genes (Zielinski et al. 1965)
Keeping quality	Quantitatively inherited (Zielinski et al. 1965)
Russeting	Quantitatively controlled in *P communis* with high heritability (Crane and Lewis 1949, Zielinski et al. 1965, Bell and Janick 1990); Two genes regulate russetting in *P. pyrifolia*, R and I (Kikuchi 1930)
Skin color	Major genes determine – yellow dominant to green, blushed recessive to non-blushed (Zielinski et al. 1965); Deep red regulated by dominant gene *C* (Zielinski 1963, Brown 1966); White flesh dominant to colored (Zielinski et al. 1965)
Shape	Quantitatively inherited with round and obovate dominant to pyriform and turbinate (Crane and Lewis 1949, Zielinski et al. 1965, Wang and Wei 1987, White and Alspach 1996); Moderate to high heritability for ratio of length to diameter (Shin et al. 1983, White et al. 2000)
Size	Quantitatively inherited (Crane and Lewis 1949, Zielinski et al. 1965, Shen et al. 1979, Wang and Wei 1987); Low heritability in some *P. pyrifolia* populations (Machida and Kozaki 1976, Shin et al. 1983)
Sugar content	Heritability of soluble solids in *P. pyrifolia* is low (Shin et al. 1983) to moderate (Machida and Kozaki 1976) depending on population

Table 10.4 (continued)

Attribute	Observations and sources
Texture	Quantitatively inherited with relatively high heritability in *P. communis* (Machida and Kozaki 1976, Kajiura and Sato 1990); Presence of stone cells dominant to stoneless in most crosses of *Pyrus* (Pu et al. 1963, Zielinski et al. 1965, Golisz et al. 1971); Grit content regulated by at least four loci, acting independently and additively (Thompson et al. 1974)

10.5.3 Fruit Quality

High fruit quality is the prime objective of all fruit breeding programs and encompasses a wide array of attributes including flavor, texture, appearance, juiciness, postharvest storage life and incidence of physiological disorders. The most important attribute is eating quality, which is largely dependent on flavor and texture. Most programs focus on fresh fruit quality, but some are also interested in the quality of canned or pureed fruit. Considerable attention is commonly given to core breakdown, bitter pit and superficial scald.

High sugar content is likely the most important factor determining a flavorful pear. The level of acidity is secondary in importance, as high and low acid pears have been characterized as good as long as they have high sugar (Visser et al. 1968). Extreme bitterness and astringency are generally not desirable in dessert pears, although they are important contributors to the bitter-sharp flavor enjoyed in perry (Luckwill and Pollard 1963). The heritability of flavor in *P. communis* has been found to be relatively low ($h^2 = 0.21$) (Bell and Janick 1990).

Subjective perceptions of sweetness and acidity, as well as physical measurements of soluble solids and acidity, are inherited as independent, quantitative traits (Table 10.4). Estimates of the heritability of soluble solids in *P. pyrifolia* cultivars have ranged from quite low (Shin et al. 1983) to moderate (Machida and Kozaki 1976). Zielinski et al. (1965) found that juiciness, depending on the parent, was regulated by either a single gene or several, with juiciness being dominant to dryness. The inheritance of flavor in hybrid populations of *P. ussuriensis* is quantitative with poor flavor showing some dominance (Lantz 1929).

It is important that pears can be held in cold storage for long periods of time without developing internal breakdown, so that they can be sold during the winter when prices tend to be the highest. Few inheritance studies have been conducted on this characteristic, although segregation patterns suggest that long term keeping quality is quantitatively inherited (Zielinski et al. 1965).

A large number of aromatic compounds are produced by pears that make important contributions to fruit flavor. In 'Bartlett', 77 volatile compounds have been identified with varying impacts on flavor (Jennings and Tressel 1974). The esters of trans-2, cis-4-decadienoic acid are highly correlated with the intensity of 'Bartlett' aroma (Jennings et al. 1964), although there are cultivars with high levels of these compounds that do not have that distinctive aroma (Bell et al. 1996). Seventeen volatile compounds, including toluene, ethyl butryate, hexanal, (E) 2-hexanal, ethyl hexanoate, a-farnesene, were found to be the most common constituents of five Asian pear cultivars (Horvat et al. 1992). 'Ya Li' of *P. ×bretschneideri* and 'Shinko', a putative interspecific hybrid, lacked five compounds found in three other *P. pyrifolia* cultivars ('Chojuro', 'Hosui', and 'Kosui'). The genetics of these compounds are unknown, although Pu et al. (1963) found that some *P. ussuriensis* cultivars had an aromatic quality that was dominant to the sweet, bland flavor of *P. ×bretschneideri* and *P. pyrifolia* cultivars.

The desired flesh texture of pears varies from region to region, in part because of species differences. In Western Europe and North America the favored type is the soft, buttery one typical of *P. communis,* while in China and Japan, the preferred texture is the crisp, breaking one of *P. pyrifolia*, *P.×bretschneideri*, and *P. ussuriensis*. A minimum of grittiness (stone cells) is appreciated in all types of pears, although more is tolerated in the Asian than European ones. Bell and Janick (1990) found heritabilities of 0.30 and 0.45 for texture and grit scores in *P. communis.* Flesh firmness has been shown to be highly heritable in populations of *P. pyrifolia* (Machida and Kozaki 1976, Kajiura and Sato 1990). Stoneless flesh was found to be dominant to stony in certain interspecific hybrids of wild *Pyrus* species (Westwood and Bjornstad 1971); however, the presence of stone cells was dominant in several studies including crosses involving *P. communis* and *P. ussuriensis* (Golisz et al. 1971), *P. ussuriensis* crossed with *P. pyrifolia* and *P. ×bretschneideri* (Pu et al. 1963) and *P. communis* (Zielinski et al. 1965). Thompson et al. (1974) suggested that the control of grit content is complex, regulated by a minimum of four loci that act in an independent and additive fashion. In most hybrids derived from crosses of Asian and European pears, it is difficult to recover pure crisp flesh selections with little grittiness (Wang 1990, Bell et al. 1996).

Several genetic studies have been conducted on skin and flesh color in pears. Zielinski et al. (1965) found that the background color of pears was under the influence of a major gene, with yellow dominant to green. They also found that the blushed vs. nonblushed character was controlled by a recessive gene, and white flesh was dominant to colored, with green and cream likely being regulated by other alleles at the same locus. Deep red colored sports of 'Bartlett' called 'Cardinal Red' and 'Max Red' were found to be regulated by a single dominant gene, *C* by Zielinski (1963). Brown (1966) found that red was dominant to white in one segregating population of 'Sanquinole' (red) × 'Conference' (white). Not all red skin mutations are transmitted sexually, as the red coloration of 'Starkrimson' only affects the epidermis and not the germlines (Dayton 1966).

Russeting of the skin is acceptable if it is smooth, uniform and light tan for the fresh market, but is not acceptable in processed puree. In *P. communis,* Zielinski et al. (1965) and Crane and Lewis (1949) found that the control of russeting was under quantitative control. Bell and Janick (1990) estimated heritability for russeting to be 0.52. Wellington (1913) originally suggested that russeting in *P. pyrifolia* was under the control of a single gene; however, Kikuchi (1930) proposed that two loci were involved, *R* and *I*. In his model, RR$_{--}$ genotypes are completely russeted and Rrii genotypes are always partially russetted. RrI$_-$ genotypes have a partial russetting that is environmentally sensitive (under humid conditions they have more russetting than under dry conditions). Wang and Wei (1987) suggested that russetting in *P. pyrifolia* was recessive to non-russetting.

There is considerable variability for fruit size in *Pyrus*. Many genotypes of *P. betulifolia* and *P. calleryana* are as small as 1 cm in diameter, compared to genotypes of *P. communis* and *P. pyrifolia* which can exceed 12 cm in diameter (Bell et al. 1996). The size of pears is under polygenic control with variable levels of heritability depending on breeding population (Zielinski et al. 1965, Shen et al. 1979, Wang and Wei 1987, White and Alspach 1996). White et al. (2000) found heritability for length and width ratios to be more than 0.5 in families of European and Asian pear parentage, while Machida and Kozaki (1976) and Shin et al. (1983) found much lower values in their *P. pyrifolia* breeding populations. Many environmental factors influence pear size including water availability, fruit set and yield.

The most common shape of European pear cultivars is pyriform and most Asian pears are round (Bell et al. 1996). The inheritance of fruit shape is under polygenic control (Crane and Lewis 1949, Zielinski et al. 1965). Round and obovate shapes are most common in segregating populations of European and Asian types, suggesting that this shape is dominant to pyriform and turbinate shapes (Zielinski et al. 1965, Wang and Wei 1987). Shin et al. (1983) found that the ratio of fruit length to diameter had a moderately low heritability of 0.23.

Several molecular studies have been conducted on the expression patterns of genes during fruit ripening and storage. El-sharkawy et al. (2004) found several 1-Aminocyclopropane-1-carboxylic acid (ACC) synthase genes in late ripening pears that were differentially expressed during cold treatment, and they discovered that cold dependent and independent cultivars had variant allelic assemblages. Fonseca et al. (2005) followed the transcript accumulation of seven genes encoding cell wall modifying enzymes during fruit growth, ripening and senescence. They found that induction of the genes for xyloglucan endotransglucosylase/hydrolase and expansin 2 was likely associated with cell wall maintenance, while expansin 1, polygalacturonase 1 and 2, ß-galactosidase and ß-xylosidase most likely played a role in cell wall disassembly and loosening. Itai et al. (2000) isolated 30 cDNA clones of genes corresponding to mRNAs up-regulated during fruit ripening of Japanese pear. The cDNAs were sequenced and were found to be associated with stress response, protein catabolism and pathogenesis. Several of the genes were inhibited by 1-methylcyclopropane (MCP), an inhibitor of ethylene action.

A number of genes have been isolated from pear which are associated with fruit development and/or ripening. Leliévre et al. (1997) cloned ACC synthase and ACC oxidase from *P. communis*. El-Sharkawy et al. (2003) isolated and characterized four ethylene perception elements and found them to be differentially expressed after cold and ethylene treatment. Itai et al. (1999a,b) characterized *ACC synthase* and *ß-D-xylosidase* from *P. pyrifolia,* the latter likely playing a role in senescence. They found an association between the expression of two specific ACC synthase genes and ethylene production in 35 Asian pear cultivars. Itai et al. (2003) developed a rapid method for analyzing fruit storage potential by utilizing CAPS (cleaved-amplified polymorphic sequences) markers of the two ACC synthase genes. One marker was associated with high ethylene producers, while a second marker was associated with moderate ethylene producers. Low ethylene producers had neither of these markers.

Sekine et al. (2006) cloned thirteen cDNAs that encode cell-wall hydrolases in the European pear and followed their expression patterns during cold storage. Hiwasa et al. (2003) found seven α-expansin genes to be differentially expressed during growth and ripening of pear fruit. They identified seven genes that might be associated with the melting texture. Yamada et al. (2006) cloned two isoforms of soluble acid invertase of Japanese pear and followed their development during ripening; one was particularly active in young fruit. Tateishi et al. (2001) isolated a cDNA fragment of the fruit softening enzyme, ß-galactosidase. Several genes encoding membrane bound proteins have been cloned including arabinogalactan proteins in *P. communis* (Mau et al. 1995) and H(+)-pyrophosphatase in *P. pyrifolia* (Suzuki et al. 1999). Additional genes that have been characterized in pear include alcohol dehydrogenase (Chervin et al. 1999), polyphenol oxidase (Haruta et al. 1999) and α-L-arabinofuranosidase (Tateishi et al. 2005).

10.5.4 Rootstocks

Historically, little breeding effort has been devoted to improving pear rootstocks, although Chevreau and Bell (2006) suggest that 'deliberate evaluation and selection of parents and hybridization has become more common'. *Pyrus communis* is the species most widely employed as rootstocks in North America and Europe. In Asia, *P. pyrifolia*, *P. betulifolia*, *P. calleryana* and *P. pashia* provide the primary rootstocks, while in Asia Minor and the Mediterranean region, *P. elaeagrifolia*, *P. syriaca*, *P. amygdaliformis* and *P. longipes* are the species most commonly utilized.

Pyrus communis seedlings produce vigorous trees that are adapted to a broad range of climates and soil types, and they are generally resistant to pear decline and Armillaria root rot [*Armillaria mellea* (Vahl:Fr.) P. Kumm.], but are often susceptible to fire blight (Bell et al. 1996). Fire blight resistant types have been found in progeny of the *P. communis* cross, 'Old Home' × 'Farmingdale' (OH F rootstock series) and individuals from this family provide some size reduction (Brooks 1984, Lombard and Westwood 1987) and precocity, but sometimes have problems with suckering (Raese 1994). Other dwarfing *P. communis* rootstocks include 'Pyrodwarf' and 'BU 2-33' coming from 'Old Home' × 'Bonne Louise d'Avranches' (Jacob 1998 and 2002), the Blossier series (Brossier 1977, Michelesi 1990) and the Rètuziére series (Michelesi 1990, Simard and Michelesi 2002). Seedlings of 'Barlett' predominate as rootstocks in older plantings, with seedlings of 'Winter Nelis' probably being the next most widely used seedling rootstock.

Among the other species utilized as rootstocks, *Pyrus calleryana* has been used in the southern U.S.A., China and Australia because of its fire blight, pear decline and wooly aphid resistance, drought tolerance and it's control of vigor, although the species is very susceptible to winter injury (Cole 1966, Batjer et al. 1967, Lombard and Westwood 1987). *Pyrus betulifolia* is used in local areas of the United States, central China, northern Italy and Israel where temperatures are not very cold; it is particularly useful where clay soils and poor drainage restrict vigor and it is resistant to salinity, black end, pear decline and the wholly pear aphid (Lombard and Westwood 1987, Matsumoto et al. 2006). Dwarfing root stocks have been selected from both *P. calleryana* and *P. betulifolia* (Robbani et al. 2006). *Pyrus ussuriensis* has proven useful where extreme cold hardiness is required in northeast China and the Great Plains of North America (Morrison et al. 1965). It is resistant to fire blight (i.e. Reimer selections) and the woolly pear aphid, but is sensitive to excessive moisture, black end and pear decline. *Pyrus pyrifolia* is utilized widely in Japan, China and Korea because of its tolerance to a wide range of soil textures and soil moistures, even though it is prone to the physiological disorder black end (hard end) and pear decline (Bell et al. 1996). *Pyrus pashia* is used in northern India and southern China; it is resistant to black end, although it is sensitive to lime-induced chlorosis and is susceptible to the woolly pear aphid (Bell et al. 1996). *Pyrus amygdaliformis* has high tolerance to salinity stress (Matsumoto et al. 2006).

Cydonia oblonga L. has long been used in Europe as a powerful dwarfing rootstock for milder climate pear production. It reduces size by 30–60% compared to standard *P. communis* seedling rootstocks, shortens time to fruiting and increases fruit size. However, quince rootstocks suffer from several problems including susceptibity to fire blight and Armillaria root rot, poor winter hardiness, low tolerance to wet soils, insufficient soil anchorage and poor graft compatibity with many common pear cultivars (Millikan and Pieniazek 1967, Lombard and Westwood 1987). Quince selections are being developed with improved cold hardiness and greater tolerance to high pH (Loreti 1994, Bassi et al. 1996, Webster 1998).

In addition to *Cydonia*, selected clones of *Amelanchier* and *Crataegus* possess graft-compatibility with *Pyrus* and can be used as dwarfing rootstocks (Lombard and Westwood 1987, Lombard 1989).

10.6 Crossing and Evaluation Techniques

10.6.1 Breeding Systems

As mentioned previously, the pear is likely an ancient allopolyploid but it behaves as a diploid with disomic inheritance (Crane and Lewis 1940). Most *Pyrus* cultivars are diploid, although there are polyploid cultivars, especially in *P. communis* (Bell et al. 1996). A number of cultivars are triploid ($2n = 3x = 51$), a few are tetraploid ($2n = 4x = 68$), and hexaploid forms have been produced, but not commercialized ($2n = 6x = 102$).

All pear species are self-infertile with the gametophytic self-incompatibility system (Crane and Lewis 1942, Westwood and Bjornstad 1971). Therefore, straight selfing to increase homozygosity can not be employed, although crosses can be made where inbreeding coefficients do not exceed 0.25. A significant, albeit small, association has been observed between inbreeding and improved flavor, grit, and texture, suggesting that limited inbreeding can aid in selection for homozygous recessive genotypes (Bell et al. 1981).

A considerable amount of work has been done to characterize the genes associated with the incompatibility locus (S) locus in European (Sanzol et al. 2006) and Asian pears (Sassa and Hirano 1997, Ushijima et al. 1998, Ishimizu et al. 1999, Kim et al. 2004). PCR-RFLP and CAPS (cleaved amplified polymorphic sequences) marker systems have been developed to genotype the S alleles of cultivars (Kim et al. 2002, Takasaki et al. 2006, Moriya et al. 2007). One mutated gene of S4-RNase has been described that confers self-compatibility in 'Osa-Nijiisseiki' (Norioka et al. 1996, Wu et al. 2007). Two other genes have been cloned in pear that are associated with pollination - the gene for uridine diphosphate (UDP)-glucose pyrophosphorylase that plays a role in pollen tube wall synthesis (Kiyozumi et al. 1999) and a pollen- and seed-transmitted RNA-dependent RNA polymerase (Osaki et al. 1998).

10.6.2 Pollination and Seedling Culture

Pollen is generally collected several weeks before the bloom period. Branches 1–1.5 m in length are collected and forced in the laboratory or greenhouse with their cut ends in water. About 2 weeks are required before the blossoms are ready for collection if the branches are gathered while the buds are still dormant. If they are collected at the tight cluster stage, only 2–3 days are usually necessary.

The anthers are generally harvested just before they dehisce, by rubbing them on a wire mesh screen over paper. The anthers are allowed to dry on the paper for 24 hours and then the pollen is poured into glass vials that are placed in a desiccator with anhydrous $CaSO_4$ and kept at about 5°C. Pear pollen remains viable for two to three weeks at room temperature, but can be stored for over a year with refrigeration. The pollen is generally transported to the field on ice in larger vials containing desiccant. The vials of pollen are removed from the cooler when required for pollination and put back immediately after use.

Emasculation is done using a variety of tools including fingernails, scalpels, tweezers, or scissors specially modified with a notch in the blades and an adjustable screw to control the amount of closure (Bell et al. 1996). A cut is made below the sepals and the flower is pulled off leaving the pistil. One to three blossoms per cluster are generally emasculated at the balloon stage of development and all other flowers are removed. The larger basal flowers tend to set more fruit than the smaller terminal ones. Pollen is applied to the stigmas using a variety of objects including glass rods, fingers or camel hair brushes. Bell et al. (1996) suggests that approximately two flowers must be emasculated and pollinated to produce one seed. Bees do not visit flowers without corollas (Visser 1951), so special precautions to prevent pollen contamination are not necessary for routine crosses.

Pear seeds require stratification to overcome their internal dormancy requirement. This is commonly done by holding them for 60–90 days in moist, finely ground peat moss placed in polyethylene bags at temperatures a little above freezing (Hartmann et al. 1990). Fungicides are often added to the water to prevent damping off. After stratification, the seeds are planted into flats or small pots in a sterilized mixture of equal parts sand and peat moss. Germination generally occurs within 9–10 days, and the seedlings are transferred to larger pots filled with commercial soil mix when they are 15 cm tall.

10.6.3 Evaluation Techniques

Pear seedlings have a long juvenile period and remain unfruitful for 4 years or longer (Visser et al. 1976, Bell 1991). The length of this juvenile period appears to be dependent on how rapidly the seedlings grow (Zimmerman 1972). Therefore, growing conditions are generally optimized as much as possible. Other factors have been evaluated to hasten flowering such as grafting onto bearing trees or quince rootstocks, root pruning and sprays with growth retardants, but these practices have proved generally ineffective (Zimmerman 1972, Verhaegh et al. 1988).

There are some seedling characters, such as disease and insect resistance, that can be conducted in the greenhouse to allow for early seedling selection (Dayton et al. 1983). These preliminary screens can be followed with an additional screening in the nursery at close spacing, to verify resistance ratings and eliminate escapes. Some juvenile characters are correlated with economically important adult characteristics. Bell et al. (1996) list some of these in *P. communis* such as seedling vigor with precocity (Zimmerman 1972), juvenile period with precocity of propagated trees (Visser and De Vries 1970), juvenile period with red leaf coloration (Zielinski 1963), early flowering before leaf break with high yield (Moruju and Slusanschi 1959), seed size with germinability and seedling vigor (Schander 1955), and fruit color with foliage color.

When the plants do fruit, selection is undertaken on such characters as precocity and productivity, fruit quality, harvest dates and ripening uniformity. The storage and processing qualities of the most promising types may also be evaluated if there is enough fruit. The selected genotypes are then evaluated as grafted trees at normal orchard spacing in one location, and then the most promising are further evaluated as grafted trees at multiple locations.

10.7 Biotechnological Approaches to Genetic Improvement

10.7.1 Genetic Mapping and QTL Analysis

A wide array of molecular markers have been developed for pear including isozymes, restriction fragment length polymorphisms (RFLPs), randomly amplified polymorphic DNA (RAPDs), amplified fragment length polymorphisms (AFLPs) and inter simple sequence repeats (ISSRs) (Chevreau and Bell 2006). This work has primarily focused on fingerprinting cultivars and measuring diversity patterns within and among *Pyrus* species (Monte-Corvo et al. 2000, Yamamoto et al. 2002a,b, Katayama and Uematsu 2006, Katayama et al. 2007, Inoue et al. 2007). In general, cultivars have been readily distinguished through these approaches and species have clustered according to traditional taxonomic classifications. Many SSRs isolated from apple have been shown to be transferable to pear (Yamamoto et al. 2001, 2002c, Pierantoni et al. 2004), and vice versa (Fernández-Fernández et al. 2006).

Physical maps of chloroplast DNA (cpDNA) have also been generated using restriction analysis. In a comparison of Asian and Occidental pear species, Iketani et al. (1998) found four cpDNA haplotypes in Asian pears, but only one in occidental pears. They argued that the two groups have evolved separately, and that considerable hybridization and introgression has occurred between species. Katayama and Uematsu (2003) developed a cpDNA map of *Pyrus ussuriensis* var. *hondoensis* and then carried out an RFLP analysis on cpDNAs from representatives of *Pyrus pyrifolia*, *P. ussuriensis*, *P. calleryana*, *P. elaeagrifolia* and *P. communis*. Two mutations, a recognition-site mutation and a length mutation (deletion), were found in the cpDNA of *P. pyrifolia* cultivars.

Several genetic linkage maps have been developed of pear. Isozyme loci were used to identify three linkage groups in pear that shared a high level of synteny with apple (Chevreau et al. 1997). Iketani et al. (2001) employed RAPD markers to construct linkage groups of 18 and 22 for the two Asian pears, 'Kinchaku' and 'Kosui'. The linkage map for 'Kinchaku' contained 120 loci in 18 linkage groups across 768 cM, while that for 'Kosui' had 78 loci in 22 linkage groups spanning over 508 cM.

Yamamoto et al. (2002c) constructed a genetic linkage map of an interspecific cross between European (*P. communis* cv. 'Bartlett') and Asian (*P. pyrifolia* cv. 'Housui') pears using isozymes, AFLPs, SSRs and morphological traits from pear, apple, peach and cherry. In the map of the female parent, 'Bartlett', 226 loci were identified on 18 linkage groups over a total length of 926 cM. In the male parent, 'Housui', 154 loci were represented on 17 linkage groups encompassing a genetic distance of 926 cM. The position of 14 SSRs from apple could be placed on the pear map, along with a few SSRs from *Prunus*.

Pierantoni et al. (2004) used 100 apple SSRs to develop linkage maps of two European pear families. A total of 41 markers were positioned on the cross of 'Passe Crassane' × 'Harrow Sweet', and 31 were placed on a map of 'Abbè Fetél' × 'Max Red Bartlett'. Considerable colinearity was observed in the linkage relationships of the apple and pear genomes. Dondini et al. (2004) expanded the map of 'Passe Crassane' × 'Harrow Sweet' with a wide array of markers including SSRs, MFLPs (microsatellite-anchored fragment length polymorphisms), AFLPs and RGAs (resistance gene analogs). They placed 155 loci on the 'Passe Crassane' map consisting of 18 linkage groups with a coverage of 912 cM. On the Harrow Sweet map they identified 156 loci on 19 linkage groups with a coverage of 930 cM.

DNA markers have been linked to a number of genes of horticultural importance in pear. Banno et al. (1999) identified a RAPD marker that was closely linked to the gene *A*, conferring susceptibility to black spot disease in Asian pear.

Inoue et al. (2006) found a RAPD marker linked to fruit skin color in Japanese pear. Iketani et al. (2001) found four RAPD markers that were loosely associated with the pear scab resistance gene, *Vnk*; they were able to place the resistance allele for pear scab and the susceptible gene for black spot on their linkage map. Terakami et al. (2006) also found an SSR marker closely linked to the scab resistance gene, *Vnk*, and converted the sequence into a sequence tagged site (STS) marker, along with the four RAPD markers previously identified by Iketani et al. (2001). Dondini et al. (2004) found four putative QTL for fire blight resistance (Fig. 10.3).

10.7.2 Somatic Cell Genetics and Genetic Manipulation

Haplo-diploidization through gametic embryogenesis has been employed in pears to obtain homozygous lines from heterozygous parents (Germaná 2006). Bouvier et al. (1993) found haploid plants ($2n = x = 17$) among seedlings of 12 crosses of European pear and were able to induce in situ parthenogenesis using irradiated pollen. The immature embryos were cultured in vitro and 1 haploid, two misoploid and several diploids with the maternal phenotype were recovered. In subsequent work, Bouvier et al. (2002) treated haploids with oryzalin *in vitro* to generate doubled haploids. These were confirmed to be diploid and homozygous using isozyme and microsatellite markers. Kadotat and Niimi (2002, 2004) have produced triploid plants of Japanese pear by anther culture, and tetraploids using in vitro colchicine treatment.

Regeneration techniques have been developed for a wide array of elite pear cultivars and genotypes, including those of *P. communis* (Chevreau et al. 1997, Matsuda et al. 2005, Yancheva et al. 2006), *P. pyrifolia* (Lane et al. 1998); *P. syriaca* (Shibli et al. 2000), *P. pyraster* (Caboni et al. 1999) and quince (Dolcet-Sanjuan et al. 1991, Baker and Bhatia 1993). Leaves from in vitro grown plants were used as explants in most work, with a brief amount of callus growth at the wounding site before bud regeneration in about 3–6 weeks. Explants were generally exposed to dark and light periods of 2–4 weeks. The most common hormones employed were thidiazuron (TDZ) and naphthaleneacetic acid (NAA). Maximum rates of regeneration varied from <20% in *P. pyrifolia* (Lane et al. 1998) to >80% in *P. communis* (Chevreau and Leblay 1993) and *C. oblonga* (Baker and Bhatia 1993).

Palombi et al. (2007) used in vitro regeneration of wild pear (*P. pyraster*) to generate somaclonal variants for higher adaptability to calcareous soils. Selective treatments involved Murashige and Skoog (MS) medium with Fe-EDTA replaced by equimolar amounts of $FeSO_4$ with $KHCO_3$ or $NaHCO_3$. Eleven putatively tolerant lines were obtained from vegetatative shoot apices.

The first transformed pears were the European cultivars 'Passe Crassane', 'Conference' and 'Doyenne du Comice' (Mourgues et al. 1996). Since then, several other cultivars of *P. communis* have been transformed including 'Vyzhnitsa' (Merkulov

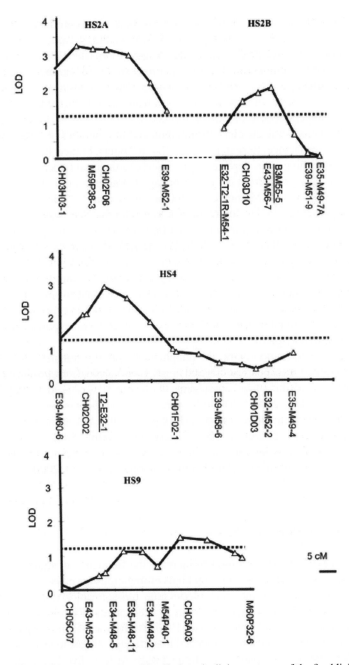

Fig. 10.3 QTL for fire blight resistance identified on the linkage groups of the fire blight tolerant cultivar 'Harrow Sweet'. The probability of association of markers is indicated by the LOD score. The black line indicates a LOD threshold value of 1.3, which is the 95% probability value (Dondini et al. 2004)

et al. 1998), 'Beurre Bosc' (Bell et al. 1999), 'Barlett' (Bommineni et al. 2001), wild *P. pyraster* (Caboni et al. 1998), *P. betulifolia* seedlings (Kaneyoshi et al. 2001) and the rootstocks GP217 (Lebedev and Dolgov 2000) and BP100030 (Zhu and Welander 2000). A number of different disarmed strains of *Agrobacterium tumifaciens* were utilized in these studies, with kanamycin or hygromycin being used as the selectable markers. In most cases, leaves were used as explants, with the only exception being cotyledons of *P. betulifolia*.

Transformation strategies have been employed in a number of instances to improve disease resistance. Researchers at INRA in Angers, France, have incorporated transgenes encoding lytic peptides from insects (attacin and cecropin) (Reynoird et al. 1999), lysozymes from T4 bacteriophage (Mourgues et al. 1998), the lactoferrin gene from bovin (Malnoy et al. 2003a) and a viral EPS-depolymerase gene (Malnoy et al. 2002) to develop resistance to fire blight. The insertion of the lytic peptide gene (*D5C*1) was shown to be partially effective against pear psylla (Puterka et al. 2002). Lebedev et al. (2002a) introduced plant defensin genes into pear to enhance pathogen resistance. Two genes have been cloned that may play a role in defense against pathogens, polygalacturonase inhibitor protein (PGIP) (Stotz et al. 1993) and thaumatin/PR5-like protein (Sassa and Hirano 1998). Malnoy et al. (2003b) searched for pathogen-inducible promoters from tobacco that would work in pear and found two, *str*246C and *sgd*24, which were responsive to inoculation by *Erwinia amylovora*.

In other transformation work, the phophinotricin acetyl transferase (PAT) gene was incorporated into pear rootstocks to produce herbicide resistance (Lebedev et al. 2002b), and the super sweet gene thaumatin II was used to modify fruit taste (Lebedev et al. 2002c). The rolC gene from *Agrobacterium rhizogenes* was introduced into pear to cause dwarfing (Bell et al. 1999) and enhance rooting (Zhu and Welander 2000, Zhu et al. 2003). The gene encoding S-adenosylmethionine hydrolase (*sam-k*) was incorporated into 'Bartlett' to improve postharvest quality and self-life by modifying ethylene synthesis (Bommineni et al. 2000).

10.7.3 Mutation Breeding

Irradiation (X-rays) has been used to increase the frequency of mutations in fruit trees (Ahloowalia et al. 2004). Several kinds of mutations have been identified after irradiation in *P. communis* including bloom time, blossom color, ripening time, fruit color (Decourtye 1971, Roby 1972a and b, Predieri and Zimmerman 2001) and growth habit (Visser et al. 1971, Lacey 1975, Predieri and Zimmerman 1997). In *P. pyrifolia*, mutations have been induced that effected disease resistance (Masuda et al. 1997) and self-compatibility (Hirata 1989). At least five European and four Japanese pears have been developed through mutation breeding. One of them, 'Gold Nijisseiki', has had a substantial impact on the Asian pear industry, and two new self-compatible, black spot resistant varieties show high promise (Ahloowalia et al. 2004).

References

Ahloowalia BS, Maluszynski M, Nichterlein K (2004) Global impact of mutation-derived varieties. Euphytica 135:187–204

Anjou K (1954) Winter injury of apples and pears at Balsgard, 1953 (in Swedish). Sverig Pomol Foren Arsskr 54:139–147

Baker BS, Bhatia SK (1993) Factors affecting shoot regeneration from leaf explants of quince (*Cydonia oblonga*). Plant Cell Tissue Organ Cult 35:273–277

Baldini E (1949) Spring frost damage to fruit trees in the spring of 1949 (in Italian). Riv Ortoflorofruttic Ital 33:78–86

Banno K, Ishikawa H, Hamauzu Y, Tabira H (1999) Identification of a RAPD marker linked to the susceptible gene of black spot disease in Japanese pear. J Jap Soc Hortic Sci 68:476–481

Bao L, Chen K, Zhang D, Cao Y, Yamamoto T, Teng Y (2007) Genetic diversity and similarity of pear (*Pyrus* L.) cultivars native to East Asia revealed by SSR (simple sequence repeat) markers. Genet Resour Crop Evol 54:959–971

Barbosa W, Pommer CV, Tombolato AFC, Meletti LMM, Veiga RF de A, Moura MF, Pio R (2007) Asian pear tree breeding for subtropical areas of Brazil. Fruits-Paris 62: 21–26

Bassi D, Marangoni B, Tagliavini M (1996) 'Fox', nuova serie di portinnest clonai per il pero. Frutticoltura 3:55–59

Batjer LP, Schomer HA, Newcomer EJ, Coyier DL (1967) Commercial pear growing. USDA Handb. 330

Bell RL (1991) Pears (*Pyrus*). Acta Hortic 290:657–700

Bell RL (1992) Additional East European *Pyrus* germplasm with resistance to pear psylla nymphal feeding. HortScience 27:412–413

Bell RL, Hough LF (1986) Interspecific and intergeneric hybridization of *Pyrus*. HortScience 21:62–64

Bell RL, Janick J (1990) Quantitative genetic analysis of fruit quality in pear. J Am Soc Hortic Sci 115:829–834

Bell RL, Quamme HA, Layne REC, Skirvin RM (1996) Pears In: Janick J, Moore JN (eds) Fruit Breeding vol 1: Tree and Tropical Fruits. John Wiley and Sons, NewYork, pp 441–514

Bell RL, Scorza R, Srinivasan C, Webb K (1999) Transformation of 'Beurre Bosc' pear with the *rolC* gene. J Am Soc Hortic Sci 124:570–574

Bell RL, Stuart LC (1990) Resistance in Eastern European *Pyrus* germplasm to pear psylla nymphal feeding. HortScience 25:789–791

Bell RL, van der Zwet T (2005) Host resistance in *Pyrus* to Fabraea leaf spot. HortScience 40:21–23

Bell RL, van der Zwet T, Blake RC (2002) 'Blake's Pride' pear. HortScience 37:711–713

Bell RL, Zimmerman RH (1990) Combining ability analysis of juvenile period in pear. HortScience 25:1425–1427

Bell RL, Janick J, Zimmerman RH, van der Zwet T (1977) Estimation of heritability and combining ability for fire blight resistance in pear. J Am Soc Hortic Sci 102:133–138

Bell RL, Janick J, Zimmerman RH, van der Zwet T, Blake RC (1981) Response of pear to inbreeding. J Am Soc Hortic Sci 106:584–589

Bellini E, Sansavini S, Lugli S, Nin S, Rivlta L (2000) Obiettivi innovatori del miglioramento genetico del pero. Rivista Frutt Ortoflor 62:56–69

Bommineni VR, Mathews H, Clenennen SK, Wagoner W, Dewey V, Kellogg J, Peters S, Matsumura W, Pieper M, Kramer M, Wagner DR (2000) Genetic engineering of fruits and vegetables with the ethylene control gene encoding S-adenosylmethionine hydrase (SAMase). Proceedings Inter Sym Plant Genetic Engineering, Havana, Cuba, pp 206–214

Bommineni VR, Mathews H, Samuel SB, Kramer M, Wagner DR (2001) A new method for rapid in vitro propagation of apple and pear. HortScience 36:1102–1106

Bouvier L, Guérif P, Djulbic M, Durel CE, Chevreau E, Lespinasse Y (2002) Chromosome doubling of pear haploid plants and homozygosity assessment using isozyme and microsatellite markers. Euphytica 123:255–262

Bouvier L, Zhang Y-X, Lespinasse Y (1993) Two methods of haploidization in pear, *Pyrus communis* L.: Greenhouse seedling selection and in situ parthenogenesis induced by irradiated pollen. Theor Appl Genet 87:229–232

Braniste N (2002) The pear industry in Eastern Europe. Acta Hortic 596:83–85

Brooks L (1984) History of the Old Home × Farmingdale pear rootstocks. Fruit Varieties J 38:126–128

Brossier J (1977) La recherche de nouveau porte-greffes du Poirier dans le genre *Pyrus communis* L. Acta Hort 69:41–47

Brown AG (1966) Genetical studies in pears, 5: red mutants. Euphytica 15:425–429

Caboni E, Tonelli MG, Lauri P, D'Angeli S, Damiano C (1999) In vitro shoot regeneration from leaves of wild pear. Plant Cell Tissue Organ Cult 59:1–7

Caboni E, Tonelli MG, Lauri P, Damiano C (1998) Regeneration and transformation of pear rootstocks. In: Abstracts XXV Inter Hortic Cong, August 2–7, Brussels

Challice J, Westwood MN (1973) Numerical taxonomic studies of the genus *Pyrus* using both chemical and botanical characteristics. Bot J Linn Soc 67:121–148

Chervin C, Spiers J, Truett J (1999) Alcohol dehydrogense expression and the production of alcohols during pear fruit ripening. J Am Soc Hortic Sci 124:71–75

Chevreau E, Bell R (2004) *Pyrus* spp. pear and *Cydonia* spp. Quince. In: Litz RE (ed) Biotechnology of fruit and nut crops. CABI Publishing, Wallingford, UK, pp 543–565

Chevreau E, Leblay C (1993) The effect of mother plant pretreatment and explant choice on regeneration from in vitro pear leaves. Acta Hortic 336:263–268

Chevreau E, Leuliette S, Gallet M (1997) Inheritance and linkage of isozyme loci in pear (*Pyrus communis* L.). Theor Appl Genet 94:498–506

Cole CE (1966) The fruit industry of Australia and New Zealand. Proc 17th Int Hor Congr 3:321–368

Crane MB, Lewis D (1940) Genetical studies in pears, 2: classification of culture varieties. J Pomol 18:52–60

Crane MB, Lewis D (1942) Genetical studies in pears, 3: incompatibility and sterility. J Genet 43:31–43

Crane MB, Lewis D (1949) Genetical studies in pears, 5: vegetative and fruit characters. Heredity 3:85–97

Dayton DF (1966) The pattern and inheritance of anthocyanin distribution in red pears. Proc Am Soc Hortic Soc 89:110–116

Dayton DF, Bell RL, Williams EB (1983) Disease resistance breeding. In: Moore JN, Janick J (eds) Methods in fruit breeding. Purdue University Press, West Layfayette, IN, pp 119–159.

Decourtye L (1967) Etude de quelques caracteres a controle genetique simple chez le pommier (*Malus* sp.) et le poirier (*Pyrus communis* L.). Ann Amel Plantes 17:243–265

Decourtye L, Lantin B (1971) Methodology in induced mutagenesis for apple and pear trees. Ann Amél Plantes 21:29–44 (in French)

Dolcet-Sanjuan R, Mok DWS, Mok MC (1991) Plantlet regeneration from cultured leaves of *Cydonia oblonga* L. (quince). Plant Cell Rep 10:240–242

Dondini L, Pierantoni L, Gaiotti F, Chiodini R, Tartarini S, Bazzi C, Sansavina S (2004) Identifying QTLs for fire blight resistance via European pear (*Pyrus communis* L.) genetic linkage map. Mol Breeding 14:407–418

El-Sharkawy I, Jones B, Gentzbittel L, Lelièvre J-M, Pech JC, Latché A (2004) Differential regulation of ACC synthase genes in cold-dependent and -independent ripening in pear fruit. Plant Cell Environ 27:1197–1210

El-Sharkawy I, Jones B, Li ZG, Lelièvre J-M, Pech JC, Latché A (2003) Isolation and characterization of four ethylene perception elements and their expression during ripening in pears (*Pyrus communis* L.) with/without cold requirement. J Exp Bot 54:1615–1625

FAOSTAT (2007) Food and Agricultural Organisation of the United Nations. http:faostat.fao.org/site/336/default.aspx

Fernández-Fernández F, Harvey NG, James CM (2006) Isolation and characterization of polymorphic microsatellite markers from European pear (*Pyrus communis* L.). Mol Eco Notes
6:1039–1041

Fisher DF (1922) An outbreak of powdery mildew (*Podosphaera leucotricha*) on pears. Phytopathology 12:103

Fisher M, Mildenberger (2000) New Naumburg/Pillnitz pear breeding results. Acta Hortic
538:735–739

Fonseca S, Monteiro L, Barreiro MG, Pais M (2005) Expression of genes encoding cell wall modifying enzymes is induced by cold storage and reflects changes in pear fruit texture. J Exp Bot
56:2029–2036

Germaná MA (2006) Doubled haploid production in fruit crops. Plant Cell Tissue Org Cult
86:131–146

Ghariani K, Stebbins RL (1994) Chilling requirements of apple and pear cultivars. Fruit Varieties
J 48:215–222

Golisz A, Basak A, Zagaja SW (1971) Pear cultivar breeding. In: S. A. Pieniazd (ed) Studies on
some local Polish fruit species, varieties and clones and on the most recently introduced to
Poland with respect to their breeding value and other characters Nov. 1, 1966, to Oct. 31, 1971.
Res Inst Pomology, Skierniewice, Poland

Harris MK, Lamb RC (1973) Resistance to the pear psylla in pears with *Pyrus ussuriensis* lineage.
J Am Soc Hortic Sci 98:378–381

Hartman H (1957) Catalog and evaluation of the pear collection at the Oregon Agriculture Experiment Station. Oregon Agric Exp Sta Tech Bull 41

Hartmann HT, Kester DE, Davies FT (1990) Plant propagation: principles and practices, 5th ed
Prentice-Hall, Englewood Cliffs, NJ

Haruta M, Murata M, Kadokura H, Homma S (1999) Immunological and molecular comparison
of polyphenol oxidase in Rosaceae fruit trees. Phytochem 50:1021–1025

Hedrick UP, Howe GH, Taylor OM, Francis EH, Tukey HB (1921) The pears of New York. New
York, Department Agriculture 29th Annual Report vol. 2, part 2

Hevesi M, Göndör M, Kása K, Honty K, Tóth MG (2004) Traditional and commercial apple and
pear cultivars as sources of resistance to fireblight. EPPO/OEPP Bull 34:377–380

Hibino H, Schneider H (1970) Mycoplasmalike bodies in sieve tubes of pear trees affected with
pear decline. Phytopathology 60:499–501

Hiroe I, Nishimura S, Sato M (1958) Pathochemical studies on *Alternaria kikuchiana*: on toxins
secreted by the fungus (in Japanese). Trans Tottori Soc Agric Sci 11:291–299

Hirata N (1989) Self-compatible mutant in Japanese pear. Gamma Field Symp 28:
71–80

Hiwasa K, Rose JKC, Nakano R, Inaba A, Kubo Y (2003) Differential expression of seven
α-expansin genes during growth and development. Physiol Plant 117:564–572

Hooker W (1818) Account of new pear, called Williams' Bon Chretien. Trans Hortic Soc London
2:250–251

Horvat RJ, Senter SD, Chapman GW, Payne JA (1992) Volatiles of ripe Asian pears (*Pyrus serotina*
Rehder). J Essent Oil Res 4:645–646

Hsu H-T, Lin S-C (1987) Oriental pear breeding for high quality and adaptation to subtropical
lowlands of Taiwan. In: The breeding of horticultural crops. Food and Fertilizer Technology
Center for the Asian and Pacific Region, Taipei , pp 95–100

Human JP (2005) Progress and challenges of the South African pear breeding. Acta Hortic
671:185–190

Hunter DM, Kappel F, Quamme HA, Bonn WG (2002a) 'AC Harrow Gold' pear. HortScience
37:224–226

Hunter DM, Kappel F, Quamme HA, Bonn WG (2002b) 'AC Harrow Crisp' pear. HortScience
37:227–229

Jennings WG, Creveling RK, Heintz DE (1964) Volatile esters of Bartlett pear. IV. Esters of
Trans:2-cis:4-decadienoic acid. J Food Chem 29:730–734

Iketani H, Abe K, Yamamoto T, Kotobuki K, Sato Y, Saito T, Terai O, Matsuta N, Hayashi T (2001) Mapping of disease-related genes in Japanese pear using a molecular linkage map with RAPD markers. Breeding Sci 51:179–184

Iketani H, Manabe T, Matsuta N, Akihama T, Hayashi T (1998) Incongruence between RFLPs of chloroplast DNA and morphological classification in east Asian pear (*Pyrus* spp.). Genet Res Crop Evol 45:533–539

Inoue E, Kasumi M, Sakuma F, Anzai H, Amano K, Hara H (2006) Identification of RAPD marker linked to fruit skin color in Japanese pear (*Pyrus pyrifolia* Makai). Sci Hortic 107: 254–258

Inoue E, Matsuki Y, Anzai H, Evans K (2007) Isolation and characterization of microsatellite markers in Japanese pear (*Pyrus pyrifolia* Nakai). Mol Ecol Notes 7:445–447

Ishimizu T, Inoue K, Shimonaka M, Saito T, Terai O, Norioka S (1999) PCR-based method for identifying the S-genotype of Japanese pear cultivars. Theo Appl Genet 98:961–967

Itai A, Kawata T, Tanabe K, Tamura F, Uchiyama M, Tomomitsu M, Shiraiwa N (1999a) Identification of 1-aminocyclopropane-1-carboxylic acid synthase genes controlling the ethylene level of ripening fruit in Japanese pear (*Pyrus pyrifolia* Nakai). Mol Gen Genet 261:42–49

Itai A, Kotaki T, Tanabe K, Tamura F, Kawaguchi D, Fukuda M (2003) Rapid identification of 1-aminocyclopropane-1-carboxylate (ACC) synthase genotypes in cultivars of Japanese pear (*Pyrus pyrifolia* Nakai) using CAPS markers. Theor Appl Genet 106:1266–1272

Itai A, Tanabe K, Tamura F, Tanaka T (2000) Isolation of cDNA clones coresponding to genes expressed during fruit ripening in Japanese pear (*Pyrus pyrifolia* Nakai): Involvement of the ethylene signal transduction pathway in their expression. J Exp Bot 51:1163–1166

Itai A, Yoshida K, Tanabe K, Tamura F (1999b) A beta-D-xylosidase-like gene is expressed during fruit ripening in Japanese pear (*Pyrus pyrifolia* Nakai). J Exp Bot 50:877–878

Jacob H (1998) Pyrodwarf, a new clonal rootstock for high density pear orchards. Acta Hortic 475:169–177

Jacob H (2002) New pear rootstocks from Geisenheim, Germany. Acta Hortic 596:337–344

Jennings WG, Tressel R (1974) Production of volatile compounds in the ripening of Bartlett pear. Chem Mikrobiol Technol Lebensm 3:52–55

Jingxian J, Shenpu F, Lanchen C (1988) Evaluation and utilization of compact pear germplasm resources. In: Int. Symp. Hort. Germplasm, Cultivated and Wild. (Abstracts), Beijing, China, Sept. 1988. Int Acad Pub, Beijing, China, p 44

Kadota M, Niimi Y (2002) In vitro induction of tetraploid plants from a diploid Japanese pear cultivar (*Pyrus pyrifolia* N. vc. Hosui). Plant Cell Rep 21:282–286

Kadota M, Niimi Y (2004) Production of triploid plants of Japanese pear (*Pyrus pyrifolia* Nakai) by anther culture. Euphytica 138:141–147

Kajiura I (1992) Nashi (Japanese pear) production in Japan. Chron Hortic 32:57–58

Kajiura I, Sato Y (1990) Recent progress in Japanese pear (*Pyrus pyrifolia* Nakai) breeding, and descriptions of cultivars based on literature review. Bull Fruit Tree Res Sta, Extra N. 1

Kajiura M (1966) The fruit industry of Japan, South Korea and Taiwan. Proc 17th Int Hortic Congr 4:403–425

Kanato K, Kajiura I, McKenzie DW (1982) The ideal Japanese pear. In: van der Zwet T, Childers NF (eds) The pear. Horticultural Publ, Gainesville, FL, pp 138–155

Kaneyoshi J, Wabiko H, Kobayashi S, Tsuchiya T (2001) *Agrobacterium tumefaciens* AKE10-mediated transformation of an asian pear, *Pyrus betulaefolia* Bunge: host specificity of bacterial stains. Plant Cell Rep 20:622–628

Katayama H, Adachi S, Yamamoto T, Uematsu C (2007) A wide range of genetic diversity in pear (*Pyrus ussuriensis* var. *aromatica*) genetic resources from Iwate, Japan revealed by SSR and chloroplast DNA markers. Genet Resour Crop Evol (Online First)

Katayama H, Uematsu C (2003) Comparative analysis of chloroplast DNA in *Pyrus* species: Physical map and gene localization. Theor Appl Genet 106:303–310

Katayama H, Uematsu C (2006) Pear (*Pyrus* species) genetic resources in Iwate, Japan. Genet Res Crop Evol 53:483–498

Khan KA (1955) Studies on stocks immune to woolly aphids of apple (*Eriosoma lanigerum* Hausm.). Punjab Fruit J 19:28–35

Kikuchi A (1930) On skin color of the Japanese pear, and its inheritance (in Japanese). Contr Inst Plant Ind 8:1–50

Kikuchi A (1946) Speciation and taxonomy of Chinese pears. Collected Records of Hortic Res 3:1–8, Kyoto University

Kim H-T, Hirata Y, Nou I-S (2002) Determination of S-genotypes of pear (*Pyrus pyrifolia*) cultivars by S-RNase sequencing and PCR-RFLP analysis. Mol Cells 13:444–451

Kim H-T, Hirata Y, Shin Y-U, Hwang H-S, Hwang J-H, Shin I-S, Kim D-I, Kang S-J, Kim H-J, Shin D-Y, Nou I-S (2004) A molecular technique for selection of self-compatible varieties of Japanese pear (*Pyrus pyrifolia* Nakai). Euphytica 138:73–80

Kim W-C, Ko K-C (1991) Studies on the inheritance of major characters in oriental pear cultivars (*Pyrus pyrifolia* Nakai and *P. pyrifolia* × *P. ussuriensis*). 1. The inheritance of physiological characteristics (colors of leaf, fruit skin, blooming time, maturity and diseases). Res Rep Rural Dev Admin 33:76–84

Kim W-C, Ko K-C (1992) Studies on the inheritance of major characters in oriental pear cultivars (*Pyrus pyrifolia* Nakai and *P. pyrifolia* ×*P. ussuriensis*). 2. The inheritance of characteristics concerned with fruit shape, fruit volume and fruit weight. Res Rep Rural Dev Admin 34: 51–59

Kiyozumi D, Ishimizu T, Nakanishi T, Sakiyama F, Norioka S (1999) Molecular cloning and nucleotide sequencing of a cDNA encoding UDP-glucose pyrophosphorylase of Japanese pear (*Pyrus pyrifolia*). Plant Physiol 119:364

Knight RL (1963) Abstract bibliography of fruit breeding and genetics to 1960 *Malus* and *Pyrus*. E. Malling, Commonwealth Agric Bur Tech Commun 29

Kovalev NV (1963) Leaf blight of pears (in Russian). Zasc Rast Vred Bolez 8:58 (abst.)

Kozaki I (1973) Black spot disease resistance in Japanese pear. I: Inheritance of disease resistance (in Japanese). Bull Hortic Res Sta Jap A 12:17–27

Kuznetzov PV (1941) The role of *Pyrus salicifolia* Pall. in the development of fruit growing in arid regions (in Russian). Sovetsk Bot 1–2:103–107

Lacey CND (1975) Induction and selection of mutant form of fruit plants. Long Ashton Ann Rep 22–24

Lane WD, Iketani H, Hayashi T (1998) Shoot regeneration from cultured leaves of Japanese pear (*Pyrus pyrifolia*) Plant Cell Tissue Organ Cult 54:9–14

Lantz HL (1929) Pear breeding: an inheritance study of *Pyrus communis* × *P. ussuriensis* hybrid fruits. Proc Am Soc Hortic Sci 26:13–19

Layne REC (1968) Breeding blight-resistant pears for southwestern Ontario. Can Agric 13:28–29

Lebedev VG, Dolgov SV (2000) The effect of selective agents and a plant intron on transformation efficiency and expression of heterologous genes in pear *Pyrus communis* L. Rus J Genet 36:650–655

Lebedev VG, Lavrova N, Lunin VG, Dolgov SV (2002a) Plant-defensin genes introduction for improvement of pear phytopathogen resistance. Acta Hortic 596:167–172

Lebedev VG, Skriabin KG, Dolgov SV (2002b) Transgenic pear clonal rootstocks resistant to the herbicide 'Basta'. Acta Hortic 596:193–197

Lebedev VG, Taran SA, Dolgov SV (2002c) Pear transformation by gene of supersweet protein thaumatin II for fruit taste modification. Acta Hortic 596:199–202

Leliévre JM, Tichit L, Dao P, Fillion L, Nam YM, Pech JC, Latche A (1997) Effects of chilling on the expression of ethylene biosynthesis genes in Passe Crassane pear (*Pyrus communis* L.) fruits. Plant Mol Biol 33:847–855

Li ZL, Shen DX, Zheng SQ (1981) On the 'juvenile span', fruiting and inheritance of pear seedlings. 1-Zhejiang Agric Univ 7(3): December

Lombard PB (1989) Dwarfing rootstocks for European pear. Compact Fruit Tree 22:74

Lombard PB, Westwood MN (1987) Pear rootstocks. In: Rom RC, Carlson RF (eds) Rootstocks for fruit crops. Wiley, New York, pp 145–183

Loreti F (1994) Attuali conoscenze sui principali portinnesti degli alberi da futto. Rivista Frutti coltura 9:21–26

Luby JJ, Bedford DS, Hoover EE, Munson ST, Gray WH, Wildung DK, Stushnoff C (1987) 'Surnmercrisp' pear. HortScience 22:964

Luckwill LC, Pollard A (1963) Perry pears. Univ. Bristol Press, Bristol, UK

Ludin Y (1942) Hardiness of fruit trees in the winter of 1941–42 (in Swedish). Fruktodlaren 6:168–171

Machida Y, Kozaki I (1976) Quantitative studies on the fruit quality of Japanese pear (*Pyrus seratina* Rehder) breeding, II: statistical analysis of a hybrid seedling population (in Japanese). J Jap Soc Hortic Sci 44:325–329

Magness JR (1937) Progress in pear improvement. In: USDA Yearb. Agr USDA, Washington, DC, pp 615–630

Malnoy M, Brisset MN, Chevreau E (2002) Expression of a depolymerase gene in transgenic pears increased only slightly their fire blight resistance. Acta Hortic 590:401–405

Malnoy M, Venisse JS, Brisset MN, Chevreau E (2003a) Expression of bovine lactoferrin cDNA confers resistance to *Erwinia amylovora* in transgenic pear. Mol Breed 12:231–244

Malnoy M, Venisse JS, Reynoird JP, Chevreau E (2003b) Activation of three pathogen-inducible promoters of tobacco in transgenic pears (*Pyrus communis* L.) after biotic and abiotic elicitation. Planta 216:802–814

Masuda T, Yoshioka T, Inoue K, Murata K, Kitagawa K, Tabira H, Yoshida A, Kotobuki K, Sanada T (1997) Selection of mutants resistant to black spot disease by chronic irradiation of gama-rays in Japanese pear 'Osanijisseiki' J Japapan Soc Hort Sci 66:85–92

Matsuda N, Gao M, Isuzugawa K, Takashina T, Nishimura K (2005) Development of an *Agrobacterium*-mediated transformation method for pear (*Pyrus communis* L.) with leaf-section and axillary shoot-meristem explants. Plant Cell Rep 24:45–51

Matsumoto K, Tamura F, Chun J-P, Tanabe K (2006) Native Mediterranean *Pyrus* rootstock, *P. amygdaliformis* and *P. elaeagrifolia* present higher tolerance to salinity stress compared with Asian natives. J Jap Soc Hortic Sci 75:450–457

Mau SL, Chen CG, Pu ZY, Mority RL, Simpson RJ, Bacic A, Clark AE (1995) Molecular cloning of cDNAse, coding the protein backbones of arabinogalactan-proteins from the filtrate of suspension-cultured cells of *Pyrus communis* and *Nicotiana alata*. Plant J 8:269–281

Merkulov SM, Bartish IV, Dolgov SV, Pasternak TP, McHugen A (1998) Genetic transformation of pear *Pyrus communis* L. mediated by *Agrobacterium tumefaciens*. Genetica Moskva 34:373–378

Michelesi JC (1990) Les porte-griffes du poirer. L'Arboriculture Fruitiere 427:19–27

Millikan DF, Pieniazek SA (1967) Superior quince rootstocks for pear from east Europe. Fruit Varieties Hortic Dig 21:2

Monte-Corvo L, Cabrita L, Oliveira C, Leitão J (2000) Assessment of genetic relationships among *Pyrus* species and cultivars using AFLP and RAPD markers. Genet Res Crop Evol 47:257–265

Moriya Y, Yamamoto K, Okada K, Iwanami H, Bessho H, Nakanishi T, Takasaki T (2007) Development of a CAPS marker system for genotyping European pear cultivars harboring 17 S alleles. Plant Cell Rep 26:345–354

Morrison JW, Cumming WA, Temmerman HJ (1965) Tree fruits for the prairies. Can Dept Agric Publ 1222

Moruju G, Slusanschi H (1959) The study of the correlation between the processes of growth and fruiting at the commencement of shooting in some apple and pear varieties (in Romanian). In: Lucar Sti Inst Cerut Hort-Vit Baneasa-Bucuresti, 1957, pp 317–330

Mourgues F, Chevreau E, Lambert C, de Bondt A (1996) Efficient Agrobacterium-mediated transformation and recovery of transgenic plants from pear (*Pyrus communis* L.). Plant Cell Rep 16:245–249

Mourgues F, Brisset MN, Chevreau E (1998) Activity of different antibacterial peptides on *Erwinia amylovora* growth, and evaluation of the phytotoxicity and stability of cecropins. Plant Sci 139:83–91

Norioka N, Norioka S, Ohnishi Y, Ishimizu T, Oneyama C, Nakanishi T, Sakiyama F (1996) Molecular cloning and nucleotide sequences of cDNAs encoding S-allele specific stylar RNases in self-incompatible cultivar and its self-compatible mutant of Japanese pear, *Pyrus pyrifolia* Nakai. J Biochem 120:335–345

Osaki H, Kudo A, Ohtsu Y (1998) Nucteotide sequence of seed- and pollen-transmitted double stranded RNA, which encodes a putative RNA-dependent RNA polymerase, detected from Japanese pear. Biosci Biotech Biochem 62:2101–2106

Palombi MA, Lombaro B, Caboni E (2007) In vitro regeneration of wild pear (*Pyrus pyraster* Burgsd) clones tolerant to Fe-chlorosis and somaclonal variation analysis by RADP markers. Plant Cell Rep 26:489–496

Perraudin G (1955) The susceptibility of fruit trees to late frosts. (in Italian). Rev Romande Agric Vitic 11:87–88

Peterson RM, Waples JR (1988) 'Gourmet' pear. HortScience 23:633

Pieniazek SA (1966) Fruit production in China. Proc 17th Int Hortic Congr 4:427–456

Pierantoni L, Cho K-H, Shin I-S, Chiodini R, Tartarini S, Dondini L, Kang S-J, Sansavini S (2004) Characterization and transferability of apple SSRs to two European pear F_1 populations. Theor Appl Genet 109:1519–1524

Predieri S, Zimmerman RH (1997) Pear mutagenesis: In vitro treatment with gamma-rays and field selection for vegetative form traits. Euphytica 93:227–237

Predieri S, Zimmerman RH (2001) Pear mutagenesis: In vitro treatment with gamma-rays and field selection for productivity and fruit traits. Euphytica 117:217–227

Pruss AG, Eremeev GN (1969) Drought resistance in pear varieties of diverse geographical origin (in Russian). Trudy Prikl Bot Genet Selek 40:56–67

Pu FS, Zing XP, Xu HY, Jia IX, Fu ZC (1963) The genetic analysis of commercial characteristics of Chinese varieties (in Chinese). Ann Scientific Rep, Res Inst Pomol, CAAS:1–15

Pu FS, Lin SH, Chen RY, Song WQ, Li XL (1986) Studies on the karyotype of *Pyrus* in China. II. Acta Hortic Sin 13:87–90

Puterka GJ, Bocchetti C, Dang P, Bell RL,Scorza R (2002) Pear transformed with a lytic peptide gene for disease control affects non-target organisms, pear psylla (Homoptera: Psyllidae). J Econ Ent 95:797–802

Quamme HA (1984) Observations of psylla resistance among several pear cultivars and species. Fruit Varieties J 38:34–36

Quamme HA, Kappel F, Hall JW (1990) Efficacy of early selection for fire blight resistance and the analysis of combining ability for fire blight resistance in several pear progenies. Can J Plant Sci 70:905–913

Quamme HA, Spearman GA (1983) 'Harvest Queen' and 'Harrow Delight' pear. HortScience 18:770–772

Raese JT (1994) Fruit disorders, mineral composition and tree performance influenced by rootstocks of 'Anjou' pears. Acta Hortic 367:372–379

Reimer FC (1925) Blight resistance in pears and characteristics of pear species and stocks. Oregon Agric Exp Stat Bull 214:99

Reynoird JP, Mourgues F, Norelli J, Aldwinckle HS, Brisset MN, Chevreau E (1999) First evidence for improved resistance to fire blight in transgenic pear expressing the attacin E gene from *Hyalophora cecropia*. Plant Sci 149:13–22

Rivalta L, Dradi M, Rosati C (2002) Thirty years of pear breeding activity at Instituto Sperimentale per la Frutticoltura of Forli: A review. Acta Hortic. 596:233–238

Robbani M, Banno K, Yamaguchi K, Fujisawa N, Liu JY, Kakegawa M (2006) Selection of dwarfing rootstock clones from *Pyrus betulaefolia* and *P. calleryana* seedlings. J Jap Soc Hortic Sci 75:1–10

Roby F (1972a) Doce mutaciones en el peral Williams obtenidas por injertos de ramitas irradiadas. Rev Invest Agropec Ser 2, 9:55–64

Roby F (1972b) Mutaciones inducida por irradiación en el peral Packham's Triumph. In: Induced mutation and plant improvement, International Atomic Energy Agency, Vienna. pp 475–483

Ronald WG, Temmerson HJ (1982) Tree fruits for the Prairie Provinces. Agric Can Pub 1672E

Rubtsov GA (1944) Geographical distribution of the genus *Pyrus* and trends and factors in its evolution. Am Nat 78:358–366

Sansavini S (1967) Studies on cold resistance in pear varieties (in Italian). Riv Ortoflorofruttic Ital 51:407–416

Sanzol J, Herrero M (2002) Identification of self-incompatibility alleles in pear cultivars (*Pyrus communis* L.). Euphytica 128:325–331

Sanzol J, Sutherland BG, Robbins TP (2006) Identification and characterization of genomic DNA sequences of the S-ribonuclease gene associated with self-incompatibility alleles S_1 to S_5 in European pear. Plant Breed 125:513–518

Sassa H, Hirano H (1997) Nucleotide sequence of a cDNA encoding S5-RNase from Japanese pear (*Pyrus serotina* Red.) Plant Physiol 113:306

Sassa H, Hirano H (1998) Style-specific and developmentally regulated accumulation of a gly-cosylated thaumatin/PR5-like protein in Japanese pear (*Pyrus serotina* Red.). Planta 205: 514–521

Sax K (1931) The origin and relationships of the pomoideae. 1. Arnold Arbor 12:3–22

Schander H (1955) On the causes of differences in weight in the seeds of pome fruits (apple and pear), I: the relationship between seed and fruit (in German). Z Pflanzenz 34:255–306

Sekine D, Munemura I, Gao M, Mitsuhashi W, Toyomasu T, Murayama H (2006) Cloning of cDNAs encoding cell-wall hydrolases from pear (*Pyrus communis*) fruit and their involvement in fruit softening and development of melting texture. Physiol Plant 126:163–174

Serdani M, Spotts RA, Calabro JM, Postman JD, Qu AP (2006) Evaluation of USDA National clonal *Pyrus* germplasm collection for resistance to *Podosphaera leucotricha*. HortScience 41:717–720

Shabi E, Rotem J, Loebenstein G (1973) Physiological races of *Venturia pirina* on pear. Phy-topathology 63:41–43

Shaw PW, Brewer LR, Wallis DR, Bus VGM, Alspach PA (2003) Susceptibility of seedling *Pyrus* clones to pear sawfly (*Caliroa cerasi*)(Hymenoptera: Tenthredinidae) damage. NZ J Crop Hor-tic Sci 31:9–14

Shaw PW, Wallis DR, Alspach PA, Brewer LR, Bus VGM (2004) Pear sawfly (*Caliroa cerasi*) (Hymenoptera: Tenthredinidae) host preference and larval development on six *Pyrus* geno-types. NZ J Crop Hortic Sci 32:257–262

Shen DX, Li ZL, Zheng SQ (1979) Inheritance of fruit characteristics in pears. J Zhejiang Agric Univ 5:83–94

Shen DX, Li ZL, Zheng SQ, Chen HQ, Lin JB (1982) Studies on the correlations between juvenile period and growth in pear seedlings (in Chinese). Acta Hortic Sin 7:25–30

Shen T (1980) Pears in China. HortScience 15:13–17

Shibli RA, Ajlouni MM, Obeidat A (2000) Direct regeneration from wild pear (*Pyrus syriaca*) leaf explants. Adv Hortic Sci 14:12–18

Shin YU, Yim YJ, Cho HM, Yae BW, Kim MS, Kim YK (1983) Studies on the inheritance of fruit characters of Oriental pear, *Pyrus serotina* Rehder var. culta (in Korean). Res Rep Office Rural Dev 25 (Hort):108–117

Simard MH, Michelesi JC (2002) 'Pyriam', a new rootstock for pear. Acta Hortic 596: 351–355

Simovski K, Ristevski B, Spirovska R (1968) Effects of negative temperatures and temperature fluctuations on pears (in Macedonian). Annu Fac Agrie Sylvie Skopjl Agric 21:125–129

Spiegel-Roy P, Alston FH (1979) Chilling and post-dormant heat requirement as selection criteria for late-flowering pears. J Hortic Sci 54:115–120

Stotz HU, Powell ALT, Damon SE, Greve LC, Bennett AB, Labavitch JM (1993) Molecular char-acterization of a polygalacturonase inhibitor from *Pyrus communis* L. cv. Bartlett. Plant Physiol 102:133–138

Stushnoff C, Garley B (1982) Breeding for cold hardiness. In: van de Zwet T, Childers NF (eds) The pear. Horticultural Publ, Gainesville, FL, pp 189–199

Suzuki Y, Maeshima M, Yamaki S (1999) Molecular cloning of vaculor H(+)-pyrophosphatase and its expression during the development of pear fruit. Plant Cell Physiol 40:900–904

Takasaki T, Moriya Y, Okada K, Yamamoto K, Iwanami H, Besso H, Nakanishi T (2006) cDNA cloning of nine S alleles and establishment of a PCR-RFLP system for genotyping European pear cultivars. Theor Appl Genet 112:1543–1552

Tateishi A, Inoue H, Shiba H, Yamaki S (2001) Molecular cloning of β-galactosidase from Japanese pear (*Pyrus pyrifolia*) and its gene expression with fruit ripening. Plant Cell Physiol 42:492–493

Tateishi A, Mori H, Watari J, Nagashima K, Yamaki S, Inoue H (2005) Isolation, characterization, and cloning of α-L-Arabinofuranosidase expressed during fruit ripening of Japanese pear. Plant Physiol 138:1653–1664

Terakami S, Shoda M, Adachi Y, Gonai T, Kasumi M, Sawamura Y, Iketani H, Kotobuki K, Patocchi A, Gessler C, Hayashi T, Yamamoto T (2006) Genetic mapping of the pear scab resistance gene *Vnk* of Japanese pear cultivar Kinchaku. Theor Appl Genet 113:743–752

Thibault B, Hermann L, Belouin A, Mangin B (1988) Inheritance of some agronomical traits in pear. Acta Hortic 224:199–209

Thibault B, Lecomte P, Hermann L, Belouin A (1987) Assessment of the susceptibility of *Erwinia amylovora* of 90 varieties or selections of pear. Acta Hortic 217:305–309

Thompson JM, van der Zwet T, Ditto WA (1974) Inheritance of grit content in fruits of *Pyrus communis* L. J Am Soc Hortic Sci 99:141–143

Thompson JM, Zimmerman RH, van der Zwet T (1975) Inheritance of fire blight resistance in pear, I: a dominant gene, Se, causing sensitivity. J Hered 66:259–264

Tufts WP, Day LH (1934) Nematode resistance of certain deciduous fruit tree seedlings. Proc Am Soc Hortic Sci 31:75–82

Tukey HB (1964) Dwarfed fruit trees. Macmillan, New York

Tuz AS (1972) The inheritance of the dwarf growth factor in pear, *Pyrus domestica* Medii (in Russian). Genetika 8:16–20

Ushijima K, Sassa H, Hirano H (1998) Characterization of the flanking regions of the S-RNase genes of Japanese pear (*Pyrus serotina*) and apple (*Malus × domestica*). Gene 211:159–167

van der Zwet T, Bell RL (1990) Fire blight susceptibility in *Pyrus* germplasm from Eastern Europe. HortScience 25:566–568

van der Zwet T, Keil HL (1979) Fire blight: a bacterial disease of Rosaceous plants. Agric Handb 510 USDA, Washington, DC

van der Zwet T, Oitto WA, Brooks HJ (1970) Scoring system for rating the severity of fire blight in pear. Plant Dis Rep 54:835–839

Vavilov NI (1951) The origin, variation, immunity and breeding of cultivated plants. Translated by K Start Chron Bot 13:1–366

Vavra M, Orel V (1971) Hybridization of pear varieties by Gregor Mendel. Euphytica 20:60–67

Verhaegh JJ, Visser T, Kellerhals M (1988) Juvenile period of apple seedlings as affected by rootstock, bud origin and growth factors. Acta Hortic 224:133–139

Visser T (1951) Floral biology and crossing technique in apples and pears (in Dutch). Meded Dir Tuinb 14:707–726

Visser T (1967) Juvenile period and precocity of apple and pear seedlings. Euphytica 16:319–320

Visser T (1976) A comparison of apple and pear seedlings with reference to the juvenile period, II: mode of inheritance. Euphytica 16:339–342

Visser T, De Vries DP (1970) Precocity and productivity of propagated apple and pear seedlings as dependent on the juvenile period. Euphytica 19:141–144

Visser T, Schaap AA, De Vries DP (1968) Acidity and sweetness in apple and pear. Euphytica 17:153–167

Visser T, Verhaegh JJ, De Vries DP (1971) Pre-selection of compact mutants induced by x-ray treatment of apple and pear. Euphytica 20:195–207

Visser T, Verhaegh JJ, De Vries DP (1976) A comparison of apple and pear seedlings with reference to the juvenile period, I: seedling growth and yield. Euphytica 25:343–351

Vondracek J (1982) Pear cultivars resistant to scab. In: van der Zwet T, Childers NF (eds) The pear. Horticultural Publ, Gainesville, FL, pp 420–424

Wang YL (1990) Pear breeding in China. Plant Breed Abstr 60:877–879

Wang YL, Wei WD (1987) Studies on the inheritance of commercial characteristics in pear crossed seedlings (in Chinese). 1. Decid Fruits 2:1–4

Webster AD (1998) A brief review of pear rootstock development. Acta Hortic 475:135–141

Wellington R (1913) Inheritance of the russet skin in the pear. Science 37:156

Westigard PH, Westwood MN, Lombard PB (1970) Host preference and resistance of *Pyrus* species to the pear psylla, *Psylla pyricola* Foerster. J Am Soc Hortic Sci 95:34–36

Westwood MN (1976) Inheritance of pear decline resistance. Fruit Var J 30:63–64

Westwood MN (1982) Pear germplasm of the new national clonal repository: Its evaluation and uses. Acta Hortic 124:57–65

Westwood MN, Bjornstad HO (1971) Some fruit characteristics of interspecific hybrids and extent of self-sterility in Pyrus. Bull Torrey Bot Club 98:22–24

Westwood MN, Westigard PH (1969) Degree of resistance among pear species to the woolly pear aphid, *Eriosoma pyricola*. J Am Soc Hortic Sci 94:91–93

White AG, Alspach PA (1996) Variation in fruit shape in three pear hybrid progenies. NZ J Crop Hortic Sci 24:409–413

White AG, Alspach PA, Weskett H, Brewer LR (2000) Heritability of fruit shape in pears. Euphytica 112:1–7

Whitesides SK, Spotts RA (1991) Susceptibility of pear cultivars to blossom blast caused by *Pseudomonas syringae*. HortScience 26:880–882

Wu H-Q, Zhang S-L, Qu H-Y (2007) Molecular and genetic analysis of S_4^{SM} RNase allele in Japanese pear 'Osa-Nijisseiki' (*Pyrus pyrifolia* Nakai). Plant Breed 126:77–82

Yamada K, Kojima T, Bantog N, Shimoda T, Mori H, Shiratake K, Yamaki S (2006) Cloning of two isoforms of soluble acid invertase of Japanese pear and their expression during fruit development. J Plant Physiol 164:746–755

Yamamoto T, Kimura T, Sawamura Y, Kotobuki K, Ban Y, Hayashi T, Matsuta N (2001) SSRs isolated from apple can identify polymorphism and genetic diversity in pear. Theor Appl Genet 102:865–870

Yamamoto T, Kimura T, Sawamura Y, Manabe T, Kotobuki K, Hayashi T, Ban Y, Matsuta N (2002a) Simple sequence repeats for genetic analysis in pear. Euphytica 124:129–137

Yamamoto T, Kimura T, Shoda M, Ban Y, Hayashi T, Matsura N (2002b) Development of microsatellite markers in the Japanese pear (*Pyrus purifolia* Nakai). Mol Biol Notes 2:14–16

Yamamoto T, Kimura T, Shoda M, Imai T, Saito T, Sawamura Y, Kotobuki K, Hayashi T, Matsuta N (2002c) Genetic linkage maps constructed by using an interspecific cross between Japanese and European pears. Theo Appl Genet 106:9–18

Yancheva SD, Shlizerman LA, Golubowicz S, Yabloviz Z, Perl A, Hanania U, Flaishman MA (2006) The use of green florescent protein (GFP) improves *Agrobacterium*-mediated transformation of 'Spadona' pear (*P. communis* L.). Plant Cell Rep 25:183–189

Yu DY, Zhang P (1979) Sinkiang pears, a new series of cultivars of pears in China. Acta Hortic 6:27–32

Zavoronkov PA (1960) Breeding winter-hardy pear varieties (in Russian). Sadovodstvo Hortic 11:28–31

Zeven AC, Zhukovsky M (1975) Dictionary of cultivated plants and their centres of diversity. Centre for Agricultural Publishing and Documentation, Wageningen

Zhejiang Agricultural University (1977) Inheritance of some characters in pear seedlings. Proc Nat Acad Conf Pears 1976:114–121

Zhejiang Agricultural University (1978) Studies on the inheritance of the precocity in pears (in Chinese). Acta Genet Sin 5:220–226

Zhu LH, Ahlman A, Welander W (2003) The rooting ability of the dwarfing rootstock BP10030 (*Pyrus communis*) was significantly increased by introduction of the *rolB* gene. Plant Sci 135:829–835

Zhu LH, Welander W (2000) Adventitious shoot regeneration of dwarfing pear rootstocks and the development of a transformation protocol. J Hortic Sci Biotechnol 75:745–752

Zielinski QB (1963) Precocious flowering of pear seedlings carrying the Cardinal Red color gene. J Hered 54:75–78

Zielinski QB, Reimer FC, Quackenbush VL (1965) Breeding behavior of fruit characteristics in pears, *Pyrus communis* L. Proc Am Soc Hortic Sci 86:81–87

Zielinski QB, Thompson MM (1967) Speciation in *Pyrus*: Chromosome number and meiotic behavior. Bot Gazette 128:109–112

Zimmerman RH (1972) Juvenility and flowering in woody plants: a review. HortScience 7:447–455

Zimmerman RH (1976) Transmittance of juvenile period in pears. Acta Hortic 56:219–224

Zimmerman RH (1977) Relation of pear seedling size to length of the juvenile period. J Am Soc Hortic Sci 102:443–447

Chapter 11
Plums

W.R. Okie and J.F. Hancock

Abstract Most of the plums grown commercially are either the hexaploid, *Prunus domestica* (European) or the diploid, *P. salicina* (Asian or Japanese). Common goals of European plum breeders are cold hardiness, modest tree size, self fertility and productivity. Some of the key abiotic problems confronting Japanese plum production are susceptibility to spring frosts, insufficient winter hardiness and limited soil adaptations. Fruit quality and disease resistance are important goals in all plum breeding projects. The genetics of only a few traits have been investigated in plum; however, significant progress has been made in identifying horticulturally useful germplasm. A Myrobalan plum clone was crossed with an almond-peach hybrid to generate a microsatellite genetic linkage map and a resistance gene to the root knot nematode (*Ma*) was identified. A transgenic European plum clone was produced that carries the plum pox virus coat protein gene (PPV-CP) and has strong resistance to all four major serotypes of PPV.

11.1 Introduction

Plums contain a hard pit, and thus are classified with other stone fruits in the genus *Prunus* of the Rosaceae. Most of the plums grown commercially fall into one of two groups: European (hexaploid) or Japanese (diploid) types (Fig. 11.1). European plums (primarily *Prunus domestica*) are generally better adapted to cooler regions than Japanese types.

Within *P. domestica*, several groups of cultivars are recognized such as Green Gage (or Reine Claude) types and prunes. The *insititia* subspecies of *P. domestica* includes bullaces, damsons, mirabelles and St. Julien types. The greengage, mirabelle and damson plums are used a great deal in the food processing industry. They are processed into jams, jellies, canned fruit, juices (prune juice, for example), and alcoholic drinks such as brandies and cordials. European plums with a sugar content high enough so that they can be dried with the pit intact are referred to as

W.R. Okie
USDA-ARS, SE Fruit and Tree Nut Research Lab, 21 Dunbar Rd, Byron, Georgia, 31008 USA
e-mail: william.okie@ars.usda.gov

Fig. 11.1 Bearing habit of a typical Japanese-type shipping plum

prunes. In some countries the term 'prune' refers primarily to the dried product; elsewhere the term refers to the fresh fruit as well. Nearly all prune production in California, and much of the world, is of 'French Prune' and its clones, under such names as 'Prune D'Agen', 'Petite Prune' and 'Prune D'Ente'.

Leading countries in European plum production are the former U.S.S.R., Romania, Yugoslavia, Germany, United States and Hungary. Part of this production is processed into dried fruit. Production of prunes is concentrated in the United States, primarily California, followed by France, Yugoslavia, Chile and Argentina. In California there are nearly 35,000 hectares of prunes concentrated in the Sacramento, Santa Clara, Sonoma, Napa and San Joaquin Valleys. Currently, these farms produce more than twice as many dried plums as the rest of the world combined; about 99% of the U.S.A. supply and 70% of the world supply. Lesser amounts of European plums are grown in Idaho, Washington, Oregon, Michigan, and New York for both fresh and canned use.

The term 'Japanese plum' originally was applied to *Prunus salicina* (formerly *P. triflora*), but now includes all the fresh-market plums developed by intercrossing various diploid species with the original species. These plums were initially improved in Japan and later, to a much greater extent, in the United States. Most of these plums are consumed as fresh fruit and in many areas of the world these are the predominant plum found in the grocery store. A wide array of skin and flesh colors are available (Fig. 11.2), but in recent years the market has been dominated by black skin with light yellow or red flesh. Production of Japanese plums is led by China, followed by the United States. Production in the United States is concentrated in

Fig. 11.2 Range in fruit color and size for diploid plum species and breeding lines

California. Substantial production also comes from Mexico, Italy, Spain, Chile, Pakistan, Republic of Korea, Egypt, Australia, South Africa, and Argentina. Production of Japanese plums has been increasing in Europe and Asia. Although most U.S. Japanese plum production is in California, Japanese plums are grown in small quantities in many states of the U.S.A.

11.2 Evolutionary Biology and Germplasm Resources

Within Rosaceae, the sub-family Prunoideae is distinguished by having simple leaves and a 1- carpelled, drupaceous fruit with a deciduous calyx. Plums are separated from cherries by lack of a terminal bud, presence of a suture and a waxy bloom on the fruit, and a flatter pit. Plums are placed in the Prunophora sub-genus to separate them from peaches, almonds and cherries based on having sutured fruit with a waxy bloom, solitary axillary buds, and no terminal buds. Peaches and almonds differ in having three axillary buds with usually sessile, solitary flowers, and conduplicate (rolled) leaves in a terminal bud. Within Prunophora, section Euprunus contains the Asian and European species of plums, distinguished by 1–2 flowers per bud, stone often sculptured, and leaves rolled in the bud. Section Prunocerasus contains the American plum species, that have three or more flowers per bud, smooth stone and leaves folded in the bud. However, Asian species appear to fit better in Prunocerasus both taxonomically and horticulturally. In the real world, these characteristics are not definitive, there being exceptions to most of them.

There are 20–40 plum species depending on authority, many of which intergrade from one to another in the wild (Table 11.1). There are three independently

Table 11.1 Important sources of germplasm in plum breeding ($2x = 2n = 16$)

Species	Chromosome number	Location+	Useful characters
P. allegehaniensis Porter	16	Conn. To Penn.	Resistance to crown gall
P. americana Marsh.	16	Eastern U.S.A. to Rocky Mountains	Tough skin; very winter hardy
P. angustifolia Marsh.	16	New Jersey to Florida, west to Illinois and Texas	Resistance to bacterial leaf spot; limited tolerance to plum leaf scald
P. besseyi Bailey	16	Manitoba to Wyoming, south to Kansas and Colorado	Late bloom; high heat threshold; very winter hardy; resistant to crown gall
P. cerasifera Ehrh.	16 (24,32,48)	Western Asia, Caucasus, Balkan	Earliness; nematode resistance
P. domestica L.	48	Europe	High flavor and fruit quality
P. hortulana Bailey	16	Midwest and S.E. U.S.A.	Resistance to bacterial leaf spot
P. maritima Marsh.	16	Maine to Virginia	Late bloom; high heat threshold
P. mexicana S.Wats.		Southern U.S.A. to Texas	Large tree; low suckering
P.munsoniana Wight & Hedr.	16	Midwest U.S.A.	Good fruit; productive
P. nigra Ait.	16	New Brunswick to Assiniboine Mts, south to New York, Ohio and Wisconsin	Very winter hardy
P. salicina Lindl.	16 (32)	China	Good size, color and attractiveness; exceptional firmness and keeping quality at high temperatures; very winter hardy
P. simonii Carr.	16	China	Firmness; upright tree
P. spinosa L.	32	Europe	Disease resistance
P. subcordata Benth.	16	California, Oregon	Drought tolerance; high chill requirement
P. umbellata Ell.	16	North Carolina to Florida, Alabama, Mississippi and Texas	Resistance to crown gall

Source: Okie and Weinberger 1996, Ramming and Cociu 1991

domesticated groups of plums. The hexaploid ($2n = 6x = 48$) European plum, *P. domestica*, is the most commonly grown species in cooler regions. The diploid ($2n = 2x = 16$) Asian or Japanese plum, *P. salicina*, originated in China. The North American plums, such as *P. americana*, *P. angustifolia*, *P. maritima* and *P. subcordata*, were grown at a number of locations across the U.S.A. and Canada by native peoples and early Americans (Fig. 11.3).

The European plum, *P. domestica,* represents all the hexaploid plums, including those formerly classified as *P. insititia* but now given subspecies rank. That subspecies is characterized by a smaller tree, smaller leaf, and smaller fruit that is generally processed in some way rather than eaten fresh. Crane and Lawrence (1956) thought *P. domestica* originated in Asia Minor as a triploid hybrid between *P. cerasifera* (Myrobalan plum) and the tetraploid *P. spinosa* L., which then doubled to produce a fertile hexaploid. Newer cytological work, indictes that *P. spinosa* itself carries the genome from *P. cerasifera* plus a second one from an unknown ancestor (Reynders-Aloisi and Grellet 1994). Thus, *P. domestica* may be descended from polyploid forms of *P. cerasifera*, which has a long history of local use and selection across the continent, and has a range of fruit color and palatability.

Fresh and dried fruit of *Prunus cerasifera* have been used for centuries in West Asia from the Tien Shan and Pamir mountains over to the Caucasus Mountains. Many local cultivars have been selected for fruit. Myrobalan is also widely used worldwide as a rootstock for plum. Yoshida suggests *P. cerasifera* is the progenitor of all plum species, because of its native range, and cross- and graft-compatibility with many other species (Okie and Weinberger 1996).

Fig. 11.3 Range in leaf shape and size for North American plum species

Although sometimes used for drying and processing, most wild *P. spinosa* fruit are bitter. This species ranges from Scandinavia across Europe to Asia Minor. In Soviet Georgia, natural *P. spinosa* have been found with $2n = 16, 32, 48, 64$ or 96. Natural hybrids ($2n = 48$) between *P. cerasifera* and *P. spinosa* have also been found.

Low-chilling types of the Japanese plum, *P. salicina,* are located in southern China and Taiwan. Cold-hardy plums in northern China have been classified as *P. ussuriensis* and *P. gymnodonta,* but are otherwise very similar to *P. salicina.* Modern breeding programs, especially in the areas of the former U.S.S.R., have utilized this source of hardiness. Western taxonomists have described other Chinese species such as *P. thibetica,* and *P. consociiflora,* but these are not listed in Chinese taxonomic references as distinct species and probably represent variants within *P. salicina. Prunus simonii* was described by Western botanists based on cultivated specimens. This species (probably the same clone each time) was used in developing California cultivars because of its firm flesh and strong flavor. Chinese botanists describe it as native to north China, and occasionally cultivated. It has some characters reminiscent of apricot and was thought by some to have descended from a natural hybrid, but more likely is just an upright variant of *P. salicina* (Okie and Weinberger 1996).

Collections of plum germplasm consist primarily of local selections and cultivars, plus a small amount of wild accessions. Because most plum breeding programs are for cultivar development and use primarily adapted, improved parents, there is little systematic evaluation of the wild germplasm. A major collection of *P. salicina* is at the Research Institute of Pomology, Chinese Academy of Agricultural Sciences, Xingcheng, Liaoning, China. Several European research institutions have large collections of European plums, including the Institute of Plant Genetics and Crop Plant Research Fruit Genebank, Dresden, Germany and the Swedish University of Agricultural Sciences, Balgard Department of Horticultural Plant Breeding, Kristianstad, Sweden. Large collections of both diploid and hexaploid plums are found at the Institut National de la Recherche Agronomique, Bordeaux and Avignon, France as well as at the United States Department of Agriculture – Agricultural Research Service, National Clonal Germplasm Repository, Davis, California, U.S.A. Unfortunately, most of the wild plum species and relatives are poorly represented in these collections.

11.3 History of Improvement

European plums have been a commonly grown garden tree in Europe since the first century A.D. Several cultivars known in 1597 are still grown, such as 'Reine Claude'. One of the earliest plum breeders was Thomas Andrew Knight in England, whose work encouraged nurseryman Thomas Rivers who released 'Early Rivers' in 1834, followed by 'Early Transparent Gage', 'Czar', 'Monarch' and 'President'. By the early 1900s, plum breeding was being carried out at Long Ashton (later East Malling) and John Innes research stations (Roach 1985). In other European countries, local selections of the older cultivars were made and became established, but little formal breeding was done until later. In Eastern Europe, intentional breeding goes back over 50 years, with many cultivars released.

Early settlers to North America brought European plums with them, but the plums thrived only in more northern areas. A few selections were made from this germplasm base, although improvements were minor. Luther Burbank developed cultivars of European plums, but only 'Giant', 'Sugar' and 'Standard' became important commercially. The first public breeding program for European plums was established at Geneva, New York in 1893. This program released 'Stanley' in 1926, which is still important in many countries. Breeding began at Vineland, Ontario in 1913 and they have released numerous fresh market cultivars adapted to northern North America, including 'Valor' (1967), 'Verity' (1967), 'Vision' (1967), 'Veeblue' (1984) and 'Voyageur' (1987), all of which are commercially planted in Ontario.

Stones from plums have been found in Japan dating back to the Yayoi Era, about 2,300 years ago. Japanese books that are 1,500 years old mention cultivated plums. Plums have been common garden plants in Japan for centuries, but improvement efforts have only occurred in the last century. Plum culture in Japan and also Korea is so ancient that it is not possible to tell if the countries were ever part of the native range for plums. Trees of improved *P. salicina* cultivars 'Kelsey' and 'Abundance' were introduced into the United States from Japan over 100 years ago. Luther Burbank intercrossed these and other imports with *P. simonii* and North American species, resulting in 'Beauty', 'Burbank', 'Duarte', 'Eldorado', 'Formosa', 'Gaviota', 'Santa Rosa', 'Satsuma', 'Shiro', and 'Wickson'. These plums formed the basis for the world's shipping plum industry, and some are still widely grown. Pure *P. salicina* and related species have been little used as parents since Burbank's early hybridizations and few pure *P. salicina* clones are available outside of China. Most of Burbank's plums are thought to have descended from *P. salicina*, *P. simonii* and *P. americana*. In general, *P. salicina* contributed size, flavor, color and keeping ability; *P. simonii* contributed firmness and acidity; whereas the American species gave disease resistance, tough skin and aromatic quality. Burbank was fortunate in having improved native material available to supply these characters (Okie and Weinberger 1996).

With the advent of Burbank's improved plums that were large and firm enough to ship long distances, a new industry developed in California that caused industries in other states to mostly die out. As local industries declined, breeding programs were closed. California-bred plum cultivars were tried around the world, but with the exception of a few places like Chile and some parts of Italy, they have not thrived as well as they did in California. As a result they were crossed with the local plums of the particular area. In the northern U.S.A., cold-hardy species such as *P. americana*, *P. nigra* and *P. besseyi* were crossed to the most adapted Japanese types to improve the plums that could be grown there. In the southeastern U.S., the Japanese plums were crossed with the local *P. angustifolia* to enhance disease resistance, resulting in plums such as 'Bruce' and 'Six Weeks'. Unfortunately for modern breeders, only a few of the improved native American selections are still available, since their cultivation is obsolete.

While *P. cerasifera* is a progenitor of European plums, it is a diploid species that is cross-fertile with Asian and American diploid species. These 'cherry plums' have not been used much in modern breeding, although two cultivars were selected from chance hybrids with *P. cerasifera*, 'Methley' in South Africa and 'Wilson' in Australia. *Prunus cerasifera* is a source of earliness, cold-hardiness and probably self-fertility, but fruit size is small.

11.4 Current Breeding Efforts

European plum breeding is naturally concentrated in Europe. Similar goals are important in former Yugoslavia, Romania, Czech Republic, and Bulgaria. Since much of the plum production is dried or processed into brandy and other products, high soluble solids are essential. Releases include both prunes and improved fresh market types. There are at least 10 breeding programs in the former U.S.S.R. They require cold hardiness, modest tree size, self fertility, and productivity. In the more southern zones, larger size (>1 oz, 30 g), higher sugar content (>13%), purple fruit, and earliness are desired. Breeding efforts in Western Europe have increased in recent years. At INRA in Bordeaux, France, goals have been to develop a series of drying prunes and dessert plums that are adapted to French conditions. Cross fertile prunes are needed that produce fruit with similar traits to improve cross pollination and maximize fruit set. Fresh plums are required that ripen before and after 'Reine Claude', with equal or better flavor and firmness, and high productivity. In the last 20 years programs have started or restarted in Germany, Switzerland, Sweden and Norway. Most of these efforts are aimed at developing better fresh market plums, with emphasis on disease resistance, particularly plum pox. 'Stanley' has been a good parent to transmit tolerance to this disease. In Italy, breeding began at Florence in 1970 to develop early ripening dessert plums with large, high quality fruit and vigorous productive trees. 'Ruth Gerstetter' has been the most important parent (Okie and Ramming 1999).

Objectives at Vineland, Ontario are to develop high quality dessert plums to complete a sequence of ripening dates from July to October. Selection criteria are cold hardiness, productivity and blue color. Despite the predominance of the Californian industry in prune production, little intentional breeding has been done there until recently. The University of California at Davis has reinstated their plum breeding project to develop prunes ripening before and after 'Improved French'. New cultivars must resemble and perform like 'Improved French' in order to fit standard production practices for dried fruit. Self pollinated seedlings of 'French Prune' display uniformly poor fruit quality, thus it is being crossed with other parents. Over the years other minor breeding programs have existed, the most important of which was USDA breeding at Prosser and later Beltsville, resulting in the recent release of 'Bluebyrd'. Currently work at USDA-Kearneysville, W.Va. centers on developing bio-engineered plums highly resistant to plum pox virus.

Japanese plum breeding in California has historically focused on size and firmness for shipping. Black skin color became very popular with the introduction of 'Friar' because it did not show bruises and was very productive. However, large, firm, highly colored fruit can be harvested prematurely resulting in reduced consumer quality. Plums showing some ground color may be easier to pick at the proper stage of maturity. Low prices and over-production of black plums have increased interest in other colors. Current objectives include a wider range of skin color and better eating quality. Red or black skin color and yellow or red flesh color appear to be most acceptable although green-skinned plums are shipped to Asian markets. Storage ability, particularly at the end of the season, is also important.

Japanese plum breeding by the USDA at Fresno, California resulted in the releases of 'Frontier' (1967); 'Friar' (1968), the predominant plum in the industry; 'Queen Rosa' (1972); 'Blackamber' (1980), another widely grown plum; and 'Fortune' (1990). 'Fortune', a red plum, represents a shift away from the dark-skinned plums which now predominate in the shipping market. New releases include 'Owen T' and 'John W', a high-quality plum ripe in September notable for its self-fertility (Okie and Ramming 1999).

Private breeders and growers in California have selected many important commercial Japanese plums. Many of the cultivars grown in California were found as chance or open-pollinated seedlings or as mutations, rather than planned hybridizations. Fred Anderson released 'Red Beaut', 'Black Beaut', and 'Grand Rosa'. John Garabedian developed 'Angeleno', still the major late plum. Floyd Zaiger released 'Joanna Red', 'Betty Anne', 'Hiromi Red', and 'Autumn Beaut', as well as 'Citation' rootstock, an interspecific hybrid, and numerous 'plum-apricot hybrids' under the trademarked terms 'pluot' and 'aprium' (some controversy exists over how much apricot blood some of these have in them). Breeders at Sunworld International (formerly Superior Farms) developed 'Black Diamond', 'Black Flame', 'Black Gold', 'Black Torch' and 'Sweet Rosa'. Their program is the largest of the private breeders, and as with most private programs, the releases are patented.

Japanese plum breeding in Europe is relatively new, but will become increasingly important, as demand grows for the large-fruited Japanese plums. Breeders at Rome and Forli are seeking smaller trees to reduce production costs in combination with large size, dark skin, and good eating quality. At Florence, goals are to develop self-fertile, late-blooming plums with high quality, particularly yellow-skinned types. Recently a breeding program has been established near Avignon for southern France where poor weather during pollination is a major problem and Sharka resistance is important. Brazil has three Japanese plum breeding programs aimed at developing lower chill red-fleshed plums with resistance to leaf scald and bacterial spot. Other programs in the Southern Hemisphere are found in South Africa and Australia. Their goals are development of large-fruited, high quality plums with resistance to bacterial spot and bacterial canker, and the ability to store without internal breakdown. Storage ability of four weeks is crucial to exporting the fruit by ship (Okie and Ramming 1999).

The main southern U.S.A. Japanese plum breeding program is USDA-ARS at Byron, Georgia. Their current breeding objectives include those of California plus additional disease resistance. Fruit firmness is somewhat less important because many local markets are available. Resistance is required to three primary diseases: bacterial leaf, fruit spot and twig canker [*Xanthomonas campestris* pv. *pruni* (Smith) Dye], bacterial canker (*Pseudomonas syringae* pv. *syringae* van Hall), and plum leaf scald (*Xylella fastidiosa* Wells et al.). The first two diseases are problems in many other countries that are trying to grow Japanese plums, such as Australia, New Zealand, Italy, and South Africa. Leaf scald is also a serious problem in Argentina and Brazil. In general, later bloom is more desirable but regions such as Florida and parts of Texas, Australia, and Brazil require even lower chilling requirements than those common in Japanese plums.

11.5 Genetics of Economically Important Traits

Plums genetics have been little studied relative to other crops, because of fewer breeding programs, and the self-incompatibility which makes selfed populations difficult to obtain. However, significant progress has been made in identifying horticulturally useful germplasm.

11.5.1 Pest and Disease Resistance

Plum production is limited by a number of fungal species (Okie and Weinberger 1996, Ramming and Cociu 1991). Brown rot [*Monilinia laxa* (Aderh. & Ruhl.) Honey] is the primary fruit disease in plums and is most important when it is rainy during bloom and fruit-ripening. Cankers caused by several pathogens can affect the longevity of plum trees, particularly in the southeastern United States where black knot [*Apisporina morbosa* (Schw.) ARK.] is a major problem. Leaf blotch (*Polystigma rubrum* Pers.) is an important problem in Europe, while rust [*Tranzschelia discolor* (Fckl.) Tranz. & Litv.] causes significant damage in warm production regions. Other fungal pathogens that are important worldwide include Phytophthora root rot, Armillaria root rot [*Armillaria mellea* (Vahl.:Fr.) P. Kumm], Verticillium wilt (*Verticillium dahliae* Kleb), Powdery mildew [*Podosphaera oxycanthae* (DC.) de Bary], rose mildew [*Sphaerotheca pannosa* (Wallr.:Fr.) Lev.], silver leaf or heart rot [*Stereum purpureum* (Pers.:Fr.) Fr.], plum pockets or bladder plum (*Taphrina communis* and *T. pruni* Tul) and peach scab [*Fusicladium carpophilum* (Thuem.) Oudem]. Shot hole is another common foliar problem that usually is caused by a fungus, *Stigmina carpophila* (Lev.) Ellis, but which can also be a manifestation of a genetic defect that can be overcome by breeding and selection (Weinberger and Thompson 1962).

Resistant genotypes have been identified for Armillaria root rot, black knot, Phytophthora root rot, silver leaf (Ramming and Cociu 1991), fruit spot, twig canker (Okie and Weinberger 1996), leaf blotch, leaf rust (Paunovic 1988) and stem cankers (Norton and Boyhan 1991). Resistance to stem cankers was associated with the spreading growth tree characteristic (Popenoe 1959). Tolerance has been described for plum leaf scald, and powdery mildew (Ramming and Cociu 1991). No source of resistance has been published for plum pockets (Atkinson 1971), bladder plum, peach scab, rose mildew and Verticillium wilt.

Two bacterial diseases are widespread on plums, bacterial canker (*Pseudomonas syringae* pv. *syringae* van Hall) and bacterial leaf spot [*Xanthomonas campestris* pv. *pruni* (Erw. Smith) Dow.]. In the southeastern U.S.A., susceptible cultivars often die before fruiting due to defoliation and dieback. Crown gall [*Agrobacterium tumefaciens* (Smith and Townsend)] is also widespread. Plum leaf scald (*Xylella fastidiosa* Wells) is an important problem in the southeastern U.S.A. Resistant or tolerant genotypes have been identified for all these diseases (Norton et al. 1991, Okie et al. 1992, Okie and Weinberger 1996).

Several viruses have significant negative impacts on plum productivity. Probably the most important is sharka (plum pox), which is found all across Europe and is transmitted by the aphid *Anurophis helicrissi*. Prune brown line, caused by tomato ring spot virus, is an important problem in North America, particularly on 'Stanley'. Other important viruses are Peach mosaic, Prunus ringspot and Rosette. High levels of tolerance are available for all these viruses, but complete resistance has only been described to the tomato ringspot virus (Albrechtova et al. 1989, Hartmann 1994, Ramming and Cociu 1991). A QTL for resistance to sharka has been identified (Dirlewanger et al. 2004a).

Among the most important insect pests on plums are plum curculio (*Conotrachelus nenuphar* Herbst), scale (*Aspidiotus perniciosus* Comstock), mites [*Panonychus silmi* (Koch)] and [*Tetranychus pacificus* (McGregor)], borers [*Synanthedon pictipes* (Grote & Robinson), *Sanninoidea exitiosa* (Say), *Scolytus rugulosus* (Ratzeburg) and *Anarsia lineatella*], fruit flys [*Ceratitis capitata* (Wiedemann), *Dacus dorsalis* Hendel, *Anastrepha ludens* (Loew)], aphids [*Myzus persicae* (Sultzer)] and thrips (Order *Thysanoptera*). There are no published reports of tolerance or resistance to these pests.

Root-knot, lesion and ring nematodes are also major problems wherever *Prunus* are grown. Resistance/tolerance has been reported in rootstocks to all these nematode pests (Okie 1987, Ramming and Cociu 1991). A gene for resistance to the root-knot nematode, *Ma*, has been mapped (Claverie et al. 2004, Dirlewanger et al. 2004b).

11.5.2 Morphological and Physiological Traits

The narrow genetic base of European plum cultivars has limited production to specific areas and made them highly susceptible to the vagaries of nature (Ramming and Cociu 1991). The European plums are restricted to areas with high numbers of chilling hours, cool summers and moderate winters. Many cultivars are self-unfruitful and as a result have poor fruit set during cool, wet pollination seasons. Rain-induced fruit cracking can also be a significant problem, along with shatter pits in some cultivars such as 'Stanley'. Japanese plums have a more diverse background than European ones, although inbreeding has restricted the genetic variability of the most widely grown cultivars (Ramming and Cociu 1991). Some of the key abiotic problems confronting Japanese plum production are susceptibility to spring frosts, insufficient winter hardiness and limited soil adaptations.

Considerable variability has been observed in the chilling requirement, cold hardiness, season of flowering and harvest date of European and Japanese plums, suggesting quantitative inheritance (Table 11.2). Other characteristics that are important in plum breeding are biennial bearing, branching habit and self-fruitfullness. The ability to set buds under high fruit load has been shown to be highly heritable and the spreading habit in *P. domestica* is dominant.

Most of the plum cultivars grown in the world are propagated on rootstocks that are selected for local soil characteristics and vigor requirements. A broad array of

Table 11.2 Genetics of adaptation, productivity, plant habit and fruit quality in plums

Attribute	Observations and sources
Adaptation	
Chilling requirement	Considerable variability exists suggesting quantitative inheritance; low and high chill cultivars have been identified (Okie and Weinberger 1995, Wilson et al. 1975)
Cold hardiness	Considerable variability exists suggesting quantitative inheritance; cold hardy genotypes have been identified (Okie and Weinberger 1995)
Season of flowering	Early and late blooming types have been identified (Okie and Weinberger 1995)
Harvest date	Quantitatively inherited in *P. salicinia* and *P. domestica* (Hansche et al. 1975, Vitanov 1972)
Productivity and habit	
Biennial bearing	Ability to set buds under high fruit load is heritable (Couranjou 1989)
Incompatibility	Most cultivars are self-incompatible, but compatible ones exist; many hybrid plums are poor pollen producers (Okie and Weinberger 1995); S-RNases were identified in plum that shared 84–94% nucleotide identity with other *Prunus* S-RNases (Sutherland et al. 2004)
Spreading habit	Spreading habit in *P. domestica* is dominant (Olden 1965)
Stamen length	Short stamens are dominant in *P. domestica* (Olden 1965)
Fruit quality	
Bloom	Thick bloom on fruit is dominant over thin in *P. domestica* (Okie and Weinberger 1995)
Firmness	Genotypes of *P. salicina* with exceptional firmness have been identified (Yamaguchi and Kyotani 1986)
Flavor	High flavored cultivars have been identified (Okie and Weinberger 1995)
Freestone character	Recessive gene in *P. salicinia* and *P. domestica*; interacts with fruit maturity and firmness (Okie and Weinberger 1995)
Skin color	Generally quantitatively inherited in *P. salicina* and *P. domestica*, although yellow skin is a single recessive gene (Hurter 1962, Weinberger and Thompson 1962, Vitanov 1972)
Shape	Quantitatively inherited in *P. salicinia* (Weinberger and Thompson 1962); single locus in *P. domestica* with oval > round > oblong (Okie and Weinberger 1995)
Size/weight	Quantitatively inherited in *P. salicinia* and *P. domestica* (Hansche et al. 1975, Weinberger and Thompson 1962)
Sugar content	Quantitatively inherited in *P. domestica* (Hansche et al. 1975)

rootstocks are now available to deal with waterlogging, alkalinity and hardiness (Okie 1987, Ramming and Cociu 1991). The most common rootstocks for European plums include 'Myro 29C', 'Marianna 2624', 'Marianna GF8-1', 'Brompton', 'Damas' and 'St. Julien'. In some areas Japanese plums are also grown on the adapted peach rootstocks, such as 'Lovell', 'Nemaguard' or 'Guardian®'. In other cases, Japanese plums are grown on myrobalan or marianna clonal or seedling stocks.

In cold regions, lack of winter bud hardiness often limits production. *Prunus* from northern climates carry the highest levels of cold-hardiness (Quamme et al. 1982). *Prunus americana* and *P. besseyi* native to the northern states, *P. nigra* native to Canada, and *P. ussuriensis* from northern China carry factors for winter hardiness. In recent years these species have been used more extensively in the former U.S.S.R. to develop hardy plums than in their home countries.

11.5.3 Flower Characters

Bloom time is determined by temperatures throughout the winter and spring. Most European plums require relatively high numbers of chilling hours (probably >1000 hours), whereas most Japanese plums need much fewer ($\sim$ 500–800 hours), meaning Japanese plums usually bloom before European types. The chilling requirements of plums have been little studied, but they seem to respond similarly to peaches. Yields can be reduced in warm regions if chilling hours are insufficient to break the rest period of both flower and leaf buds.

Very low-chill Japanese plums have been developed in Florida, California and Taiwan. Plum cultivars requiring <450 chill hours (below 7°C) include 'Gema de Ouro', 'Golden Talisma', 'Kelsey Paulista', 'Roxa de Itaquera', 'Sanguinea', 'Amerelinha', 'Carmesim' and 'Pluma-7' from Brazil; 'Salad', 'Donsworth', and 'Narrabeen' from Australia, and seedlings from Taiwan (Okie and Weinberger 1996). Other low-chilling plums are 'Reubennel' and 'Harry Pickstone' from South Africa; 'Gulfblaze', 'Gulfbeauty' and 'Gulfrose' from the United States; and 'Remolacha de Capuseo', 'Estrela Purpura', and 'Gigaglia' from Argentina.

Late blossoming can result in greater productivity in an area subject to spring frosts. Breeders in England have attempted to select for high heat requirement to prolong dormancy and delay bloom (Wilson et al. 1975). Northern U.S.A. plum species *P. besseyi* and *P. maritima* bloom very late in Byron, Georgia but fruit well, suggesting they have a higher heat requirement, higher heat threshold, or both, relative to other plums. *Prunus besseyi* has been shown to have a higher heat accumulation threshold than peach (Werner et al. 1988).

Most plum cultivars are self-incompatible and many are poor pollen producers, but highly self fertile ones have been identified (Okie and Weinberger 1996). The S-RNases that regulate self incompatibility in plums have been cloned and sequenced and have been shown to share 84–94% homology with other *Prunus* S-RNases (Sutherland et al. 2004). Since many plum cultivars are self-unfruitful, compatibility with other cultivars can strongly influence productivity. Olden (1965) found short stamens to be dominant in *P. domestica*. Low productivity has reduced the popularity of cultivars derived from native American species in Minnesota (Andersen and Weir 1967). Patterns of compatibility between cultivars are unpredictable, although they have been widely studied (Alderman and Weir 1951, Flory 1947, Tehrani 1991). The general trend is towards self-incompatibility, but compatibility within a ploidy level does occur. Some diploids such as 'Beauty', 'Climax', 'Methley', 'Friar', 'Simka', and 'Santa Rosa' are relatively self-fruitful. Many of the hybrid plums produce little viable pollen, which limits their ability to pollinate regardless of compatibility.

11.5.4 Tree Characters

Tree productivity, essential for successful cultivars, is associated with tree vigor, disease resistance, hardiness and other characteristics. A spreading type tree is easier to manage in the conventional orchard than an upright growing tree. Popenoe (1959) noted that resistance to *Xanthomonas* stem cankers in *P. salicina* cultivars was associated with the spreading growth tree character, perhaps because *P. simonii* as a parent imparted both upright habit and disease susceptibility. Many of the California cultivars have upright growth and bear primarily on spurs. This tree form is preferred for high-density plantings. In the Southeast, vigorous growth can compensate to some extent for the effects of plum leaf scald. Some cultivars such as 'Harry Pickstone' and 'Byrongold' also bear well on year-old shoots. However, healthy trees of cultivars such as 'Robusto' and 'Segundo' may be too vigorous, and require both winter and summer pruning to keep the tree open. Many wild plums and some Japanese seedlings, especially juvenile trees, have sharp spurs or thorns, which are unacceptable on commercial plums because they injure both the fruit and the picker, and can even puncture tractor tires.

Crane and Lawrence (1956) reported that purple color of leaves and fruit in *P. cerasifera pissardi* Bailey was controlled by a single pair of genes with heterozygous individuals having intermediate color intensity (as in peach). In *P. domestica* spreading character of growth appears to be recessive. Hairiness of growth and leaves was dominant to sub-glabrous (Olden 1965).

11.5.5 Fruit Quality

A number of factors are key in consumer acceptance of European plums including attractive fruit appearance, large size, firmness, good flavor and texture (Okie and Weinberger 1996). Processing quality is also increasing in importance for drying and brandy making. In the Asian plums, the most important factors are dark skin color and firmness. Several factors associated with plum appearance have been shown to be highly heritable including degree of bloom, color, shape and size (Table 11.2). Thick bloom has been shown to be dominant over thin in *P. domestica*. Skin color is largely quantitatively inherited in *P. domestica*, although yellow skin is regulated by a single recessive gene. Shape has been shown to be quantitatively inherited in *P. salicinia*, while a single locus regulates shape in *P. domestica*.

The fruit quality traits firmness and flavor are also highly heritable and genotypes have been identified that are exceptional for these (Table 11.2). Sugar content has been shown to be quantitatively inherited in *P. domestica* (Hansche et al. 1975). The freestone character in *P. domestica* has been found to be under the regulation of a single, recessive allele, although there is a strong interaction with fruit maturity and firmness. Oval fruit shape was reported to be dominant over round fruit shape which was dominant to oblong; yellow or green skin color recessive to red, purple, blue and black; thick bloom on fruit dominant over thin bloom; and freestone recessive to cling (Okie and Weinberger 1996, Renaud 1975). Hansche et al. (1975) reported fruit size, ripening date, and soluble solids to be highly heritable, in contrast to yield. Vitanov (1972) reported polygenic inheritance for skin color, stone freeness, and ripening date. Genetic factors also affect ability to set buds under heavy crop loads thus avoiding biennial bearing (Couranjou 1989).

The mode of transmission of a few *P. salicina* plum characters has been determined by Weinberger and Thompson (1962). Time of ripening is quantitatively inherited. The average ripening dates of progeny was close to the mid-parent value with some individuals ripening earlier or later than either parent. Size of fruit is also quantitatively inherited. When both parents had large-sized fruit approaching the extreme size, the progeny fruit average smaller in size than that of the parents. Shape of fruit is controlled by multiple factors with neither round nor ovate shape dominant. Yellow skin color is a single gene recessive to red, black, or purple, which intergrade and appear to be quantitatively inherited. Red flesh color is dominant over yellow, and a single factor is involved. The intensity of the anthocyanin color is controlled by multiple genes. Hurter (1962) also found red flesh color dominant over yellow, with monofactorial inheritance. The freestone character is apparently recessive, as occasional seedlings with freestone fruits were found in progeny from clingstone parents. The maturity of the fruit and the firmness of the flesh affects the degree of clinginess. Some plums are 'air-free', such that an air pocket surrounds the pit inside the flesh.

Several plum species have fruiting characteristics that would be valuable in a breeding program. *Prunus salicina* hybrids have good size, color, and attractiveness. Some have exceptional firmness and keeping quality. *Prunus cerasifera* plums transmit earliness. They produce progeny which are quite variable in hardiness, fruit form, and other characters even when selfed (Murawski 1959). Tough skin is often carried by *P. americana* which can be used to improve shipping quality.

11.6 Crossing and Evaluation Techniques

11.6.1 Breeding Systems

The European plums (*P. domestica*) are hexaploid ($2n = 48$, $x = 8$). *Prunus spinosa* plums are tetraploids ($2n = 32$). The Japanese plums (*P. salicina*) are diploid ($2n = 16$) as are *P. cerasifera*, *P. americana* and most other species. Since the two leading groups of commercial plums have different chromosome numbers, hybridization between them often gives poor results. Where both parents have the same number of chromosomes, interspecific hybridization is generally successful. Many hybrids have been made, particularly with *P. cerasifera*, *P. salicina*, *P. simonii*, *P. besseyi*, *P. americana*, *P. angustifolia*, *P. hortulana*, *P. munsoniana*, and *P. nigra*. Hybrids have also been made with *P. japonica*, Chinese bush cherry. Hybrids between the first four species and apricot (*P. armeniaca* and *P. mume*) have also been successful, but many are not very productive.

Several plumcots have been introduced ('Red Velvet', 'Royal Velvet', 'Flavor Supreme', 'Flavor Delight', 'Flavor Queen', 'Rutland', 'Plum Parfait', 'Dapple Dandy', 'Spring Satin', 'Yuksa', *P. × blireiana*). Others have been grown for generations in southwest Asia as *P. × dasycarpa* ('Irani Olju' and 'Tlor Csiran'). Fruit set has been a problem with plumcots, but has improved with some of the newer releases. After backcrossing and intercrossing these plumcots, it becomes difficult to distinguish hybrids from plums. The California industry has recently begun using the term 'interspecific' to describe these plumcot derivatives although the term overlooks the fact that most Japanese plums are already just that. Confusion over what is legally a plum affects marketing orders, grade standards, monetary box assessments, and pesticide usage (Okie and Ramming 1999).

11.6.2 Pollination and Seedling Culture

During the summer, reproductive buds develop in leaf axils. Buds contain either a vegetative axis, or a flower primordia. Initially all reproductive parts are enclosed in the five petals. The five sepals are fused to form a cup at the base of the 5 petals. Plum flowers range in diameter from 5–30 mm, with most commercial plums about 2–3 cm across, with petals opening flat atop the cup-shaped corolla. The 20–30 anthers are attached along with petals to the rim of the calyx cup. The single pistil protrudes above the corolla, but the stigma is positioned only slightly beyond the anthers at full bloom. Some plums form many spurs which carry flower buds, others flower mainly on previous seasons extension growth. For European plums there are usually 1–2 flowers per flower bud, whereas Japanese plum has 3 or more. Some species have as many as 5–6 flowers per bud.

European plums are often self-fertile in contrast to Japanese plums, which usually require ample bees to cross-pollinate the flowers. In practice bees are helpful to increase fruit set in European plums as well, since the often pistil extends above the stamens, and the pollen is little moved by the wind. Since most Japanese plums are self-incompatible, different cultivars blooming at the same time in close proximity are necessary for good fruit set. In years when spring weather inhibits bee flight, fruit set is often inadequate.

For hand pollination by breeders, pollen may be collected by gathering unopened blossoms just before the petals separate. They are placed in a wire screen sieve. Maceration of the flowers forces the anthers through the screen to be collected underneath. Alternately the blossoms can be clipped just above the top of the sepals so that petals and anthers fall into the screen. With gentle tapping the anthers fall through. Pollen produced this way will be cleaner. The anthers are dried overnight at room temperature or with slight extra heating. Pollen is easily shed from the dried anthers by manipulation with a camel's hair brush. Pollen of *P. domestica* can be stored for 550 days at 2°C and 25% humidity, and still produce high percentages of germination. Short-term storage in liquid nitrogen has also been successful. Emasculation of the larger flowers may be performed by grasping the calyx cup with the fingernails and tearing away the unopened corolla. The stamens are attached to the rim and are removed with it. The bare pistil and part of the calyx cup remain. Many breeders have found that fruit set is low after emasculation, perhaps because the pistil is damaged either during emasculation or by weather conditions after pollination. Cultivars of plums which are self-unfruitful do not need to be emasculated before pollination. Using them as seed parents reduces the work of making crosses and improves the chances of obtaining a good set of fruit. Some 'self-unfruitful' cultivars will set a small percentage of blossoms with their own pollen. For breeding purposes this is negligible, but for cytological and genetic studies emasculation is necessary. Cheesecloth, screens or row covers can also be used to enclose a honeybee hive with the tree without using a framework. If the tree is self-fertile, selfed seed can be obtained. Alternately, blooming potted trees or bouquets in buckets of water can be placed inside the cage to provide pollen for the bees to transfer (Okie and Weinberger 1996).

In situations where genetic purity is not essential, breeders often collect open-pollinated seed, particularly from commercial blocks where two desirable parents are interplanted. Some breeders have established small isolated blocks of elite lines. Open-pollinated polycross seed from these trees can be collected in a type of recurrent selection. This approach is particularly useful in producing large progenies where poor adaptation due to climate or disease susceptibility will eliminate many seedlings. Genes for improved adaptability will be concentrated using this approach with less effort expended making hand pollinations. Since many plums are self-infertile, bees can be used to make interspecific pollinations. Species of interest can be planted in the midst of a block of the second species and allowed to open-pollinate. Alternately, caged trees can be used with a beehive and bouquets. Bees are able to effect many more pollinations with less damage than humans.

Seed of early ripening cultivars usually germinate poorly. It is advisable to culture these seed on sterile nutrient agar after removing the endocarp and integuments. Current technology allows the successful culture of ovules as small as 0.6 mm. After-ripening of freshly cultured seed is not necessary, but may be of some help.

Seed of midseason and later-ripening cultivars usually give satisfactory germination when prevented from drying out. The rehydration of dry, stored seed often introduces bacterial and fungal infections.

11.6.3 Evaluation Techniques

Plum breeding work is best accomplished in the regions where the plums will be grown. Each region has its own climatic distinctions in which a new cultivar must be tested to prove its adaptability. High summer temperatures in the San Joaquin Valley of California, for example, can cause internal browning of flesh of some *P. domestica* cultivars. Testing of potential cultivars is more or less a local problem. The 12 most important fruit characters of interest to breeders are: time of maturity, size, shape, crop load, skin color, attractiveness of ground color, color of flesh, firmness, freeness of pit, texture, quality, and resistance to disease. These are not listed in order of importance but rather in sequence in which observations are usually made. Attractiveness is perhaps the most important feature, for a fruit must have consumer appeal to be successful in the markets. It must also be firm enough to arrive in markets in good condition, and must have adequate quality to assure repeat sales.

Plums are generally evaluated in the field when they first fruit 3–5 years after planting. Most breeders only take detailed evaluation notes on seedling trees that meet minimum requirements for vigor and disease resistance, and fruit size. A preliminary field rating for fruit characters identifies trees worthy of follow-up for post-harvest tests. After several years of cropping, seedlings will be propagated by budding and tested in a semi-commercial setting, followed by replicated or commercial trials.

11.7 Biotechnological Approaches to Genetic Improvement

11.7.1 Genetic Mapping and QTL Analysis

Genetic diversity among European *Prunus* rootstocks was assessed using RAPD markers. There was more diversity among *P. domestica* clones than *P. cerasifera* (myrobalan) stocks (Casas et al. 1999). Hexaploid and diploid plum cultivars were also studied, and found to be distinguishable via RAPD markers (Ortiz et al. 1997).

A microsatellite genetic linkage map of myrobalan plum clone P. 2175 and an almond-peach hybrid have been generated (Dirlewanger et al. 2004a,b). The linkage map of plum is composed of 93 markers that cover 524.8 cM. SSR markers from almond-peach as well as apricot were found to be widely transportable across various Prunus species, including plum (Messina et al. 2004). The resistance gene *Ma* was located on linkage group 7. Claverie et al. (2004) was able to use two closely flanking markers to identify a single BAC clone encompassing *Ma*.

11.7.2 Regeneration and Transformation

Adventitious shoots have been regenerated from European plum (*P. domestica*) using cotyledons (Mante et al. 1989), hypocotyls (Mante et al. 1991) and leaves (Novak and Miczynski 1996). Hypocotyl sections were used to produce a transgenic European plum clone, C5, which carried the plum pox virus coat protein gene (PPV-CP) and had strong resistance to all four major serotypes of PPV (Scorza et al. 1994, Ravelonandro et al. 1997). After 5–6 years of natural aphid vectored inoculation, trees of clone C5 remained virus free (Scorza et al. 2003). Other transgenic plum plants carrying the papaya ringspot virus coat protein gene (PRV-CP), delayed symptoms to PPV but the plants eventually became diseased (Scorza et al. 1995).

The expression patterns of the PPV-CP gene has been well studied and hybrids with C5 have been shown to carry the gene and have resistance (Ravelonandro et al. 1998, 2002, Scorza et al. 1998). The resistance of the C5 plants appeared to be RNA-mediated through post-transcriptional gene silencing (PTGS), where transgene mRNA is degraded in the cytoplasm soon after synthesis (Scorza et al. 2001).

References

Albrechtova L, Karesova R, Pluhar Z (1989) Evaluation of resistance of plum cultivars and hybrids to plum pox virus (in German). Z Pflanzenkr Pflanzenschutz 96:455–463

Alderman WH, Weir TS (1951) Pollination studies with stone fruits. Minn Agr Expt Sta Tech Bull, 198:1–16

Andersen ET, Weir TS (1967) Prunus hybrids, selections and cultivars at the University of Minnesota Fruit Breeding Farm. Minn Agr Expt Sta Tech Bull, 252:1–49

Atkinson JD (1971) Diseases of tree fruits in New Zealand. AR Shearer, Govt. Printer, Wellington, New Zealand

Casas AM, Igartua E, Balaguer G, Moreno MA (1999) Genetic diversity of *Prunus* rootstocks analyzed by RAPD markers. Euphytica 110:139–149

Claverie M, Bosselut N, Voisin R, Esmenjaud D, Chalhoub B, Direwanger E, Kleinhentz M, Laigret F (2004) High resolution map of the *Ma* gene for resistance to root-knot nematodes in myrobaslan plum (*Prunus cerasifera*). Acta Hortic 663:69–74

Couranjou J (1989) A second cultivar factor of biennial bearing in *Prunus domestica* L: the sensitivity of flower bud formation to fruit load. Sci Hortic 40:189–201

Crane MB, Lawrence WJC (1956) The genetics of garden plants (4th ed). London, Macmillan

Dirlewanger E, Cosson P, Howard W, Capdeville G, Bosselut N, Claverie M, Voisin R, Poizat C, Lafargure B, Baron O, Laigret F, Kleinhentz M, Arús P, Esmenjaud D (2004b) Microsatellite genetic linkage maps of myrobalan plum and an almond-peach hybrid – location of root-knot nematode resistance genes. Theor Appl Genet 109:827–838

Dirlewanger E, Graziano E, Joobeur T, Garriga-Caldere F, Cosson P, Howard W, Arús P (2004a) Comparative mapping and marker-assisted selection in Rosaceae fruit crops. Proc Natl Acad Sci USA 101:9891–9896

Flory WS (1947) Crossing relationships among hybrid and specific plum varieties, and among several Prunus species which are involved. Am J Bot 34:330–335

Hansche PE, Hesse CO, Beres V (1975) Inheritance of fruit size, soluble solids and ripening date in *Prunus domestica* cv. Agen. J Am Soc Hortic Sci 100:522–524

Hartmann W (ed) (1994) Fifth international symposium on plum and prune genetics, breeding and pomology. Acta Hortic 359:1–295

Hurter N (1962) Inheritance of flesh color in the fruit of the Japanese plum *Prunus salicina*. S Afr J Agric Sci 5:673–674

Mante S, Morgens P, Scorza R, Cordts JM, Callahan A (1991) Agrobacterium-mediated transformation of plum (*Prunus domestica* L.) hypocotyls slices and regeneration of transgenic plants. Biotechnology 9:853–857

Mante S, Scorza R, Corts JM (1989) Plant regeneration from cotyledons of *Prunus persica, Prunus domestica* and *Prunus cerasus*. Plant Cell Tissue Organ Cult 19:1–11

Messina R, Lain O, Marrazzo MT, Huang WG, Cipriani G, Testolin R (2004) Isolation of microsatellites from almond and apricot genomic libraries and testing for their transportability. Acta Hortic 663:79–82

Murawski H (1959) Contributions to breeding research on plums. II. Further investigations on the breeding value of seedlings (in German). Zuchter 29:21–36

Norton JD, Boyhan GE (1991) Inheritance of resistance to black knot in plums. HortScience 26:1540

Norton JD, Boyhan GE, Smith DA, Abrahams BR (1991) AU-Cherry plum. HortScience 26:1091–1092

Novak B, Miczynski K (1996) Regeneration capacity of *Prunus domestica* L. cv. Wegierka Zwykla from leaf explants of in vitro shoots using TDZ. Folia Hortic 8:41–49

Okie WR (1987) Plum Rootstocks. In: Rom RC, Carlson RF (eds) Rootstocks for fruit crops. Wiley, New York

Okie WR, Ramming DW (1999) Plum breeding worldwide. HortTechnology 9:162–176

Okie WR, Thompson JM, Reilly CC (1992) Segundo, Byrongold and Rubysweet plums and BY69-1637P plumcot: Fruits for the Southeastern United Sates. Fruit Varieties J 46:102–107

Okie WR, Weinberger JH (1996) Plums. In: Janick J, Moore JN (eds) Fruit Breeding, vol. 1. Tree and tropical fruits. John Wiley and Sons, Inc., NewYork

Olden EJ (1965) Interspecific plum crosses. Research Report 1. Balsgard Fruit Breeding Institute. Fjalkestad, Sweden

Ortiz A, Renaud R, Calzeda I, Ritter E (1997) Analysis of plum cultivars with RAPD markers. J Hortic Sci 72:1–9

Paunovic AS (1988) Plum genotypes and their improvement in Yugoslavia. Fruit Varieties J 42:143–151

Popenoe J (1959) Relation of heredity to incidence of bacterial spot on plum varieties in Alabama. Proc. Assoc. So. Agric. Workers 56th Annu Conv Memphis 1959:176–177

Quamme HA, Layne REC, Ronald WG (1982) Relationship of supercooling to cold hardiness and the northern distribution of several cultivated and native *Prunus* species and hybrids. Can J Plant Sci 62:137–148

Ramming DW, Cociu V (1991) Plums In: Moore JN, Ballington, JR (eds) Genetic resources of temperate fruit and nut crops. Acta Hortic 290(1):233–287

Ravelonandro M, Briard P, Monsion M, Scorza R (2002) Stable transfer of the plum pox virus (PPV) capsid transgene to seedlings of two French cultivars 'Prunier D'Ente 303' and 'Quetsche 2906' and preliminary results of PPV challenge assays. Acta Hortic 577:91–96

Ravelonandro M, Scorza R, Bachelier JC, Labonne G, Levy L, Damsteegt V, Callahan A, Dunez J (1997) Resistance of transgenic *Prunus domestica* to plum pox virus infection. Plant Dis 81:1231–1235

Ravelonandro M, Scorza R, Renaud R, Salesses G (1998) Transgenic plums resistant to plum pox virus infection and preliminary results of cross-hybridization. Acta Hortic 478:67–71

Renaud R (1975) The study of inheritance in the plum intraspecific cross-breeding (in French). Acta Hortic 48:79–82

Reynders-Aloisi S, Grellet E (1994) Characterization of the ribosomal DNA units in two related Prunus species (P. Cerasifera and P. Spinosa). Plant Cell Reports 13:641–646

Roach FA (1985) Cultivated fruits of Britain: Their origin and history. Basil Blackwell, New York

Scorza R, Callahan AM, Levy L, Damsteegt V, Ravelonandro M (1998) Transferring potyvirus coat protein genes through hybridization of transgenic plants to produce plum pox virus resistant plums (*Prunus domestica* L.). Acta Hortic 472:421–427

Scorza R, Callahan A, Levy L, Damsteegt V, Webb K, Ravelonandro M (2001) Post-transcriptional gene silencing in plum pox virus resistant transgenic European plum containing the plum pox potyvirus coat protein gene. Transgenic Res 10:201–209

Scorza R, Levy L, Damsteegt V, Yepes LM, Cordts JM, Hadidi A, Gonsalvez D (1995) Transformation of plum with the papaya ringspot virus coat protein gene and reaction of transgenic plants to plum pox virus. J Am Soc Hortic Sci 120:943–952

Scorza R, Ravelonandro M, Callahan A, Cordts JM, Fuchs M, Dunez J, Gonsalvez D (1994) Transgenic plums (*Prunus domestica*) express the plum pox virus coat protein gene. Plant Cell Rep 14:18–22

Scorza R, Ravelonandro M, Malinowski T, Minoliu N, Cambra M (2003) Potential use of trasgenic plums resistant to plum pox virus field infection. Acta Hortic 622:119–122

Sutherland BG, Tobutt KR, Robbins TP (2004) Molecular genetics of self-incompatibility in plums. Acta Hortic 663:557–562

Tehrani G (1990) Seventy-five years of plum breeding and pollen compatibility studies in Ontario. Acta Hortic 283:95–103

Vitanov M (1972) Inheritance of some traits and properties of fruits from the hybridization of plum varieties of *Prunus domestica* L.II. Time of ripening, stone adherence to flesh, fruit skin and flesh colour, correlations (in Bulgarian). Gen Syst 5:341–356

Weinberger JH, Thompson LA (1962) Inheritance of certain fruit and leaf characters in Japanese plums. Proc Am Soc Hortic Sci 81:172–179

Werner DJ, Mowrey BD, Young E (1988) Chilling requirements and post-rest heat accumulation as related to difference in time of bloom between peach and Western sand cherry. J. Amer. Soc. Hort. Sci. 113:775–778

Wilson D, Jones RP, Reeves J (1975) Selection for prolonged winter dormancy as a possible aid to improving yield stability in European plum (*Prunus domestica* L.). Euphytica 24:815–819

Yamaguchi M, Kyotani H (1986) Differences in fruit ripening patterns of Japanese plum cultivars under high (30°C) and medium (20°C) temperature storage. Bull Fruit Tree Res Sta A13:1–19

Chapter 12
Raspberries

C.E. Finn and J.F. Hancock

Abstract All raspberry breeders are interested in improving fruit quality and increasing the efficiency of fruit production. The development of primocane fruiting cultivars with excellent shipping quality has allowed major raspberry industries to emerge in non-traditional areas such as California. An increased interest in fruit chemistry, particularly anthocyanins, has led to many studies determining the inheritance of these compounds. Progress towards resistance to major diseases such as Phytophthora root rot has been made through greater understanding of the inheritance of these traits, and the use of novel and traditional germplasm resources. Black raspberry breeding efforts have been greatly increased in the early 21st Century in response to increased disease pressure and raised consumer awareness of the high levels of antioxidants in their fruit. A genetic linkage map of red raspberry ('Glen Moy' × 'Latham') has been constructed and used to search for QTL associated with cane spininess, root sucker density and root sucker spread. Transformation was used to develop a red raspberry cultivar with resistance to *Raspberry bushy dwarf virus,* although it was not commercialized.

12.1 Introduction

The most popular raspberry species grown commercially in temperate climates are *Rubus idaeus* L. (red raspberries) and *R. occidentalis* L. (black raspberries). There are also limited acreages of yellow raspberries grown, which are mutations of red raspberries, and purple ones, which are hybrids of red and black raspberry genotypes. Another domesticated species, *R. arcticus* L., is important in Scandinavia. Several native species have a small niche in the world market including *R. chamaemorus* L. in Scandinavia, *R. parvifolius* L., *R. niveus* Thunb. and *R. coreanus* Miq. in China and *R. phoenicolasius* Maxim. in Japan (Finn 1999).

C.E. Finn
USDA-ARS, Horticultural Crops Research Unit, 3420 NW Orchard Avenue, Corvallis, Oregon 97330, USA
e-mail: finnc@hort.oregonstate.edu

Raspberries are most productive in regions with mild winters and long, moderate summers. The major production areas of red raspberries in North America are the Pacific Northwest (Oregon, Washington and British Columbia), California, the eastern U.S. (New York, Michigan, Pennsylvania and Ohio) and rapidly expanding industries in Mexico and Guatemala. In Europe, red raspberries are grown to the largest extent in Serbia, Russia and Poland, with commercial production scattered all across the European Union. The value of the early season fresh market production in Spain is incredibly high, but their acreage is much less than the other countries listed. In the southern hemisphere, red raspberries are most widely grown in Chile and New Zealand.

Raspberry canes are biennial with the first year canes being called primocanes and the second year canes floricanes. Within the red raspberries there are two types of cultivars, the primocane (fall) fruiting and floricane (summer) fruiting. Floricane fruiting raspberries produce canes that are vegetative in the first year and in the second year they flower, fruit, die (floricanes) and are pruned out. Therefore, in a given planting, in a given year, there will be vegetative canes that will produce next year's crop and fruiting canes. Some of the more popular cultivars of this type are 'Tulameen', 'Glen Ample', 'Meeker' and 'Willamette'. The primocane fruiting red raspberry cultivars produce fruit in the fall at the top of the current season's primocanes and then again in the second year, if they are not pruned out. Some of the more popular cultivars of this type include 'Heritage', 'Caroline', 'Josephine', 'Amity', and 'Autumn Bliss' and the proprietary cultivars from companies like Driscoll's (Watsonville Cal.). While it is easiest to cut the canes of these cultivars off at ground level each winter after recovering just the late-summer primocane crop, the canes are sometimes left to over-winter and produce a very early spring crop. Because these primocane fruiting types can be double cropped in this way, they are sometimes called 'everbearing raspberries'.

Black raspberry cultivars are typically floricane fruiting. The primocanes that emerge from the crown are tipped in commercial plantings to about 1 m tall to encourage branching. During the winter the branches are cut back to about 45 cm. The following year these canes become floricanes and produce the crop. In the Northwest, where there is a strong but small (600–700 ha) industry, nearly all the commercial crop is planted in 'Munger', a cultivar released in 1890. A couple of primocane fruiting black raspberries exist; the very old cultivar 'Ohio Everbearer' (Hedrick 1925) and 'Explorer' (U.S. Plant Patent 17,727) which was patented in 2007.

Purple raspberries tend to have a great deal of 'hybrid vigor' and are crown forming and floricane fruiting with large, soft fruit. They are generally considered to have only fair quality fresh but truly shine when they are processed. 'Brandywine' and 'Royalty' are mostly commonly listed by commercial nurseries.

12.2 Evolutionary Biology and Germplasm Resources

Raspberries are in the genus *Rubus* of the Rosaceae. There are 15 subgenera recognized within *Rubus* (USDA 2007) with the domesticated raspberries being found in the subgenus *Idaeobatus. Idaeobatus* contains about 200 wild species with nine sections. Almost all of the raspberry species are diploid ($2n = 14$), with a few triploid and tetraploid types (Thompson 1995a,b, Thompson 1997). *Idaeobatus* species are concentrated in northern Asia, but are also located in Africa, Australia, Europe and North America (Jennings 1988). The greatest diversity is found in southwest China, the likely center of origin of the subgenus.

The major commercial taxa of raspberries share a considerable amount of inter-fertility. *Rubus idaeus* and *R. strigosus* are completely inter-fertile (Darrow 1920) and are often considered two subspecies of the same species. The cross of *R. occidentalis* × *R. idaeus* is only successful if *R. occidentalis* is used as the female parent, although bud pollination and heat treatment can help overcome this unilateral incompatibility (Hellman et al. 1982). At least 40 additional species in *Idaeobatus* have also been used in raspberry breeding, along with a few species in the *Cylactis, Anoplobatus, Chamaemorus, Dalibardastrum, Malachobatus,* and *Rubus* (Table 12.1).

Finn and Knight (2002) found that almost all raspberry breeding programs devote energy to the evaluation and incorporation of species germplasm. In Europe, at least 16 species have been evaluated and used as sources of new traits. In North America, at least 58 species have been evaluated and used in breeding. The program at East Malling has long been particularly active in incorporating genes from European species (Jennings 1988, Knight 1993).The USDA-ARS program in Oregon has more recently focused on Asian species, as have the programs in Maryland, North Carolina, Washington and British Columbia (Finn 1999, 2002a,b).

12.3 History of Improvement

The European red raspberry, *R. idaeus* was first mentioned in the historical record by Pliny the Elder. He described it as 'ida' fruit grown by the people of Troy at the base of Mount Ida. However, it is likely that these plants originally came from the Ide Mountains of Turkey, as raspberries were not native to Greece (Jennings 1988). Raspberries gradually grew in popularity over the centuries and by the 1500s, *R. idaeus* was cultivated all over Europe. In 1829, 23 cultivated varieties were listed by George Johnson in his 'History of English Gardening'. The North American *R. strigosus* was introduced into Europe in the early 19th century and natural hybrids with *R. idaeus,* resulted in much advancement. In fact, most red raspberry cultivars dating from this period are hybrids of these two species (Daubeny 1983, Dale et al. 1989, 1993).

The first formal breeding work on raspberries was begun in North America; Darrow (1937) cites Dr. Brinkle of Philadelphia, Pennsylvania as the 'first successful raspberry breeder of this country'. The most enduring cultivar from this early breeding period was 'Latham' which was introduced in 1914 by the Minnesota Fruit Breeding Farm and is still grown. Five early European cultivars played the dominant role in the breeding of red raspberries including, 'Pruessen', 'Cuthbert' and 'Newburgh', which are hybrids between the North American and European species, and 'Lloyd George' and 'Pyne's Royal', which are pure *R. idaeus.*

Table 12.1 Sources of germplasm that breeders have attempted to incorporate into their breeding material. $2x = 2n = 14$

Subgenus	Species	Ploidy	Location	Important traits
Idaeobatus	*R. biflorus* Buch.-Ham ex Sm.	2x	China	Low chilling requirement; resistant to drought, high temperature, leaf spot, cane spot
Idaeobatus	*R. chingii* Hu	2x	China	High yield, vigorous, fresh and processed fruit quality
Idaeobatus	*R. cockburnianus* Hemsl.	2x	China	High fruit numbers per lateral; ease of harvest; late ripening
Idaeobatus	*R. corchorifolius* L.	2x	China	Earliness; disease resistance; good flavor
Idaeobatus	*R. coreanus* Miq.	2x	China	Earliness; vigor; Range of fruit colors (orange-black); resistant to aphids, cane blight, midge blight, spur blight, cane *Botrytis*, anthracnose, European raspberry beetle, powdery mildew, leaf spot, root rot
Idaeobatus	*R. crataegifolius* Bunge	2x	China	Firm fruit with a bright, non-darkening red color; early ripening; resistant to fruit rot, cane *Botrytis*, cane midge, cane beetle, root lesion nematode, strong laterals; winter tolerance
Idaeobatus	*R. ellipticus* Sm	2x	China	Vigor, low chilling requirement
Idaeobatus	*R. eustephanos* Focke ex Diles	2x	China	Very high drupelet count; Vigor
Idaeobatus	*R. flosculosus* Focke	2x	China	High fruit numbers per lateral; condensed fruit ripening; erect habit; vigorous; cane disease resistance
Idaeobatus	*R. glaucus* Benth.[z]	4x	S. America	Low chilling requirement; vigor; excellent fruit quality particularly aroma; large fruit size; small seeds and drupelets; extended production season; root rot resistance
Idaeobatus	*R. innominatus* var. *kuntzeanus* (Hemsl.) L.H. Bailey (= *R. kuntzeanus*)	2x	China	Low chilling requirement; resistant to drought, high temperature, leaf spot, cane spot, cane beetle
Idaeobatus	*R. hirsutus* Thunb	2x	China	Large size and bright red color; heat and high humidity tolerance; tolerant of fluctuating winter temperatures
Idaeobatus	*R. idaeus* L.	2x	Europe	As the primary species in red raspberry background a tremendous source of untapped diversity for most traits

Table 12.1 (continued)

Subgenus	Species	Ploidy	Location	Important traits
Idaeobatus	*R. strigosus* Michx.	2x	N. America	As a primary species in red raspberry background a tremendous source of untapped diversity for most traits
Idaeobatus	*R. innominatus* S. Moore	2x	China	Vigor; late ripening; heat and humidity tolerance; high fruit numbers per lateral; productivity; erect plant habit; excellent fruit size
Idaeobatus	*R. lasiostylus* Focke	2x	China	Vigor; high drupelet count; large fruit size; ease of harvest; fruit cohesiveness/pubescence; foliar disease resistance; yellow rust resistance
Idaeobatus	*R. leucodermis* Douglas ex Torr. & A. Gray	2x	W.N. America	Productive; large fruit; vigor; potential source RBDV resistance; resistant to cane and leaf rust
Idaeobatus	*R. mesogaeus* Focke	2x	China	Resistant to cane blight, cane midge
Idaeobatus	*R. niveus* Thunb.	2x	India, Asia	Vigor; fruit firmness; tolerance to heat and humidity, and cane and leaf disease; orange rust resistance, erect; good flavor; primocane fruiting; high number fruit/lateral; fruit rot resist, late ripening
Idaeobatus	*R. occidentalis* L.	2x	E.N. America	Progenitor species for black raspberry cultivars so wide degree of diversity may be available for black raspberry improvement. For red raspberry improvement: tolerance to heat and humidity; resistant to aphids, bud moth, leaf rollers, cane beetle, two-spotted spider mite, fruit rot; firm fruit; late-ripening floricane fruit
Idaeobatus	*R. parvifolius* L.	2x,4x	Japan, China, Australia	Low chilling requirement; resistant to drought, high temperature, high humidity, leaf spot, cane spot, spider mite, root rot; some tolerance fluctuating winter temperatures; productive; fruit size
Idaeobatus	*R. phoenicolasius* Maxim.	2x	Japan	Very early; resistant to cane and Japanese beetle, powdery mildew, root rot

Table 12.1 (continued)

Subgenus	Species	Ploidy	Location	Important traits
Idaeobatus	R. pileatus Focke	2x	Europe	Fruit flavor; resistant to cane blight, cane midge, cane Botrytis, spur blight, fruit rot, root rot; low chilling
Idaeobatus	R. pungens Cambess.	2x	Indonesia	Early ripening floricane fruit; winter hardiness; resistant to spur blight
Idaeobatus	R. rosifolius Sm.	2x	Asia, Australia	High drupelet count; tolerant to high temperature and humidity
Idaeobatus	R. sachalinensis Leveille	4x	E. Asia	Hardiness, flavor; vigor, drupelet size
Idaeobatus	R. spectabilis Pursh.	2x	W.N. America	Early floricane and primocane fruiting; condensed fruit ripening; fruit with a bright, non-darkening, red color; ease of harvest; resistant to root rot and aphids; erect growth
Idaeobatus	R. sumatranus Miq	2x	Asia	High drupelet count; tolerant to root rot in greenhouse trials; primocane fruiting
Idaeobatus	R. trifidus Thunb.	2x	Japan	Foliar disease resistance; black fruit color
Anoplobatus	R. deliciosus Torr	2x	W.N. America	Upright growth habit, drought tolerance; cold hardiness
Anoplobatus	R. odoratus L.	2x	E. N. America	Early primocane ripening; self-supporting canes; resistant to raspberry midge, cane blight; winter hardiness
Anoplobatus	R. parviflorus Nutt.	2x	W.N. America	Upright habit; large, well formed fruit; veinbanding mosaic virus resistance
Chamaemorus	R. chamaemorus. L.	8x	Circumpolar/ Sub-arctic	Excellent, aromatic flavor; high ascorbic acid content; thornlessness, winter hardiness
Rubus, Ursini	R. ursinus Cham. et Schlecht	7–13x	W.N. America	Good fruit quality; early ripening; resistant to Verticillium wilt and Phytophthora root rot

[z] While classified by USDA-ARS in *Idaeobatus*, has also been designated as 'natural inter-subgeneric hybrid' (Williams et al. 1949, Thompson 1995a) and has been more successfully used in hybrids with blackberry than raspberry (Finn et al. 2002b, HK Hall pers. comm.).
Jennings 1988, Ying et al. 1989, Jennings et al. 1991, Swartz et al. 1993, Thompson 1995a, Daubeny 1996, Finn et al. 2002a,b

'Lloyd George' has been a particularly important parent, being in the direct ancestry of 32% of the North American and European cultivars in 1970 (Oydvin 1970). This cultivar contributed several important traits including primocane fruiting, large fruit size and resistance to the American aphid. Jennings (1988) speculates that the success of 'Lloyd George' hybrids 'was possibly achieved because they combined the long-conical shape of "Lloyd George" receptacle with the more rounded shape of the American raspberries'. A key example of such a hybrid is 'Willamette', which is a cross of 'Newburgh' × 'Lloyd George', and dominated the industry in western North America for over a half century.

Many programs released red raspberry cultivars in the latter half of the 20th Century and into the 21st Century, but only a few programs stood out as particularly active. In the United Kingdom, the program at East Malling was responsible for the 'Malling series'. A number of selections were made prior to World War II and released in the 1950s, 'Malling Promise', 'Malling Exploit' and the most successful, 'Malling Jewel' (Jennings 1988). This program continued to have an impact with the later release of 'Malling Admiral' as a late, high yielding genotype and most recently 'Octavia' (Finn et al. 2007). In addition to these floricane cultivars, the program has developed a number of very important primocane fruiting cultivars, with 'Autumn Bliss' being the most important.

Further north at the Scottish Crop Research Institute the 'Glen series' was developed, the first being 'Glen Clova' in 1969. 'Glen Moy' and 'Glen Prosen', released in 1981, were the first spineless raspberries and both offered great improvements in fruit size and flavor. 'Glen Ample' released in 1994 became a standard for quality and yield throughout much of Europe and the program continues to be active with the recent release of 'Glen Doll' (Fig. 12.1).

The breeding programs in the Pacific Northwest of North America at Washington State University (WSU; Puyallup, Wash.), Agriculture and Agri-Foods Canada (AAFC; Agassiz, BC) and the U.S. Dept. of Agriculture-Agricultural Research Service in Oregon (USDA-ARS; Corvallis) benefited from many years of collaboration among one another and with the U.K. programs. The USDA-ARS's releases from the mid 1900s, 'Willamette' and 'Canby', are still commercially important floricane cultivars. The recent release 'Coho' from that program has been widely planted for its high yields of IQF fruit (Finn et al. 2001). The USDA-ARS primocane fruiters 'Summit' and 'Amity' have been very important since their release and 'Summit' has found new life in the developing Mexican industry. 'Meeker', developed by WSU and released in the 1960s, is still the processing industry standard (Finn 2006). This program continues to be active and the newest releases 'Cascade Delight' and 'Cascade Bounty' are likely to become the standards for root rot tolerant cultivars (Moore 2004, 2006, Moore and Finn 2007).

The AAFC program has been one of the most prolific and important programs over the past 30 years. The breeders there took full advantage of germplasm exchanges with the U.K. and were very successful at identifying outstanding selections out of crosses between British Columbia selections and some of the 'Glen series' particularly 'Glen Prosen' (Finn 2006). The 1977 releases 'Chilcotin', 'Skeena' and 'Nootka' had excellent fruit quality and high yields for a fresh market berry.

Fig. 12.1 'Glen Doll', bred at the Scottish Crop Research Institute

The program followed these releases with 'Chilliwack' in the mid 1980s and the incredibly important 'Tulameen' in 1989. 'Tulameen' set new standards for fresh market quality particularly flavor. This program remains active and the recent releases 'Esquimalt', 'Chemainus', 'Cowichan', and 'Saanich' are being widely planted (Kempler et al. 2005a,b, 2006, 2007).

Elsewhere in the U.S., the New York Agricultural Experiment Station (Geneva) used their own primocane fruiting germplasm in combination with material such as 'Durham', developed in New Hampshire, to produce an excellent primocane fruiting germplasm pool that culminated with the release of 'Heritage' in 1969 and 'Ruby' ('Watson') in 1988 (Daubeny 1997). First viewed as a novelty, the primocane fruiting types became the standard in regions where cold winter temperatures caused considerable winter damage to canes of floricane fruiting raspberries. Later, private companies in California, such as Driscoll's Strawberry Associates (Watsonville, Cal.) developed cultivars and whole new production systems where the plants were only in the ground 18 months to fuel the rapid expansion of the enormous California raspberry industry (Finn and Knight 2002).

The importance of the University of Minnesota release 'Latham' has already been mentioned, but other releases from that program such as 'Chief' have been valuable in breeding programs for their root rot resistance. While the program was discontinued in the early 2000s, 'Redwing', released in 1987 has been a popular primocane cultivar where other cultivars cannot mature their crops in time before fall frosts (Daubeny 1997).

The cooperative program centered at the University of Maryland, in cooperation originally with Virginia Tech University, Rutgers University, and the University of Wisconsin – River Falls, really hit their stride in the late 1990s and early 2000s with the release of the primocane fruiting 'Caroline', 'Anne', and 'Josephine'.

The eastern North America black raspberry (*R. occidentalis*) was not cultivated until the 19th century, probably because of its abundance in the wild and the public's preference for red raspberry (Jennings 1988). Purple raspberry cultivars were actually grown earlier in the 1820s as hybrids of black and red raspberries. The first known pure black raspberry cultivar was 'Ohio Everbearer' that was selected for its propensity to produce a significant fall crop (Jennings 1988). Hedrick (1925) listed 193 black raspberry cultivars in his 'The Small Fruits of New York', although most were wild selections.

The breeding of black raspberries was slow to develop, with the first active breeding work being initiated in the late 1800s at the New York Agricultural Experiment Station at Geneva. The station continued as the primary center of research for much of the 20th Century (Slate 1934, Slate and Klein 1952, Ourecky and Slate 1966, Ourecky 1975), although significant work was also done by Drain (1952, 1956) in Tennessee. In the late 20th Century, breeding efforts ebbed and only three cultivars were released. In the early 21st Century, black raspberry breeding efforts were renewed at the New York Agriculture Experiment Station and the U.S. Department of Agriculture-Agricultural Research Service (USDA-ARS) in Corvallis, Oregon. In Corvallis, Dossett (2007) evaluated black raspberry families from sibling families from crosses among cultivars and a North Carolina selection to assess variation and inheritance of vegetative, reproductive and fruit chemistry traits in black raspberry. In New Zealand, spinelessness from red raspberry has been transferred to black raspberry and resulted in the recent release of the spineless 'Ebony' (H.K. Hall, pers. comm.).

12.4 Current Breeding Efforts

There are now 38 active red raspberry breeding programs in 21 countries, found mostly in Europe and North America. These programs have released at least 160 red raspberry cultivars over the last 30 years (Tables 12.2 and 12.3). In a survey conducted in 2001 by Finn and Knight (2002), raspberry breeders were found to be optimistic about their programs financial support, as most were able to maintain or

Table 12.2 Red raspberry breeding programs worldwide

Country	Location
Australia	Inst. Horticultural Develop., Knoxfield, Victoria
Bulgaria	Kostinbrod
Canada	
British Columbia	Agriculture and Agri-Food Canada, Agassiz
Nova Scotia	Agriculture and Agri-Foods Canada, Kentville
Ontario	University of Guelph, Guelph
Chile	
Metropolitan Region	Hortifrut, Santiago
Metropolitan Region	VBM, Santiago
VII Region	INIA
China	Beijing Institute of Pomology & Forestry, Beijing
Germany	Freising-Weinhenstephan
Hungary	Small Fruit Research Station, Fertod
Italy	University of Ancona
Latvia	Dobele Horticultural Plant Breeding Experimental Station
Mexico	Universidad Michoacana de San Nicolás de Hidalgo, Uruapan
Norway	Norwegian Crop Research Institute, Planteforsk Njoes
New Zealand	HortResearch, Inc., Motueka
Poland	Research Institute of Pomology, Brzezna
Romania	Research Institute for Fruit Growing, Pitesti
Russia	VIR, St. Petersburg
Serbia	Fruit and Viticulture Research Centre, Cacak
	IPTCH Vilamet, Cacak
Sweden	Balsgard, Kristianstad
Turkey	Atatürk Central Horticultural Research Institute, Yalova
United Kingdom	
England	Driscoll's Assoc., E. Malling
England	East Malling Research, formerly Hort. Research Int.; Malling
England	Redeva Ltd.
Scotland	Scottish Crop Research Inst., Dundee
U.S.A.	
California	Driscoll's Assoc., Watsonville
California	Plant Sciences Inc., Watsonville
Florida	Florida A&M, Tallahassee
Maryland-Virginia	Univ. of Maryland, Virginia Tech
Wisconsin Coop Program	Univ. of Wisconsin
Maryland	USDA-ARS, Beltsville
North Carolina	N.C. State University
Oregon	USDA-ARS/Ore. St. Univ., Corvallis
Washington	Wash. State Univ., Puyallup
Washington	Northwest Plants Co., Lynden
Washington	Driscoll's Assoc., Lynden

expand their programs. Support came from a varied mix of federal, state, commodity and royalty support, with the government support generally decreasing.

All *Rubus* breeding programs emphasize the development of cultivars with dependable yields of high quality fruit, suitability for shipping if for the fresh market and for machine harvestability if for the processed market, adaptation to the local environment and improved pest and disease resistance. Resistance to *Phytophthora*

Table 12.3 Red raspberry cultivars released since the 1970s

Location of releasing program	Cultivar
Australia	Alkoopina, Bogong, Dinkum, Glen Yarra
Bulgaria	Raliza, Samodiva, Ljlin, Essenna Poslata
Canada	
British Columbia	Chemainus, Chilliwack, Comox, Cowichan, Esquimalt, Kitsilano, Malahat, Nanoose Qualicum, Saanich, Tulameen
Manitoba	Double Delight, Red River, Souris
Nova Scotia	Nova
Ontario	OAC Regal, OAC Regency
Québec	Perron's Red
Czech Republic	Granat
Denmark	Zenith
Estonia	Aita, Alvi, Helkal
Finland	Jatsi, Jenkka, Ville
France	Comtesse, Favorite, Galante, Meco, Princess, Wawi
Germany	Rusilva (Rrabant), Resa (Lucana), Rubaca (Naniane), Weirula
Hungary	Fertodi aranyfurt, Fert. karmin (Marla), Fert. ketszertermo, Fert. rubina, Fert. Venus, Fert. Zamatos, Fert. Zenit
Mexico	Gina
New Zealand	Clutha, Kaituna, Kiwigold, Motueka, Moutere Rakaia, Selwyn, Waiau, Waimea, Tadmor
The Netherlands	Marwe
Norway	Balder, Borgund, Frosta, Hitra, Stiora, Tambar, Varnes, Vene
Poland	Benefis, Beskid, Laska, Nawojka, Polana, Polka, Pokusa, Poranno Rosa
Romania	Citria, Gustar, Opal, Ruvi, Star
Russia	Approximately 35 floricane and 17 primocane cultivars; however information is not available in English leaving a great deal of uncertainty on timeframe of releases and accuracy of information (HK Hall, pers. comm.)
Serbia (Cacak)	Gradina, Krupna Dvorodna, Podgorina,
Sweden	Ariadne, Boheme, Carmen
Switzerland	Elida, Framita, Himbo Star, Himbo Top (Rafzaqu)
U.K.	
England	Autumn Bliss, A. Britten, A. Cascade, A. Cygnet, A. Byrd, Brice, Gaia, Joan Irene, Joan J., Joan Squire, Julia, Malling Augusta, M. Hestia, M. Joy, M. Juno, M. Minerva, Marcela, Octavia, Terri-Louise, Valentina
Scotland	Glen Ample, G. Doll, G. Garry, G. Lyon, G. Magna, G. Moy, G. Prosen, G. Rosa, G. Shee
U.S.A.	
California	AnnaMaria, Bababerry, Driscoll Cardinal, D. Carmelina, D. Dulcita, D. Francesca, D. Madonna, D. Maravilla, Gloria, Godiva, Graton Gold, Hollins, Holyoke, Isabel, Joe Mello, Lawrence, PSI 79, PSI 114, PSI 127, PSI 168, PSI 737, PSI 744, PS 1070, PS 1764, PS-1703, Stonehurst, Sweetbriar, Tola, Wilhelm
Minnesota	Nordic, Redwing
NJ/Md/Virg/Wisc	Alice, Anne, Caroline, Claudia (Christmas Tree), Deborah, Emily, Esta (Esther), Georgia, Jaclyn, Josephine, Lauren
New York	Encore, Prelude, Ruby (Watson), Titan
Oregon	Amity, Chinook, Coho, Lewis, Summit
Washington	Centennial, Cascade Bounty, Cascade Dawn, Cascade Delight, Cascade Nectar

fragariae C.J. Hickman var. *rubi* Wilcox & Duncan is a universal goal. European programs were very concerned with cane Botrytis (*Botrytis cinerea* Pers.: Fr.), spur blight (*Didymella applanata* [Niessl] Sacc.), and anthracnose (*Elsinoë veneta* [Burkolder] Jenk.). American programs are particularly concerned with resistance to *Raspberry bushy dwarf virus* (RBDV).

While molecular genetics, including transgenic technologies, has been cautiously added to the *Rubus* breeders toolbox, there is currently no interest in the industry for transgenic cultivars, although a transgenic RBDV resistant 'Meeker' has been developed (Martin, pers. comm.). The programs that were using molecular tools were most commonly dealing with marker assisted selection, mapping, and genetic fingerprinting (see section on biotechnological approaches to genetic improvement).

Modern raspberry breeders must be aware of industry changes so that they can effectively address these needs. Breeders must decide what kind of cultivars will be most appropriate for the increasing amount of tunnel production worldwide, the necessity of machine harvesting for processing, the demand for nutraceuticals by the consumer, and the fact that globalization means the product grown in one region can be shipped just about anywhere. Breeding programs must also recognize where their industries physically will be in the future. China, which has never had a significant commercial caneberry industry, is rapidly planting raspberries, and they will develop cultivars that are specifically adapted to their production regions (Wu et al. 2006). Regional cultivars will be needed that more efficiently produce high yields of high quality fruit in order to keep their industries competitive.

Other changes facing all public breeding efforts are the rise in importance of private programs and the protection of intellectual property rights. Private companies with their own breeders and proprietary cultivars or public breeders with some part of their program privatized are no longer an anomaly and no longer regional. Patents, breeder's rights, and licenses, while not new, are becoming the standard (see Chapter 13.7.3). While plant protection and the potentially associated royalty stream have reduced the level of germplasm exchange between programs, it has allowed for the stabilization or survival of breeding programs.

12.5 Genetics of Important Traits

12.5.1 Disease and Pest Resistance

One of the primary factors associated with low yields, poor quality fruit, and degeneration of raspberry cultivars are viruses (Jennings 1988). The inheritance of virus resistance is often not known, however differences in susceptibility and resistance have often been noted among different species and genotypes (Jennings and Jones 1986, Converse 1991, Jennings et al. 1991, Jones and McGavin 1998).

The pollen-born RBDV is the most serious raspberry virus disease worldwide in the commercial industry (Martin 2002) and it is common in the wild (Finn and Martin 1996, Chamberlain et al. 2003). While plant growth and fruit yield are often not affected by RBDV, on susceptible cultivars the fruit is crumbly making it worthless for the fresh market or for the higher value IQF processed market (Converse 1991). *Rubus leucodermis*, the western North American counterpart to *R. occidentalis,* does not appear to carry RBDV in wild populations (Finn and Martin 1996) and will likely be a potential source of variability for breeding improved *R. occidentalis* genotypes (Finn et al. 2003). *Rubus leucodermis* was successfully incorporated into *R. occidentalis* germplasm to produce 'Earlysweet' (*R. leucodermis* is a grandparent) (Galletta et al. 1998).

While symptomless, *Tobacco streak virus* (TSV) is another common pollen borne virus. Virus-free planting stock is an important component of control; however, genetic resistance is the only long term viable control option (Converse 1991). Resistance to RBDV is based on a single dominant gene *Bu* (Jones et al. 1982, Knight and Barbara 1999). 'Willamette', an industry standard, as well as 'Chilcotin', 'Nootka', 'Haida' and 'Heritage' carry this gene (Daubeny 2002a, Kempler et al. 2002). In addition, some genotypes such as 'Cowichan' have shown long term resistance and may carry this gene (Stahler et al. 1995, Kempler et al. 2005b). Despite this, it has been difficult to develop new commercial cultivars with this trait, and in fact, the low percentage of selections with immunity suggest that potentially a negative linkage exists between some traits important to commercial quality and disease resistance (P. Moore, pers. comm.). A strain of RBDV capable of infecting genotypes with the *Bu* gene has been found in Europe and Asia; resistance to this strain has been identified in some genotypes (Jennings et al. 1991).

The aphid-born viruses that make up the raspberry mosaic complex [*Rubus yellow net virus* (RYNV), *Black raspberry necrosis virus* (BRNV), *Raspberry leaf mottle virus* (RLMV) and *Raspberry leaf spot virus* (RLSV)] continue to be major problems and can be devastating in red and black raspberries (Stace-Smith 1956, Halgren et al. 2007). In Europe, these viruses are vectored by *Amphorophora idaei* Börner and in North America by *A. agathonica* Hottes. While there is no immunity to the raspberry mosaic complex, the *Ag1* resistance gene to the aphid vector has been successfully incorporated into many cultivars and programs continue to actively screen seedling populations for this resistance (Knight et al. 1960, Daubeny and Stary 1982). The use of *A. idaei* resistant cultivars in the U.K. has been effective for a long time; however, cultivars have been found to vary in their response to the aphid despite having the same source of resistance (Jones et al. 2000). The effectiveness of the gene for resistance seems to be compromised in plants grown with partial shade and there appear to be minor genes affecting susceptibility (Jones et al. 2000). Dossett (2007) found that populations from crosses among commercial black raspberry cultivars grown in the Pacific Northwest under substantial virus pressure generally had poorer vigor than did the populations derived from crosses between cultivars and a wild selection of *R. occidentalis* from North Carolina. This suggests that broadening the germplasm pool for disease tolerance and for other traits could be very important for genetic improvement (Weber 2003).

Nematode borne viruses are commonly a problem with *Tomato ringspot virus* (TmSV) and *Tobacco ringspot virus* (ToSV) of primary concern in North America and *Raspberry ringspot virus* (RRSV), *Tomato black ring virus* (TBRV), *Arabis mosaic virus* and *Strawberry latent ringspot virus* (SLRV) of greatest concern in Europe. The primary vectors of these viruses are *Xiphinema americanum* Cobb. in North America and *Longidorus elongates* de Man and *Xiphinema diversicaudatum* Micoletsky in Europe. While sources of resistance to the nematode borne viruses, through resistance to the vector, have been identified in red raspberry and *Rubus crataegifolius* Bunge, they have not been pursued vigorously (Jennings 1964, Vrain and Daubeny 1986).

The most important cane diseases of raspberries in Europe are midge and cane blights. Midge blight is a disease complex instigated by damage from the raspberry midge, *Resseliella theobaldi* Barnes. Cane blight is caused by *Leptosphaeria coniothyrium* (Fuckel) Sacc., which generally enters through wounds caused by mechanical harvesting (Williamson and Jennings 1992). These two diseases are generally not important in North America. Several other fungal diseases cause significant damage. Grey mold, *B. cinerea,* is the most important fungal disease of raspberry fruit worldwide. Spur blight is a very important disease in the Pacific Northwest and eastern Europe, where it affects a large portion of the canes surface area, while in western Europe its effects are less severe, generally limited to individual buds. Problems with cane spot or anthracnose (*E. veneta*) are widespread. Leaf and cane spot (*Sphaerulina rubi* Demaree and Wilcox) often effect raspberries grown in warmer climates. Yellow rust [*Phragmidium rubi-idaei* (DC.) P. Karst] has variable, but sometimes severe effects across Europe and Australasia. Late yellow rust [*Pucciniastrum americanum* (Farl.) Arth.] can be severe in the eastern parts of North America.

Sources of resistance have been found for many of these fungal diseases (Table 12.4). Gene *H*, for hairy vs. glabrous canes has been found to be associated with resistance to spur blight (Jennings 1983, 1988) and recently has been added to the raspberry linkage map (Graham et al. 2006). Keep et al. (1977a) found that *R. coreanus* imparted strong resistance to not only spur blight, but cane blight, anthracnose and the leaf disease powdery mildew [*Sphaerotheca macularis* (Fr.) Jaczewski]. Daubeny (1987) proposed that resistance to cane Botrytis was due to two gene pairs with the presence of at least two dominant genes necessary to give resistance. He also listed a number of resistant and susceptible cultivars. Sources of resistance to yellow rust have been identified by Anthony et al. (1986).

Ramanathan et al. (1997) cloned two genes for polygalacturonase-inhibiting protein (*PGIP1* and *PGIP2*) that could play a role in grey mold resistance. Plant PGIPs inhibit fungal endopolygalacturonases, which are released by fungi to degrade plant cell walls. Activity levels of *PGIP* were found to decline during floral and fruit development, but expression of *PGIP* was stable throughout development from closed flower to ripe fruit.

Root rot (*P. fragariae* var. *rubi*) is a devastating problem throughout red raspberry production areas worldwide (Kennedy and Duncan 1993, Wilcox et al. 1993, Daubeny 2002b). Fungicides, particularly metalaxyl, were effective at controlling the root rot organism from the 1980s–1990s and while it has not lost all effectiveness, breeding programs are being pressed to develop resistant cultivars. Some success has been accomplished with the release of resistant 'Cascade Bounty' and

Table 12.4 Genetics of pest and disease resistance in raspberry

Attribute	Observations and source
Bacteria	
Crown gall	'Willamette' is resistant (Daubeny 1996)
Fungi	
Cane blight	'Latham' is resistant (Jennings et al. 1991); resistance is additive (Williamson and Jennings 1992)
Cane Botrytis	High levels of additive variation for resistance (Jennings 1983); the gene *H* controlling cane pubescence provides resistance (Williamson and Jennings 1992)
Cane spot	Resistant genotypes identified (Jennings et al. 1991); major resistance genes may exist (Williamson and Jennings 1992)
Grey mold	Resistant genotypes identified (Jennings 1988, Jennings et al. 1991)
Late yellow rust	Resistant genotypes identified (Jennings et al. 1991)
Leaf spot	Good germplasm sources identified (Jennings et al. 1991)
Midge blight	Resistant genotypes identified (Jennings et al. 1991)
Phytophthora root rot	'Latham', 'Winkler's Samling', and 'Cascade Bounty' are resistant as are several species (Bristow et al. 1988, Jennings et al. 1991, Pattison and Weber 2005, Moore and Finn 2007)
Powdery mildew	Multiple resistance genes identified in red raspberry (Keep 1968, Williamson and Jennings 1992)
Raspberry yellow rust	Major gene for resistance (Williamson and Jennings 1992)
Spur blight	Resistant genotypes identified (Jennings et al. 1991); high levels of additive variation for resistance (Jennings 1988); the gene *H* controlling cane pubescence provides resistance (Williamson and Jennings 1992)
Verticillium wilt	Primarily additive resistance, 'Willamette' and 'Southland' tolerant (Fiola and Swartz 1994)
Yellow rust	Single gene for resistance (Anthony et al. 1986)
Insects	
Amphorophora idaei	Multiple genes for resistance identified (Daubeny 1996, Daubeny and Stary 1982); resistant genotypes identified (Daubeny 1996)
Aphis idaei	Resistant genotypes identified; multiple genes identified for various races (Knight et al. 1960, Knight et al. 1972)
Raspberry midge	Cultivars with few splits in canes are resistant (Jennings et al. 1991)
Raspberry beetle	Primocane fruiting cultivars are resistant (Jennings et al. 1991)
Raspberry budmoth	Resistant cultivars identified (Wilde et al. 1991)
Raspberry fruitworm	'Royalty' purple raspberry has resistance (Schaefers et al. 1978)
Spider mite	Resistant genotypes identified (Shanks and Moore 1996)
Nematodes	
Pratylenchus penetrans	'Nootka' is resistant (Vrain and Daubeny 1986)
Viruses	
Arabis mosaic	Single gene for immunity (Jennings 1964); resistant genotypes identified (Jones and McGavin 1998)
Black raspberry necrosis	Most cultivars are tolerant (Jennings et al. 1991, Jones and McGavin 1998)
Raspberry bushy dwarf	Single gene for immunity (Jones et al. 1982); 'Willamette' is immune. Genotypes with long term resistance have been identified (Stahler et al. 1995, Kempler et al. 2005b); resistant cultivars identified for isolate RBDV-D200 but not RBDV-RB (Jones and McGavin 1998).

Table 12.4 (continued)

Attribute	Observations and source
Raspberry ringspot	Single gene for immunity (Jennings 1964); resistant cultivars identified (Jennings et al. 1991, Jones and McGavin 1998)
Raspberry leaf mottle	All tested varieties susceptible(Jones and McGavin 1998)
Raspberry leaf spot	All tested varieties susceptible(Jones and McGavin 1998)
Raspberry yellow net	Most cultivars are tolerant (Jennings et al. 1991, Jones and McGavin 1998)
Raspberry yellow spot	No published source of resistance
Raspberry vein chlorosis	All tested varieties susceptible(Jones and McGavin 1998)
Strawberry latent ringspot	Resistant genotypes identified (Jones and McGavin 1998)
Tomato black ring	Single gene for immunity; resistant genotypes identified (Jennings et al. 1991, Jones and McGavin 1998)

the tolerant 'Cascade Dawn' and 'Cascade Delight' (Moore 2004, 2006, Moore and Finn 2007). Hydroponic culture methods have been developed that can classify phenotypes of individuals for their resistance to root rot (Pattison et al. 2004, Pattison and Weber 2005). The root disease Verticillium wilt (*Verticillium albo-atrum* Reinke & Berthier and *V. dahliae* Kleb.) is also a common problem that is locally severe. While sources of resistance have been identified and the quantitative inheritance of tolerance documented (Fiola and Swartz 1989, 1994) for Verticillium wilt they have not been actively pursued by breeders.

The bacterial disease crown gall, caused by *Agrobacterium tumefaciens* (E.F. Smith & Townsend) Conn, is common wherever caneberries are grown, and fireblight (*E. amylovora* [Burr.] Winslow et al.) and Pseudomonas blight (*Pseudomonas syringae* van Hall) can occasionally be a problem. While genetic resistance to fireblight has been identified (Stewart et al. 2005), no raspberry breeding program is actively breeding for resistance to bacterial diseases.

Insect and mite problems are usually specific to regions or environments. In monocultures, insecticides/acaricides are often applied as needed for specific problems such as raspberry crown borer (*Pennisetia marginata* [Harris]), red-necked caneborer (*Agrilus ruficollis* [Fabricius]), strawberry bud weevil (*Anthonomus signatus* Say), brown and green stink bugs (*Euschistus* spp. and *Acrosternum hilare* Say, respectively), Japanese beetle (*Popillia japonica* Newman), thrips (eastern and western flower thrips, *Frankliniella tritici* Fitch and *F. occidentalis* Pergande, respectively), grass grub (*Costelytra zealandica* White), raspberry fruitworms (*Byturus tomentosus* Degeer in Europe and *B. unicolor* Say in North America), root weevils (*Otiorhynchus singularis* L., *O. sulcatus* Fab, *O. ovatus* L. and *Sciopithes obscurus* Horn), and foliar nematode (*Aphelenchoides ritzemabosi* [Schwartz] Steiner). Insecticides are also used as a 'knockdown' to remove insects such as orange tortrix (*Argyrotaenia citrana* Fernald) that can be a contaminant in machine harvested fruit (Jennings 1988, Daubeny 1996, Clark et al. 2007, HK Hall pers. comm.). In New Zealand, caneberries are attacked severely by raspberry bud moth (*Heterocrossa rubophaga* Dugdale) and/or blackberry bud moth (*Eutorna phaulacosma* Meyrick) and the leaf roller species (*Epiphyas postivittana* Walker,

Planotortrix exessana Walker, *P. octo* Dugdale, *Ctenopseustis obliquana* Walker, *C. herana* Felder, and Rogenhofer and *Cnephasia jactatana* Walker). *Rubus occidentalis* has been found to be a source of resistance to both groups and has been carried through four generations of breeding improvement into red raspberries (H.K. Hall, pers. comm.).

As crops are moved to new regions or environments, nuisance or minor pests can become severe. Red raspberries grown in warm, dry environments are generally very susceptible to the two-spotted spider mite (*Tetranychus urti*cae Koch), and therefore as glasshouse and tunnel production become more important parts of commercial production they are becoming more of a problem (Finn and Knight 2002). Resistance to this pest has been identified in red raspberry and related species (Shanks and Moore 1996).

Numerous sources of disease and pest resistance have been identified in the wild raspberries (Table 12.4) and a good summary is presented by Jennings et al. (1991). Of particular note are a number of species within the Idaeobatus that carry multiple resistances including: (1) *R. crataegifolius* (fruit rot, cane Botrytis, cane midge, cane beetle and root lesion nematode), (2) *R. coreanus* (aphids, cane blight, spur blight, cane Botrytis, cane spot, cane beetle, powdery mildew, leaf spot and root rot), (3) *R. mesogaeus* Focke (cane *Botrytis*, cane blight and cane midge), (4) *R. parvifolius* (leaf spot, cane spot, spider mite and root rot, (5) *R. occidentalis* (aphids, bud moth, leaf rollers, cane beetle, two-spotted spider mite and fruit rot), and (6) *R. pileatus* Focke (cane blight, cane midge, cane Botrytis, spur blight, fruit rot and root rot).

12.5.2 Environmental Adaptation

Adaptation to low winter temperatures, high summer temperatures and low chilling are three of the most important characteristics sought by breeders for continued expansion of the raspberry industry. Lack of winter cold tolerance limits the range of successful raspberry cultivation in the continental climates of central and eastern Europe, and eastern and central North America (Daubeny 1996).

The inheritance of winter hardiness is under complex genetic control. Winter hardy caneberries have four key characteristics: (1) rapid hardening in the fall before severe temperatures occur, (2) long rest or deep dormancy making them resistant to temperature fluctuations in the spring, (3) the ability to re-harden if initial cold tolerance is lost, and (4) late bud break (Warmund and George 1990, Daubeny 1996). It has been difficult to combine cold hardiness with early flowering and fruiting (Jennings 1988).

Attempts have been made to predict relative hardiness in red raspberry using individual characteristics such as when leaf drop occurs, rates of bud development and bud water content, but hardiness has been most accurately assessed by evaluating reproductive performance in the field after 'test' winters. Numerous raspberry cultivars with good winter hardiness have been identified in North America, Scandinavia, Eastern Europe, the former Soviet Union and the U.K. (Daubeny 1995, 1996,

1997,1999). Strong winter hardiness has also been described in native populations of the raspberries *R. idaeus*, *R. sachalinensis* H. Lev. ($2n = 4x = 28$), *R. chamaemorus*, *R. crataegifolius*, *R. arcticus* and *R. arcticus* subsp. *stellatus* (Table 12.1). The development of primocane fruiting raspberries probably had the most dramatic effect on the winter hardiness of any berry crop. Severe winter cold damage can be avoided in primocane fruiting raspberries by removing the canes after they are harvested in the fall and therefore, only those extremely cold winters that damage crowns are a problem.

Raspberry production can be limited by fluctuating spring and fall temperatures, and low temperatures during fruiting. There is considerable variability among raspberry genotypes in their ability to set high proportions of their drupelets under cool conditions during flowering. In general, the raspberry cultivars developed in the U.K. are better adapted to cool temperatures than those developed in the Pacific Northwest, with some exceptions. The northwestern cultivar 'Meeker' has been shown to have good drupelet set in both locations (Dale and Daubeny 1985).

Heat and drought are limiting in southern Europe, southeastern North America and much of the southern hemisphere. Since fruit quality standards are high in the commercial market, irrigation is becoming the standard practice and heat and ultraviolet (UV) damage are more of a concern than drought. Considerable variability has been found among raspberry cultivars for adaptation to high summer temperatures. The wild Asiatic raspberry species have been most widely used as sources of high temperature tolerance (Williams 1961, Stafne et al. 2000) and a low chilling requirement, although genes regulating a low chilling requirement have also been found in cool weather adapted *R. idaeus* (Rodriguez-A and Avitia-G 1989). While breeders have attempted to use *R. parvifolius* as a source of heat tolerance and cultivars such as 'Dormanred', 'Southland' and 'Mandarin' have been produced that are purported to be either *R. idaeus* × *R. parvifolius* hybrids or second 2nd generation hybrids, this route has never proven as successful/easy as it would appear to be in theory (Daubeny 1997, Stafne et al. 2000).

In red raspberry, while there remains interest in low chilling cultivars, the use of long cane production techniques and primocane fruiting genotypes has largely circumvented this challenge. Long cane production is the practice where floricane fruiting cultivars are grown in northerly climates (e.g. Scotland), dug after they go dormant, refrigerated and then replanted with the entire cane intact in a warmer climate (e.g. Spain) where they quickly break bud and flower. For those breeders interested in reducing chilling requirement, the wild raspberry species *R. biflorus* Buch.-Ham. ex Sm., *R. innominatus* S. Moore, *R. glaucus*, *R niveus* and *R. parvifolius* are excellent sources of a low chilling requirement, while *R. biflorus*, *R. coreanus*, *R. niveus*, *R. occidentalis*, *R. parvifolius* and *R. innominatus* are sources of resistance to high temperatures.

Much genetic variability exists for season extension among raspberry cultivars and species (Jennings et al. 1991). Fruiting season appears to be highly heritable trait, with some genotypes showing considerable environment interaction (Hoover et al. 1988). The fastest ripening cultivars have an early bloom date and a rapid developmental rate, although early flowering can be a problem in frosty areas. Among the wild raspberry species, *R. corchorifolius* L., *R. crataegifolius*, *R.* pungens Cambess. and *R. spectabilis* Pursh are generally early ripening, while *R. innominatus*, *R. coreanus*, and *R. occidentalis* are generally late. *Rubus glaucus* has an extended production season, while the season of *R. flosculosus* Focke is concentrated.

12.5.3 Plant Characteristics

Daubeny (1999) describes the ideal floricane raspberry cultivar as follows: '... has erect canes with few or no spines and adequate but not excessive cane numbers and cane heights. Fruiting laterals should be upright, strongly attached but flexible, and of moderate length with fruit well spaced.' He suggests the ideal primocane fruiting type 'produces abundant canes that branch to produce higher numbers of fruiting nodes. Cane height is moderate, which in some environments will eliminate the need of supports.' All these characters are heritable and genetic variability exists for them among cultivars (Knight and Keep 1960, Keep et al. 1977b, Daubeny 1996), although no raspberry cultivar is perfect for all these traits. Breeding programs have used a variety of species to alter architectures when not available in cultivar quality material (Yeager 1950, Finn et al. 2002a,b).

Ideally red raspberry cultivars are erect and sparsely-spined, with adequate but not excessive cane numbers and cane heights (Jennings et al. 1991). Spines are not generally a significant commercial issue for red raspberry as the spines are small and seldom noticed; however genes for spinelessness and its inheritance have been identified (Jennings 1984, 1988, Jennings and Brydon 1990, Daubeny 1996). Most raspberry cultivars are not completely spine-free, but spineless ones do exist and many only have spines on the basal portions of canes making them spineless from a commercial production standpoint, whether hand or machine harvested. The *s* gene originally found by Lewis (1939) in segregates of 'Burnetholm' has been widely used in breeding.

Black raspberry spines can be a commercial issue in that they are large and 'aggressive'. However, since black raspberries are primarily mechanically pruned, machine harvested and processed as puree or juice, where the product is pressed through a sieve, thornlessness has not been a high breeding priority. If the fresh black raspberry market is to be developed, thornlessness will become important and raspberry might be the most appropriate source of thornlessness, since no thornless black raspberries mutations have been identified.

12.5.4 Fruit Quality

The critical traits associated with high fruit quality in raspberry include size, shape, color, firmness, skin strength, texture, seed (botanically pyrene) size, flavor, and nutritional/nutraceutical content and ease of harvest. Obviously, whether the fruit is being grown for the fresh or processing market determines which traits rise or drop in importance (Fig. 12.2). Machine harvested fruit are frozen or otherwise processed within hours of harvest and therefore do not need the same level of firmness that is essential for fruit for fresh shipping. Fruit that is processed needs high soluble solids, high titratable acidity levels, and relatively low pH in order that they have long shelf stability. Since fruit for processing is often only a small portion of a product, it is essential that they have intense flavor and color.

A series of papers from Washington State University examined the relationships among raspberry fruit characteristics including firmness (Barritt 1982, Robbins and Sjulin 1989, Robbins and Moore 1990a, 1991, 1993, 1998). Barritt (1982) found very high heritability for fruit firmness in a diverse breeding population of rasp-berries containing genotypes from the Pacific Northwest and the United Kingdom. Daubeny (1996) attributes improvements in the firmness of modern cultivars to the incorporation of *R. occidentalis* through such cultivars as 'Glen Prosen' and 'Bur-netholm', likely a selection from indigenous *R. idaeus*.

Ease of harvest at maturity has been essential since the advent of viable commer-cial harvesters in the 1950s and 1960s for the processing industry and for the most efficient hand harvest for the fresh market (Hall et al. 2002). Important character-istics associated with machine harvesting are: (1) fruiting laterals that are flexible but firmly attached, (2) easy fruit detachment, (3) concentrated ripening, and (4) firm fruit (Moore 1994). Since raspberries have the dynamic where each individual drupelet must form an abscission zone at their connection to the torus, there is a great range in ease of harvest from those genotypes whose berries fall at the slightest shake to those whose fruit dry on the torus and cannot be shaken off. While various tools have been used to try to objectively measure ease of harvest (Mason 1976, Sjulin and Robbins 1987), these methods have not proven practical enough to be adopted in breeding programs. Breeders most commonly estimate ease of harvest

Fig. 12.2 Two different but acceptable ideals for red raspberry – Fruit on left has ideal color and acceptable fruit size for processing market; fruit on right is larger and 'brighter' and is more appropriate for fresh market

using subjective evaluation such as scoring how easily fruit can be removed by hand or noting how readily overripe fruit drop to the ground. When critical, breeding programs have incorporated commercial machines into their program. Most typically after a selection is made, it is put into machine harvest trials even before it is evaluated in an advanced replicated trial (Moore and Kempler pers. comm.). Some programs feel they can do an excellent job of evaluating seedlings for machine harvest by driving the machine over the seedling field (Sjulin, pers. comm.). The ease of harvest trait can be greatly affected by the environment (Daubeny 1996). In addition to the processing industry standards 'Meeker' and 'Willamette', all successful newer processing cultivars have this trait (e.g., 'Cascade Bounty', 'Chemainus', 'Chilliwack', 'Coho', 'Cowichan', and 'Saanich').

Appropriate fruit color is essential for the success of a new cultivar. Red raspberries for the fresh market must be bright and glossy red colored whereas those for processing need much greater color intensity (Sjulin and Robbins 1987, Robbins and Moore 1990b). Black raspberries, which are often sold for their natural colorant properties, need to be dark black. Blackberries and black raspberries, and to a lesser extent raspberries, naturally have a very intense color and high anthocyanin levels.

Several major genes have been described that control fruit color in red raspberry including R for the rhamnose containing anthocyanins (Barritt and Torre 1975); T, which when recessive yellow fruit are produced (Crane and Lawrence 1931); Bl, which is epistatic to T but which when dominant gives black or purple fruit (Britton et al. 1959); P, which is also epistatic to T but which give apricot/orange fruit color (Crane 1931); Y in $R.$ $phoenicolasius$ and its counterpart which suppresses yellow color Ys; and $Ycor$, which has a similar effect in $R.$ $coreanus$ (Jennings and Carmichael 1980); however, there is some contention as to whether all these genes are valid (Keep 1984, Jennings 1988). The effects of processing and environment on the anthocyanin content of red raspberry juices made from various genotypes has been examined (Boyles and Wrolstad 1993, Rommel and Wrolstad 1993).

In a breeding program, genotypes are objectively scored for color and then if they become potential processing cultivars their anthocyanin content is determined. As a compromise, Moore (1997a) found that the reflectance readings (a^*/b^*) from a tri-stimulus color meter correlated well ($r = 0.73$) with anthocyanin concentrations and required much less effort than anthocyanin extraction. A considerable amount of new research has been performed on variation patterns in the antioxidant capacity of $Rubus$ species and crosses. The fact that anthocyanins and polyphenolics are powerful antioxidants has led a number of investigators to look at the nutraceutical/antioxidant levels of raspberries (Moyer et al. 2002, Perkins-Veazie and Kalt 2002, Wada and Ou 2002, Beekwilder et al. 2005, Anttonen and Karjalainen 2005, Moore 2007).

Conner et al. (2005a,b) estimated narrow-sense heritabilities for antioxidant capacity (AA), total phenolic content (TPH) and fruit weight from progeny of a factorial mating design of seven female and six male red raspberry genotypes. A rapid response to selection appears possible, as heritability estimates were all high, at 0.54, 0.48 and 0.77 for AA, TPH and fruit weight, respectively. AA and TPH were only weakly correlated with fruit weight, suggesting that selection for high antioxidant capacity and large fruit weight is possible. In further work evaluating individual anthocyanin (ACY) content with total anthocyanin content and antioxidant capacity in the same families, Conner et al. found high values of h^2 for individual ACYs (0.54–0.90), but ACY content and profile information were 'inefficient proxies and predictors of AA in red raspberry fruit'. The inclusion of a pigment-deficient *R. parvifolius* × *R. idaeus* hybrid resulted in significant female and male contributions to variation, but its removal from the analysis made female × male interaction negligible. The overall conclusion from all of these studies is that the traits related to anthocyanin content and nutraceutical value are heritable and improvement should be expected with a recurrent mass selection breeding approach. The greater challenge for many of these traits is, when the human eye is not the best selection tool, are there tools that allow for these traits to be effectively, efficiently and cost effectively selected in seedling or parental populations?

Dossett (2007) examined variation in fruit chemistry properties including pH, titratable acids, soluble solids, anthocyanin profiles, and total anthocyanins in 26 black raspberry families from a partial diallel cross among eight cultivars and a selection of *R. occidentalis*. For each of the traits general combining ability (GCA) effects were significant and larger than specific combining ability (SCA) effects. Narrow-sense heritability estimates were generally moderate to high when year effects were excluded from the analysis, indicating the potential for progress from selection within the examined families.

Daubeny (1996) stated that when breeding caneberries, 'Flavor, the most difficult of the quality traits to define, is becoming more important...'. While the statement that flavor is becoming more important is still accurate, our ability to define the trait and therefore successfully select for it is improving rapidly. The increased sophistication of flavor chemists instrumentation (Klesk et al. 2004), combined with the sensory evaluation and the knowledge of the germplasm provided by a breeder is allowing a much greater understanding of the genetics of flavor and how to most efficiently select for it in seedling populations.

Raspberry flavor has had considerable effort devoted to it over the years (Jennings 1988, Daubeny 1996) but recent research has looked at genetic, environmental, and treatment effects, such as freezing and thawing, on flavor (Casabianca and Graff 1994, Morel et al. 1999, Sewenig et al. 2005). One of the major challenges faced by breeders is to make selections based on fresh fruit quality in the field for a market where either the fruit are harvested immature and refrigerated for several days or frozen and later thawed for processed applications. The potential for the use of either molecular markers (Paterson et al. 1993) or some in-field tool that objectively, reliably and quickly determines flavor profiles are in the process of becoming a reality (Qian, pers. comm.). Overall post harvest fresh fruit quality, which is affected by treatment, and cultivar characteristics have been well studied and useful variability identified (Jennings 1988, Crandall and Daubeny 1990, Jennings et al. 1991, Perkins-Veazie et al. 1996, 1999, 2000).

Research is beginning to emerge on the genes associated with fruit ripening in *Rubus*. Jones et al. (2000) profiled changes in gene expression during raspberry fruit ripening and identified 34 up-regulated genes. Genes have been cloned from ripening fruit that are similar to major latex proteins and endo-polygalacturonases (Jones et al. 1998). L-phenylalanine-lyase (PAL) and 4-Coumarate:CoA ligase (4CL), were found to be encoded by gene families in raspberry. Four classes of 4CL genes were identified that had distinct temporal patterns of expression during flower and fruit development (Kumar and Ellis 2003). PAL was found to be encoded by two similar genes in raspberry, *RiPAL1* which was associated with early fruit ripening, and *RiPAL2* which was more associated with later stages of development (Kumar and Ellis 2001).

Twenty genes that play a rol in fruit ripening were identified by Jones (1998) in the red raspberry 'Glen Clova'. Most of these genes were associated with cell wall hydrolysis and ethylene biosynthesis. Iannetta et al. (2000) cloned two putative endo-β-1,4 glucanase genes (*RI-EGL1* and 2) from ripe receptacle mRNA. The expression of these genes were limited to ripe-fruit receptacles, and the application of 1-methylcyclopropene (1-MCP) to green fruit indicated that ethylene accelerates raspberry abscission and increases EGase activity.

12.5.5 *Yield*

Yield in raspberries is a complex trait that is quantitatively inherited and is the sum of many components (Hoover et al. 1986, Dale 1989, Daubeny 1996, Pritts 2002). Yield in floricane fruiting types is dependent on the factors influencing the growth and development of canes in the first and second year. The most critical parameters in the first year are cane number, height and diameter, the number of nodes and root growth. In the second year, fruiting laterals per cane and fruit numbers per lateral are especially important to yield, along with fruit weight (composed of ovule number, drupelet set and drupelet size). The highest yielding plants have abundant numbers of intermediate sized canes, dense node numbers in the cropping area and vigorous root growth. Dense node numbers can be produced by a compact growth habit or short internodes. High yields have been obtained by selecting for fruit size, lateral numbers and fruits per lateral, although excesses in any of the yield components can lead to negative component interactions (Jennings 1980, Dale and Daubeny 1985). In red raspberry breeding plots, a single year's data is fairly predictive of fruit size and firmness but not incidence of Botrytis fruit rot or yield (Moore 1997b).

All the yield components are inherited additively in raspberries with significant genetic interactions (Daubeny 1996). While single genes have been identified that influence fruit size – L_1 which enhances fruit size, and l_2 which results in 'miniature' fruit (Jennings 1961, 1966a,b), the L_1 gene has proven to be very unstable and most breeding programs have actively worked to eliminate it from their programs. Among the wild species, *R. cockburnianus* Hemsl., *R. flosculosus* and *R. innominatus* have high fruit numbers per lateral; large fruit size is found in *R. glaucus*, *R. lasiostylus* Focke, and *R. nubigenus* (= *R. macrocarpus*) (Knight et al. 1989, Finn et al. 2002a,b).

Yield in primocane fruiting raspberries is most dependent on cane number and amount of branching, which directly influence the number of fruiting laterals (Hoover et al. 1988). Fruit size is negatively associated with high cane numbers in some raspberry cultivars, but not all. Earliness can also be an important component of yield in primocane fruiting types, particularly where the growth season is short. Considerable genetic variability exists for fruiting season among primocane raspberry cultivars, and several wild species have been used to breed early primocane fruiting types including *R. arcticus*, *R. odoratus* L. and *R. spectabilis* (Howard 1976, Keep 1988).

12.6 Crossing and Evaluation Techniques

The crossing and evaluation of raspberries is very similar to that of blackberries, which is thoroughly outlined in Chapter 3. Raspberries are self compatible and in many cases, interspecific crosses are possible, even across ploidies (see section on germplasm resources). Raspberry flowers are typical for the *Rosaceae* and their emasculation and pollination techniques are similar to those for others in the family. *Rubus* seeds generally require scarification and stratification. The standard germination-to-field protocol consists of an acid scarification (concentrated sulfuric acid), water and sodium bicarbonate rinse, a 5–6 days calcium hypochlorite soak, another rinse, overnight warm stratification, 6–10 weeks cold stratification, 1–4 weeks germination and transplanting, six weeks as greenhouse plugs, one week acclimation to outdoor conditions and, finally, field planting.

While most seed lots are germinated using this basic procedure, in vitro procedures are used for small seed lots or for seeds from wide or challenging crosses. The in vitro germination protocol involves surface sterilization with ethanol and bleach, a 6–10 weeks cold stratification, repeat surface sterilization, dissection, 1–2 weeks germination on media, transplanting, six weeks as greenhouse plugs and one week of acclimation prior to field planting.

For field testing, a minimum of 100 seedlings per cross are typically planted with each plant at 0.8–0.9 m apart within the row. The primocanes produced in the second year are intensively managed so that all of the seedlings can be evaluated in the third year, two years after planting. Primocane fruiting seedlings can be evaluated in the planting year in many climates but are often left for a second year to ensure that the assessment of the flowering and fruiting habit is accurate and not affected by any juvenility. In northern and southern California, the evaluation of primocane fruiting seedlings would roughly parallel the production systems. For northern California, seedlings are winter planted; the primary evaluation is done in the fall, followed by a second spring evaluation. For southern California, the seedlings are started and grown to 10–20 cm tall, chilled, summer planted and evaluated just once, the following winter. Only 0.5–1% of the seedlings are selected, primarily based on the perceived vigor, yield and fruit quality with few notes or detailed evaluations made. The most elite selections are propagated for more trialing, in either single, multiple plant observation plots or in replicated trials. Decisions about release as a cultivar are generally made 8–12 years after the initial crosses.

12.7 Biotechnological Approaches to Genetic Improvement

12.7.1 Genetic Mapping, QTL Analysis and Genomic Resources

A wide array of molecular markers has been developed in *Rubus* (Antonius-Klemola 1999, Hokanson 2001, Stafne et al. 2005). Genomic in situ hybridization (GISH) and fluorescence *in situ* hybridization (FISH) have been utilized to distinguish between raspberry and blackberry chromosomes, and identify translocations (Lim et al. 1998). Weber (2003) used RFLP markers to assess genetic diversity in black raspberry and found that on the whole the group of genotypes they evaluated had a collective marker similarity of 92% compared to 70% reported for red raspberry (Graham et al. 1994).

Graham et al. (2004) constructed a genetic linkage map of red raspberry ('Glen Moy' × 'Latham') using AFLP, genomic-SSR and EST-SSR markers. The SSR markers were developed from genomic and cDNA libraries of 'Glen Moy'. A total of 273 markers were mapped in nine linkage groups, covering 789 cM (Fig. 12.3). This map was used to search for QTL associated with cane spininess, root sucker density and root sucker spread. Two QTL for cane spininess were mapped to linkage group 2, while one QTL for root sucker density and two for root sucker spread were mapped to linkage group 8. Most of these QTL explained in excess of 50% of the phenotypic variability. Graham et al. (2006) have most recently added gene *H* to their linkage map, which determines whether canes are pubescent (*HH* or *Hh*) or glabrous (*hh*). *H* has been found to be closely associated with cane botrytis and spur blight resistance, but not to cane spot or rust.

12.7.2 Regeneration and Transformation

Regeneration and transformation systems have been developed for raspberries utilizing leaves, cotyledons and internodal stem segments (Swartz and Stover 1996, Kokko and Karenlampi 1998). A number of factors have been shown to play critical roles in determining regeneration and transformation rates including environmental conditions (Palonen and Buszard 1998), leaf orientation (McNicol and Graham 1990), type of hormone (Fiola et al. 1990, Millan-Mendoza and Graham 1999), and most importantly genotype (Reed 1990, Owens et al. 1992, Graham et al. 1997).

Mathews et al. (1995) transformed 'Canby', 'Chilliwack' and 'Meeker' red raspberries with the gene for S-adenosylmethionine (SAMase), as a potential strategy to delay fruit decay. Leaf and petiole explants were inoculated with

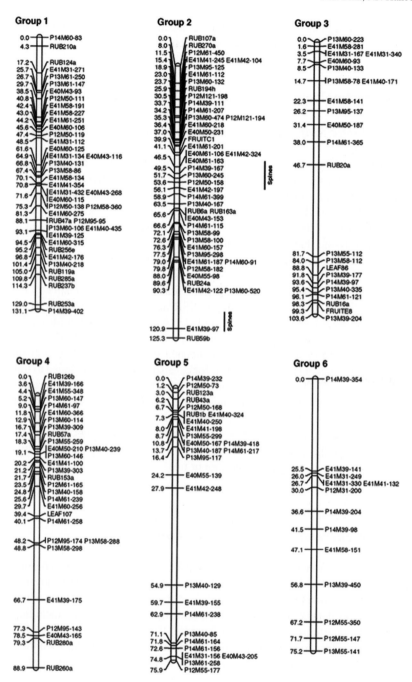

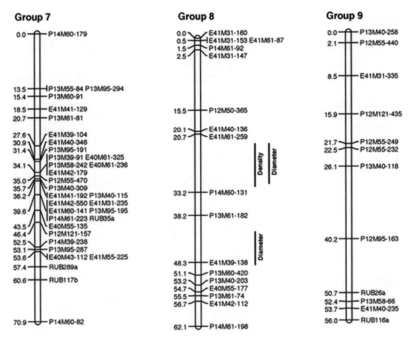

Fig. 12.3 The genetic linkage group of red raspberry built with AFLP, genomic-SSR markers and EST-SSR markers. OTL regions for spines, root sucker spread and root sucker density are also noted (Graham et al. 2004)

Agrobacterium strain EHA 105 carrying the binary vectors pAG1452 or pAG1552 encoding SAMase under control of the wound and fruit specific E4 promoter. Petiole explants produced the highest rates of transformation, and more transformants were recovered using hygromycin phospotransferase (*HPT*) as the selective agent rather than neomycin phospotransferase (*NPT11*). They reported on establishment of the transformants in soil, but have not published information on levels of resistance to decay.

There have been two attempts to generate resistance to raspberry bushy dwarf (RBDV) using *Agrobacterium*–mediated transformation. Jones et al. (1998) isolated the coat protein gene (*cp*) from a resistance-breaking strain of RBDV and transformed plants with it in the sense and anti-sense orientation. Some of their transformants were partially resistant. Taylor and Martin (1999) sequenced the *cp* gene, mutations of the movement protein and non-translatable RNA of RBDV and transformed 'Meeker' red raspberry with each of these constructs (Martin and Mathews 2001). Transformed 'Meeker' from this work, has been successfully field trialed and the processed fruit quality compared to wild-type 'Meeker' with no discernable differences, although it is not commercially grown (Martin and Qian, pers. comm.). 'Ruby' was successfully transformed with the DEfH9-iaaM gene which produces parthenocarpic fruit (Mezzetti et al. 2002).

References

Anthony VM, Williamson B, Jennings DL, Shattock RC (1986) Inheritance of resistance to yellow rust (*Phragmidium rubi-idaei*) in red raspberry. Ann Appl Biol 109:365–374

Antonius-Klemola K (1999) Molecular markers in *Rubus* (Rosaceae) research and breeding. J Hortic Sci Bio 74:149–160

Anttonen MJ, Karjalainen RO (2005) Environmental and genetic variation of phenolic compounds in red raspberry. J Food Comp Anal 18:759–769

Barritt BH (1982) Heritability and parent selection for fruit firmness in red raspberry. HortScience 17:648–649

Barritt BH, Torre LC (1975) Inheritance of fruit anthocyanins pigments in red raspberry. HortScience 10:526–528

Beekwilder J, Hall R, de Vos CHR (2005) Identification and dietary relevance of antioxidants from raspberry. Biofactors 23:197–205

Boyles MJ, Wrolstad RE (1993) Anthocyanin composition or red raspberry juice: Influences of cultivar, processing, and environmental factors. J Food Sci 58:1135–1141

Bristow PR, Daubeny HA, Sjulin TM, Pepin HS, Nestby R, Windom GE (1988) Evaluation of *Rubus* germplasm for reaction to root rot caused by *Phytophthora erythroseptica*. J Am Soc Hortic Sci 113:588–591

Britton DM, Lawrence FJ, Haut IC (1959) The inheritance of apricot fruit color in raspberries. Can J Genet Cytol 1:89–93

Casabianca H, Graff JB (1994) Enantiomeric and isotopic analysis of flavour compounds of some raspberry cultivars. J Chromatogr A 684:360–365

Chamberlain CJ, Kraus J, Kohnen PD, Finn CE, Martin RR (2003) First report of Raspberry bushy dwarf virus in *Rubus multibracteatus* from China. Plant Dis 87:603

Clark JR, Stafne ET, Hall HK, Finn CE (2007) Blackberry breeding and genetics. Plant breeding rev 29:19–144

Conner AM, McGhie TK, Stephens MJ, Hall HK, Alspach PA (2005a) Variation and heritability estimates of anthocyanins and their relationship to antioxidant activity in a red raspberry factorial design. J Am Soc Hortic Sci 130:534–542

Conner AM, Stephens MJ, Hall HK, Alspach PA (2005b) Variation and heritabilities of antioxidant activity and total phenolic content estimated from a red raspberry factorial experiment. J Am Soc Hortic Sci 130:403–411

Converse RH (1991) Diseases caused by viruses and viruslike agents. In: Ellis MA, Converse RH, Williams RN, Williamson B (eds) Compendium of raspberry and blackberry diseases and insects. APS Press, St. Paul, Minn, pp 42–58

Crandall PC, Daubeny HA (1990) Raspberry management. In: Galletta GJ, Himelrick DG (eds) Small fruit crop management. Prentice Hall, Englewood Cliffs NJ, pp 157–213

Crane MB, Lawrence WJC (1931) Inheritance of sex, colour, and hairiness in the raspberry, *Rubus idaeus* L. J Genet 24:243–255

Dale A (1989) Productivity in red raspberries. Hortic Rev 11:185–228

Dale A, Daubeny HA (1985) Genotype-environmental interactions involving British and Pacific Northwest raspberry cultivars. HortScience 20:68–69

Dale A, McNicol RJ, Moore PP, Sjulin TM (1989) Pedigree analysis of red raspberry. Acta Hortic 262:35–39

Dale A, Moore PP, McNicol RJ, Sjulin TM, Burmistrov LA (1993) Genetic diversity of red raspberry varieties throughout the world. J Am Soc Hortic Sci 118:119–129

Darrow GM (1920) Are our raspberries derived from American or European species? J Hered 11:179–184

Darrow GM (1937) Blackberry and raspberry improvement. pp. 496–533. U.S. Dept. of Agr. Yearbook of Agriculture Yrbk. 1937

Daubeny HA (1983) Expansion of genetic resources available to red raspberry breeding programs. Proc 21st Int Hortic Cong 1:150–155

Daubeny HA (1987) A hypotheses for inheritance of resistance to cane *Botrytis* in red raspberry. HortScience 22:116–119

Daubeny HA (1995). In: Cummins JN (ed) Register of new fruit and nut varieties Brooks and Olmo list no. 37: Blackberries and hybrid berries. HortScience 30:1136–1137

Daubeny HA (1996) Brambles. In: Janick J, Moore JN (eds) Fruit Breeding. vol. II. Vine and small fruits. John Wiley & Sons, NewYork

Daubeny HA (1997) Raspberry. The Brooks and Olmo register of fruit and nut varieties, 3rd edn. ASHS Press, Alexandria, Va, pp 635–662

Daubeny HA (1999) Raspberry. In: Okie WR (ed) Register of new fruit and nut varieties List 39. HortScience 34:196–197

Daubeny HA (2002a) Raspberry. In: Okie WR (ed) Register of new fruit and nut varieties List 41. HortScience 37:264–266

Daubeny H (2002b) Raspberry breeding in the 21st century. Acta Hort 585:69–72

Daubeny HA, Stary D (1982) Identification of resistance to *Amphorophora agathonica* in the native North American red raspberry. J Am Soc Hortic Sci 91:593–597

Dossett M (2007) Variation and heritability of vegetative, reproductive and fruit chemistry traits in black raspberry (*Rubus occidentalis* L.) M.S. Thesis, Oregon State University, Corvallis, Ore

Drain BD (1952) Some inheritance data with black raspberries. Proc Am Soc Hortic Sci 60:231–234

Drain BD (1956) Inheritance in black raspberry species. Proc Am Soc Hortic Sci 68:169–170

Finn CE (1999) Temperate berry crops. In: Janick J (ed) Perspectives on new crops and new uses. ASHS Press, Alexandria, Virg, pp 324–333

Finn CE (2006) Caneberry breeders in North America. HortScience 41:22–24

Finn CE, Knight VH (2002) What's going on in the world of *Rubus*breeding? Acta Hortic 585:31–38

Finn CE, Lawrence FJ, Strik BC, Yorgey B, DeFrancesco J (1999) 'Siskiyou' trailing blackberry. HortScience 34:1288–1290

Finn CE, Lawrence FJ, Yorgey B, Strik BC (2001) 'Coho' red raspberry. HortScience 36:1159–1161

Finn CE, Martin RR (1996) Distribution of tobacco streak, tomato ringspot, and raspberry bushy dwarf viruses in *Rubus ursinus* and *R. leucodermis* collected from the Pacific Northwest. Plant Dis 80:769–772

Finn CE, Moore PP, Kempler C (2007) Raspberry cultivars: What's new? What's succeeding? Where are breeding programs headed? Acta Hortic (In press)

Finn CE, Swartz HJ, Moore PP, Ballington JR, Kempler C (2002a) Use of 58 *Rubus* species in five North American breeding programs-breeders notes. Acta Hortic 585:113–119

Finn CE, Swartz HJ, Moore PP, Ballington JR, Kempler C (2002b) Breeders experience with *Rubus* species. http://www.ars-grin.gov/cor/rubus/rubus.uses.html

Finn CE, Wennstrom K, Link J, Ridout J (2003) Evaluation of *Rubus leucodermis* populations from the Pacific Northwest. HortScience 38:1169–1172

Fiola JA, Hassan MA, Swartz HJ, Bors RH, McNichols R (1990) Effect of thidiazuron, light fluence rates, and kanamycin on in vitro shoot organogenesis from excised *Rubus* cotyledons and leaves. Plant Cell Tissue Organ Cult 20:223–228

Fiola JA, Swartz HJ (1989) Screening raspberry (Idaeobatus) hybrids for resistance to *Verticillium albo-atrum*. Acta Hortic 262:181–187

Fiola JA, Swartz HJ (1994) Inheritance of tolerance to *Verticillium albo-atrum* in raspberry. HortScience 29:1071–1073

Galletta GJ, Maas JL, Enns JM (1998) 'Earlysweet' black raspberry. Fruit Varieties J 52:123

Graham J, Iasi L, Millam S (1997) Genotype-specific regeneration from a number of *Rubus* cultivars. Plant Cell Tissue Organ Cult 48:167–173

Graham J, McNichols R, Greig K, Van de Ven WTG (1994) Identification of red raspberry cultivars and an assessment of their relatedness using fingerprints produced by random primers. J Hortic Sci 69:123–130

Graham J, Smith K, MacKenzie K, Jorgenson L, Hackett C, Powell W (2004) The construction of a genetic linkage map of red raspberry (*Rubus idaeus* subsp. *idaeus*) based on AFLPs, genomic-SSR and EST-SSR markers. Theor Appl Genet 109:740–749

Graham J, Smith K, Tierney I, MacKenzie K, Hackett C (2006) Mapping gene H controlling cane pubescence in raspberry and its association with resistance to cane botrytis and spur blight, rust and cane spot. Theor Appl Genet 112:818–831

Halgren A, Tzanetakis IE, Martin RR (2007) Identification, characterization and detection of Black raspberry necrosis virus. Phytopathology 97:44–50

Hall HK, Stephens MJ, Alspach P, Stanley CJ (2002) Traits of importance for machine harvest of raspberries. Acta Hortic 585:607–610

Hedrick UP (1925) The small fruits of New York. J.B. Lyon. Albany, New York

Hellman EW, Skirvin RM, Otterbacher AG (1982) Unilateral incompatibility between red and black raspberries. J Am Soc Hortic Sci 107:718–784

Hokanson SC (2001) SNiPs, chips, BACs, and YACs: are small fruits part of the party mix. HortScience 36:859–871

Hoover E, Luby J, Bedford D (1986) Yield components of primocane-fruiting red raspberries. Acta Hortic 183:163–166

Hoover E, Luby J, Bedford D, Pritts M (1988) Vegetative and reproductive yield components of primocane-fruiting red raspberries. J Am Soc Hortic Sci 113:824–826

Howard GS (1976) 'Pathfinder' and 'Trailblazer' everbearing raspberries released. Fruit Varieties J 3:94

Iannetta PPM, Wyman M, Neelam A, Jones C, Taylor MA, Davies HV, Sexton R (2000) A causal role for ethylene and endo-β-1,4-glucanase in the abscission of red-raspberry (*Rubus idaeus*) druplets. Physiol Plant 110: 535–543

Jennings DL (1961) Mutation for larger fruit in the raspberry. Nature 191:302–303

Jennings DL (1964) Studies on the inheritance in the red raspberry of immunities from three nematode-borne viruses. Genetica 34:152–164

Jennings DL (1966a) The manifold effects of genes effecting fruit size and vegetative growth in the raspberry, I. gene L_1. New Phytol 65:176–187

Jennings DL (1966b) The manifold effects of genes effecting fruit size and vegetative growth in the raspberry, II. gene l_2. New Phytol 65:188–191

Jennings DL (1980) Recent progress in breeding raspberries and other *Rubus* fruits at the Scottish Horticulture Research Institute. Acta Hortic 112:109–116

Jennings DL (1983) Inheritance of resistance to *Botrytis cinerea* and *Didymella applanata* in canes of *Rubus idaeus*, and relationships between these resistances. Euphytica 32:895–901

Jennings DL (1984) A dominant gene for spinelessness in *Rubus*, and its use in breeding. Crop Res 24:45–50

Jennings DL (1988) Raspberries and blackberries: their breeding, diseases and growth. Academic Press, London

Jennings DL, Brydon E (1990) Variable inheritance of spinelessness in progenies of a mutant of the red raspberry cv. Willamette. Euphytica 46:71–77

Jennings DL, Carmichael E (1980) Anthocyanin variation in the genus *Rubus*. New Phytol 84:505–513

Jennings DL, Daubeny HA, Moore JM (1991) Blackberries and raspberries (*Rubus*). In: Moore JN, Ballington JR (eds) Genetic resources of fruit and nut crops, vol 1. International Society for Horticultural Science, Wageningen, pp 329–320

Jennings DL, Jones AT (1986) Immunity from raspberry vein chlorosis virus in raspberry and its potential for control of the virus through plant breeding. Ann Appl Biol 108:417–422

Jones AT, McGavin WJ (1998) Infectibility and sensitivity of U.K. raspberry, blackberry and hybrid berry cultivars to *Rubus* viruses. Ann Appl Biol 132:239–251

Jones AT, McGavin WJ, Birch ANE (2000) Effectiveness of resistance genes to the large raspberry aphid, Amphorophora idaei Börner, in different raspberry (*Rubus idaeus* L.) genotypes and under different environmental conditions. Ann Appl Biol 136:107–113

Jones AT, Murant AF, Jennings DL, Wood GA (1982) Association of raspberry bushy dwarf virus with raspberry yellows disease: Reaction of *Rubus* species and cultivars, and the inheritance of resistance. Ann Appl Biol 100:135–147

Jones CS, Davies HV, McNicol RJ, Taylor MA (1998) Cloning of three genes up-regulated in ripening raspberry fruit (*Rubus idaeus* cv. Glen Clova). J Plant Physiol 153:643–648

Jones CS, Davies HV, Taylor MA (2000) Profiling of changes in gene expression during raspberry (*Rubus idaeus*) fruit ripening by application of RNA fingerprinting techniques. Planta 211:708–714

Keep E (1968) The inheritance of accessory buds in Rubus idaeus L. Genetica 39:209–219

Keep E (1984) Inheritance of fruit color in a wild Russian red raspberry seedling. Euphytica 33:507–515

Keep E (1988) Primocane (autumn)-fruiting raspberries: A review with particular reference to progress in breeding. J Hortic Sci 63:1–18

Keep E, Knight RL (1968) Use of the black raspberry (*Rubus occidentalis* L.) and other Rubus species in breeding red raspberries. Rep E Malling Res Stn for 1967 pp 105–107

Keep E, Knight VH, Parker JH (1977a) The inheritance of flower color and vegetative characters in *Rubus coreanus*. Euphytica 26:185–192

Keep E, Knight VH, Parker JH (1977b) *Rubus coreanus* as donor or resistance to cane disease and mildew in red in red raspberry breeding. Euphytica 26:505–510

Kempler C, Daubeny HA, Frey L, Walters T (2006) 'Chemainus' red raspberry. HortScience 41:1364–1366

Kempler C, Daubeny H, Harding B (2002) Recent progress in breeding red raspberries in British Columbia, Canada. Acta Hortic 585:47–50

Kempler C, Daubeny HA, Harding B, Baumann T, Finn CE, Moore PP, Sweeney M, Walters T (2007) 'Saanich' red raspberry. HortScience 42:176–178

Kempler C, Daubeny HA, Harding B, Finn CE (2005a) 'Esquimalt' red raspberry. HortScience 40:2192–2194

Kempler C, Daubeny HA, Harding B, Kowalenko CG (2005b) 'Cowichan' red raspberry. HortScience 40:1916–1918

Kennedy DM, Duncan JM (1993) Occurrence of races of *Phytophthora fragariae* var Rubi on raspberry. Acta Hortic 352:555–559

Klesk K, Qian M, Martin RR (2004) Aroma extract dilution analysis of cv. Meeker (*Rubus idaeus* L.) red raspberries from Oregon and Washington. J Agric Food Chem 52:5155–5161

Knight VH (1993) Review of *Rubus* species used in raspberry breeding at East Malling. Acta Hortic 352:363–371

Knight VH, Barbara BJ (1999) A review of raspberry bushy dwarf virus at HRI-East Malling and the situation in a sample of commercial holdings in England in 1995 and 1996. Acta Hortic 505:263–271

Knight VH, Briggs JB, Keep E (1960) Genetics of resistance to *Amphorophora rubi* (Kalt.) in the raspberry, II: the genes A2-A7 from the American variety, Chief. Genet Res 1: 319–331

Knight VH, Jennings DL, McNicol RJ (1989) Progress in the U.K. raspberry breeding programme. Acta Hortic 262:93–103

Knight VH, Keep E (1960) The genetics of suckering and tip rooting in the raspberry. In: Report of East Malling Research Station for 1959, pp 57–62

Knight VH, Parker JH, Keep E (1972) Abstract bibliography of fruit breeding and genetics, 1956–1969: Rubus and Ribes. In: Tech. Commun. 32. Commonwealth Bur. Hort. Plantation Crops, East Malling

Kokko HI, Karenlampi SO (1998) Transformation of arctic bramble (*Rubus arcticus* L.) by *Agrobacterium tumefaciens*. Plant Cell Rep 17:822–826

Kumar A, Ellis BE (2001) The phenylalanine ammonia-lyase gene family in raspberry. Structure, expression and evolution. Plant Physiol 127:230–239

Kumar A, Ellis BE (2003) 4-Coumarate:CoA ligase gene family in *Rubus idaeus*: cDNA structures, evolution and expression. Plt Mol Biol 51:327–340

Lewis D (1939) Genetical studies in cultivated raspberries. I. Inheritance and linkage. J Genet 38:367–379

Lim KY, Leitch IJ, Leitch AR (1998) Genomic characterization and the detection of raspberry chromatin in polyploid *Rubus*. Theo Appl Gene 97:1027–1033

Mathews H, Wagoner W, Cohen C, Kellogg J, Bestwick R (1995) Efficient genetic transformation of red raspberry, *Rubus idaeus* L. Plant Cell Rpt 14:471–476

Martin RR, Mathews H (2001) Engineering resistance to raspberry bushy dwarf virus. Acta Hortic 551:33–38

Martin RR (2002) Virus diseases of *Rubus* and strategies for their control. Acta Hortic 585:265–270

Mason DT (1976) Changes in the fruit retention strength of the red raspberry (*Rubus idaeus* L.) during ripening and their relevance to the selection of raspberry clones suitable for mechanical harvesting. Acta Hortic 60:113–122

McNicol RJ, Graham J (1990) In vitro regeneration of *Rubus* from leaf and stem segments. Plant Cell Tissue Organ Cult 21:45–50

Mezzetti B, Landi L, Spena A (2002) Biotechnology for improving *Rubus* production and quality. Acta Hortic 585:73–78

Millan-Mendoza B, Graham J (1999) Organogenesis and micropropagation in red raspberry using florchlorfenuron (CPPU). J Hortic Sci Bio 74:219–223

Moore PP (1993) Variation in drupelet number and drupelet weight among raspberry clones in Washington. Acta Hortic 352:405–412

Moore PP (1994) Yield compensation of red raspberry following primary bud removal HortScience 29:701

Moore PP (1997a) Estimation of anthocyanin concentration from color meter measurements of red raspberry fruit. HortScience 32:135

Moore PP (1997b) Year-to-year consistency of harvest data in raspberry breeding plots. J Am Soc Hortic Sci 122:211–214

Moore PP (1998) Variation in drupelet number and weight in Pacific Northwest red raspberries. Fruit Varieties J 2:103–106

Moore PP (2004) 'Cascade Delight' red raspberry. HortScience 39:185–187

Moore PP (2006) 'Cascade Dawn' red raspberry. HortScience 41:857–859

Moore PP, Finn CE (2007) 'Cascade Bounty' red raspberry. HortScience 42:393–396

Moore PP, Perkins-Veazie P, Weber CA, Howard L (2007) Environmental effect on antioxidant content on ten raspberry cultivars. Acta Hortic (In press)

Morel S, Harrison RE, Muir DD, Hunter EA (1999) Genotype, location and harvest date effects on the sensory character of fresh and frozen red raspberries. J Am Soc Hortic Sci 124:19–23

Moyer R, Hummer K, Finn C, Frei B, Wrolstad R (2002) Anthocyanins, phenolics and antioxidant capacity in diverse small fruits: *Vaccinium*, *Rubus* and *Ribes*. J Agric Food Chem 50:519–525

Ourecky DK (1975) Brambles. In: Janick J, Moore JN (eds) Advances in Fruit Breeding. Purdue University Press, West Lafayette, Indiana, pp 98–120

Ourecky DK, Slate GL (1966) Hybrid vigor in *Rubus occidentalis-R. leucodermis* seedlings. In: Proc 17th Int Hortic Cong 1: Abstr 277

Owens Y, de Novoa C, Conner AJ (1992) Comparison of in vitro shoot regeneration protocols from *Rubus* leaf explants. New Zealand J Crops and Hortic Sci 20:471–476

Oydvin J (1970) Important breeding lines and cultivars in raspberry breeding. St Forsokag Njos 1970:1–42

Palonen P, Buszard D (1998) In vitro screening for cold hardiness of raspberry cultivars. Plant Cell Tissue Organ Cult 53:213–216

Paterson A, Piggott JR, Jiang J (1993) Approaches to mapping loci that influence flavor quality in raspberries. In: Spanier AM, Okai H, Tamura M (eds) Food flavor and safety. ACS Symposium Series 528:110–115

Pattison JA, Weber CA (2005) Evaluation of red raspberry cultivars for resistance to Phytophthora rot root. J Am Pom Soc 59:50–56

Pattison JA, Wilcox WF, Weber CA (2004) Assessing the resistance of red raspberry (*Rubus idaeus* L.) genotypes to *Phytophthora fragariae* var. *rubi* in Hydroponic culture. HortScience 39:1553–1556

Perkins-Veazie P, Collins JK, Clark JR (1996) Cultivar and maturity affect postharvest quality of fruit from erect blackberries. HortScience 31:258–261

Perkins-Veazie P, Collins JK, Clark JR (1999) Cultivar and storage temperature effects on the shelflife of blackberry fruit. Fruit Varieties J 53:201–208

Perkins-Veazie P, Collins JK, Clark JR (2000) Shelflife and quality of 'Navaho' and 'Shawnee' blackberry fruit stored under retail storage conditions. J Food Qual 22:535–544

Perkins-Veazie P, Kalt W (2002) Postharvest storage of blackberry fruit does not increase antioxidant levels. Acta Hortic 585:521–524

Pritts MP (2002) From plant to plate: How can we redesign *Rubus* production systems to meet future expectations. Acta Hortic 585:537–543

Ramanathan V, Simpson CG, Thow G, Iannetta PPM, McNicol RJ, Williamson B (1997) cDNA-cloning and expression of polygalacturonase-inhibiting proteins (PGIPs) from red raspberry (*Rubus idaeus*). J Exp Bot 48:1185–1193

Reed BM (1990) Multiplication of *Rubus* germplasm in vitro: a screen of 256 accessions. Fruit Varieties J 44:141–148

Robbins J, Moore PP (1990a) Relationship of fruit morphology and weight to fruit strength in 'Meeker' red raspberry. HortScience 25:679–681

Robbins JA, Moore PP (1990b) Color change in fresh red raspberry fruit stored at 0, 4.5 or 20° C. HortScience 25:1623–1624

Robbins J, Moore PP (1991) Fruit morphology and fruit strength in a seedling population of red raspberry. HortScience 26:294–295

Robbins JA, Sjulin T (1989) Fruit morphology of red raspberry and its relationship to fruit strength. HortScience 24:776–778

Rodriguez AJ, Avitia GE (1989) Advances in breeding low-chill red raspberries in central Mexico. Acta Hortic 262:127–132

Rommel A, Wrolstad RE (1993) Composition of flavonols in red raspberry juice as influenced by cultivar, processing, and environmental factors. J Agric Food Chem 41:1941–1950

Schaefers GA, Labnowska BH, Brodel CF (1978) Field evaluation of eastern raspberry fruitworm damage to varieties of red raspberry. J Econ Entomol 71:566–569

Sewenig S, Bullinger D, Hener U, Mosandl A (2005) Comprehensive authentication of (*E*) – $\alpha(\beta)$-ionone from raspberries, using constant flow MDGC-C/P-IRMS and enantio-MDGC-MS. J Agric Food Chem 53:838–844

Shanks CH Jr, Moore PP (1996) Resistance of red raspberry and other *Rubus* species to Twospotted spider mite (Acari: Tetranychidae). J Econ Entomol 89:771–774

Sjulin TS, Robbins JA (1987) Effects of maturity, harvest date, and storage time on postharvest quality of red raspberry fruit. J Am Soc Hortic Sci 112:481–487

Slate GL (1934) The best parents in purple raspberry breeding. Proc Am Soc Hortic Sci 30:108–112

Slate GL, Klein LG (1952) Black raspberry breeding. Proc Am Soc Hortic Sci 59:266–268

Stace-Smith R (1956) Studies on *Rubus* virus diseases in British Columbia. III. Separation of components of raspberry mosaic. Can J Bot 34:435–442

Stafne ET, Clark JR, Rom CR (2000) Leaf gas exchange characteristics of red raspberry germplasm in a hot environment. HortScience 35:278–280

Stafne ET, Clark JR, Weber CA, Graham J, Lewers KS (2005) Simple sequence repeat (SSR) markers for genetic mapping of raspberry and blackberry. J Am Soc Hortic Sci 103:722–728

Stahler MM, Lawrence FJ, Martin RR (1995) Incidence of raspberry bushy dwarf virus in breeding plots of red raspberry. HortScience 30:113–114

Stewart PJ, Clark JR, Fenn P (2005) Sources and inheritance of resistance to fire blight [*Erwinia amylovora*] in eastern U.S. blackberry genotypes. HortScience 40:39–42

Swartz HJ, Naess SK, Yongping Z, Cummaragunta J, Luchsinger L, Walsh CS, Stiles H, Turk BA, Fordham I, Zimmerman RH, Fiola JA, Smith B, Popenoe J (1993) Maryland/Virginia/New Jersey/Wisconsin *Rubus* breeding program. Acta Hortic 352:484–492

Swartz HJ, Stover EW (1996) Genetic transformation in raspberries and blackberries (*Rubus* species). In: Bajaj YPS (ed) Biotechnology in Agriculture and Forestry, vol 38. Springer-Verlag, Berlin, pp 297–307

Taylor S, Martin RR (1999) Sequence comparison between common and resistance breaking strains of raspberry bushy dwarf virus. Phytopathology 89:576 (Abstr.)

Thompson MM (1995a) Chromosome numbers of *Rubus* species at the National Clonal Germplasm Repository. HortScience 30:1447–1452

Thompson MM (1995b) Chromosome numbers of *Rubus* cultivars at the National Clonal Germplasm Repository. HortScience 30:1453–1456

Thompson MM (1997) Survey of chromosome numbers in *Rubus* Rosaceae: Rosoideae. Ann Rpt Mo Bot Garden 84:128–163

USDA, ARS, National Genetic Resources Program (2007) Germplasm Resources Information Network – (GRIN) [Online Database]. National Germplasm Resources Laboratory, Beltsville, Md. URL: http://www.ars-grin.gov/cgi-bin/npgs/html/genus.pl? 10574 (Jan. 2007)

Vrain TC, Daubeny HA (1986) Relative resistance of red raspberry and related genotypes to the root lesion nematode. HortScience 21:1435–1437

Wada L, Ou B (2002) Antioxidant activity and phenolic content of Oregon caneberries. J Agric Food Chem 50:3495–3500

Warmund MR, George MF (1990) Freezing survival and supercooling in primary and secondary buds of *Rubus* spp. Can J Plant Sci 70:893–904

Weber CA (2003) Genetic diversity in black raspberry detected by RAPD markers. HortScience 38:269–272

Wilcox WF, Scott PH, Hamm PB, Kennedy DM, Duncan JM, Brasier CM, Hansen EM (1993) Identity of a Phytophthora species attacking raspberry in Europe and North America. Mycol Res 97:817–831

Wilde G, Hall HK, Thomas WP (1991) Resistance to raspberry bud moth (Lepidoptera: Carposinidae) in raspberry cultivars. J Econ Entomol 84:247–250

Williams CF (1961) Raspberry and blackberry breeding. North Carolina Agricultural Experiment Station

Williams CF, Smith BW, Darrow GM (1949) A pan-America blackberry hybrid. J Hered 40:261–265

Williamson B, Jennings DL (1992) Resistance to cane and foliar diseases in red raspberry (*Rubus idaeus*) and related species. Euphytica 63:59–70

Wu W, Chen Y, Lu L, Li W, Sun Z (2006) Blackberry and raspberry introduction in Nanjing. J Jiangsu Forestry Sci Tech 33:13–20

Yeager AF (1950) Breeding improved horticultural plants II Fruits, nuts and ornamentals. New Hampshire Agric Expt Sta Bull 383

Ying G, Sun ZG, Cai JH, Huang YS, He SE (1989) Introduction and utilization of small fruits in China with special reference to Rubus species. Acta Hortic 262:47–55

Chapter 13
Strawberries

J.F. Hancock, T.M. Sjulin and G.A. Lobos

Abstract The major cultivated strawberry species, *Fragaria × ananassa*, is a hybrid of two native species, *F. chiloensis* and *F. virginiana*. Strawberry breeders are focused on improving local adaptations, fruit quality, productivity and disease resistance, and many are interested in developing new day-neutral cultivars. Some of the major pathogens worldwide are *Botrytis cinerea, Colletotrichum* spp., *Phytophthora cactorum, Phytophthora fragariae* and *Verticillium albo-atrum*. The genetics of many of the horticulturally important traits have been investigated in strawberry and a number of genes have been characterized and cloned that are highly expressed during fruit ripening and maturation. Marker systems have been developed in strawberry for genetic linkage mapping and QTL have been identified for the day neutrality trait and several other fruit characteristics. Transgenic strawberries have been produced with herbicide and pest resistance and an effective marker-free transformation process has been developed. Two major EST libraries have been generated as genomic resources.

13.1 Introduction

The most popular cultivated strawberry is the dessert strawberry, *Fragaria× ananassa*. Annual world production of this species has steadily grown through the ages, with quantities doubling in the last 20 years to over 2.5 million tones (FAO Production Statistics). Most of the production is located in the northern hemisphere (98%), but there are no genetic or climatic barriers preventing greater expansion into the southern hemisphere. There are two primary types of strawberries now grown commercially, day-neutral and short day plants. Long day ('everbearing') plants are also available, but they are only commercially important in southern California.

J.F. Hancock
Department of Horticulture, 342C Plant and Soil Sciences Building, Michigan State University, East Lansing, Michigan 48824, USA
e-mail: hancock@msu.edu

There are two major production systems utilized in the world – matted rows and hills. The matted row system employs runners as the primary yield component. Both mother and daughter plants are allowed to runner freely, with periodic training into narrow rows. The hill or 'plasticulture system' relies on crowns as the primary yield component, and any runners that form are removed. The hill system is used primarily in areas having warm winters and either hot or moderate summers such as California, Florida, Italy and Spain. Matted rows are used to grow short day cultivars in climates with short summers and cold winters such as continental Europe and northern North America (Hancock 1999). In general, cultivars perform best in one or the other cultural system, although some are more flexible than others.

The most widely planted cultivar in the world is 'Camarosa', released from the University of California (UC) breeding program. It is important in all climates with mild winters (Florida, Southern U.S.A., Australia, Italy, New Zealand, South America, South Africa, Turkey, Mexico and Spain). The newer UC release 'Ventana' appears likely to be the next dominate cultivar in the warmer regions. 'Honeoye' has gained the broadest foothold in world climates with cold winters, followed by 'Earliglow' which dominates eastern U.S.A. crop area and Dutch bred 'Elsanta' which predominates in Europe.

Three other strawberry species are of minor importance in the world, *Fragaria chiloensis*, *F. vesca* and *F. moschata*. *Fragaria chiloensis* is currently grown to a small extent in Chile, but was widely grown there until the late 1800s when it was replaced by *F. ×ananassa*. It was domesticated 1,000 years ago by the indigenous Chilean Mapuches and was spread widely by the Spanish during the colonization period. *Fragaria vesca* was probably cultivated by the ancient Romans and Greeks, and by the 1300s, it was being grown all across Europe (Darrow 1966). *Fragaria vesca*, the alpine strawberry or fraise de bois, had its widest popularity in the 1500s and 1600s in Europe before the introduction of strawberry species from the New World. It is now generally restricted to home gardens where its small, aromatic fruits are considered a delicacy; most of the varieties grown are everbearers. The musky-flavored *F. moschata* (Hautbois or Hautboy) was also planted in gardens by the late 15th century, along with the green strawberry, *F. viridis*. *Fragaria viridis* was used solely as an ornamental all across Europe, while *F. moschata* was utilized for its fruit by the English, Germans and Russians. Neither of these two species is of current commercial importance.

13.2 Evolutionary Biology and Germplasm Resources

The strawberry belongs to genus *Fragaria* in the Rosaceae. Its closest relatives are Duchesnea and Potentilla. There are four basic fertility groups in *Fragaria* that are associated primarily with their ploidy level or chromosome number (Table 13.1). The most common native species, *F. vesca*, has 14 chromosomes and is considered a diploid. The most important cultivated strawberry, *F. ×ananassa*, is an octoploid with 56 chromosomes. It is an accidental hybrid of *F. chiloensis* and *F. virginiana*

Table 13.1 Strawberry species of the world and their important horticultural traits

Species	Ploidy	Location	Important traits
F. vesca L.	2x	Worldwide	Bright red, aromatic, soft fruit; long ovate-variable; cold, heat and drought tolerant; multiple disease resistances; self-compatible
F. viridis Duch.		Europe and Asia	Firm, green-pink fruit; spicy, cinnamon like flavor; self-incompatible
F. nilgerrensis Schlect.		Southeastern Asia	Pink, tasteless to unpleasant fruit; subglobose; immune to aphids and several leaf diseases; self-compatible
F. daltoniana J. Gay		Himalayas	Shiny red, tasteless fruit; ovoid to cylindrical; self-compatible
F. nubicola Lindl.		Himalayas	Fruit resembles *F. vesca*; self-incompatible
F. iinumae Makino		Japan	Spongy, nearly tasteless fruit; elongate; cold tolerant; self-compatible
F. yesoensis Hara.[1]		Japan	Fruit resembles *F. nipponica*; self-compatible
F. mandshurica Staudt		North China	Very acid fruit; subglobose to obovoid; self-incompatible
F. nipponica Makino.[1]		Japan	Unpleasant flavored fruit; cold tolerant; globose to ovoid; self-incompatible
F. gracilisa A. Los.		North China	Elongated and ovate fruit; self-incompatible
F. pentaphylla Losinsk		North China	Bright red, firm fruit with little flavor; ovoid-globose; multiple leaf disease resistances; self-incompatible
F. corymbosa Losinsk		North China	Seeds in deep pits; dioecious
F. orientalis Losinsk	4x	Russian Far East/ China	Soft fruit with slight aroma; obovoid; trioecious
F. moupinensis (French.) Card		North China	Resembles *F. nilgerrensis*; orange red, spongy fruit; nearly tasteless; dioecious
F. ×bringhurstii Staudt	5x	California	Intermediate to *F. vesca* and *F. chiloensis*; dioecious
F. moschata Duch.	6x	Euro-Siberia	Light to dark dull purplish red fruit, soft, irregular to ovoid; musky flavored and aromatic; tolerant to shade, cold and water-logged soil; immune to powdery mildew; trioecious
F. chiloensis (L.) Miller	8x	Western N. America and Chile	Dull red brown, white flesh, mild, firm, round to oblate; very broad range of adaptations and traits; trioecious

Table 13.1 (continued)

Species	Ploidy	Location	Important traits
F. virginiana Miller		North America	Soft to deep red or scarlet fruit; white flesh, tart, aromatic; very broad range of adaptations and traits; trioecious
F. iturupensis Staudt		Iturup Island	Spherical, bright red; trioecious
F. ×*ananassa* Duchesne ex Lamarck		Worldwide	Very large, red, fruit; variable in all traits

[1] According to Staudt (1989), *F. nipponica* and *F. yesoensis* are the same species.
Staudt 1989, 1999, Galletta and Bringhurst 1991, Hummer 1995, Bors and Sullivan 1998

that arose in the mid-1700s when plants of *F. chiloensis* from Chile were planted in France next to *F. virginiana* from the eastern seaboard of the United States (see more details below).

An accurate taxonomy of the native strawberry species is still emerging. Diploid, tetraploid and hexaploid species are found in Europe and Asia (Table 13.1), but octoploids are restricted to the New World and perhaps Iturup Island northeast of Japan (Staudt 1989). Only one diploid species, *F. vesca*, is located in North America. The genomic complement of the octoploids is likely AAA'A'BBB'B' (Bringhurst 1990), with *F. vesca* probably being the A genome donor. The B genome donor has not been clearly elucidated, although molecular evidence is accumulating that Japanese *F. iinumae* may be it (Davis 2004).

The most likely scenario is that the octoploids originated in northeastern Asia when *F. vesca* combined with other unknown diploids, and the polyploid derivatives then migrated across the Bering Strait and dispersed across North America (Hancock 1999). It is possible that *F. chiloensis* and *F. virginiana* are extreme forms of the same biological species, separated during the Pleistocene, which subsequently evolved differential adaptations to coastal and mountain habitats. The two species are completely inter-fertile, carry similar cpDNA restriction fragment mutations (Harrison et al. 1997) and have very similar nuclear internal transcribed spacer (ITS) regions (Potter et al. 2000).

Polyploidy in *Fragaria* probably arose through the unification of 2*n* gametes, as several investigators have noted that unreduced gametes are relatively common in *Fragaria* (Hancock 1999). Staudt (1984) observed restitution in microsporogenesis of a F_1 hybrid of *F. virginiana* × *F. chiloensis*. In a study of native populations of *F. chiloensis* and *F. vesca*, Bringhurst and Senanayake (1966) found frequencies of giant pollen grains to be approximately 1% of the total. Over 10% of the natural hybrids generated between these two species were the result of unreduced gametes.

The inheritance patterns of the octoploids are in dispute. Lerceteau-Köhler et al. (2003) concluded that *F.* ×*ananassa* has mixed segregation ratios using AFLP markers, as they found the ratio of coupling vs. repulsion markers fell between the fully disomic and polysomic expectations. However, two other studies evaluating isozyme, SSR and RFLP segregation observed predominantly disomic ratios, indicating that the octoploid strawberry is completely diploidized (Arulsekar and Bringhurst 1981, Ashley et al. 2003).

Most commercial strawberries have been selected to be strict hermaphrodites, but sex is regulated as a single gene trait in *F. vesca*, *F. chiloensis*, *F. virginiana* and *F. ×ananassa* (Ahmadi and Bringhurst 1991). Female (F) (pistillate) is dominant to hermaphrodite (H), which is dominate to male (M) (staminant). Females are heterogametic (F/H or F/M), while hermaphrodites can be homo- or heterogametic (H/H or H/M) and males are homogametic (M/M). A range in fertility can be found in hermaphrodites ranging from self infertility to complete fruit set (Stahler et al. 1990, 1995, Luby and Stahler 1993). In *F. orientalis* and *F. moschata*, Staudt (1967) found tetrasomic inheritance for sex and he described the alleles for sex as male suppressor *SuM* (F) dominant to male inducer *Su+* (H) and to the female suppressor *SuF* (M). *SuF* was dominant to *Su+*.

While there appear to be some barriers to interfertility among the diploid strawberries, they all can be crossed to some extent, and meiosis in the hybrids is regular, even in cases where the interspecific hybrids are sterile (Federova 1946, Staudt 1959, Fadeeva 1966). There are at least three overlapping groups of diploid species that are inter-fertile (Bors and Sullivan 1998, 2005): (1) *Fragaria vesca*, *F. viridis*, *F. nubicola* and *F. pentaphylla* (2) *F. vesca*, *F. nilgerrensis*, *F. daltoniana* and *F. pentaphyta* (3) *F. pentaphyta*, *F. gracilis* and *F. nipponica*. *Fragaria iinumae* may belong in group 3 or in an additional group, as no fertile seeds have been recovered when it was crossed with either *F. vesca*, *F. viridis* or *F. nubicola*, but it has not been crossed with enough other species to accurately classify it. *Fragaria iinumae* does, however, have a glaucous leaf trait that is unique among the diploids, and its chloroplast RFLPs cluster it with *F. nilgerrensis* in a group that is isolated from the rest (Harrison et al. 1997).

Numerous valuable characteristics exist in the lower ploidy species that could be of value in the cultivated species (Darrow 1966, Hancock 1999). A particularly excellent comparison of the quantitative and qualitative differences between the diploids can be found in Sargent et al. (2004b). *Fragaria iinumae*, *F. vesca* and *F. nipponica* are likely highly cold tolerant as they are located on cold, alpine meadows. *F. vesca* have high tolerance to heat and drought, and high aroma along with resistance to Verticillium wilt (Arulsekar 1979), powdery mildew (Harland and King 1957) and crown rot (*Phytophthora cactorum*) (Gooding et al. 1981). *Fragaria moschata* is found under heavy shade and is immune to powdery mildew (Maas 1998). *Fragaria viridis* tolerates alkaline soils. In a comprehensive study of diploid species in Ontario, Bors and Sullivan (1998) found *F. nilgerrensis* to have immunity to aphids and leaf diseases. *Fragaria iinumae* produced unusual tap roots from runners. *F. moschata* survived a particularly cold winter in water-logged soil and displayed excellent leaf disease resistance. *Fragaria pentaphylla* was extremely vigorous, with unusually bright red, firm fruit and leaf disease immunity.

The incorporation of traits from a number of lower ploid species has been accomplished through pollinations with native unreduced gametes or by artificially doubling chromosome numbers. The utility of this approach has been shown for a wide range of species in *Fragaria* and in the related genus *Potentilla* (Hancock et al. 1996).

Particular success in incorporating lower ploidies into the background of *F.* ×*ananassa* has come through combining lower ploidy species and then doubling to the octoploid level (Sangiacomo and Sullivan 1994, Bors and Sullivan 1998).

Native clones of *F. chiloensis* and *F. virginiana* also offer a rich genetic storehouse and may be more useful in improving *F.* ×*ananassa* than the lower ploidies, as they cross readily with the cultivated types and offer as much if not more genetic diversity. Some of the wild clones have particularly interesting flavors and aromas that have not yet been characterized, and they possess resistance to extreme environments, as well as a number of disease and pest problems. In addition, variability exists in several yield-related physiological traits including: (1) heat and cold tolerance (2) rates and patterns of CO_2 fixation (3) the levels of dry matter allocated to reproduction (4) the number of flowering cycles, and (5) the length of the floral induction period.

In many cases, important components of yield can be combined with known disease resistances (Hancock et al. 2001). Most reports of pest resistance are limited to one disease or insect, but there are some genotypes that have been identified with multiple resistances. Clones of *F. chiloensis* have been described that carry resistance to aphids, 2-spotted spider mites, red stele, leaf spot, powdery mildew and root lesion nematodes. Two clones in particular from California, RCP 37 and CA 11, stand out as they are resistant to most of the pests described above, have very high photosynthetic rates (Hancock et al. 1989), and originated on dry, salty dunes.

Recently an elite group of 38 strawberry accessions was selected to represent the diversity found in *F. chiloensis* and *F. virginiana*, and was evaluated for plant vigor, flower number per inflorescence, flowering date, runner density, fruit set, fruit appearance and foliar disease resistance (Hancock et al. 2001). This collection is available at the National Clonal Germplasm Repository at Corvallis, Oregon, U.S.A. (http://www.ars.usda.gov/main/site_main.htm?modecode = 53581500). Among the individual taxa of the octoploid species, the largest fruit sizes were observed in the cultivated land races of *F. chiloensis* from South America, although the native clones of North American *F. chiloensis* ssp. *pacifica* were much more vigorous (Fig. 13.1). *Fragaria virginiana* ssp. *platypetala* had by far the largest fruit of any *F. virginiana* subspecies, rivaling the native clones of *F. chiloensis*. Day neutrals were found among northern *F. virginiana* ssp. *virginiana* and *F. virginiana* ssp. *glauca* accessions. Northern *F. virginiana* ssp. *virginiana* flowered the earliest and longest of any taxa, and had the most deeply colored fruit. By far the greatest winter hardiness was found in *F. virginiana* ssp. *glauca* and ssp. *virginiana*. Considerable genotype × location interaction was observed for many of the traits measured, indicating that individual site analyses cannot always be used to predict the broad range performance of individual genotypes; however, a few genotypes were impressive at all locations including CFRA 368 (California) with its unusually large, early fruit, and NC 95191 (North Carolina), Frederick 9 (Ontario) and RH 30 (Minnesota), which were very vigorous and had unusually good color.

A much larger sample of 270 genotypes of wild *F. virginiana* and *F. chiloensis* from the National Clonal Germplasm Repository at Corvallis, Ore. has also been compared in a greenhouse at Michigan State University for variation in fourteen

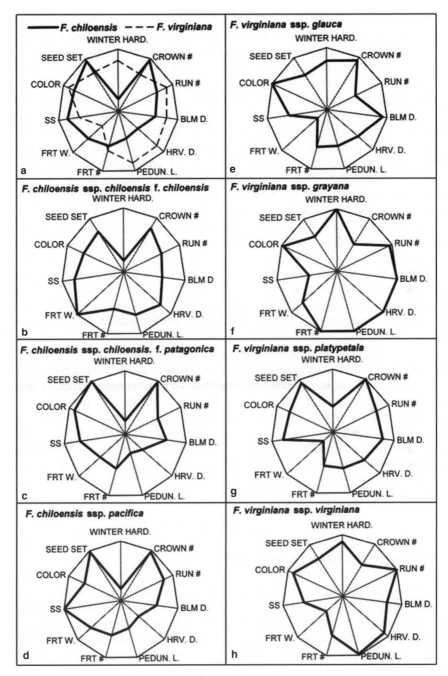

Fig. 13.1 Pictograms contrasting the horticultural traits of *F. virginiana* and *F. chiloensis* (**a**), and the various octoploid subspecies (**b–h**) (Hancock et al. 2003). The outer circumference represents the highest mean value of any of the subspecies. Each axis was normalized by dividing each trait by the highest overall value

horticultural traits (Hancock et al. 2003). Significant levels of variation were found for all but a few of the traits at the species, subspecies, regional and genotypic level, with the highest amount of variation generally being partitioned among genotypes. *Fragaria chiloensis* was superior to *F. virginiana* for crown number, fruit weight, soluble solids and seed set, while *Fragaria virginiana* was superior for runner production, peduncle length, fruit number, fruit color and winter hardiness. *Fragaria chiloensis* ssp. *pacifica* had the highest soluble solids and among the earliest bloom dates, highest crown numbers and highest seed set. *Fragaria chiloensis* ssp. *chiloensis* f. *chiloensis* produced the largest fruit and some of the earliest bloom dates and longest peduncles. *Fragaria chiloensis* ssp. *chiloensis* f. *patagonica* had some of the highest crown numbers and the highest percentage seed set. *Fragaria virginiana* ssp. *platypetala* produced the most crowns and its fruit ripened earliest. *Fragaria virginiana* ssp. *glauca* were the latest flowering, had the darkest fruit color and the most flowering cycles. *Fragaria virginiana* ssp. *virginiana* displayed the most winter dieback, longest peduncles, and the highest flower and runner numbers.

Among the most impressive individual genotypes, CFRA 0024 (*F. chiloensis* ssp. *chiloensis* f. *chiloensis*) possessed unusually high crown numbers, was extremely early blooming and displayed multiple fruiting cycles. CFRA 1121 (*F. chiloensis* ssp. *chiloensis* f. *chiloensis*) had unusually long peduncles and much higher than average values for fruit weight, soluble solids, fruit color and seed set. CFRA 0094 (*F. chiloensis* ssp. *pacifica*) was extremely early flowering and had much darker fruit color than most other *F. chiloensis* genotypes. CFRA 0368 (*F. chiloensis* ssp. *lucida*) flowered unusually early and had among the largest fruit. CFRA 0366 (*F. chiloensis* ssp. *lucida*) possessed unusually long peduncles and the largest fruit of any North American genotype. CFRA 0560 and CFRA 1369 (*F. virginiana* ssp. *glauca*) had an unusual combination of multiple flowering cycles and high runner production. CFRA 1170 and 1171 (*F. virginiana* ssp. *virginiana*) were unusually late fruiting and had high numbers of large fruit on long peduncles. CFRA 1385 and JP 9531 (*F. virginiana* ssp. *virginiana*) had extremely high flower numbers, long peduncles and large fruit.

13.3 History of Improvement

The wood strawberry, *F. vesca*, dominated strawberry cultivation in Europe for centuries, until *F. virginiana* from eastern Canada and Virginia began to replace it in the 1600s. All of the clones that found their way to Europe were wild in origin, as the aboriginal peoples of North America did little gardening with strawberries. The early cultivar development of *F. virginiana* was primarily conducted by growers who found raising seed imported from North America often resulted in horticulturally important variations.

One of the domesticated Chilean clones of *F. chiloensis* found its way into Europe in the 1700s compliments of a French spy, Captain Amédée Frézier (Darrow 1966, Wilhelm and Sagen 1974). Unfortunately, early reports on the

Chilean strawberry were negative, as the plants were largely barren because Frézier had inadvertently brought back staminate plants. French horticulturalists solved the problem when they discovered that the 'Chili' would produce fruit when pollinated by *F. moschata* or *F. virginiana*. The Chilean strawberry reached its highest acclaim in Brittany, and by the mid-1800s, there was probably more *F. chiloensis* cultivated in France than its native land.

Unusual seedlings began to appear in Brittany and other gardens with unique combinations of fruit and morphological characteristics. While the origin of these seedlings was initially clouded, the great French Botanist Antoine Nicholas Duchesne determined in 1766 that they were hybrids of *F. chiloensis* × *F. virginiana* and he named them *Fragaria* × *ananassa* to denote the perfume of the fruit that smelled like pineapple (*Ananas*). It is not clear where the first hybrids of The Pineapple or Pine strawberry appeared, but they must have arisen early in the commercial fields of Brittany and in botanical gardens all across Europe such as the Trianon, the Royal Garden at Versailles where Duchesne studied.

Formal strawberry breeding was initiated in England in 1817 by Thomas A. Knight (Darrow 1966, Wilhelm and Sagen 1974). He was one of the first systematic breeders of any crop, and used clones of both *F. virginiana* and *F. chiloensis* in his crosses. He produced the famous 'Downton' and 'Elton' cultivars, noted for their large fruit, vigor and hardiness. Michael Keen, a market gardener near London, also became interested in strawberry improvement about this time and developed 'Keen's Imperial' whose offspring, 'Keen's Seedling' is in the background of many modern cultivars (Fig. 13.2). This variety dominated strawberry acreage for close to a century.

Thomas Laxton of England was the most active breeder during the later part of the 18th century and released a number of important varieties including 'Noble' and 'Royal Sovereign'. These two varieties were grown on both sides of the Atlantic Ocean, and were popular until the middle of the 20th century. 'Nobel' was known for its earliness, cold hardiness and disease resistance. 'Royal Sovereign' was popular because of its earliness, productivity, flavor, attractiveness and hardiness.

Charles Hovey, of Cambridge, Massachusetts, produced the first important North American cultivar, 'Hovey', by crossing the European pine strawberry, 'Mulberry' with a native clone of *F. virginiana* in 1836. It was the first variety of any fruit to come from an artificial cross in America and for awhile made the strawberry the major pomological product in the country (Hedrick 1925).

Albert Etter of California developed dozens of varieties around the turn of the century utilizing native *F. chiloensis* clones (Wilhelm and Sagen 1974, Fishman 1987). His most successful variety was 'Ettersburg 80' (1910), which was widely grown in California, Europe, New Zealand and Australia. Renamed as 'Huxley', it was still popular in England as late as 1953. Ettersburg 80 was extremely drought resistant, of outstanding dessert and jam quality due to its solid bright red color, and was unusually hardy for a California type. Other outstanding Etter varieties were 'Ettersburg 121', 'Fendalcino' and 'Rose Ettersburg'. While his releases were very successful as cultivars, they may have had their greatest impact as breeding parents. Most California cultivars (and many others) have an Ettersburg variety in their background (Darrow 1937 and 1966, Sjulin and Dale 1987).

Fig. 13.2 The strawberry 'Keen's Seedling', which was a sensation in England in the late 1800s. It is in the pedigree of many modern cultivars

In the middle of the 20th century, a number of particularly active breeding programs emerged in Scotland, England, Germany and Holland. In Scotland, Robert Reid developed a series of red stele resistant varieties utilizing American 'Aberdeen' as a source of resistance. His variety 'Auchincruive Climax' dominated acreage in Great Britain and northern Europe until its demise due to June yellows in the mid 1950s. He then released 'Redgauntlet' (1956) and 'Talisman' (1955), which served as suitable replacements. In England, D. Boyle produced a large series of varieties with the prefix 'Cambridge'; 'Cambridge Favorite' (1953) became the most important of the group and dominated the acreage in Great Britain by the 1960s. In

Germany, R. von Sengbusch's produced a 'Senga' series, of which 'Senga Sengana' (1954) became paramount. 'Senga Sengana' was widely planted for its processing quality and is still important in Poland and other eastern European countries. In Holland, H. Kronenberg and L. Wassenaar's released several cultivars, of which 'Gorella' (1960) made the greatest impact. It was noted for its size, bright red glossy skin and red flesh.

George Darrow came to the USDA in Beltsville, Maryland in the 1920s and began his illustrious career with the release of 'Blakemore' in 1929 and 'Fairfax' in 1933. 'Blakemore' became the major southern U.S. variety in the mid-1930s and 'Fairfax' was widely planted in the middle of this century from southern New England to Maryland and westward to Kansas. These two cultivars were used extensively in breeding, finding their way into the ancestry of a diverse array of cultivars grown in all parts of the U.S. Other important releases from Darrow were 'Pocahontas', 'Albritton', 'Surecrop' and 'Sunrise'. Donald Scott took over the program in the 1950s and released 'Midway', 'Redchief', 'Guardian' and 'Earliglow'. An active USDA breeding program was also conducted at Corvallis, Oregon in the middle of the 20th century by Darrow, G.F. Waldo and F.J. Lawrence. Some of the more important cultivars emerging from this program were 'Narcissa' (1932), 'Brightmore' (1942), 'Hood' (1965) and 'Benton' (1974).

H. Thomas and E. Goldsmith's of the University of California released the important cultivars 'Lassen' and 'Shasta' in 1945. 'Shasta' was widely grown in the central coast of California in the 1950s and 1960s because of its large size, firmness and long season. 'Lassen' was grown extensively in southern California about the same period, prized for its short rest period and high productivity. Thomas and Goldsmith ultimately left the University and founded the highly successful breeding program of Driscoll Associates in Watsonville, California (Sjulin 2006). Royce Bringhurst and Victor Voth took over the Cal-Davis program in the 1950s and generated an amazing succession of internationally important, Mediterranean adapted cultivars including 'Tioga' (1964), 'Tufts' (1972), 'Aiko' (1975), 'Pajaro' (1979), 'Chandler' (1983), 'Selva' (1983), 'Camarosa' (1992) and 'Seascape' (1991).

Several significant breeding programs were conducted by various Agricultural Experiment Stations in the early to mid-1900s. A.N. Brooks in Florida selected 'Florida 90' (1952) from an open pollinated population of 'Missionary'. This variety had excellent flavor, very high yields and found an important seasonal niche in March and early April as other southern production diminished. Miller and Hawthorn released 'Klonmore' (1940), 'Headliner' (1957) and 'Dabreak' (1961) in Louisiana with leaf resistance to leaf spot and scorch, and good shipping quality. A number of important cultivars came out of New Jersey including 'Pathfinder' (1937) and 'Sparkle' (1942) introduced by J. H. Clark and 'Jerseybelle' (1955) developed by F. A. Gilbert. In New York, George Slate released 'Catskill' in 1933 for its large attractive berries and high productivity.

The greatest concentration of breeding activity in the world was centered in the U.S.A. and Europe until the modern period, although the Japanese produced two important varieties: Dr. H. Fukuba's 'Fukuba' (1899), noted for its large size and high flavor (Darrow 1966), and K. Tamari's 'Kogyoku' (1940), respected for its vigor, earliness and fruit size (Mochizuki 1995). 'Fukuba' was the most important variety in forcing culture until the early 1970s. 'Kogyoku' was one of the leading field grown cultivars after World War II, until it lost importance to the American import 'Donner' in the 1950s (Darrow 1966).

13.4 Current Breeding Efforts

There are numerous public and private breeding programs across the world that focus on improvement of *F. ×ananassa*. The largest European efforts are found in France, Italy, the Netherlands, Spain and the U.K. The most active North American programs are located in British Columbia, California, Florida, Maryland, New York, Nova Scotia, Ontario, Oregon and Quebec. In Asia, the largest number of public and private breeding programs is found in Japan. *Fragaria chiloensis* is being bred at the Universidad de Talca in Chile.

The Centre Interrégional de Recherche et d'Expérimentation de la Fraise (CIREF) directed by Philippe Chartier in France is active in developing short-day and day neutral cultivars with superior fruit appearance and flavor as well as soil disease tolerance. The most important recent cultivars from this program are 'Ciflorette' and 'Cirafine'. Two other important French breeding efforts are Darbonne which produced 'Darselect' and Marionnet SARL which released 'Mara des Bois' and the new 'Matis'.

In Italy, there is a national program 'Frutticoltura' funded by the Minister of Agriculture, with a number of institutions being involved. G. Baruzzi and W. Faedi lead this effort. Concentration is being placed on developing new dessert varieties with adaptations to the south, Po Valley and north mountain regions including disease resistance and tolerance to alkaline soils. Significant new cultivars from this program are 'Patty', 'Granda' and 'Queen Elisa'. A private company, Consorzio Italiano Vivaisti (C.I.V.) directed by A. Martinelli is also active in producing dessert varieties for both north and south Italy, and have generated 'Marmolada', 'Clery' and 'Miranda'.

The breeding effort of Fresh Forward, directed by B. Meulenbroek in the Netherlands, is concerned with developing types with broad adaptations, high yields and large fruit size. The most important European cultivar, 'Elsanta' (1981) came from the public progenitor of this program and their newest variety, 'Sonata', is increasing in popularity.

In Spain, a breeding program at the Universidad de Malaga conducted by J.M. Lopez-Aranda and C. Soria is concentrating on producing cultivars for Huelva. Their goals are to breed high yielding, early cultivars for the fresh market with large, high quality fruit. Their most important new cultivars are 'Andana' and 'Carisma'. The private firm, Plantas de Navarra S.A. (PLANASA) directed by D. Sanchez is also very active in searching for highly productive, high quality types for Spain. This program's most recent successes are 'Tudla', 'Cartuno' and 'Candonga'.

David Simpson of East Malling Research in England is concentrating on combining excellent fruit quality with resistance to diseases, particularly *Verticillium dahliae* and *Sphaerotheca macularis*. His most important cultivars are 'Florence', 'Pegasus' and 'Flamenco'. Edward Vinson Limited also has an active program at Kent, with a focus on day-neutrals. Their recent cultivar 'Everest' has proven to be widely adapted.

The largest U.S. breeding programs reside at the University of California at Davis, and the USDA-ARS centered at Beltsville, Maryland and Corvallis, Oregon. Now under the direction of D.V. Shaw and K.D. Larson, the Cal-Davis program is known for its broadly adapted, large-fruited and high yielding cultivars. Some of the most important new releases from this program are 'Aromas', 'Diamante', 'Ventana' and 'Albion'. The USDA-ARS program in Maryland is probably the longest continually maintained program in the world. It was directed by G.J. Galletta through the latter part of the 20th century in collaboration with A.D. Draper, and most recently is under the direction of K.S. Lewers. The most important cultivars from these programs are 'Allstar', 'Tribute', 'Tristar', 'Northeaster', 'Delmarvel' and 'Ovation'. These cultivars are noted for their resistance to soil pathogens, particularly *Phytophthora fragariae*. C.E. Finn now conducts the USDA breeding effort at Corvallis, Oregon. The current emphasis in this program is on developing high quality types for the processed market, although effort is shifting towards developing fresh market cultivars. Recent cultivars include 'Redcrest', 'Redgem', 'Pinnacle' and 'Tillamook'.

State supported programs in the U.S.A. are located in Washington (P. Moore), Michigan (J. Hancock), Minnesota (J. Luby), New Jersey (G. Jelenkovic), New York (C. Weber), North Carolina (J. Ballington), Wisconsin (B. Smith), Maryland (H. Swartz) and Florida (C. Chandler). The current goals of these programs are: (1) Florida – high fresh fruit and shipping quality including resistance to water damage, high late November to mid-March yields, anthracnose resistance (2) Michigan – day neutral types with higher heat tolerance and resistance to soil pathogens, germplasm development using wild octoploids (3) and (4) Minnesota and Wisconsin – winter hardiness, high quality, disease resistance, germplasm development using wild octoploids (5) New Jersey – early cultivars with excellent fruit flavor and size that are adapted to matted row and hill culture (6) New York – fruit quality including size, symmetry; high, steady yields; black root rot resistance (7) and (8) North Carolina and Maryland – superior genotypes that are resistant to anthracnose and are adapted to annual hill plasticulture systems, and (9) Washington – June bearing types with firm, easily harvested fruit for processing and fresh outlets, increased disease and insect resistance (particularly fruit rots and aphid transmitted viruses). Some of the most important cultivars that have come out of these programs over the years are 'Honeoye' and 'Jewel' (New York), 'Mesabi' (Minnesota), 'Raritan' (New Jersey), 'Strawberry Festival' and 'Sweet Charlie' (Florida).

There are several large breeding programs in Canada that are federally funded. At Agriculture and Agri-Food Canada in Kentville (Nova Scotia), A. Jamison is making wide use of European cultivars to produce early season, red stele resistant types. He is building on the previous decades of work of D.L. Craig and L.E. Aalders. Some of the most significant cultivars to come out of this program are 'Bounty', 'Annapolis', 'Glooscap', 'Kent' and 'Cavendish'. An Ontario program led by A. Dale at the University of Guelph is concentrating on large-fruited, firm types and he is actively exploiting native germplasm. Recent releases from his program are 'Startyme', 'Sapphire' and 'Serenity'. S. Khanizadeh at the Horticultural Research and Development Centre in Quebec is searching for large fruited, pale skin colored and firm types with resistance to red stele. Numerous cultivars have been released from this program including the most recent 'Harmonie', 'St-Jean d'Orléans', 'St-Laurent d'Orléans' and 'La Clé de Champs'.

Probably the largest program in Canada is the Agriculture and Agri-Foods Canada program in Agassiz, British Columbia. It was originally run by H. Daubeny, and now C. Kempler. The predominant cultivar grown in the Pacific Northwest, 'Totem', was released from this program in 1971. 'Sumas' and 'Shuswap' were two other important cultivars from this program that has focused on June-bearing types with excellent processing characteristics (intense internal and external color, high soluble solids, high titratable acidy, low pH, and intense flavor).

R. Harrison (Production, Breeding and Research Department) and B. Mowrey (Head Plant Breeder) of the Driscoll Strawberry Associates in Watsonville, California direct the most vigorous private, breeding effort in the U.S. Their primary goals focus on consumer attributes of flavor, appearance and shelf life, coupled with the production attributes of fruit size, timing of harvest and harvestability. Other private efforts are conducted by Plant Sciences Inc (California), New West Fruit Corporation (California) and Well-Pict Inc. (California).

Significant breeding work is also being conducted in Japan. Probably the largest program in Japan is operated by the National Research Institute of Vegetables, Ornamental Plants and Tea with two branches at Kurume and Morioka. Numerous other Prefecture Experiment Stations are actively breeding strawberries including ones at Aichi, Chiba, Hyogo and Saga. Common goals are to produce large, dessert quality berries that are adaptable to forcing culture.

13.5 Genetics of Important Traits

13.5.1 Disease and Pest Resistance

Several soil pathogens damage strawberry roots, resulting in vigor declines and ultimately death (Table 13.2). Two very common problems across the world are red stele or red core caused by *Phytophthora fragariae* Hickman and Verticillium wilt caused by *Verticillium albo-atrum* Reinke & Berth. and *V. dahlia*. Black root rot is also widespread and is caused by a complex of organisms including *Pythium*, *Rhizoctonia* and the root lesion nematode (*Pratylenchus penetrans* Cobb). Fusarium wilt or Fusarium yellows (*Fusarium oxysporum* Schl. f. sp. *fragariae* Winks and Williams) is of major importance in Japan, Korea and Australia.

Table 13.2 Inheritance patterns of disease resistance in strawberries

Disease	Observations	Selected references
Bacteria		
Angular leaf spot – *Xanthomonas fragariae*	Resistant types identified; three or four unlinked loci regulate resistance; isolate × cultivar interactions do not exist; *Fragaria* species vary in their levels of resistance	Lewers et al. 2003, Hildebrand et al. 2005, Xue et al. 2005
Fungi		
Alternaria leaf spot (black leaf spot) – *A. alternate*	Resistant types identified; race variation in pathogen; single dominant locus for susceptibility	Yamamoto et al. 1985, Takahashi 1993
Anthracnose – *Colletotrichum fragariae*	Resistant types identified; race variation in pathogen; types resistant to *C. fragariae* tend to be resistant to *C. acutatum*	Delp and Milholland 1981, Smith and Black 1987, Smith et al. 2007
Anthracnose – *C. acutatum*	Resistant types identified; major and minor genes are associated with resistance; QTL identified for resistance	Gimenez and Ballington 2002, Denoyes-Rothan et al. 2005, Lerceteau-Köhler et al. 2005, Smith et al. 2007
Black root rot – Several Species	Tolerant genotypes identified	Wing et al. 1995, LaMondia 2004, Particka and Hancock 2005
Cactorum crown rot – *Phytophthora cactorum*	Resistant genotypes identified, *F. vesca* may have stronger resistance	van der Scheer 1973, Gooding et al. 1981, Bell et al. 1997
Fusarium wilt – *Fusarium oxysporum*	Resistant genotypes identified	Kim et al. 1982, Cho and Moon 1984, Takahashi 2003, Mori et al. 2005
Grey mold – *Botrytis cinerea*	Resistant genotypes identified; resistance is quantitative and additive; fruit firmness may be related	Maas and Smith 1978, Barritt 1980, Popova et al. 1985
Leaf scorch – *Diplocarpon earliana*	Resistant types identified; race variation in pathogen; parents not useful in predicting resistance of progeny	Nemec 1971
Leaf spot – *Mycosphaerella fragariae*	Resistant types identified; moderate levels of heritability	Nemec 1971, Shaw 1988, Delhomez et al. 1995

Table 13.2 (continued)

Disease	Observations	Selected references
Powdery mildew – *Sphaerotheca macularis*	Resistant types identified; general and specific combining ability is important; resistance is highly heritable; cuticle thickness is associated with resistance	Hsu et al. 1969, Murawski 1968, Simpson 1987, Nelson et al. 1996, Davik and Honne 2005
Red stele root rot – *Phytophthora fragariae*	Resistant genotypes identified; inbreeding can concentrate resistance genes; specific combining ability is very important; data fits a gene for gene model	Daubeny 1964, Melville et al. 1980a, Van de Weg 1997, Maas 1998
Verticillium wilt – *V. albo-atrum* and *V. dahlia*	Resistant genotypes identified; additive variation is important; resistance is partially dominant	Gooding et al. 1975, Maas 1989, Shaw and Gordon 2003
Viruses		
Arabis mosaic	Little resistance identified	Murant and Lister 1987
Clover phyllody	Resistant genotypes identified	Chiykowski and Craig 1975
Raspberry ringspot	Little resistance identified	Murant and Lister 1987
Strawberry latent ringspot	Little resistance identified	Murant and Lister 1987
Tomato black ring	Resistant genotypes identified	Murant and Lister 1987
Tomato ringspot	Resistant genotypes identified	Converse 1987
Misc. virus complexes	Resistant genotypes identified to regional virus complexes including yellows; a high proportion of the genetic variability is additive	Barritt et al. 1982, Graichen et al. 1985, Sjulin et al. 1986

Resistant and/or tolerant genotypes have been found for all four of these major soil pathogens, although the underlying genetics of resistance have only been studied for red stele and Verticillium wilt (Table 13.2). Resistance to Verticillium wilt has been shown to be inherited in an additive fashion with partial dominance. Red stele resistance has been demonstrated in several studies to be regulated primarily through additive interactions, although Van de Weg (1997) has provided evidence that red stele resistance fits a gene for gene model, with five virulence and five resistance genes. Haymes et al. (1997) found molecular markers that were tightly associated with one of the resistance loci (*Rpf1*). No further molecular characterizations have been made on these genes or any other resistance genes in strawberry.

Fumigation has been widely employed to control soil pathogens, but the impending ban on methyl bromide fumigation has stimulated increased interest in developing resistant cultivars. Without fumigation, cultivars yield 50% less fruit on average. Screens of the California breeding population on fumigated and non-fumigated soil have uncovered little general resistance to the total array of soil pathogens normally found in strawberry soils (Larson and Shaw 1995a,b), although a screen of eastern breeding material did uncover some tolerant individuals (Particka and Hancock 2005).

Among the foliar diseases, three are very widespread and can cause serious damage including, leaf blight [*Phomopsis obscurans* (Ell. and Ev.) Suton], Ramularia leaf spot, [*Mycosphaerella fragariae* (Tul.) Lindau] and leaf scorch [*Diplocarpon earliana* (Ell. & Everh.) Wolf]. Alternaria leaf spot or black leaf spot (*Alternaria alternata* (FR.) Keissler) causes serious damage in Europe, New Zealand and Korea. Powdery mildew [*Sphaerotheca macularis* (Wallr. Ex Fr.) Jaez] is also found across most of the strawberry range, although it rarely does economic damage. Angular leaf spot, *Xanthomonas fragariae* Kennedy and King, is a rapidly growing problem in strawberries all across the world (Maas et al. 1998). Moderate to high levels of heritability have been found for resistance to leaf spot, leaf scorch, powdery mildew and ramularia leaf spot (Table 13.2). Black leaf spot resistance has been reported to be controlled at a single locus.

Anthracnose is a common problem in strawberries, causing a wide array of symptoms including fruit rot, crown rot, and lesions of the stolons, petioles and leaves. Anthracnose diseases of strawberry are caused by *Colletotrichum fragariae* A.N. Brooks, *C. acutatum* J.H. Simmonds, and *C. gloeosporioides* (Penz.) Penz. & Sacc. in Penz. *Colletotrichum acutatum* is the primary pathogen causing crown rot in Europe (Denoyes and Baudry 1995), while *C. fragariae* is the most common cause of crown rot in the southeastern U.S.A. (Howard et al. 1992). *C. acutatum* is the primary pathogen in Israel and California. Sources of resistance to anthracnose fruit and crown rots exist in strawberry (Table 13.2); however, the genetic factors conditioning host resistance to crown and fruit infection differ and only a few genotypes are resistant to both fruit and plant infection. Strong environmental × genotype interactions affect the expression of resistance, and multiple isolates of *C. acutatum*, *C. gloeosporioides* and *C. fragariae*, vary in pathogenicity to *Fragaria* genotypes (Smith and Black 1990). Denoyes-Rothan et al. (2005) found both major and minor gene resistance to *C. acutatum*, with the major gene common in the germplasm evaluated. MacKenzie et al. (2006) discovered that resistance to *C. fragariae* and *C. gloeosporoides* was nonspecific, and the major gene resistance described by Denoyes-Rothan et al. (2005) to *C. acutatum* may be effective against these other 2 *Colletotrichum* species.

Phytophthora cactorum (Leb. & Cohn) Schroet also causes widespread incidences of severe crown (Cactorum crown rot) and fruit rots (leather rot), particularly in warm climates. Other important fungal fruit rots are: (1) Botrytis fruit rot or gray mold (*Botrytis cinerea* Pers. ex. Fr.), which is a worldwide problem (2) Mucor fruit rot (*Mucor mucedo* L. ex Fries), sometimes important in the U.S.A. and U.K. (3) Rhizopus leak [*Rhizopus* (spp.)] a particular problem in the U.K. but worldwide in scope (4) Tan-brown rot, [*Discohainesia oenotherae* (Cook & Ellis)], a major problem in humid strawberry regions (5) *Phomopsis obscurans* (Ellis & Everh.) Sutton in Florida, and (6) Septoria hard rot (*Septoria fragariae*) rarely a problem in the U.S.A. but common in Europe and Australia. Moderate to high levels of heritability have been found to grey mold and leather rot (Table 13.2).

A gene encoding a polygalacturonase-inhibiting protein (PGIP) has been cloned that shows developmental regulation and pathogen-induced expression in strawberry and likely plays a role in defense against fruit rots (Mehli et al. 2004, 2005, Schaart et al. 2005). After inoculation with *Botrytis cinerea*, fruit of five cultivars ('Elsanta', 'Korona', 'Polka', 'Senga Sengana' and 'Tenira') showed a significant induction in PGIP expression and the most resistant one, 'Polka', had the highest constitutive expression. Work is ongoing to produce transgenic strawberries that over-express PGIP sequences and screen them for resistance to *B. cinerea*.

Strawberries across their range are hosts to numerous viruses and phytoplasma. Among the most important are the aphid-borne viruses involved in the yellows complex (mottle, mild yellow-edge, crinkle and vein banding viruses) and the nematode-borne viruses (raspberry ringspot virus, tomato black ring, strawberry latent ringspot virus and arabis mosaic virus) (Maas 1998). There are also two important phytoplasma diseases spread by leaf hoppers, Aster yellows which are caused by a variety of species and green petal or clover phyllody. Resistance has been identified to the regional virus complexes found in the Pacific Northwest, but no formal genetic analyses have been performed (Table 13.2).

The nematodes causing the most widespread problems include the Northern root-knot nematode (*Meloidogyne hapla* Chitwood), root lesion nematode and the needle nematode (*Longidorus elongatus* de Man). Resistance has been described for all these pests, although no genetic studies have been conducted (Table 13.3).

Two aphids are widespread that damage strawberries, the strawberry root aphid, *Aphis forbesi* Weed and the strawberry aphid *Chaetosiphon fragaefolii* Cockerell. The strawberry aphid is found all across the range of cultivation, while the strawberry root aphid is restricted to east of the Rockies in the U.S.A. The latter is most important as a vector of virus disease. Resistant genotypes have been identified for the strawberry aphid, and resistance has been shown to be regulated by more than one locus with partial dominance and additive action.

Other important strawberry pests are plant bugs (*Lygus* spp.), root weevils (*Otiorhynchus* spp.), strawberry weevil (*Anthonomus signatus* Say), two-spotted spider mite (*Tetranychus urticae* Koch.) and the cyclamen mite [*Steneotarsonemus pallidus* (Banks)]. Moderate levels of resistance have been identified to two-spotted spider mite, strawberry aphid and black vine weevil (Table 13.3). Both leaf volatile and essential oil content have been examined as possible inhibitors to two-spotted spider mite attack (Hamilton-Kemp et al. 1988, Khanizadeh and Bélanger 1997).

Breeding for resistant types has been frequently complicated by negative correlations between resistance and horticulturally important traits (Maas and Galletta1989, Hancock et al. 1990). For example, Bringhurst et al. 1967 found Verticillium wilt resistance was negatively correlated with yield. Breeding for disease resistance has been further complicated by the presence of eco-or biotypes of the pathogen. In the

Table 13.3 Inheritance patterns of pest resistance in strawberries

Disease	Observations	Representative studies
Insects		
Black vine weevil – *Otiorhynchus sulcatus*	Tolerant genotypes identified; probable quantitative inheritance; trichome density associated with resistance	Cram 1978, Shanks and Doss 1986, Doss and Shanks 1988
Blossom weevil – *Anthonomus rubi*	Resistance is under independent genetic control from flowering time; additive genetic variance is the most important	Simpson 1997, 2002
Cyclamen mite – *Steneotarsonemus pallidus*	Resistant genotypes identified	Oydvin 1980
Obscure root weevil – *Sciopithes obscurus*	Tolerant genotypes identified	Cram 1978
Root aphid – *Aphis forbesi*	Resistant genotypes identified	Darrow et al. 1933
Strawberry aphid – *Chaetosiphon fragaefolii*	Resistant genotypes identified; regulated by more than one locus, with partial dominance and additive action, but highly resistant types are recoverable in backcross generations	Shanks and Barritt 1974, Barritt 1980, Crock et al. 1982
Strawberry root weevil – *Otiorhynchus* sp.	Tolerant genotypes identified	Cram 1978
Tarnished plant bug – *Lygus lineolaris*	Resistant genotypes identified	Tingey and Pillemer 1977, Schaefers 1980, Handley et al. 1991
Two-spotted Spider mites – *Tetranychus urticae*	Resistant genotypes identified; resistance is biotype specific; strong additive and dominance effects; little genotype × environment interaction	Schuster et al. 1980, Barritt and Shanks 1981, Shanks et al. 1995, Medina et al. 1999
Woods weevil – *Nemocestes incomptus*	Tolerant genotypes identified	Cram 1978
Nematodes		
Root lesion – *Pratylenchus penetrans*	Resistant genotypes identified	Szczygiel 1981c, Potter and Dale 1994
Needle – *Longidorus elongates*	Resistant genotypes identified, segregation patterns suggest high heritability	Szczygiel 1981a
Northern root-knot *Meloidogyne hapla*	Resistant genotypes identified, segregation patterns suggest high heritability	Szczygiel and Danek 1974, Szczygiel 1981b, Edwards et al. 1985

case of red stele root rot, there are over ten known races in the U.S., 12 in the U.K., 6 in Canada and 6 in Japan (Table 13.2). Regional variation in cultivar susceptibility to pathogens has also been documented for Alternaria leaf spot, ramularia leaf spot, leaf scorch, Verticillium wilt and anthracnose (Hancock et al. 1996b).

13.5.2 Environmental Adaptation

Strawberries are grown across a vast environmental zone, but several environmental factors commonly limit their productivity including: (1) heat and drought (2) salinity (3) winter cold (4) spring frosts and (5) insufficient chilling hours.

One of the most important production problems of strawberries is drought, which is often associated with high temperatures. Prior to the general use of irrigation, losses due to drought were very high and less than optimum soil moisture still plagues parts of all non-irrigated production regions. While there are numerous published suggestions that cultivars vary in their resistance to heat and drought, there are few formal genetic studies (Table 13.4). Heat and drought tolerance have also been described in several native genotypes of *F. virginiana* and *F. chiloensis*.

Excess salt from irrigation water is a major production problem in many arid agricultural regions. Irrigation with water containing more than 100 ppm sodium or chloride ions results in enough salt accumulation to cause yield loss without visible plant injury (Brown and Voth 1955). Some cultivars have been shown to be more 'salt tolerant' than others ('Lassen', 'Festival naya' and 'Fresno'), but few surveys have been made. Perhaps the best source of salt tolerance will come from native genotypes of *F. chiloensis* which live alongside the ocean in Chile and California (Hancock and Bringhurst 1979).

Strawberries generally bloom in early spring, when the chance of frost is relatively high. Flower buds, open flowers and young fruit are all injured by frost. Pistils are most sensitive to damage; however, some damage is likely to all flower parts if temperatures fall to $-2°$ C (Darrow 1966). Winter freezing injury to the strawberry crown and inflorescence buds is also a serious limitation to strawberry production throughout the upper half of the Northern Hemisphere. Non-acclimated strawberry plants are usually killed when the crown temperature remains at $-3°$ C for more than 1 or 2 hours (Scott and Lawrence 1975). Acclimated strawberry plants can survive crown temperatures of $-12°$ C to $-15°$ C, although injury such as decreased vigor is visible at higher temperatures (Zurawicz and Stushnoff 1977, Marini and Boyce 1979).

Cultivars grown in the more northern regions of North America and Europe tend to be more winter hardy and this hardiness is highly heritable (Table 13.4). A wide range in bloom tolerance to frost has also been described, although regional correlations are not always apparent (Ourecky and Reich 1976). To elucidate the molecular basis of cold acclimation in strawberry, NDong et al. (1997) used differential screening to identify genes associated with low temperature acclimation. They identified three transcripts, *Fcor1 –3* (Fragaria Cold-Regulated 1–3), whose levels

Table 13.4 Genetics of adaptation, productivity, plant habit and fruit quality in strawberry

Attribute	Observations	Representative studies
Adaptations		
Chilling requirement	Quantitatively inherited; resistant genotypes identified	Darrow 1966
Concentrated ripening	Quantitatively inherited; negatively correlated with yield	Denisen and Buchele 1967, Moore et al. 1970, Barritt 1974, Moore et al. 1975
Drought and heat	Resistant genotypes identified	Hancock et al. 1990
Flowering date	Quantitatively inherited; earliness partially dominant; bloom and ripening dates closely correlated	Powers 1945, Wilson and Giamalva 1954, Zych 1966, Scott et al. 1972
Frost tolerance	Quantitatively inherited; resistant genotypes identified	Darrow and Scott 1947, Ourecky and Reich 1976
Harvest date	Quantitatively inherited; bloom and ripening dates closely correlated	Wilson and Giamalva 1954, Zych 1966
Photoperiod sensitivity	Numerous models proposed from single dominant gene to quantitative inheritance	Ahmadi et al. 1990, Serce and Hancock 2005, Shaw and Famula 2005
Salinity tolerance	Resistant genotypes identified	Hancock and Bringhurst 1979, Hancock et al. 1990
Winter cold hardiness	Quantitatively inherited; highly heritable; resistant genotypes identified	Powers 1945
Productivity		
Flower number	Quantitatively inherited through several yield components; both additive and epistatic variation is important depending on the population	Morrow et al. 1958, Watkins et al. 1970, Spangelo et al. 1971, Lal and Seth 1981
Fruit size	Quantitatively inherited, with 6 to 8 allelic pairs regulating fruit expansion; both additive and epistatic variation is important depending on the population	Comstock et al. 1958, Sherman et al. 1966, Hansche et al. 1968, Scott et al. 1972
Runner number	Quantitatively inherited; high general combining ability	Simpson and Sharp 1988
Total yield	Quantitatively inherited; both additive and epistatic variation is important depending on the population; often negative interactions between yield components	Hansche et al. 1968, Watkins et al. 1970, Spangelo et al. 1971, Webb 1974, Mason and Rath 1980, Shaw et al. 1989

Table 13.4 (continued)

Attribute	Observations	Representative studies
Fruit quality		
Acidity	Controlled with varying levels of additive and dominance control depending on the population	Duewer and Zych 1967, Lal and Seth 1979, Shaw et al. 1987, Shaw 1988
Firmness	Quantitatively inherited; flesh firmness and skin toughness often correlated positively	Hansche et al. 1968, Ourecky and Bourne 1968, Barritt 1979, Shaw et al. 1987
Ease of calyx removal	Quantitatively inherited; much additive variation; low capping force can be dominant	Brown and Moore 1975, Barritt 1976
Color	Skin and flesh color quantitatively inherited; largely additive with a few major genes; Internal and external color poorly correlated	MacLachlan 1974, Murawski 1968, Lundergan and Moore 1975, Shaw and Sacks 1995
Pedicle length	Quantitatively inherited	MacIntyre and Gooding 1978, Dale et al. 1987
Sugar content	Controlled with varying levels of additive and dominance control depending on the population; individual sugars vary more than total sugars; negative association between soluble solids and yield	Duewer and Zych 1967, Lal and Seth 1979, Wentzel 1980, Shaw et al. 1987, Shaw 1988
Vitamin C	Quantitatively inherited, with partial dominance for high levels	Hansen and Waldo 1944, Anstey and Wilcox 1950, Lundergan and Moore 1975, Lal and Seth 1979

changed dramatically after cold-acclimation. Transcript accumulation for *Fcor3* was the most closely correlated with freezing tolerance, suggesting it may be a useful marker for this trait. *Fcor3* encodes a polypeptide that shows high identity with PSI polypeptides from spinach and barley.

Insufficient chilling can result in reduced yields in many of the regions of the world with moderate winters. Cultivars vary substantially in their chilling requirements, time of bloom and ripening dates; however, few quantitative genetics studies have been performed on these characteristics (Table 13.4). Cultivars that are adapted to warm southern areas, such as the southern U.S., Mediterranean regions and Africa, appear to have the shortest rest periods and these plants are capable of growing and ripening fruit during the short days of summer. Considerable variation in bloom and ripening dates also exist within regions of adaptation, with time of bloom and ripening dates often being closely correlated. Earliness can act as a partially dominant trait (Powers 1945, Scott et al. 1972).

13.5.3 Flowering and Fruiting Habit

There are two types of octoploid plants that can produce more than one crop a year: day-neutrals and long-day; although continuums in growth habit and flowering behavior make rigid classifications difficult (Nicoll and Galletta 1987). Short day cultivars tend to have a limited harvest window, and as a result day-neutral cultivars have increased in importance. The inheritance behavior of multiple cropping in strawberries has been the subject of numerous studies. Hypothesis concerning the inheritance of day-neutrality have ranged from a single recessive gene (Darrow 1937), a single dominant gene (Ahmadi et al. 1990), two dominant complementary genes (Ourecky and Slate 1967), two or more complementary dominant genes of equal potency and at least four recessive genes (Powers 1954). The most recent studies suggest that a large portion of the variance can be explained by a dominant gene (Shaw and Famula 2005), although numerous other loci probably play a role in conditioning day-neutrality (Serce and Hancock 2005). Major QTL have been identified for photoperiod sensitivity in $F. \times ananassa$, as will be described in the section on genetic mapping, but no attempts have been made to identify the specific gene(s) responsible (Weebadde et al. 2007).

The genetics of multiple cropping in 'alpine' forms of European $F.$ $vesca$ is much simpler than that of $F. \times ananassa$, due partly to their diploid instead of octoploid nature. The everbearers 'Baron Solemacher' and 'Bush White' contain a homozygous recessive gene for day neutrality (Brown and Wareing 1965). Day neutrality has not been observed in North American populations of $F.$ $vesca$, and when California clones were crossed with alpine forms, at least three genes were identified that controlled photoperiodism (Ahmadi et al. 1990). Molecular markers have been identified that are closely linked to the seasonal flowering locus in $F.$ $vesca$ (Cekic et al. 2001).

Most day-neutral types of diploids and octoploids produce limited numbers of runners, although Simpson and Sharp (1988) found considerable variation for stolon production and yield in everbearing, octoploid types. General combining ability was the strongest component of fruit yield, but specific combining ability played a more important role in stolon production. They suggested that early fruiting and adequate stolon production could be combined in an everbearing type. Yu and Davis (1995) found a tight genetic linkage between runnering and a phosphoglucoisomerase locus in diploid strawberry.

13.5.4 Fruit Quality

Several factors restrict consumer acceptance of strawberry fruit including size, flavor, nutrition and color. Other important factors are flesh firmness and skin toughness. Size of fruit is inherited quantitatively, with 6 to 8 allelic pairs controlling fruit expansion (Table 13.4). There is a decline in size of fruits from the primary to inferior positions, and the relative decline varies substantially among genotypes. Several genetic studies have shown that a large part of the genetic variance for fruit size is epistatic, although there is still considerable additive variability, depending on the parents. Much genetic variability has also been identified in breeding populations for firmness. Flesh firmness and skin toughness are often correlated positively and are generally inherited quantitatively.

A number of recent molecular studies have searched for the genes in strawberry that are involved in cell wall modification during ripening and therefore influence fruit-firmness. Harrison et al. (2001) identified and characterized a number of expansin genes (*FaExp2* to *FaExp7*), which likely induce cell wall extension in vitro. Messenger RNA from most of these were present in leaves, roots and fruit, except for *FaExp5*, which showed fruit specific expression. Castillejo et al. 2004 isolated four pectin esterases genes from strawberry (*FaPE1* to *FaPE4*). *FaPE1* was specifically expressed in fruit and was up-regulated by auxin treatment in green fruit and down regulated by exogenous applied ethylene in ripe and senescing fruits. The repression of *FaPE1* may be involved in textural changes during fruit senescence. Blanco-Portales et al. (2004) identified a fruit-specific gene encoding for a HyPRP protein involved in the anchoring of polyphenols to cell membranes. Salentijn et al. (2003) found the expression of two genes associated with lignin metabolism (cinnamoyl CoA reductase and cinnamyl alcohol dehydrogense) to vary dramatically between soft fruited ('Gorella') and firm-fruited ('Holiday') cultivars.

Three full length cDNAs encoding ß-galactosidases (*Faßgal1*, *Faßgal2* and *Faßgal3*) were isolated from a library representing red fruit transcripts by Trainotti et al. (2001). Two of the genes had a C-terminus domain that was structurally related to known animal peptides with sugar-binding ability. Galactose is released during the dismantling of cell walls and the galactosidases are thought to play an important role in the mobilization of galactose. In a study of salt extractable proteins from the cell walls of immature and ripe strawberry, Iannetta et al. (2004) identified seven abundant polypeptides; two of which were thought to be important determinants in the regulation of the sugar:acid balance (mitochondrial malate dehydrogenase and mitochondrial citrase synthase).

Soluble solids and acidity are controlled with varying levels of additive and dominance control (Shaw et al. 1987, Shaw 1988). Shaw (1988) found little difference in the soluble solids and total sugars in his breeding population, although he did observe significant genotypic variation in sucrose, glucose, fructose and acidity levels. Wenzel (1980) found a negative association between soluble solids concentration and yield. Vitamin C content has also been shown to be polygenic, with some parents displaying partial dominance for high levels and some progeny having higher levels than their parents (Hancock et al. 1996b). Several studies have described variation in the flavor of progeny families suggesting additive quantitative control (Darrow 1966), although few formal genetic studies on this character have been conducted. Zubov and Stankevich (1982) found significant seedling variation in fruit consistency, anthocyanin content and vitamin C, but not flavor. GCA was greater than SCA for all the other traits except vitamin C.

An NADPH-dependent D-galacturonic acid reductase gene (*GalUR*) was isolated and characterized from strawberry to determine its role in vitamin C content (Agius et al. 2003). Expression of *GalUR* correlated closely with ascorbic acid levels during strawberry fruit ripening and GalUR protein levels were found to be associated with ascorbic acid content in four species of *Fragaria* (*F. ×ananassa*, *F. chiloensis*, *F. virginiana* and *F. moschata*). The gene was not engineered into strawberry, but overexpression of *GalUR* in Arabidopsis enhanced vitamin C content two-to three-fold.

Skin and flesh color have been shown to be largely under the control of additive variation, although a few genes appear to have much larger effects than others (Table 13.4). Internal and external colors are probably regulated by separate sets of genes as correlations between these two parameters are small. No molecular studies to date have attempted to associate specific genes with segregation for fruit color in *F. ×ananassa*, but Wilkinson et al. (1995) has identified a gene for chalcone synthase (CHS) that is highly expressed in ripening strawberry fruit and is likely a key enzymatic step in flavonoid biosynthesis. Aharoni et al. (2001) has also cloned the transcription factor *FaMYB1* from ripening fruit, which plays a key role in the biosynthesis of anthocayanins and flavonols. Deng and Davis (2001) found a polymorphism in the flavone 3-hydroxylase gene to be associated with a yellow fruit color in *F. vesca*.

Hoffman et al. (2006) used RNAi-induced silencing to reduce activity of CHS in strawberry fruits. They used a construct containing the partial sense and corresponding antisense sequences of CHS separated by an intron (from a strawberry quinone oxidoreductase gene). An *Agrobacterium* suspension containing the gene was injected into 14-day-old fruit still attached to the plant. Almost white fruit were produced when the injection was repeated three days in a row.

DNA microarrays have been utilized to identify and clone genes associated with strawberry flavor and aroma. Aharoni et al. (2000) found a novel strawberry alcohol acyltransferase (SAAT) in *F. ×ananassa* cultivar 'Elsanta' that is critical in flavor biogenesis in ripening fruit. This gene combines acyl-CoA and alcohol to generate the esters, the most important class of volatile compounds in fruit. Aharoni et al. (2004) also cloned the gene, *F. ananassa Nerolidol Synthase 1* (*FaNES1*), which was found in all three octoploid species, but not in *F. vesca* and *F. moschata*. It generates linalool and nerolindol when supplied by geranyl disphosphate or farnesyl diphosphate. They also found *F. vesca* to carry an insertion mutation in a terpene synthesase gene that differs from the one in the cultivated strawberry (*F. ×ananassa Pinene Synthase*). This insertion limited its expression and further altered aroma by reducing quantities of pinene and myrcene.

In a study of the catalytic properties of AAT in different strawberry species and cultivars, Olías et al. (2002) found that heptanol was the best straight-chain substrate for three European varieties, while hexanol was the prefered alcohol for two American cultivars; a genotype of *F. vesca* had the highest activity with pentanol. The cultivars had generally similar patterns of activity on straight chain acyl-CoAs, except for 'Eros' which had much higher activity than the others for pentanoyl-CoA. *F. vesca* also showed much lower activities for pentanoyl-CoA than most of the cultivars, and much higher levels of activity for acetyl-and propionyl-CoA.

A number of other genes have been characterized that are highly expressed during fruit ripening and maturation. Manning (1998) generated a cDNA library from messenger RNA isolated from ripe fruit, and identified a number of genes encoding enzymes of phenylpropenoid metabolism, and genes for cellulase, expansins, cysteine proteinase and acyl carrier protein (Manning 1998). Three mRNAs with fruit specific, ripening-enhanced expression have also been identified in ripening fruit using polymerase chain reaction (PCR) differential display (Wilkinson et al. 1995). When sequenced, they had high homology with known proteins including: (1) an annexin which may play a role in membrane function and cell wall structure (2) chalcone synthase which is a key enzymatic step in flavonoid biosynthesis, and (3) a ribosomal protein, most likely a 40S subunit. In addition, a gene (*njjs4*) has been identified which is associated with the process of seed maturation and fruit ripening, and is related to the class-I LMW heat-shock-protein-like genes (Medina-Escobar et al. 1998). Two auxin-induced and one auxin-repressed mRNAs from unknown genes have been cloned from receptacles of immature green fruit (Reddy et al. 1990, Reddy and Poovaiah 1990). Yubero-Serrano et al. (2003) identified a gene encoding a lipid transfer protein (*Fxaltp*) in strawberry fruit that responds to ABA, wounding and cold stress. Aharoni et al. (2001) cloned the transcription factor *FaMYB1* from ripening fruit, which plays a key role in the biosynthesis of anthocayanins and flavonols.

DNA microarrays have also been used to profile cosmic patterns of gene expression during ripening. Aharoni and O'Connell (2002) found 441 transcripts to differ significantly between the achene and receptacle tissues. The most common transcripts found in achenes were those for signal and regulation cascades associated with achene maturation, and stress tolerance. Representatives included phosphatases, protein kinases, 14-3-3 proteins and transcription factors. Several genes were identified in the receptacle that encode proteins related to stress, the cell wall, DNA/RNA protein and primary metabolism.

13.5.5 Yield

Yield is the product of a combination of characters, such as number and size of fruit, plant vigor, hardiness, and disease resistance of the plant. Crown number per row area is often the factor most strongly associated with yield, although flower number and fruit size are also important components. High crown numbers can be achieved through either high levels of stolon production or branch crown production.

Strong compensatory interactions have often been found between the various yield components (plant density, crowns per plant, trusses per crown, fruit per truss, etc.), indicating that breeding for high fruit numbers or individual fruit size by themselves will not necessarily increase productivity (Hancock et al. 1996b). However, outlier types do exist with both large fruit and high fruit numbers (Hancock and Bringhurst 1988).

Considerable levels of genetic variability have been described for most yield components, although the relative levels of additive and non-additive variation have varied greatly from study to study (Table 13.4). In most studies, sufficient levels of additive variation were considered available for rapid improvement of yield.

Hansche et al. (1968) found extensive levels of genetic associated with fruit size, firmness and yield in the University of California (UC-Davis) breeding program, but not appearance. A significant genetic correlation existed between fruit size and yield, indicating that plants with large berries have a genetic potential for high yield. When the UC-Davis breeding population was evaluated 20 years later, heritability estimates were not significantly different from the ancestral population (Shaw et al. 1989). In a few breeding populations, non-additive gene influences have appeared to be more important than additive ones (Watkins et al. 1970, Spangelo et al. 1971), suggesting that crosses should be designed to exploit all the genetic variance, whether it be additive, dominant, or epistatic.

13.5.6 Adaptability to Mechanical Harvesting

A recurring objective in strawberry breeding has been to produce types adapted to mechanical harvesting, although few cultivars have been developed that produce consistently profitable yields when machine harvested. Paramount are concentrated ripening for once over harvest, long pedicles and either easy calyx removal or long necks for machine decapping. There is considerable variation for concentrated ripening (Denisen and Buchele 1967, Moore et al. 1970, Barritt 1974), although concentrated ripeners are often lower yielding than longer season types (Moore et al. 1975). Ease of calyx removal shows considerable additive genetic action (Barritt 1976), and in some parents, low capping force is dominant (Brown and Moore 1975). Considerable variation in pedicle and fruit neck length has also been reported (MacIntyre and Gooding 1978, Dale et al. 1987). Unfortunately, few cultivars have been released for mechanical harvesting (Daubeny et al. 1980).

13.6 Crossing and Evaluation Techniques

13.6.1 Breeding Systems

The dessert strawberry is an outcrossed crop that is relatively sensitive to inbreeding (Morrow and Darrow 1952, Melville et al. 1980b), and it can be asexually propagated by runners, so most varietal improvement programs have been based on pedigree breeding where elite parents are selected each generation for inter-crossing. If adequate population sizes are maintained, changes in levels of homozygosity across generations appear to be minimal (Shaw and Sacks 1995). Since highly heterozygous genotypes can be propagated as runners, few breeding programs have developed hybrid varieties using inbred lines, although some cultivars have been developed this way.

Selfing has been successfully employed in a number of instances to concentrate genes of interest (Hancock et al. 1996a) and backcrossing has been used occasionally to incorporate specific traits. Barritt and Shanks (1980) moved resistance to the strawberry aphid from native *F. chiloensis* to *F. ×ananassa*. Bringhurst and

Voth (1976, 1984) transferred the day neutrality trait from native *F. virginiana* spp. *glauca* to *F. ×ananassa*. Approximately 3 generations were necessary to restore fruit size and yield to commercial levels.

Numerous studies have been designed to test the effects of temporal, spatial and developmental variation on production traits, with the final intent to develop efficient selection strategies (Gooding et al. 1975, Hortynski 1989). Shaw and coworkers (1987) found that within a single year, the distribution of genetic and environmental variance components for a single trait vary continuously, with heritabilities for yield and fruit size being highest in the middle of the season. They also found that seedling location has a large effect on the expression of genetic variation (Shaw 1989, Shaw et al. 1989). Shaw (1991) observed that nursery treatments induced large interactions for production traits in annual systems, especially those that condition variable levels of plant development and chilling. His studies indicated that crossing among parents chosen for breeding value may be more effective than simple clonal performance in generating superior seedling populations. In fact, the performance of seedlings may be very different when propagated as runner plants or when grown in different environments.

13.6.2 Pollination and Seedling Culture

The blossom of strawberry is composed of many pistils, each with its own style and stigmata, attached to a receptacle that on fertilization of the pistils develops into a fleshy 'fruit'. The true fruits of the strawberry are the achenes which carry one seed and are found on the surface of the swollen receptacle. A single blossom may have 20–400 pistils that develop into seeds, depending on the size of the blossom and its position on the cluster. Primary flowers have the largest number of pistils, with secondary, tertiary and later flowers having progressively fewer (Hancock 1999).

Strawberry flowers are usually emasculated using a scalpel, tweezers or thumb nails, by carefully removing the ring of sepals, petals and anthers surrounding the receptacle. Care is taken not to rupture any anthers. Emasculation is usually done 1 to 3 days before anthesis to prevent selfing; this is frequently done when the first white of the petals begins to show as the sepals separate. Emasculated flowers must be protected from foreign pollen either by bagging or isolation. Cotton gauze is often used to cover individual flowers or clusters. Much of the hybridization work is done using potted plants in the greenhouse as it is easier to prevent contamination and control the environment.

Pollen is collected by removing individual anthers from the blossoms 1 to 2 days before anthesis and placing them in vials to dehisce, or by detaching flowers from the clusters, removing petals and sepals, and placing them overnight in paper lined shallow vessels such as petri plates. Pollen will remain viable for several days if stored at room temperature, and for several years if stored at 4° C under low humidity. Pollen can be transferred to stigmas with a small camel's hair brush, a rubber-tipped rod or a finger tip. Alcohol is commonly used to sterilize the transfer vehicles between crosses.

Fruits ripen 25–30 days after pollination at 18–25° C. Large quantities of fruit can be processed by threshing them in water with a 10 to 15 second spin in a food blender. The pulp floats, while the seeds sink. Smaller quantities of fruit can be mashed on absorbent paper and the seeds scraped off after the residue has dried. For long term storage, seeds are generally placed in coin envelopes and held under low humidity at 1–4° C. Under these conditions seed remain viable for over 20 years, depending on genotype (Scott and Draper 1970).

For germination, seeds should be spread on the surface of the soil and held under light. Without pre-treatment, seedling emergence is irregular, with some seeds germinating within 10 days, while others can take up to 90 days. Time of emergence can be normalized by after-ripening the seed for 2.5–3 months at 1–4° C (Bringhurst and Voth 1957), or scarifying it for 10–15 minutes in concentrated sulfuric acid. Seedlings are generally allowed to grow for six weeks in the seed trays until they have a few true leaves and then they are transplanted into pots, where they are grown for another 6 to 8 weeks before being planted in the field.

13.6.3 Evaluation Techniques

Field plantings of first-test seedlings intended for matted row culture are generally planted in the spring at 45–60 cm spacing and are allowed to form small matted blocks that are about 25 cm wide. Hybrids intended for annual hill systems are planted in the fall in plastic covered ridges at 20 cm spacing. Elite clones are selected from both systems in the summer of the second year and runners are collected in the fall for trial planting in the third year. A randomized block design is then used to evaluate the elite clones in replicates of 5–10 plants maintained at commercial spacing. The hybrids are evaluated for one to two years, and then runners from these are sent to collaborators for further testing in randomized designs. Decisions about release are generally made 6–8 years after the initial crosses.

13.7 Biotechnological Approaches to Genetic Improvement

13.7.1 Genetic Mapping and QTL Analysis

Numerous marker systems have been developed in strawberry for genetic linkage mapping and QTL analysis (Hadonou et al. 2004, Sargent et al. 2004a, Cipriani et al. 2006). These have been shown to be broadly applicable across all strawberry species, although SSRs developed from other Rosaceae species have only limited utility.

Davis and Yu (1997) provided the first diploid map of *F. vesca*, using RAPD markers and isozymes, plus some morphological traits. They crossed the cultivar Baron Solemacher of *F. vesca* f. *semperflorens* and a wild clone of *F. vesca* ssp. *vesca* collected in New Hampshire, and developed an 80-marker map in the

F_2 population that represented all seven linkage groups and was 445 cM long. Unusually high levels of segregation distortion was noted (47%) that was skewed toward the maternal grandparent, 'Baron Solemacher'. Davis and Yu speculated that the segregation distortion was caused by the maternal cytoplasm favoring maternal genes.

Deng and Davis (2001) used a candidate gene approach to determine the molecular basis of the yellow fruit color locus (c) in diploid strawberry. They employed PCR and degenerate primer pairs to examine segregation patterns in intron length polymorphism's of a number of genes involved in the anthocyanin biosynthetic pathway. They studied F_2 progeny populations of a wild clone of northern California *F. vesca* 'Yellow Wonder' × an *F. nubicola* genotype from Pakistan, and were able to place five genes into their previously published map. They found *F3H*, the gene encoding flavanone 3-hydrolase, to be the likely candidate for the yellow fruit color locus.

Most recently, a diploid map of 78 markers was constructed from a hybrid population of *F. vesca* ssp. *vesca* f. *semperflorens* × *F. nubicola* (Sargent et al. 2004a). They authors used a combination of SSRs, SCARs, gene specific markers and morphological markers that came from the GenBank data base and other studies. All seven linkage groups were identified in their map that covered 448 cM. Segregation distortions were noted at 54% of the loci that were skewed toward the paternal parent *F. nubicola*. They speculated that the segregation distortions were due to meiotic irregularities or the self-incompatible nature of *F. nubicola*.

Lerceteau-Köhler et al. (2003) used a total of 727 AFLP markers and 119 individuals to build both a female map and a male map from the cross of 'Capitola' × CF1116 ['Pajaro' × ('Earliglow' × 'Chandler')]. The female map was built with 235 markers and was 1604 cM long, while the male map was 1496 cM long with 280 markers. Only 3.2% of the markers displayed distorted segregation ratios. They detected 30 linkage groups on the female side and 28 on the male side, but did not develop a consensus map of the two parents. The female genome size was estimated to be 2870 cM, while the male genome size was 1861 cM.

Viruel et al. (2002) used 300 SSR and RFLP markers and 86 progeny to build a consensus linkage map with 17 linkage groups and a total distance of 627 cM. 120 markers were unlinked or linked to only one marker, suggesting the need for more markers to build a complete map. Only 10% of the markers showed distorted segregation ratios.

Weebadde et al. (2007) genotyped sixty-seven individuals of the cross 'Tribute' × 'Honeoye' with AFLP markers. Out of 611 polymorphic bands obtained using 52 primer combinations, 410 single dose restriction fragments (SDRFs) were identified and 23 linkage groups. Most of the markers (255/410) remained unlinked, indicating the need for more markers and larger population sizes to build a map with wide genome coverage.

Only a few quantitative trait loci (QTL) analyses have been conducted in strawberry. In the study of Weebadde et al. (2007), two AFLP markers were significantly associated with segregation of the day-neutrality trait at a 0.01% level and five at a 0.1% level (Fig. 13.3). Several of these markers were not linked, indicating

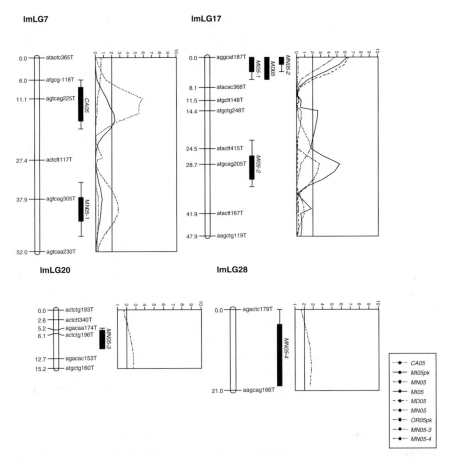

Fig. 13.3 QTL for the day-neutral flowering response detected in a segregating population of 'Tribute' (day-neutral) × 'Honeoye' (short day) strawberries evaluated in Michigan (MI), Minnesota (MN), Maryland (MD), California (CA) and Oregon (OR). All the QTL associated with day-neutrality were derived from the cultivar 'Tribute' (Webbadde et al. 2007)

that day-neutrality is a quantitative trait in the octoploids. Haymes found AFLP markers linked to three red stele resistance genes (Hokanson and Maas 2001). Lerceteau-Köhler et al. (2004) found fourteen QTL associated with seven characters (fruit height, ratio fruit height/diameter, fruit color, firmness, malate content, glucose content and ratio fructose to glucose). The percentages of phenotypic variance explained by the QTL ranged from 12% to 20%.

13.7.2 Regeneration and Transformation

Strawberries were one of the first crops to be routinely proliferated through micropropagation (Zimmerman 1991), and regeneration systems for *F. ×ananassa* have been developed with disarmed strains of *Agrobacterium tumefaciens* using anthers, callus, flower buds, leaf discs, protoplasts, petioles, stems, stipules, roots and runners (Hokanson and Maas 2001, Passey et al. 2003). Callus, petiole sections and leaf discs have also been used in combination with *A. tumefaciens* to regenerate *F. vesca* (Haymes and Davis 1998, Alsheikh et al. 2002). Genotypes vary widely in the success of the various techniques, and some are quite recalcitrant to all techniques. A genetic line of *F. ×ananassa*, LF9, has been developed that produces transformed shoots in as few as 15 days (Folta et al. 2006).

Most of the transgenic strawberries have been generated using *Agrobacterium*-mediated tranformation systems. *Agrobacteriun* strains LBA4404 and EHA105 have been most commonly employed, with pBIN19 derivates as the binary vector (Graham 2005). In most cases, leaf or stem based systems have been utilized in Murashige and Skoog (MS) medium. The hormones have been BA plus 2,4-D or IAA. Regenerates have generally been selected using 25 mg/L kanamycin, and contamination after inoculation has been limited by using cefotaxime, carbenicillim and ticaricillin. Biolistics have been employed in three instances with strawberry. Cordero de Mesa et al. 2000 bombarded leaf discs with *Agrobacterium* coated gold microprojectiles as a means to enhance stable transformation of GUS. Wang et al. (2004) bombarded strawberry calli with tungsten particles coated with the pBY520 plasmid using PDS-1000/He. Agius et al. (2005) used particle bombardment to effect transient transformation of strawberry fruit.

An effective marker-free transformation process has recently been successfully tested in strawberry (Schaart et al. 2004). In it, a vector was constructed in which site-specific recombination left only the 35S promoter of cauliflower mosaic and a GUS encoding sequence. The system incorporated an inducible site-specific recombinase to eliminate the selectable marker. Fully marker-free transgenic 'Calypso' plants were obtained through this procedure.

Two types of herbicide resistance have been engineered into strawberry through *Agrobacterium* mediated transformation. The phosphinothricin acetyl transferase gene (PAT) was incorporated into the 'Selekta' strawberry using *Agrobacterium* mediated gene transfer (du Plessis et al. 1995, 1997). Putatively transformed shoots were rooted and established in the greenhouse and most transgenic plants were found to be resistant to the herbicide glufosinate-ammonium (Ignite®). The CP4.EPSP synthase gene, which confers resistance to glyphosate (Roundup®) was introduced into 'Camarosa' (Morgan et al. 2002), and when 73 independent transformations were sprayed with Roundup in the nursery, a range of responses were noted from complete resistance to death. Expression levels of the *CP4.EPSPS* gene was strongly correlated with phenotype. The best lines were subsequently tested in the field and appeared to produce good quality fruit.

To provide insect resistance, the cowpea protease trypsin inhibitor gene (*CpTi*) (Agricultural Genetics Company, Cambridge, England) was incorporated into strawberry via *Agrobacterium* mediated transformation using the *NPTII* marker (Graham et al. 1995, Graham et al. 1997, 2002). The insertion of *CpTi* into strawberry cultivars 'Melody' and 'Symphony' was found to reduce vine weevil (*Otiorhynchus sulcatus*) damage in both greenhouse and field trials. The transgenic lines showed increased root growth, less larval feeding and fewer pupae. In other work, strawberries transformed with the lectin *Galanthus nivalis agglutin gene* (*GNA*), did not show any significant reduction in weevil feeding (Graham 2005).

Enhanced resistance to *Verticillium dahliae* was observed in transgenic 'Joliette' strawberry plants expressing a *Lycopersicon chilense* chitinase gene (*pcht28*) under the control of the CAMV 35S promoter (Chalavi et al. 2003). A stipule regeneration system was used with *Agrobacterium*-mediated gene transfer. Constitutive expression of the chitinase gene was demonstrated by northern analysis, and in growth chamber studies, transgenic strawberry plants had significantly higher resistance than controls, based on rates of crown infection and leaf wilting symptoms.

The antisense of strawberry pectate lyase was incorporated into 'Chandler' strawberry under control of the 35S promoter to increase fruit firmness (Jiménez-Bermúdez et al. 2002). At full ripening, no differences in color, shape and weight were noted between the transgenic and control plants, but the transgenics were significantly firmer. Pectase lyase activity was 30% lower in ripe transgenic fruit than the control. In another study, Agius et al. (2003) found that expression of an antisense sequence of a strawberry pectate lyase gene reduced ascorbic acid content, presumably through reduced pectin solubilization in cell walls of transgenic plants.

Two endo-β-1,4-glucanase (EG) genes, *cel1* and *cel2*, have been isolated from strawberry that are closely related to tomato genes influencing softening. *Cel1* were expressed specifically in ripening fruit (Manning 1998), while *cel2* mRNA was found primarily in young vegetative tissues and early green fruit (Trainotti et al. 1999). *Cel1* has been cloned into strawberry in the antisense orientation via *Agrobacterium* – mediated transformation using the plant binary vector pBIN-PLUS (Woolley et al. 2001). In the transgenic strawberries, mRNA was strongly suppressed in ripe fruit; however, EG activity and firmness were not affected. The incorporation of *cel1* had no effect on the transcription *cel2*.

The S-adenosylmethionine hydrolase gene (*SAMase*) has been incorporated into strawberry which controls ethylene biosynthesis and presumably effects fruit softening (Mathews et al. 1995). Strawberries are not climacteric fruit, but do have a limited response to ethylene and it is possible that reductions in ethylene biosynthesis during the post harvest period could slow down softening.

De la Fuente et al. (2006) has cloned *FaGAST* from strawberry, which encodes a small protein with 12 cysteine residues conserved in the C-terminal region that is similar to a group of proteins in other plant species that regulate cell division and elongation. Expression of *FaGAST* in transgenic *F. vesca*, under the control of CaMV-35S, resulted in delayed fruit growth, reduced fruit size, late flowering and low sensitivity to gibberellin. Apparently, *FaGAST* plays a role in arresting fruit elongation during strawberry fruit ripening.

The acidic dehydrin gene *WCOR410* from wheat was transferred to strawberry in an attempt to improve freezing tolerance (Houde et al. 2004). The WCOR410 protein has been associated with the plasma membrane in wheat and its levels have been correlated with freezing tolerance. After acclimation, transgenic strawberry leaves had a 5° C improvement in freezing tolerance compared to controls. However, there was not a difference in the freezing tolerance of non-acclimated transgenics and controls, suggesting that another factor induces its expression during cold acclimation.

Transgenic strawberries have also been developed that expressed anti-freeze protein gene (*AFP*) isolated from white flounder (Firsov and Dolgov 1998), but no data from freezing trials has been published.

In an earlier attempt to increase the freezing tolerance of strawberries, the transcription factor *CBF1* from Arabadopsis was overexpressed in the strawberry 'Honeoye' (Owens et al. 2002). The *CBF* genes are part of a family of cold and drought inducible transcription factors that bind to promoters containing a C-Repeat/DehydrationResponsive Element (CRT/DRE). This element is found in many cold-induced plant genes. *Agrobacterium*-mediated transformation with a CaMV35S-CBF1 construct was utilized to produce transgenic strawberries. The freezing tolerance of strawberry leaf-discs from non-acclimated plants was significantly increased by 3–5° C. However, the freezing tolerance of floral tissue was not, despite expression of the *CBF1* transgene in receptacles at levels similar to developing leaves. A putative ortholog [*Fragaria ×ananassa CBF1* (*FaCBF1*)] was identified with 48% amino acid identity to *CBF1* from Arabidopsis.

The late embryogenesis abundant protein gene (*LEA3*) from barley (*Hordeum vulgare*) was used to transform the 'Toyonaka' strawberry (Wang et al. 2004) in hopes of increasing the resistance of strawberry to salt stress. Calli from anthers were transformed by particle bombardment with plasmid pBY520. In vitro plants of transgenic strawberry had significantly less wilting than controls under 50 mmol (19% vs. 62%) and 100 mmol NaCl (43% vs. 96%).

Mezzetti et al. (2004) developed transgenic strawberries and raspberries carrying the *defH9-iaaM* auxin-synthesizing construct, composed of the regulatory region of the *DefH9* gene from snapdragon and the *iaaM* coding region from *Pseudomonas syringae*. The *defH9-iaaM* gene was found to promote parthenocarpy in emasculated flowers of both strawberry and raspberry, and to increase fruit size, weight and yield.

The *FBP7* promoter (*floral binding protein7*) from Petunia was found to be active in floral and fruit tissues of strawberry, using the ß-glucuronidase gene as a reporter (Schaart et al. 2002). GUS activity was found in floral and fruit tissues, but not vegetative ones, although gus-derived mRNAs were found in roots and petioles. The *35S* promoter was found to be sixfold stronger than the *FBP7* promoter.

Agius et al. (2005) used a transient expression system to conduct a functional analysis of homologous and heterologous promoters in strawberry fruit. The *CaMV 35S* promoter was fused to the *LUC* gene to optimize the transient assay. The *GalUR* promoter from strawberry was found to be active in fruit and under light regulation. Slight activity in fruit was found for the pepper fibrillin promoter, but not for the tomato polygalacturonase promoter.

A MADS box gene from strawberry, *STAG1*, has been cloned and characterized in transgenic plants (Rosin et al. 2003). *STAG1* shares 68–91% sequence homology with *AGAMOUS* from numerous plant species. Analysis of the expression patterns of a GUS marker gene driven by the *STAG1* promoter revealed that *STAG1* was active in stamens, receptacles, petals, central pith and vascular cells during floral development and achenes, pith and cortical cells during fruit ripening.

13.7.3 Genomic Resources

Two major EST libraries of have been generated as genomic resources in strawberry. A cDNA library of over 1,800 ESTs has been produced from whole plants treated with salicyclic acid by Folta et al. (2005). This effort is part of a major Rosaceae genomics project (Jung et al. 2004). Batley et al. (2005) have generated over 23,600 ESTs from a range of tissues, developmental stages and experimental conditions, and identified 11,690 single nucleotide polymorphisms (SNPs) and 4,200 SSRs for mapping. Their focus is on traits such as day neutrality and the fruit characteristics of firmness, flavor, taste, aroma and color. cDNA libraries of *F. vesca* are also being constructed by Davis (2005) and Slovin (person. comm.).

Acknowledgments Chad Finn (USDA-ARS, Corvallis, OR) and David Simpson (East Malling Research, U.K.) made valuable contributions to the manuscript.

References

Agius F, González-Lamothe R, Caballero JL, Muñoz-Blanco J, Botella MA,Valpuesta V (2003) Engineering increased vitamin C levels in plants by overexpression of a D-galacturonic acid reductase. Nat Biotechnol 21:177–181

Agius F, Amaya I, Botella MA, Valpuesta V (2005) Functional analysis of homologous and heterologous promoters in strawberry fruits using transient expression. J Exp Bot 56:37–46

Aharoni A, De Vos C, Wein M, Sun Z, Greco R, Kroon A, Mol J, O'Connell A (2001) The strawberry *FaMYB1* transcription factor suppresses anthocyanin and flavonol accumulation in transgenic tobacco. Plant J 28:319–332

Aharoni A, Giri A, Verstappen F, Bertea C, Sevenier R, Sun Z, Jongsma M, Schwab W, Bouwmeester H (2004) Gain and loss of fruit flavor compounds produced by wild and cultivated strawberry species. Plant Cell 16:3110–3131

Aharoni A, Keizer L, Bouwmeester H, Sun Z, Alvarez-Huerta M, Verhoeven H, Blaas J, van Houweilingen A, De Vos C, van der Voet H, Jansen R, Guis M, Mol J, Davis R, Schena M, van Tunen A, O'Connell A (2000) Identification of the SAAT gene involved in strawberry flavor biogenesis by use of DNA microarrays. Plant Cell 12:647–661

Aharoni A, O'Connell A (2002) Gene expression analysis of strawberry achene and receptacle maturation using DNA microarrays. J Exp Bot 53:2073–2087

Ahmadi H, Bringhurst RS (1991) Genetics of sex expression in *Fragaria* species. Am J Bot 78:504–514

Ahmadi H, Bringhurst RS, Voth V (1990) Modes of inheritance of photoperiodism in *Fragaria*. J Am Soc Hortic Sci 115:146–152

Alsheikh M, Suso H, Robson M, Battey N (2002) Appropriate choice of antibiotic and Agrobacterium strain improves transformation of antibiotic-sensitive *Fragaria vesca* and *F. v. semperflorens*. Plant Cell Rep 20:1173–1180

Anstey TH, Wilcox AN (1950) The breeding value of selected inbred clones of strawberries with respect to their vitamin C content. Sci Agric 30:367–374

Arulsekar S (1979) *Verticillium* wilt resistance in the cultivated strawberries and preliminary studies on isozymes genetics in *Fragaria*. PhD. University of California, Davis

Arulsekar S, Bringhurst RS (1981) Genetic model for the enzyme marker PGI in diploid California *Fragaria vesca*. J Hered 73:117–120

Ashley M, Wilk J, Styan S, Craft K, Jones K, Feldman K, Lewers K, Ashman T (2003) High variability and disomic segregation of microsatellites in the octoploid *Fragaria virginiana* Mill. (Rosaceae). Theor Appl Genet 107:1201–1207

Barritt BH (1974) Single harvest yields of strawberries in relation to cultivar and time of harvest. J Am Soc Hortic Sci 99:6–8

Barritt BH (1976) Evaluation of strawberry parent clones for easy calyx removal. J Am Soc Hortic Sci 101:590–591

Barritt BH (1979) Breeding strawberries for fruit firmness. J Amer Soc Hort Sci 104:663–665

Barritt BH (1980) Resistance of strawberry clones to Botrytis fruit rot. J Am Soc Hortic Sci 105:160–164

Barritt BH, Daubeny HA (1982) Inheritance of virus tolerance in strawberry. J Am Soc Hortic Sci 107:278–282

Barritt BH, Shanks CH (1980) Breeding strawberries for resistance to the aphids *Chaetosiphon fragaefolii* and *C. thomasi*. HortScience 15:287–288

Barritt BH, Shanks CH (1981) Parent selection in breeding strawberries resistant to two-spotted spider mites. HortScience 16:323–324

Batley J, Keniry A, Hopkins C, Mountford H, Logan E, Gramzow L, Morrison B, Spangenberg G, Edwards D (2005) A new genomics resource for strawberry: Towards molecular genetic markers for day neutrality traits. In: Plant & Animal Genomes XIII Conference, San Diego, CA

Bell JA, Simpson DW, Harris DC (1997) Development of a method for screening strawberry germplasm for resistance to *Phytophthora cactorum*. Acta Hortic 439:175–180

Blanco-Portales R, López-Raéz J, Bellido M, Moyano E, Dorado G, González-Reyes J, Caballero JL, Muñoz-Blanco J (2004) A strawberry fruit-specific and ripening-related gene codes for a HyPRP protein involved in polyphenol anchoring. Plant Mol Biol 55:763–780

Bors B, Sullivan JA (1998) Interspecific crossability of nine diploid *Fragaria* species. HortScience 32:439 (abst.)

Bors B, Sullivan JA (2005) Interspecific hybridization of *Fragaria vesca* subspecies with *F. nilgerrensis*, *F. nubicola*, *F. pentaphylla*, and *F. viridis*. J Am Soc Hortic Sci 130:418–423

Bringhurst R (1990) Cytogenetics and evolution in American *Fragaria*. HortScience 25:879–881

Bringhurst RS, Hansche PE, Voth V (1967) Inheritance of verticillium wilt resistance and the correlation of resistance with performance traits of the strawberry. Proc Am Soc Hortic Sci 92:369–375

Bringhurst RS, Senanayake YDA (1966) The evolutionary significance of natural *Fragaria chiloensis* × *F. vesca* hybrids resulting from unreduced gametes. Am J Bot 53:1000–1006

Bringhurst RS, Voth V (1957) Effect of stratification on strawberry seed germination. Proc Am Soc Hortic Sci 70:144–149

Bringhurst RS, Voth V (1976) Origin and evolutionary potentiality of the day-neutral trait in octoploid *Fragaria*. Genetics 83:s10

Bringhurst RS, Voth V (1984) Breeding octoploid strawberries. Iowa State University J Res 58:371–381

Brown GR, Moore JN (1975) Inheritance of fruit detachment in strawberry. J Am Soc Hortic Sci 100:569–572

Brown JG, Voth V (1955) Salt damage to strawberries. Calif Agric 9:11–12

Brown T, Wareing PF (1965) The genetical control of the everbearing habit and three other characters in varieties of *Fragaria vesca*. Euphytica 14:97–112

Castillejo C, de la Fuente J, Iannetta P, Botella M, Valpuesta V (2004) Pectin esterase gene family in strawberry fruit: study of FaPE1, a ripening-specific isoform. J Exp Bot 55:909–918

Cekic C, Battey J, Wilkinson JQ (2001) The potential of ISSR-PCR primer-pair combinations for genetic linkage analysis using the seasonal flowering locus in *Fragaria* as a model. Theor Appl Genet 103:540–546

Chalavi V, Tabaeizadeh Z, Thibodeau P (2003) Enhanced resistance to *Verticillium dahliae* in transgenic strawberry plants expressing a *Lycopersicon chilense* chitinase gene. J Am Soc Hortic Sci 128:747–753

Chiykowski LN, Craig DL (1975) Reaction of strawberry cultivars to clover phyllody (green petal) agent transmitted by *Aphrodes bicincta*. Can Plant Dis Survey 55:66–68

Cho CT, Moon BJ (1984) Studies on the wilt of strawberry caused by *Fusarium oxysporum* f.sp. *fragariae* in Korea. Korean J Plant Prot 23:74–81

Cipriani G, Pinosa F, Bonoli M, Faedi W (2006) A new set of microsatellite markers for *Fragaria* species and their application in linkage analysis. J Hortic Sci Bio 81:668–675

Comstock RE, Kelleher T, Morrow EB (1958) Genetic variation in an asexual species, the garden strawberry. Genetics 43:634–46

Cordero de Mesa M, Jiménez-Bermúdez S, Pliego-Alfaro F, Quesada MA, Mercado JA (2000) *Agrobacterium* cells as microprojectile coating: a novel approach to enhance stable transformation rates in strawberry. Aust J Plant Physiol 27:1093–1100

Cram WT (1978) The effect of root weevils (Coleoptera: Curculionidae) on yield of five strawberry cultivars in British Columbia. J Entomol Soc B C 75:10–13

Crock JE, Shanks CH Jr, Barritt BH (1982) Resistance in *Fragaria chiloensis* and *F*. ×*ananassa* to the aphids *Chaetosiphon fragaefolii* and *C*. *thomasi*. HortScience 17:959–960

Dale A, Gray VP, Miles NW (1987) Effects of cultural systems and harvesting techniques on the production of strawberries for processing. Can J Plant Sci 67:853–862

Darrow GM (1937) Strawberry improvement. In: Better plants and animals 2. USDA Yearbook of Agriculture, pp 496–533

Darrow GM (1966) The Strawberry. history, breeding and physiology. Holt, Rinehart and Winston, NewYork

Darrow GM, Scott DH (1947) Breeding for cold hardiness of strawberry flowers. Proc Am Soc Hortic Sci 50:239–242

Darrow GM, Waldo GF, Schuster CE (1933) Twelve years of strawberry breeding. A summary of the strawberry breeding work of the United States Department of Agriculture. J Hered 24:391–402

Daubeny HA (1964) Effect of parentage in breeding for red stele resistance of strawberry in British Columbia. Proc Am Soc Hortic Sci 84:289–294

Daubeny HA, Lawrence FJ, Martin LW, Barritt BH (1980) 'Tyee', a new strawberry cultivar suited to machine harvest. Sta Bull, Agric Exp Sta, Oregon State Univ 645:40–42

Davik J, Honne BI (2005) Genetic variance and breeding values for resistance to a wind-borne disease [*Sphaerotheca macularis* (Wallr. ex Fr.)] in strawberry (*Fragaria* ×*ananassa* Duch.) estimated by exploring mixed and spatial models and pedigree information. Theor Appl Genet 111:256–264

Davis R (2005) A diploid platform for strawberry genomics. In: Plant & Animal Genomes, San Diego, CA

Davis TM (2004) Identification of putative diploid genome donors to the octoploid cultivated strawberry, *Fragaria* ×*ananassa*. In: Plant & Animal Genomes XII Conference, San Diego, CA

Davis TM, Yu H (1997) A linkage map of the diploid strawberry, *Fragaria vesca*. J Hered 88:215–221

De la Fuente JI, Amaya I, Castillejo C, Sánchez-Sevilla, Quesada MA, Botella MA, Valpuesta V (2006) The strawberry gene *FaGAST* affects plant growth through inhibition of cell elongation. J Exp Bot 57:2401–2411

Delhomez N, Carisse O, Lareau M, Khanizadeh S (1995) Susceptibility of strawberry cultivars and advanced selections to leaf spot caused by *Mycosphaerella fragariae*. HortScience 30: 592–595

Delp BR, Milholland RS (1981) Susceptibility of strawberry cultivars and related species to *Colletotrichum fragariae*. Plant Dis 65:421–423

Deng C, Davis TM (2001) Molecular identification of the yellow fruit color (c) locus in diploid strawberry: a candidate gene approach. Theor Appl Genet 103:316–322

Denisen EL, Buchele WF (1967) Mechanical harvesting of strawberries. Proc Am Soc Hortic Sci 91:267–273

Denoyes B, Baudry A (1995) Species identification and pathogenicity study of French *Colletotrichum* strains isolated from strawberry using morphological and cultural characteristics. Phytopathology 85:53–57

Denoyes-Rothan B, Guerin G, Lerceteau-Köhler E, Risser G (2005) Inheritance of resistance to *Colletotrichum acutatum* in *Fragaria* ×*ananassa*. Phytpathology 95:405–412

Doss RP, Shanks Jr CH (1988) The influence of leaf pubescence on the resistance of selected clones of beach strawberry (*Fragaria chiloensis* (L.) Duchesne) to adult black vine weevils (*Otiorhynchus sulcatus* F.). Sci Hortic 34:47–54

Du Plessis HJ, Brand RJ, Glynn-Woods C, Goedhart MA (1995) Genetic engineering leads to a herbicide-tolerant strawberry. S Afr J Sci 91:218

Du Plessis HJ, Brand RJ, Glynn-Woods C, Goedhart MA (1997) Efficient genetic transformation of strawberry (*Fragaria* ×*ananassa* Duch.) cultivar Selekta. Acta Hortic 447:289–294

Duewer RG, Zych CC (1967) Heritability of soluble solids and acids in progenies of the cultivated strawberry (*Fragaria* ×*ananassa* Duch.). Proc Am Soc Hortic Sci 90:153–157

Edwards WH, Jones RK, Schmitt DP (1985) Host suitability and parasitism of selected strawberry cultivars by *Meloidogyne hapla* and *M. incognita*. Plant Dis 69:40–42

Fadeeva TS (1966) Communication 1. Principles of genome analysis (with reference to the genus *Fragaria*). Genetika 1:12–28

Federova NJ (1946) Crossibility and phylogenetic relations in the main European species of *Fragaria*. Compt Rend (Doklady) Acad Sci USSR 1:12–28

Firsov AP, Dolgov SV (1998) Agrobacterial transformation and transfer of the antifreeze protein gene of winter flounder to the strawberry. Acta Hortic 484:581–586

Fishman R (1987) Albert Etter: fruit breeder. Fruit Varieties J 41:40–46

Folta KM, Dhingra A, Howard L, Stewert P, Chandler CK (2006) Characterization of LF9, an octoploid strawberry genotype selected for rapid regeneration and transformation. Planta 224:1058–1067

Folta KM, Staton M, Stewart PJ, Jung S, Bies DH, Jesdurai C, Main D (2005) Expressed sequence tags (ESTs) and simple sequence repeat (SSR) markers from octoploid strawberry (*Fragaria* ×*ananassa*). BMC Plant Biol 5:12

Galletta GJ, Bringhurst RS (1991) Strawberry management. In: Galletta GJ, Himelrick D (eds) Small fruit crop management. Prentice Hall, Englewood Cliffs, NJ, pp 83–156

Gimenez G, Ballington JR (2002) Inheritance of resistance to *Colletotrichum acutatum* Simmonds on runners of garden strawberry and its backcrosses. HortScience 37:686–690

Gooding HJ, Jennings DL, Topham TP (1975) A genotype-environment experiment on strawberries in Scotland. Heredity 34:105–115

Gooding HJ, McNicol RJ, MacIntyre D (1981) Methods of screening strawberries for resistance to *Sphaerotheca macularis* (Wall ex Frier) and *Phytophthora cactorum* (Leb. and Cohn). J Hortic Sci 56:239–245

Graham J (2005) *Fragaria* strawberries In: Litz R.E. (ed) Biotechnology of fruit and nut crops. CABI Publishing, Wallingford, UK

Graham J, Gordon SC, McNicol RJ (1997) The effect of the *CpTi* gene in strawberry against attack by vine weevil (*Otiorhynchus sulcatus* F. Coleoptera: Curculionidae). Ann Appl Biol 131:133–139

Graham J, Gordon SC, Smith K, McNicol RJ, McNicol JW (2002) The effect of the Cowpea trypsin inhibitor in strawberry on damage by vine weevil under field conditions. J Hortic Sci Biotechnol 77:33–40

Graham J, McNicol RJ, Greig K (1995) Towards genetic based insect resistance in strawberry using the Cowpea trypsin inhibitor gene. Ann Appl Biol 127:163–173

Graichen K, Proll E, Leistner HU, Kegler H (1985) Preliminary results of purification experiments with strawberry viruses. Archiv Phytopathol Pflanzenschutz 21:499–501

Hadonou A, Sargent D, Wilson F, James C, Simpson DW (2004) Development of microsatellite markers in *Fragaria*, their use in genetic diversity analysis, and their potential for genetic linkage mapping. Genome 47:429–438

Hamilton-Kemp TR, Andersen RA, Rodriguez JG, Loughrin JH, Patterson CG (1988) Strawberry foliage headspace vapor components at periods of susceptibility and resistance to *Tetranychus urticae* Koch. J Chem Ecol 14:789–796

Hancock JF (1999) Strawberries. CABI Publishing, Wallingford, UK

Hancock JF, Bringhurst RS (1979) Ecological differentiation in perennial, octoploid species of *Fragaria*. Am J Bot 66:367–375

Hancock JF, Bringhurst RS (1988) Yield component interactions in wild populations of California *Fragaria*. HortScience 23:889–891

Hancock JF, Callow PW, Serçe S, Phan PQ (2003) Variation in the horticultural characteristics of native *Fragaria virginiana* and *F. chiloensis* from North and South America. J Am Soc Hortic Sci 128:201–208

Hancock JF, Finn C, Hokanson S, Luby JJ, Goulart B, Demchak K, Callow P, Serce S, Schilder A, Hummer K (2001) A multistate comparison of native octoploid strawberries from North and South America. J Am Soc Hortic Sci 126:579–586

Hancock JF, Flore JA, Galletta GJ (1989) Variation in leaf photosynthetic rates and yield in strawberries. J Soc Hortic Sci 64:449–454

Hancock JF, Maas JL, Shanks CH, Breen PJ, Luby JJ (1990) Strawberries (*Fragaria* ssp). In: Moore J, Ballington J (eds) Genetic Resources in Temperate Fruit and Nut Crops International Society of Horticultural Sciences, Wageningen, The Netherlands, pp 489–546

Hancock JF, Sakin M, Luby JJ, Dale A, Darnell D (1996) Germplasm resources in octoploid strawberries: Potential sources of genes to increase yield in northern climates. In: Proceedings of the IV North American Strawberry Conference, Gainesville, FL

Hancock JF, Scott DH, Lawrence FJ (1996b) Strawberries. In: Janick J, Moore JN (eds) Fruit breeding. vol II. vine and small fruits. John Wiley and Sons, NewYork, pp 419–470

Handley DT, Dill JF, Pollard JE (1991) Field susceptibility of twenty strawberry cultivars to tarnished plant bug injury. Fruit Varieties J 45:166–169

Hansche PE, Bringhurst RS, Voth V (1968) Estimates of genetic and environmental parameters in the strawberry. Proc Am Soc Hortic Sci 92:338–345

Hansen E, Waldo GF (1944) Ascorbic acid content of small fruits in relation to genetic and environmental factors. Food Res 9:453–461

Harland SC, King E (1957) Inheritance of mildew resistance in *Fragaria* with special reference to cytoplasmic effects. Heredity 11:257

Harrison EP, McQueen-Mason SJ, Manning K (2001) Expression of six expansin genes in relation to extension activity in developing strawberry fruit. J Exp Bot 52:1437–1446

Harrison RE, Luby JJ, Furnier GR (1997) Chloroplast DNA restriction fragment variation among strawberry (*Fragaria* spp.) taxa. J Am Soc Hortic Sci 122:63–68

Haymes K, Davis TM (1998) Agrobacterium–mediated transformation of 'Alpine' *Fragaria vesca*, and transmission of transgenes to R1 progeny. Plt Cell Rep 17:279–283

Haymes KM, Henken B, Davis TM, van de Weg ME (1997) Identification of RAPD markers linked to a *Phytophthora fragariae* resistance gene (*Rpf1*) in the cultivated strawberry. Theor Appl Genet 94:1097–1101

Hedrick UP (1925) The Small Fruits of New York. J.B. Lyon Company, Printers, Albany, NY

Hildebrand PD, Braun PG, Renderos WE, Jamieson AR, McRae KB, Binn M (2005) A quantitative method for inoculating strawberry leaves with *Xanthomonas fragariae*, factors affecting infection, and cultivar reactions. Can J Plant Pathol 27:16–24

Hoffman T, Kalinowski G, Schwab W (2006) RNAi-induced silencing of gene expression in strawberry fruit (*Fragaria ×ananassa*) by agroinfiltration: a rapid assay for gene function analysis. Plant J 48:818–826

Hokanson S, Maas JL (2001) Strawberry biotechnology, vol. 21. John Wiley & Sons, NewYork

Houde M, Dallaire S, N'Dong D, Sarhan F (2004) Overexpression of the acidic dehydrin WCOR410 improves freezing tolerance in transgenic strawberry leaves. Plant Biotechnol J 2:381–387

Hortynski J (1989) Genotype-environmental interaction in strawberry breeding. Acta Hortic 265:175–180

Howard CM, Maas JL, Chandler CK, Albregts EE (1992) Anthracnose of strawberry caused by the *Colletotrichum* complex in Florida. Plant Dis 76:976–981

Hsu CS, Watkins R, Bolton AT, Spangelo LPS (1969) Inheritance of resistance to powdery mildew in the cultivated strawberry. Can J Genet Cytol 11:426–438

Hummer K (1995) What's new in strawberry genetic resources: raw materials for a better berry. In: Pritts MP, Chandler CK, Crocker TE (eds) Proceedings of the IV North American Strawberry Conference, University of Florida, Orlando, pp 79–86

Iannetta P, Escobar N, Roos H, Souleyre E, Hancock R, Witte C, Davis H (2004) Identification, cloning and expression analysis of strawberry (Fragaria ×ananassa) mitochondrial citrate synthase and mitochondrial malate dehydrogenase. Physiol Plant 121:15–26

Jiménez-Bermúdez S, Redondo-Nevado J, Muñoz-Blanco J, Caballero JL, López-Aranda JM, Valpuesta V, Pliego-Alfaro F, Quesada MA, Mercado JA (2002) Manipulation of strawberry fruit softening by antisense expression of a pectate lyase gene. Plant Physiol 128: 751–759

Jung S, Jesudurai C, Staton M, Du Z, Ficklin S, Cho I, Abbott A, Tomkins J, Main D (2004) GDR (Genome Database for Rosaceae): Integrated web resources for Rosaceae genomics and genetics research. BMC Bioinformatics 5:130–138

Khanizadeh S, Bélanger A (1997) Classification of 92 strawberry genotypes based on their leaf essential oil composition. Acta Hortic 439:205–210

Kim CH, Seo HD, Cho WD, Kim SB (1982) Studies on varietal resistance and chemical control to the wilt of strawberry caused by Fusarium oxysporum. Korean J Plant Prot 21:61–67

Lal D, Seth JN (1979) Studies on genetic variability in strawberry (Fragaria ×ananassa Duch.). Progressive Hortic 11: 49–53

Lal D, Seth JN (1981) Studies on combining ability in strawberry (Fragaria ×ananassa): 1. Number of inflorescences, number of flowers, days to maturity and number of fruits. Can J Genet Cytol 23: 373–378

LaMondia JA (2004) Field performance of twenty-one strawberry cultivars in a black root rot-infested site. J Am Pom Soc 58:226–232

Larson KD, Shaw DV (1995a) Relative performance of strawberry genotypes on fumigated and nonfumigated soil. J Am Soc Hortic Sci 120:274–277

Larson KD, Shaw DV (1995b) Strawberry nursery soil fumigation and runner plant production. HortScience 30:236–237

Lerceteau-Köhler E, Guerin G, Denoyes-Rothan B (2005) Identification of SCAR markers linked to Rca2 anthracnose resistance gene and their assessment in strawberry germplasm. Theor Appl Genet 111:862–870

Lerceteau-Köhler E, Guérin G, Laigret F, Denoyes-Rothan B (2003) Characterization of mixed disomic and polysomic inheritance in the octoploid strawberry (Fragaria ×ananassa) using AFLP mapping. Theor Appl Genet 107:619–628

Lerceteau-Köhler E, Moing A, Guerin G, Renaud C, Courlit S, Camy D, Praud K, Parisy V, Bellec F, Maucourt M, Rolin D, Roudeillac P, Denoyes-Rothan B (2004) QTL analysis for fruit quality traits in octoploid strawberry (Fragaria × ananassa). Acta Hortic 663:331–335

Lewers KS, Maas JL, Hokanson SC, Gouin C, Hartung JS (2003) Inheritance of resistance in strawberry to bacterial angular leafspot disease caused by Xanthomonas fragariae. J Am Soc Hortic Sci 128:209–212

Luby JJ, Stahler MM (1993) Collection and evaluation of Fragaria virginiana in North America. Acta Hortic 345:49–54

Lundergan CA, Moore JN (1975) Inheritance of ascorbic acid content and color intensity in fruits of strawberry (Fragaria ×ananassa Duch.). J Am Soc Hortic Sci 100:633–635

Maas JL (1998) Compendium of stawberry diseases, 2nd edn. APS Press, Beltsville, Maryland

Maas JL, Galletta GJ (1989) Germplasm evaluation for resistance to fungus-incited diseases. Acta Hortic 265:461–472

Maas JL, Galletta GJ, Draper AD (1989) Resistance in strawberry to races of Phytophthora fragariae and to isolates of Verticillium from North America. Acta Hortic 265:521–526

Maas JL, Smith WL (1978) 'Earliglow', a possible source of resistance to Botrytis fruit rot in strawberry. HortScience 13:275–276

MacIntyre D, Gooding HJ (1978) The assessment of strawberries for decapping by machine. Hortic Res 18:127–137

MacKenzie SJ, Legard DE, Timmer LW, Chandler CK, Peres NA (2006) Resistance of strawberry cultivars to crown rot caused by *Colletrotrichum gloeosporioides* isolates from Florida is nonspecific. Plant Dis 90:1091–1097

MacLachlan JB (1974) The inheritance of colour of fruit and the assessment of plants as sources of colour in the cultivated strawberry. Hortic Res 14:29–39

Manning K (1998) Genes for fruit quality in strawberry. In: Cockshull KE, Gray D, Seymour GB, Thomas B (eds) Genetic and environmental manipulation of horticultural crops, vol. 51–61. CAB International, Wallingford, UK

Marini RP, Boyce BR (1979) Influence of low temperatures during dormancy on growth and development of 'Catskill' strawberry plants. J Am Soc Hortic Sci 104:159–162

Mason DT, Rath N (1980) The relative importance of some yield components in East of Scotland strawberry plantations. Ann Appl Bio 95:399–408

Mathews H, Wagoner W, Kellogg J, Bestwick R (1995) Genetic transformation of strawberry: Stable integration of a gene to control the biosynthesis of ethylene. In Vitro Cell Dev Biol 31:36–43

Medina JL, Moore PP, Shanks CH, Gil FF, Chandler CK (1999) Genotype × environment interaction for resistance to spider mites in *Fragaria*. J Am Soc Hortic Sci 124:353–357

Medina-Escobar N, Cárdenas J, Muñoz-Blanco J, Caballero JL (1998) Cloning and molecular characterization of a strawberry fruit ripening-related cDNA corresponding a mRNA for a low-molecular-weight-heat-shock protein. Plt Mol Biol 36:33–42

Mehli L, Kjellsen TD, Dewey FM, Hietala AM (2005) A case study from the interaction of strawberry and *Botrytis cinerea* highlights the benefits of comonitoring both partners at genomic and mRNA level. New Phyt 168:465–474

Mehli L, Schaart JG, Kjellsen TD, Tran DH, Salentijn EMJ, Schouten HJ, Iversen T-H (2004) A gene encoding a polygalacturonase-inhibiting protein (PGIP) shows developmental regulation and pathogen-induced expression in strawberry. New Phyt 163:99–110

Melville AH, Draper AD, Galletta GJ (1980a) Transmission of red stele resistance by inbred strawberry selections. J Am Soc Hortic Sci 105:608–610

Melville AH, Galletta GJ, Draper AD, Ng TJ (1980b) Seed germination and early seedling vigor in progenies of inbred strawberry selections. HortScience 15:49–750

Mezzetti B, Landi L, Pandolfini T, Spena A (2004) The defH9-iaaM auxin-synthesizing gene increases plant fecundity and fruit production in strawberry and raspberry. BMC Biotechnol 4:1–10

Mochizuki T (1995) Past and present strawberry breeding programs in Japan. Advances in Strawberry Research 14:9–17

Moore JN, Brown GR, Bowen HL (1975) Evaluation of strawberry clones for adaptability to once-over mechanical harvest. HortScience 10:407–408

Moore JN, Brown GR, Brown ED (1970) Comparison of factors influencing fruit size in large-fruited and small-fruited clones of strawberry. J Am Soc Hortic Sci 95:827–831

Morgan A, Baker CM, Chu JSF, Lee K, Crandell BA (2002) Production of herbicide tolerant strawberry through genetic engineering. Acta Hortic 567:113–115

Mori T, Kitamura H, Kuroda K (2005) Varietal differences in Fusarium wilt-resistance in strawberry cultivars and the segregation of this trait in F_1 hybrids. J Jap Soc Hortic Sci 75:57–59

Morrow EB, Comstock RE, Kelleher T (1958) Genetic variances in strawberries. Proc Am Soc Hortic Sci 72:170–85

Morrow EB, Darrow GM (1952) Effects of limited inbreeding in strawberries. Proc Am Soc Hortic Sci 59:269–276

Murant AF, Lister RM (1987) European nepoviruses in strawberry. In: Converse RH (ed) Virus diseases of small fruits. USDA/ARS, Washington, DC

Murawski H (1968) Studies on heritability in strawberry varieties. Height of inflorescences, mildew resistance, fruit colour, flesh colour and the shape of the berries. Arch Gartenb 16:293–318

NDong C, Quellet F, Houde M, Sarhan F (1997) Gene expression during cold acclimation in strawberry. Plt Cell Physiol 38:863–870

Nelson MD, Gubler WD, Shaw DV (1996) Relative resistance of 47 strawberry cultivars to powdery mildew in California greenhouse and field environments. Plant Dis 80:326–328

Nemec S (1971) Studies on resistance of strawberry varieties and selections to *Mycosphaerella fragariae* in southern Illinois. Plant Dis Rep 55:573–576

Nemec S, Blake RC (1971) Reaction of strawberry cultivars and their progenies to leaf scorch in southern Illinois. HortScience 6:497–498

Nicoll MF, Galletta GJ (1987) Variation in growth and flowering habits of Junebearing and everbearing strawberries. J Am Soc Hortic Sci 112:872–880

Olías R, Pérez AG, Sanz C (2002) Catalytic properties of alcohol acyltransferase in different strawberry species and cultivars. Agric Food Chem 50:4031–4036

Ourecky DK, Bourne MC (1968) Breeding and Instron evaluation of strawberry firmness. HortScience 3:92–93

Ourecky DK, Reich JE (1976) Frost tolerance in strawberry cultivars. HortScience 11:413–414

Ourecky DK, Slate GL (1967) Behavior of the everbearing characteristics in strawberries. Proc Am Soc Hortic Sci 91:236–248

Owens CL, Thomashow MF, Hancock JF, Iezzoni AF (2002) CBF1 orthologs in sour cherry and strawberry and the heterologous expression of CBF1 in strawberry. J Am Soc Hortic Sci 127:489–494

Oydvin J (1980) Records of two-spotted spider mite *Tetranychus urticae* Koch, strawberry mite *Steneotarsonemus pallidus* Banks and strawberry mildew *Sphaerotheca macularis* (Wallr.) Magn. in a progeny test of strawberries, 1976–77. Forskning og Forsok i Landbruket 31:1–9

Particka C, Hancock JF (2005) Field evaluation of strawberry genotypes for tolerance to black root rot on fumigated and nonfumigated soil. J Am Soc Hortic Sci 130:688–693

Passey AJ, Barrett KJ, James DJ (2003) Adventitious shoot regeneration from seven commercial strawberry cultivars (*Fragaria* ×*ananassa* Duch.) using a range of explant types. Plant Cell Rep 21:397–401

Popova IV, Konstantinova AE, Zekalashvili AU, Zhananov BK (1985) Features of breeding strawberries for resistance to berry molds. Sov Agric Sci 3:29–33

Potter JW, Dale A (1994) Wild and cultivated strawberries can tolerate or resist root-lesion nematode. HortScience 29:1074–1077

Potter D, Luby JJ, Harrison RE (2000) Phylogenetic relationships among species of *Fragaria* (Rosaceae) inferred from non-coding nuclear and chloroplast DNA sequences. Syst Bot 25:337–348

Powers L (1945) Strawberry breeding studies involving crosses between the cultivated varieties (*Fragaria* ×*ananassa*) and the native Rocky Mountain strawberry (*F. ovalis*). J Agric Res 70:95–122

Powers L (1954) Inheritance of period of blooming in progenies of strawberries. Proc Am Soc Hortic Sci 64:293–298

Reddy ASN, Jena PK, Mukherjee SK, Poovaiah BW (1990) Molecular cloning of cDNAs for auxin-induced mRNAs and developmental expression of the auxin-inducible genes. Plant Mol Biol 14:643–653

Reddy ASN, Poovaiah BW (1990) Molecular cloning and sequencing of a cDNA for an auxin-repressed mRNA: correlation between fruit growth and repression of the auxin-regulated gene. Plant Mol Biol 14:127–136

Rosin FM, Aharoni A, Salentijn EMJ, Schaart JG, Boone MJ, Hannapel DJ (2003) Expression patterns of a putative homolog of *AGAMOUS*, *STAG1*, from strawberry. Plant Sci 165:959–968

Salentijn EMJ, Aharoni A, Schaart JG, Boone MJ, Krens FA (2003) Differential gene expression analysis of strawberry cultivars that differ in fruit firmness. Physiol Plant 118:571–578

Sangiacomo MA, Sullivan JA (1994) Introgression of wild species into the cultivated strawberry using synthetic octoploids. Theor Appl Genet 88:349–354

Sargent DJ, Davis TM, Tobutt KR, Wilkinson MJ, Battey NH, Simpson DW (2004a) A genetic linkage map of microsatellite, gene-specific and morphological markers in diploid *Fragaria*. Theor Appl Genet 109:1385–1391

Sargent DJ, Geibel M, Hawkins JA, Wilkinson MJ, Battey NH, Simpson DW (2004b) Quantitative and qualitative differences in morphological traits revealed between diploid *Fragaria* species. Ann Bot 94:787–769

Schaart JG, Krens FA, Pelgrom KTP, Mendes O, Rouwendal JA (2004) Effective production of marker-free transgenic strawberry plants using inducible site-specific recombination and a bifunctional selectable marker gene. Plant Biotechnol 2:233–240

Schaart JG, Mehli L, Schouten HJ (2005) Quantification of allele-specific expression of a gene encoding strawberry polygalacturonase-inhibiting protein (PGIP) using Pyrosequencing. Plant J 41:493–500

Schaart JG, Salentijn EMJ, Krens FA (2002) Tissue-specific expression of the β-glucuronidase reporter gene in transgenic strawberry (*Fragaria ×ananassa*) plants. Plant Cell Rep 21:313–319

Schaefers GA (1980) Yield effects of tarnished plant bug feeding on June-bearing strawberry varieties in New York State. J Econ Entomol 73:721–725

Schuster DJ, Price JF, Martin FG, Howard CM, Albregts EE (1980) Tolerance of strawberry cultivars to twospotted spider mites in Florida. J Econ Entomol 73:52–54

Scott DH, Draper AD (1970) A further note on the longevity of strawberry seed in cold storage. HortScience 5:439

Scott DH, Draper AD, Greeley LW (1972) Interspecific hybridization in octoploid strawberries. HortScience 7:382–384

Scott DH, Lawrence FJ (1975) Strawberries. In: Janick J, Moore JN (eds) Advances in fruit breeding. Purdue University Press, pp 71–79

Serce S, Hancock JF (2005) Inheritance of day-neutrality in octoploid species of *Fragaria*. J Am Soc Hortic Sci 130:580–584

Shanks CH, Barritt BH (1974) *Fragaria chiloensis* clones resistant to the strawberry aphid. HortScience 9:202–203

Shanks CH, Chandler CK, Show ED, Moore PP (1995) *Fragaria* resistance to spider mites at three locations in the United States. HortScience 30:1068–1069

Shanks CH, Moore PP (1995) Resistance to two-spotted spider mite and strawberry aphid in *Fragaria chiloensis*, *F. virginiana*, and *F. ×ananassa* clones. HortScience 30: 596–599

Shaw DV (1988) Genotypic variation and genotypic correlations for sugars and organic acids of strawberries. J Am Soc Hortic Sci 113:770–774

Shaw DV (1989) Variation among heritability estimates for strawberries obtained by offspring-parent regressions with relatives raised in separate environments. Euphytica 44:157–162

Shaw DV (1991) Recent advances in the genetics of strawberries. In: Dale A, Luby JJ (eds) The Strawberry into the 21st Century Timber Press, Portland, Oregon, pp 76–83

Shaw DV, Bringhurst RS, Voth V (1987) Genetic variation for quality traits in an advanced-cycle breeding population of strawberries. J Am Soc Hortic Sci 112:699–702

Shaw DV, Bringhurst RS, Voth V (1988) Quantitative genetic variation for resistance to leaf spot (*Ramularia tulasnei*) in California strawberries. J Am Soc Hortic Sci 113:451–456

Shaw DV, Bringhurst RS, Voth V (1989) Genetic parameters estimated for an advanced-cycle strawberry breeding population at two locations. J Am Soc Hortic Sci 114:823–827

Shaw DV, Famula T (2005) Complex segregation analysis of day-neutrality in domestic strawberry (*Fragaria ×ananassa* Duch.). Euphytica 145:331–338

Shaw DV, Gordon TR (2003) Genetic response for reaction to *Verticillium* wilt in strawberry with two stage family and genotypic selection. HortScience 38:432–434

Shaw DV, Sacks EJ (1995) Response in genotypic and breeding value to a single generation of divergent selection for fresh fruit color in strawberry. J Am Soc Hortic Sci 120:270–273

Sherman WB, Janick J, Erickson HT (1966) Inheritance of fruit size in strawberry. Proc Am Soc Hortic Sci 89:309–17

Simpson DW (1987) The inheritance of mildew resistance in everbearing and day-neutral strawberry seedlings. J Hortic Sci 62:329–334

Simpson DW, Easterbrook MA, Bell JA (2002) The inheritance of resistance to the blossom weevil, *Anthonomus rubi*, in the cultivated strawberry, *Fragaria* ×*ananassa*. Plant Breed 121: 72–75

Simpson DW, Sharp RS (1988) The inheritance of fruit yield and stolon production in everbearing strawberries. Euphytica 38:65–74

Sjulin TM (2006) Private strawberry breeders in California. HortScience 41:17–19

Sjulin TM, Dale A (1987) Genetic diversity of North American strawberry cultivars. J Am Soc Hortic Sci 112:375–385

Sjulin TM, Robbins J, Barritt BH (1986) Selection for virus tolerance in strawberry. J Am Soc Hortic Sci 111:458–464

Smith BJ, Black LL (1987) Resistance of strawberry plants to *Colletotrichum fragariae* affected by environmental conditions. Plant Dis 71:834–837

Smith BJ, Black LL (1990) Morphological, cultural and pathogenic variation among *Colletotrichum* species isolated from strawberry. Plant Dis 74:69–76

Spangelo LPS, Hsu CS, Fejer SO, Bedard PR, Rouselle GL (1971) Heritability and genetic variance components for 20 fruit and plant characters in the cultivated strawberry. Can J Genet Cytol 13:443–456

Stahler MM, Ascher PD, Luby JJ, Roelfs AP (1995) Sexual composition of populations of *Fragaria virginiana* (Rosaceae) collected from Minnesota and western Wisconsin. Can J Bot 73:1457–1463

Stahler MM, Luby JJ, Ascher PD (1990) Comparative yield of female and hermaphroditic *Fragaria virginiana* germplasm collected in Minnesota and Wisconsin. In: Dale A, Luby JJ (eds) The Strawberry into the 21st Century Timber Press Portland, Oregon pp. 104–105

Staudt G (1959) Cytotaxonomy and phylogenetic relationships in the genus *Fragaria*. IX Int Bot Congr Proc 2:377

Staudt G (1967) The genetics and evolution of heterosis in the genus *Fragaria* II.. Species hybridization of *F. vesca* ×*F. orientalis* and *F. viridis* ×*F. orientalis* (in German). Z Pflanzenz 58:309–322

Staudt G (1984) Cytological evidence of double restitution in *Fragaria*. Plant Sys Evol 146: 171–179

Staudt G (1989) The species of *Fragaria*, their taxonomy and geographical distribution. Acta Hortic 265:23–33

Staudt G (1999) Systematics and geographic distribution of the American strawberry species: taxonomic studies in the genus *Fragaria* (Rosaceae: Potentilleae). Univ Calif Publ Bot 81:1–162

Szczygiel A (1981a) Trials on susceptibility of strawberry cultivars to the needle nematode, *Longidorus elongatus*. Fruit Sci Rep 8:127–131

Szczygiel A (1981b) Trials on susceptibility of strawberry cultivars to the northern root-knot nematode, *Meloidogyne hapla*. Fruit Sci Rep 8:115–119

Szczygiel A (1981c) Trials on suceptibility of strawberry cultivars to the root lesion nematode, *Pratylenchus penetrans*. Fruit Sci Rep 8:121–125

Szczygiel A, Danek J (1974) Pathogenicity of three species of root parasitic nematodes to strawberry plants as related to methods of inoculation. Zeszyty Problemowe Postepow Nauk Rolniczych 154:133–149

Takahashi H (1993) Breeding of strawberry cultivars resistant to Alternaria black spot of strawberry (*Alternaria alternata* strawberry pathotype). Bull Akita Prefectural Coll Agric 19: 1–44

Takahashi H, Yoshida Y, Kanda H, Furuya H, Matsumoto T (2003) Breeding of Fusarium wilt-resistant strawberry cultivar suitable for field culture in Northern Japan. Acta Hortic 626:113–118

Tingey WM, Pillemer EA (1977) Lygus bugs: crop resistance and physiological nature of feeding injury. Bull Entomol Soc Am 23:277–287

Trainotti L, Ferrarese L, Vecchia F dalla, Rascio N, Casadoro G (1999) Two different endo-β-1,4-glucanases contribute to the softening of strawberry fruits. J Plant Physiol 154:355–362

Trainotti L, Spinello R, Piovan A, Spolaore S, Casadoro G (2001) beta – Galactosidases with a lectin-like domain are expressed in strawberry. J Exp Bot 52:1635–1645

Van de Weg WE (1997) A gene-for-gene model to explain interactions between cultivars of strawberry and races of *Phytophthora fragariae* var *fragariae*. Theor Appl Genet 94:445–451

Van der Scheer HAT (1973) Susceptibility of strawberry to isolates of *Phytophthora cactorum* and *Phytophthora citricola*. Mededelingen van de Faculteit Landbouwwetenschappen, Rijksuniversiteit Gent 38:1407–1415

Viruel MA, Sánchez D, Arús P (2002) An SSR and RFLP linkage map for the octoploid strawberry (*Fragaria* ×*ananassa*) In: Plant & Animal Genome x Conference, San Diego, CA

Wang J, Ge H, Peng S, Zhang H, Chen P, Xu J (2004) Transformation of strawberry (*Fragaria* ×*ananassa* Duch.) with late embryogenesis abundant protein gene. J Hortic Sci Biotechnol 79:735–738

Watkins R, Spangelo LPS, Bolton AT (1970) Genetic variance components in cultivated strawberry. Can J Genet Cytol 12:52–59

Weebadde C, Wang D, Finn CE, Lewers KS, Luby JJ, Bushakra J, Sjulin TM, Hancock JF (2007) Using a linkage mapping approach to identify QTL for day-neutrality in the octoploid strawberry. Plant Breed (In press)

Wenzel WG (1980) Correlation and selection index components. Canadian Journal of Genetics and Cytology 13, 42–50

Wilhelm S, Sagen JA (1974) A History of the Strawberry. University of California Division of Agriculture Publication 4031, Berkeley

Wilkinson JQ, Lanahan MB, Conner TW, Klee HJ (1995) Identification of mRNAs with enhanced expression in ripening strawberry fruit using polymerase chain reaction differential display. Plant Mol Biol 27:1097–1108

Wilson WF Jr, Giamalva MJ (1954) J. Days from bloom to harvest of Louisiana strawberries. Proc Am Soc Hortic Sci 63:201–204

Wing KB, Pritts MP, Wilcox WF (1995) Field resistance of 20 strawberry cultivars to black root rot. Fruit Varieties J 49:94–98

Woolley LC, James DJ, Manning K (2001) Purification and properties of an *endo*-β-(1,4)-glucanase from strawberry and down-regulation of the corresponding gene, *cell*. Plant 214:11–21

Xue SM, Bors RH, Strelkov SE (2005) Resistance sources to *Xanthomonas fragariae* in non-octoploid strawberry species. HortScience 40:1653–1656

Yamamoto M, Namiki F, Nishimura F, Kohmoto K (1985) Studies on host-specific AF-toxins produced by *Alternaria alternata* strawberry pathotype causing Alternaria black spot of strawberry (3). Use of toxin for determining inheritance of disease reaction in strawberry cultivar Morioka-16. Ann Phytopathol Soc Jpn 51:530–535

Yu H, Davis TM (1995) Genetic linkage between runnering and phosphoglucoisomerase allozymes, and systematic distortion of monogenic segregation ratios in diploid strawberry. J Am Soc Hortic Sci 120:687–690

Yubero-Serrano EM, Moyano E, Medina-Escobar N, Munoz-Blanco J, Caballero JL (2003) Identification of a strawberry gene encoding a non-specific lipid transfer protein that responds to ABA, wounding and cold stress. J Exp Bot 54:1865–1877

Zimmerman RH (1991) Micropropagation of temperate zone fruit and nut crops. In: Debergh PC, Zimmerman RH (eds) Micropropagation: Technology and application. Kluer Academic, Dordrecht, The Netherlands, pp 231–264

Zubov AA, Stankevich KV (1982) Combining ability of a group of strawberry varieties for fruit quality characters (In Russian). Sov Genet 18:984–992

Zurawicz E, Stushnoff C (1977) Influence of nutrition on cold tolerance of 'Redcoat' strawberries. J Am Soc Hortic Sci 102:342–346

Zych CC (1966) Fruit maturation times of strawberry varieties. Fruit Varieties Hortic Dig 20:51–53

Chapter 14
Intellectual Property Rights for Fruit Crops

J.R. Clark and R.J. Jondle

Abstract The intellectual property protection of fruit cultivars has escalated in recent years and this trend will continue using various protection options, primarily plant patents (U.S.A.), plant breeders rights (worldwide), and trademarks, with minimal use of utility patents or plant variety protection (U.S.A.). Contracts and licensing are an integral part of the protection process, whereby the terms of the assignment of rights for propagation or other use of the protected cultivar are defined between the owner of the rights and licensee. In the area of testing, material transfer agreements are very important in sharing of germplasm for testing and the terms of these are extremely critical in defining the use of the test material. A potential area of expansion in fruit breeding is in formal breeding agreements, in which germplasm is shared. These agreements can be executed between public and private entities or within public institutions. Expansion of the use of these various protection, testing and breeding options will occur, allowing increased monetary income to breeding programs while increasing the complexity of commercialization and use of fruit cultivars.

14.1 Introduction

Parallel to genetic advances in fruit crop improvement in recent years has been a great expansion in the use of intellectual property rights (IPR) for protection of cultivars and other genetic advances. Intellectual property rights includes various protections for cultivars, cultivar names, genes, breeding processes, and many other inventions including protection types such as plant patents, utility patents, plant breeders rights, trademarks, and other legal designations. A number of factors have contributed to the increase in IPR in fruit crops, but the primary reason has been

J.R. Clark
Department of Horticulture, 316 Plant Science, University of Arkansas, Fayetteville, Arkansas 72701, USA
e-mail: jrclark@uark.edu

money. Protection provides legal restrictions for use of new developments without the permission of the holder of the rights of the invention and provides an avenue for royalties or other fees to be collected for the use of the invention. Use of IPR is now widely carried out by private and public fruit breeding programs worldwide. The incentive of potential IPR income has sustained many programs in recent years and likely has provided for expanded advances in many of our fruit species.

The following discussion is provided to inform those interested in IPR for fruit crops of the major forms of protection and associated issues such as licensing and germplasm sharing. It is not intended that the information be used as legal advice for IPR protection; rather appropriate legal counsel should be consulted for more detailed information and procedures.

14.2 Protection Options

14.2.1 Plant Patents

The plant patent is an option for protection in the United States, and is the most common choice for fruit cultivars and plants that can be asexually reproduced (and that cannot be reproduced by seed). The Townsend-Purnell Plant Patent Act (Title 35 of the United States Code, Chapter 15) was passed by the U.S. Congress on 13 May 1930. This was the first act in the world that granted patent rights to plant breeders. The intent of this act was to afford agriculture the same IPR rights as provided for industrial inventions. The Act stated 'Whoever invents or discovers and asexually reproduces any distinct and new variety of plant, including cultivated sports, mutants, hybrids, and newly found seedlings, other than a tuber propagated plant or a plant found in an uncultivated state, may obtain a patent therefore, subject to the conditions and requirements of this title' (Chapter 15, 35 U.S.C. 161 Patents for Plants). Upon amendment over the years the protection further stated 'In the case of a plant patent, the grant shall include the right to exclude others from asexually reproducing the plant, and from using, offering for sale, or selling the plant so re-produced, or any of its parts, throughout the United States, or from importing the plant so reproduced, or any parts thereof, into the United States' (1998 Amendment to Chapter 15, 35 U.S.C. 163 Grant). This type of protection therefore is specific for asexually propagated plants and is not allowed for seed-propagated crops that cannot be asexually reproduced.

A noteworthy amendment to the Plant Patent Act was enacted by Congress in 1998 that provided provisions restricting the importation of plant parts into the United States. The primary plant parts one would consider important for fruit crops would be the fruit, while key parts of other types of crops could include flowers, leaves or other items of commerce. Very important to recognize is that if fruits or other plant parts are produced outside the United States, the plant patent law comes into affect at the border when the item enters the country. Finally, some have suggested that 'plant parts' includes gametes and that this protection infers a breeding restriction on plant patented plants within the United States. This issue has not been legally resolved by any court decision as of this writing.

The first plant patent was issued 18 August 1931 for a rose. The first fruit plant to attain a plant patent was 'Thornless Young Dewberry' issued 20 October 1931 to Elmer L. Pollard and Jubal E. Sherrill of Chino, California. Five plant patents were issued in 1931. Plant patents issued increased substantially over the years, with the highest number achieved in 2006 (Table 14.1). Fruit plant patents issued made a major jump between 1980 and 1990 and were between 50 and 100 in most years from 1990 to 2006. From 2000 through 2006, fruit patents ranged from 7% to 13% of all plant patents (U.S. Patent and Trademark records used for this compilation).

For a plant patent to be considered for granting for an invention, the key items of patentability must be met including:

- Novelty – the invention must be new in some manner,
- Utility – must be useful, and
- Non-obvious – must be distinct from other related known cultivars.

Complete guidelines for filing a plant patent can be attained from the United States Patent and Trademark Office (www.uspto.gov). Legal assistance is commonly utilized for plant patents, although individuals can complete the process successfully.

When considering plant patents as a form of protection, several major aspects should be considered. Major characteristics of a plant patent include:

- Protection is granted for 20 years from the date of filing the application (patents that issued prior to 8 June 1995 or that were filed prior to 8 June 1995 and issued after that date have protection for 17 years or 20 years from date of filing the application, whichever is greater),

Table 14.1 Plant patents issued from 1931, the starting year of each decade, and 2001 through 2006

Year	Plant patents issued	Plant patent numbers	Fruit patents issued
1931	5	1–5	1
1940	85	352–436	7
1950	89	911–1,000	13
1960	116	1,983–2,008	24
1970	52	2,959–3,010	17
1980	117	4,491–4,607	17
1990	318	7,089–7,407	54
2000	551	11,169–11,719	71
2001	585	11,720–12,304	43
2002	1,134	12,305–13,439	88
2003	994	13,440–14,433	100
2004	1,019	14,434–15,453	100
2005	716	15,454–16,170	57
2006	1,149	16,170–17,320	84

- Only one claim is allowed, for the cultivar,
- Can be applied for by any inventor, regardless of citizenship,
- Provides for protection *only* in the United States, and,
- Cost is less than most other forms of protection and the examination procedure is simpler than utility patents

Of critical importance to a plant breeder is developing a timeline for collection of the botanical description of the proposed invention, release of the cultivar, and filing for protection. There is no specific list of botanical characteristics for each species required by the U.S. Patent and Trademark Office for a plant patent application. However, a review of the characteristics included in the description of a recent plant patent issued for the species provides an excellent guideline for data collection. One should consider the time required for this data collection (an entire season may be required for descriptive data to be attained, including observations of canes or dormant branches or buds during dormancy, budbreak and bloom characters in the springtime, fruit, shoot, and leaf development in early to mid-summer, fruit characters at ripening, and possibly late-season to beginning dormancy observations). A template form of characters to be measured or observed is commonly used to assist in keeping track of data collection. Issues such as technical labor available to collect data, potential environmental impacts on plant characteristics (crop loss or damage due to winter injury, frost, hail etc.), in addition to the timeline for preparation and filing of the plant patent application in conjunction with public release and plant sale should be carefully considered.

For a plant to be eligible for obtaining a valid plant patent, the invention must not have been sold, publicly available, or offered for sale in the U.S.A. more than one year prior to the filing date of the application. Likewise, sale, public availability, and public description in any other country in the world more than one year prior to filing the application can prevent a plant patent from being granted. The issue of public description is of increasing importance to fruit and flower breeders in that many are reluctant to share information on new developments with growers and others interested in their program accomplishments. It is suggested that if plant material is shared on new advances that it be accompanied by the statement 'breeding selection not available and not for sale.' It is also suggested that if any plant material is made available for testing or for increase, that it be accompanied by a testing agreement or a confidentiality agreement until the application for patent has been made.

14.2.2 Utility Patents

This type of protection is not as routinely used for fruit crops as plant patents in the U.S.A. The protection is much more powerful however, and may be desired if an inventor wants more thorough protection of a new plant or associated development. Utility patents can have claims which protect plants, pollen, ovules genes, promoters, selectable markers, DNA, expressed sequence tags (ESTs), quantitative trait loci (QTL), proteins, methods, bioinformatics, genomes, proteomics, and software, etc. Examples include new traits (such as disease resistance or male sterility genes), increased level of a trait (higher nutraceutical, oil, or fatty acid concentration), mutants for any trait, an improved method, process or apparatus used in breeding or genetic advancement, or simply for a variety, inbred or hybrid.

Of primary importance to plant breeders is that a utility patent provides protection for the variety against use in further breeding or research without permission of the patent holder. It is advised that to ensure protection from breeding use that utility patents be considered for varieties that are anticipated to be of substantial value when used as parents in subsequent breeding.

Examples of utility patents involving fruit crops that are not solely variety protection include infra short-day strawberry types (U.S. Patent No. 5,444,179), pathogen-resistant grape plants (U.S. Patent No. 6,995,015), regeneration system for grape and uses thereof (U.S. Patent No. 6,455,312), vitro propagation of grape via leaf disk culture (U.S. Patent No. 4,931,394), rapid recovery of shoots through thin stem slices after preconditioning of micropropagated fruit tree shoots (U.S. Patent No. 6,127,182), and method of producing vole-resistant apple trees and trees produced thereby (U.S. Patent No. 4,516,353).

Utility patents provide for the same term as plant patents (currently 20 years from date of application) and have the same requirements for filing within one year of any public disclosure or sale of the invention. The claims for a utility patent can be multiple, compared to the single claim of the variety only for a plant patent. A patent attorney is required for a utility patent filing since there are substantially more specifications and details for this type of protection compared to a plant patent. Costs for utility patents can be substantial and this might be examined closely when decisions are made to file a plant vs. utility (or both) patent application. Also, provisional utility patent applications can be filed which provide a lower-cost first patent filing. With provisional applications, an applicant can claim the benefit of a provisional utility application in a corresponding non-provisional utility application filed not later than 12 months after the provisional application filing date. Information on provisional and non-provisional patent filings can be attained from the U.S. Patent and Trademark Office (www.uspto.gov).

14.2.3 Plant Variety Protection

This type of protection is provided for in the U.S.A. for seed propagated (sexually reproduced) crops and for tuber crops. Since fruit crops are normally clonally propagated, they would not usually be considered for protection using this form. If seeds are a common plant part used for propagation then Plant Variety Protection might be considered. An example would be a peach or other species rootstock, where seed propagation may be the method of propagation. This type of protection was attained for peach rootstock BY520-9 (PVP 9400013), and also a seed-propagated peach Truegold (PVP 200400055). For more details on this type of protection, see the website for the U.S. Plant Variety Protection office http://www.ams.usda.gov/Science/PVPO/PVPO_Act/PVPA.htm.

14.2.4 Plant Breeders Rights

The most common international protection offered for varieties is termed plant breeder's rights. The term 'plant breeder's rights' (PBR) is not used in the United States, but PBR is very similar to U.S. plant variety protection. The International Union for the Protection of New Varieties of Plants was established by the UPOV (Union internationale pour la protection des obtentions vègètales) Convention in 1961 with additional acts of the Union in subsequent years. UPOV was established to provide for and promote an international system of plant variety protection to encourage new breeding developments worldwide. As of January, 2007, there were 63 member countries. Both seed and clonally propagated crops are covered by UPOV plant breeder's rights.

Conditions for the grant of breeder's rights in the UPOV system has similarities to the American Plant Variety Protection Act in that a variety must be new, distinct, uniform and stable. The novelty requirement for breeder's rights requires that a variety must not have been sold or otherwise disposed of in the territory of the country concerned for more than one year prior to application for the right, or more than four years (or six years for trees and vines) in a country other than that of the member of the Union in which the application was filed.

The acts that the breeder's rights protect the variety for include:

- Propagation,
- Conditioning for the purpose of propagation,
- Offering for sale,
- Selling or other marketing,
- Exporting,
- Importing,
- Stocking for any of the purposes listed above.

For countries under the 1991 UPOV convention, the scope of protection extends to harvested material – thus only authorized propagation of the variety allows for fruit or other products to be marketed in the territory. Also, protection can be extended to products made directly from the harvested material. Further protection can be extended by member states of the Union. Finally, protection in countries under the 1991 UPOV convention extends to:

- Varieties that are essentially derived from the protected variety, where the protected variety is not itself a protected variety,
- Varieties that are not clearly distinguishable from the protected variety, and
- Varieties whose production requires the repeated use of the protected variety.

Plant breeder's rights do not restrict breeding activity with a protected variety, nor other experimental or private/non-commercial uses. The intention of the UPOV system is to ensure that germplasm sources such as protected varieties remain accessible by plant breeders. Plant breeder's rights include protection of the variety for not less than 18 years from the date of the grant, or 15 years for trees or vines and depends on which act of the UPOV convention a country has adhered.

A major component of UPOV is to provide for cooperation among UPOV member countries in the use and approval of variety names and examination of new varieties. The concept allows for one member to conduct examination of a variety for potential granting of breeders rights and another member can choose to accept the evaluations for its grant. The intention is that this system can reduce the cost of protection by members and protection might be obtained in several territories at lower costs. Each member defines the criteria for granting, whether this includes a technical description of the new variety, or actual growing of the plants for examination.

The European Union (E.U.) provides a good example of a system administered by a group of UPOV member states whereby plant breeders rights are granted. The 'Community Plant Variety Office' has been operating since 1995 and is based in Angers, France. This system allows for application for protection for numerous countries in one act. A common protocol is used including an application form, technical questionnaire, proposed variety name, and submission of photographs. Parties outside the E.U. must appoint a procedural representative residing in the E.U. to file for plant breeders rights. A major difference in filing for plant breeders rights in the E.U. compared to U.S. Plant Patents and U.S. PVP is that plant material is submitted by the breeder (rather than a complete botanical description as provided for U.S. Plant Patent and Plant Variety Protection consideration), and the variety is grown for examination in a selected site for that crop alongside other candidates and/or standard or established varieties in the E.U. Also, in addition to a filing or application fee, there are fees for examination for the growing period of the plant along with annual fees for continued protection after approval for as long as the protection is desired or allowed.

The E.U. provides protocols for distinctness, uniformity, and stability tests which provide details of plant health status of the material of the variety being submitted for protection along with guidelines for the botanical description. For example using blackberry, growth habit, number of new canes emerged, dormant cane length, dormant cane diameter, cane branch number, presence and density of spines, leaf characteristics, and several flower and fruit characteristics must be described along with comparison to known variety examples.

As more and more fruit genotypes are used on a worldwide basis, familiarity with the UPOV system is imperative to provide for widespread protection. Costs, timing, choosing of commercial cooperators, and targeted countries for filing must all be considered when planning for broad protection.

14.2.5 Trademarks

The increasing use of trademarks as a form of protection of commercialized names of fruit crops has been noteworthy in recent years. A trademark is a word, name, symbol, or other device which is used in trade with goods to indicate the source of the goods and to distinguish them from the goods of others. Trademark rights may be used to prevent others from using a confusingly similar mark, but not to prevent others from making the same goods or from selling the same goods or services under a completely different mark. Trademarks used in interstate or international commerce may be registered with the U.S. Patent and Trademark Office.

A trademark does not restrict the propagation, use of plant parts, or breeding use of a plant. Rather, in the U.S.A., a federally registered trademark provides for the exclusive right to use the mark nationwide; in this instance a plant name used for marketing for which a trademark has been claimed and/or registered. Why are trademarks of interest to fruit breeders?

- The life of a trademark is infinite if maintained and used in commerce (does not expire after 20 years such as patents),
- Can give a 'product line' or theme to a series of patented varieties,
- Provides restrictions in marketing and use of the mark by the propagator and or seller,
- Provides the right to file a lawsuit concerning the use of the mark in U.S. federal court,
- Is a basis to obtain trademark registration in other countries under the Madrid Protocol, and,
- Provides for the ability to file with the U.S. Customs Service to prevent importation of infringing foreign goods

Fruit varieties are increasingly being protected by both plant patents and trademarks in the U.S.A. The reasoning behind this practice is that a plant patent provides for the restriction of propagation or use of the variety for 20 years, during which a trademark name can be established for the variety with the intention that this trademarked name will be the readily recognizable variety connotation when the patent restriction expires. Trademarks can also be used for a group or series of varieties, with the same name used for more than one variety. This can provide for a protected name used for marketing a set of varieties that are very similar in phenotype and ripen over a range of dates during the growing and/or marketing season.

One of the more noteworthy uses of a trademark is that for Pink Lady® apple. The variety Cripps Pink was released by the Western Australia Department of Agriculture and patented in the United States (U.S. Plant Patent 7880). Pink Lady® has had great success as a 'managed' variety in that trademark protection is widely used even where restrictions on propagation may not be in place. Pink Lady® is a registered trademark of Apple and Pear Australia Limited and marketing in the United States is managed by Pink Lady® America, L.L.C. In the United States, only 'Cripps Pink' apples that meet the quality control requirements of Pink Lady® America can be sold under the Pink Lady® Brand. The Pink Lady® 'Flowing Heart™' logo is the brand used only with these apples meeting the quality standard.

The symbol™ indicates a claim to a mark and alerts the public to this claim regardless of if the mark has been filed for federal registration with the U.S. Patent and Trademark Office. The ™ symbol is also used if registration is in progress but not complete. The symbol ® indicates that the mark is registered with the U.S. Patent and Trademark Office and carries full trademark rights protection.

In considering trademarks for fruit cultivars, one should consider: (1) the selection of a name that will be attractive to marketers, (2) the mark must be shown to be used in commerce to complete registration and be maintained, (3) if not registered, the mark can be contested by a holder of a similar mark or the rights be taken away by another applicant for the mark, and (4) requires more cost and paperwork.

It is also vital that the trademark be used correctly and consistently. This includes the use of ^{TM} or ® in association with the mark on all literature, websites, bags, tags, invoices, or other places the mark will appear. Clear language in contracts with propagators or other users of the mark should delineate proper uses.

Details of trademarks including filing and other procedures can be found at the U.S. Patent and Trademark Office (www.uspto.gov). If plants, fruit or other plant parts are sold outside the United States, then it may be desirable to have trademark protection in the countries in which those plants, fruit, or other plant parts are being sold. Depending on the countries, it may be possible to file a single application in the United States and designate other countries under a treaty called the Madrid Protocol. Currently 72 countries, including the E.U., are members of the Madrid Protocol which means that a single trademark application can be filed with the Patent and Trademark Office in the United States with up to 72 countries designated in which trademark protection is desired. There is a specific government fee for each country designated. Some countries like Canada and Mexico are not members of the Madrid Protocol and therefore separate trademark applications must be made in each of these countries. If a trademark application is filed under the Madrid Protocol with the U.S. Patent and Trademark Office, this office forwards the application to the World Intellectual Property Office (WIPO) which then forwards the application to each designated country. Each designated country examines the trademark application as though it had been filed directly with that country. WIPO transmits communications between each designated country's trademark office and the applicant; the U.S. Patent and Trademark Office does not participate in any further communications after forwarding the application to WIPO. Within a time period set by the Madrid Protocol, each designated country will examine the trademark application and communicate the results to the applicant.

14.2.6 Trade Secrets

Trade secrets are important assets to a company, university, or inventor prior to filing for intellectual property protection of a variety or invention. Agreements that are important in maintaining trade secrets and confidentiality may include using Confidentiality Agreements, Material Use/Testing Agreements, and Production Agreements.

14.2.7 Contracts and Licensing

A contract is an agreement between two or more parties which creates an obligation to do or not to do a particular thing; the term 'contract' also refers to a written document which contains the agreement, although a written agreement is sometimes not necessary to create the obligation, i.e., a verbal agreement. Licensing is the granting of the rights of the invention through contracts. Contracts are governed by state law with the terms of the license agreement agreed to by both parties.

Since the license agreement is the vehicle that grants the rights to use the patent or other IPR protection, it is paramount that the language covers a wide range of issues. Common issues addressed in fruit crop license agreements can include:

- Statements of who owns the rights to the invention (the licensor) and who the agreement is being established with (the licensee),
- The invention being licensed, including information on protection such as patent numbers and/or trademarks,
- Definitions of the rights agreed to, including exclusivity or non-exclusivity, the definition of the territory where the rights are provided (for sale or propagation), any assignment of rights.
- Licensing payments due initially (if any), royalty due (per plant, tree, quantity of fruit sold, etc.), minimum royalty due annually (if any), time of payment of royalty, sales record submissions, or other information pertaining to money due to the licensor,
- Requirements of the licensee for any registration or protection activities within the territory,
- Labeling or other use of cultivar or institution name language required to be used by the licensee,
- Termination clauses providing for the termination of the agreement by either party,
- Indemnity clauses, which can release either party from liability by any use of the plant
- Warranty clauses which disclaim any warranty of the fitness for a particular purpose or the warranty of merchantability of the plant, and,
- Resolution of dispute language defining details of the procedures and rights of each party should differences arise under the agreement.

The value of the breadth and precision of the license agreement cannot be overstated, as the agreement plus the honesty of the licensee provide for the main basis of the IPR success of the invention. Many questions arise in determining how to approach licensing. Sole source or exclusive licensing is convenient for the licensor, in that only one entity is dealt with in the agreement. This route is more often chosen for international licensees with defined territories, and multiple licensees may be used in an array of countries or territories. For domestic licensing, multiple licensees are often used so that the plant is more widely available for growers. If domestic propagators are limited, then relationships between the breeding program and the local industry can be a concern, although this would only have major potential political aspects with a public breeding program where the program is either funded by or perceived to be so by local growers and/or taxpayers.

Choosing licensees can be a substantial challenge. A major issue that must be addressed is the potential interest in and value of the invention. If a breeding program is known widely as a major source of proven developments, then potential licensees will be waiting for the new releases and generating interest is often easy. However, if a program is not known for successes in a crop, then limited licensing opportunities may be available and the development may need to be marketed or promoted by the breeding program or technical transfer officer to potential licensees.

On occasion, particularly when exclusive licensing is desired, a licensor may want to take proposals from potential licensees so that information can be provided about the licensee and how the variety will be commercialized. Items that might be asked of potential licensees include:

- What is the company's history in marketing plants of this species?
- What is the current, potential, or projected volume of plant sales (or fruit sales if the agreement is as fruit-based licensing issue) of the plant species of interest for your organization?
- What is the potential sales volume projected for the new variety?
- What nursery capability does the company have to propagate the plants?
- What experience does the company have in applying for, attaining and managing intellectual property protection in the territory? and,
- What is the company willing to pay for the rights to the variety, in addition to the per plant and/or per unit of fruit royalty?

Finally, it is common to learn about the performance of potential licensees from other technical transfer officers at universities or institutions. First-hand knowledge and experiences of others can be of great value in determining who is the best partner in commercializing a new variety. Additionally, a licensor can insert language in an agreement that outlines the progress or goals of the licensee under the agreement.

14.3 Increasing Complexity of IPR Management

The array of items discussed above provide a look at the breadth of issues that developers and users of new inventions deal with. The amount of time required for IPR management by inventing entities and licensees has escalated in recent years. David Brazelton of Fall Creek Nursery, listed the areas of greatest expanse at the ASHS Symposium 'Intellectual Property Rights for Clonally Propagated Plants: Basics to Application' in 2006 at New Orleans, LA. (http://www.ashs.org/resources/IPRsymposium.html). These were:

- Proposal, negotiation and license development increasing in complexity,
- Responsibilities of IPR management are often increasingly migrating to the licensee,
- A wide array of approaches to licensing and commercialization has provided for a lack of harmonization of performance requirements among licensors,
- Increased numbers of varieties to manage in licensing, and,
- Increasing license complexity.

14.4 Testing Agreements

14.4.1 Testing of Selections and Simple Material Transfer Agreements

Testing agreements involve a range of options for breeding programs including the simplest of agreements that provide for testing of selections from other programs or sharing material for testing at other locations. These arrangements are usually governed by a Material Transfer Agreement (MTA), which often includes the following items:

- Statements that limit the testing including restriction from sharing the selections with others along with precise limits on the number of plants that can be used for testing,
- Limitations on where the material can be tested (only experiment station sites, etc.), and requirements of security for the test site,
- Ownership statements that indicate the selections remain the property of the breeding program providing them and that no grant of ownership or proprietary right is conferred with the testing of the material,
- Restrictions on the altering of the material in any way, including the use of the selections in breeding or manipulation where the germplasm is moved into other ownership,
- Restrictions on publishing results of the testing, reporting results without written permission, describing the material in publication or public presentations and other items of potential concern in the area of disclosure about the material,
- Provisions for reporting of test results back to the provider of the material,
- Requirement of reporting any sports, noteworthy results or other findings that may have proprietary value in relation to the testing,
- Agreement that the material is used in compliance with all applicable statutes and regulations including those related to research involving the use of recombinant DNA,
- Statement indicating that the source (the breeding program providing, university, etc.) of the material shall in no event be liable for any use, loss, claim, damage, or liability, that may arise from or in connection with the use, handling, or storage of the material,
- Termination language and instructions on destroying of the material when the test is terminated, and,
- Laws of governance in effect for the agreement.

Material Transfer Agreements generally do not provide for fees or other monetary exchange for testing rights, and these are the more common type used among public breeding programs.

14.4.2 Testing Agreements for Varieties or Selections for Fee

This type of agreement provides a potentially more complex arrangement where money is exchanged for the right to test. This might be done for advanced selections where a commercial tester is interested in identifying the value of new developments prior to release, or for recently released varieties where the testing partner is interested in evaluating the genotype for commercialization in a specified territory. One might also consider allowing testing of selections that were not worthy and have been passed over for release, but may have value in a narrow market such as home gardens or other limited, non-commercial production areas. For a fee, the cooperator normally would require some sort of right to the genotype in the form a first right of refusal for licensing and an exclusive agreement for use in commercialization. The greatest value of these arrangements to a breeding program include: (1) program income can be generated from testing fees, (2) selections might be found to perform better at the cooperator's location (a more desirable genotype × environment interaction than where the genotype was developed) and commercialization might be done only in that territory (and otherwise the genotype might be discarded and no value attained for it), (3) the cooperator might be capable of testing the material using other cultural management methods not available to the breeding program, increasing the chances of finding use for the material, (4) the cooperator might have commercialization capabilities that far exceed those of the institution that developed the material and be able to provide more return for the invention and broader use, (5) protection costs might be shared or paid for by the cooperator upon commercialization, (6) the determination of commercial value might be determined much earlier, providing time for protection to be filed in the territory (for instance, prior to the expiration of the four- or six-year time limitation after first sale of plants outside of the state or territory for UPOV-member countries).

These agreements usually include similar clauses as the previously described material transfer agreement, plus items concerning fees, performance requirements of the tester, lack of warranty of the performance of the material, a defined time frame for testing after which the rights to testing expire, rights of access and possibly expenses paid for the breeder to examine the material, non-assignability of the testing rights, and other items of legal interest. One might consider multi-year testing agreements where a cooperator attains a specified number or range of selections to be provided each year, for instance a five-year period. This provides for a longer relationship and allows testing of very new developments during the agreement period.

14.5 Breeding Agreements

Largely gone are the days when breeding programs openly share breeding material freely with each other, this change due to IPR issues and the potential commercial value of germplasm. And, often the decisions concerning restrictions of sharing are not made by the breeder but rather administrative or intellectual property officials. Many feel this restriction has damaged fruit breeding efforts. However, cooperative breeding agreements can provide for sharing of material among public programs and should be considered as a way to continue germplasm sharing.

14.5.1 Public-to-Public Breeding Agreements

When fruit breeding programs were begun, mostly in the 1900s, sharing of material was more the rule than the exception. As this sharing has become more limited there has been a reduction in breadth of use of genetic advances from other programs (other than named varieties). Since genetic diversity is one of the keys to success for breeding programs, this limitation can have major negative implications on program progress. Overcoming this limitation can be achieved by formal breeding agreements between universities or other public agencies.

Reciprocal agreements allow sharing of selections, seedlings, or pollen among programs. Selection of the shared material would occur and likely commercial genotypes would result that warrant release. When these are identified, decisions on release would be made together by the programs involved, and any resulting royalties or other IPR income would be shared by the institutions. Sharing could be based on a percentage of royalty income, a per plant basis, or some other formula. Usually the program that did the initial crossing, seedling evaluation, and selection would 'own' the variety, file for protection, take steps in commercialization, collect IPR income, and handle other IPR issues. However, these issues would be best agreed to prior to the initiation of material sharing to avoid any conflicts that might arise after the 'winners' begin to be identified. Also, it is important that all administrative entities be aware of the terms of the breeding agreement so that conflicts beyond the programs do not surface later in the relationship. Another aspect of the agreement is how long does the sharing occur, as in only first-generation hybrids, or second or later generations? Most programs would require at least second if not third generation sharing of royalty proceeds, although on a reduced scale than those from the first-generation use.

What are some implications of public to public agreements? Jim Hancock of Michigan State University shared some ideas on this topic at the ASHS Symposium 'Intellectual Property Rights for Clonally Propagated Plants: Basics to Application' in 2006 at New Orleans, LA. (http://www.ashs.org/resources/IPRsymposium.html). He suggested that similar germplasm may be released by both programs, and these might 'compete' in the marketplace. This could be a positive or negative for the programs involved. The more productive programs might provide a disproportional gain to smaller programs by generating more royalty income than the smaller program. However, the genetic diversity gained from sharing along with broader testing of germplasm could still be very valuable to the larger program and in the long run be very beneficial. And, chances of program continuation might be enhanced with a broader IPR income base. Finally, the sharing of ideas, breeding strategies and resulting increased interaction among the breeders might increase the creativity and enthusiasm of those involved.

14.5.2 Public-to-Private Breeding Agreements

As increased activity in private breeding in fruit crops has developed in recent years, the consideration of sharing of germplasm among public and private programs has surfaced. Unlike public-to-public agreements, these arrangements are more often 'one-way' agreements where germplasm moves from the public to the private program only. Why might a public program consider this option? Program support is a primary reason, as the private entity would pay for access to the program's germplasm. Further, wider use of germplasm could be gained by sharing, and the resulting commercialization of varieties derived from the effort might be greater than that possible by the public program. The private program might also have diversity in environment to uncover more favorable genotype × environment interactions, and have broader testing capability (possibly on a worldwide basis).

Disadvantages to the program contributing the germplasm in this type of agreement include: (1) sharing can result in the release of competing and/or similar varieties from the two programs, (2) there could be increased potential for loss of the material if the commercial partner does not operate fairly and honestly within the terms of the agreement, and (3) the providing program could have its material genetically 'laundered' by the private program using the material, and quickly crossing in subsequent generations so that any rights or returns to the providing program are lost. However, this issue can be addressed to some extent by covering multiple generations of use in the breeding agreement. A final concern lies in the issue of any negative implications of the public-to-private relationship with the local industry, program supporters, political, or other entities that support the public program. For instance, if a breeding agreement was set up with a company located in a distant location that competed with local growers for market share, this could lead to conflict with the local industry. This is often more of a concern with a processed crop however, as the product produced could be stored for long periods and or shipped long distances by the distant producer, and introduced in the market of local growers and negatively impact price. If local growers who support the public program could potentially be unhappy with a breeding agreement, this could lead to reduced program support and other negative consequences.

From the private partner side, there are also some items to examine closely. First, it is valuable to determine if the public program is stable in funding and if it will continue to operate at a productive level during the agreement to fulfill the obligations of germplasm sharing. Also, should the current leader of the public program leave, will the program activity be continued at the same level or possible activity interrupted or discontinued due to that vacancy of the breeder's position?

Setting up agreements of this type can be quite complex. Prior to entering into discussions between the potential programs, the program contributing the germplasm should determine what private entity would be best to work with. This could be done by contact with potential individual companies, or a request for interest could be issued to a group of potential private cooperators in which the companies provide information. Items to consider in such an inquiry would include information such as: (1) the existing sales volume (fruit or plants or other product) of the company, (2) potential impact a new development would have on the company's sales, (3) where the company would market the development or product, (4) the technical expertise of the company to conduct breeding activities, (5) experience of the company in commercializing, protecting, and otherwise handling new developments, (6) nursery and propagation capability that the company would utilize in commercializing the new development, and (7) the amount of money in initial fees paid and a schedule of royalties to be paid on a development. Companies with long-standing breeding activities are very familiar with breeding activity costs and other issues, while those that have not ventured into such an endeavor lack knowledge of cost, time involved, technical expertise needed, and other major components of operating a breeding program.

The structure of a public-to-private agreement could include the following items (these might also be included in public-to-public agreements):

- Terms of location (the extent of the territory), length and exclusivity of the proposed breeding cooperation
- Security of the germplasm's use including prohibition of sharing with others, security at the site of the program, etc.
- Right of the cooperator to use developments in subsequent crossing within its program
- The type and amount (number of selections, seeds populations, etc) of material to be supplied to the cooperator, such as plants, seeds, pollen, or other item that moves the germplasm
- Ownership of subsequent developments
- Fees to be paid for initial access to the germplasm, royalties on developments (including multi-generation schedule of royalties to be paid on second or later-generation developments)
- Protection requirements for any developments
- Use or commercialization restrictions on the use of the developments outside the territory defined for the breeding activity (whether the development can be used anywhere in the world, including where the contributing program is based)
- Confidentiality among the parties concerning breeding activities, progress, or other potential proprietary issues,
- Access to the program by the germplasm contributor, such as visits by the public program breeder, and possible funding of travel to breeding sites by the breeder,
- Transfer or assignability of the agreement by the private company,
- Liability and indemnity clauses, and,
- Termination language.

Finally, one must look closely at the issue of the time required to carry out the terms of the agreement by both entities, particularly the public program, in that agreements of this sort require additional time to determine the material to send, to make crosses and produce seeds, to collect pollen or other plant parts, and to monitor the progress and other issues of the cooperative effort. Likewise, time on the private contributor's side should be examined also since breeding activity must be a primary focus by personnel, and they not be heavily involved with commercial activities of the company. When 'money is on the table' on the commercial side, such as harvesting, propagation, marketing, or other commercial activities, this will often take precedence over research activities due to the immediacy of the need for action to produce revenue and profits.

14.6 Closing Comments

The protection options available and details of IPR for fruit crops are expanding areas that are increasingly gaining attention by fruit breeders. The type of protection to be used in the U.S.A., whether plant or utility patent or trademark (or a combination of these), must be examined and planned for prior to variety release. Likewise, international protection options must be planned for and explored. The subsequent license agreements with propagators or other users of the variety must be developed, a process that can require substantial time and effort. Finally, germplasm sharing has become much more complex than in previous times, but sharing is still possible with careful considerations and agreement by the cooperating institutions. However, IPR is much like breeding technology developments that are incorporated into a breeding program; understanding of details, vision, practice, and inspiration all contribute to success.